STATISTICS

A FIRST COURSE

STATISTICS
A FIRST COURSE

EIGHTH EDITION

John E. Freund
Arizona State University

Benjamin M. Perles
Suffolk University

PEARSON
Prentice
Hall

Upper Saddle River, NJ 07458

Library of Congress Cataloging-in-Publication Data

Freund, John E.
    Statistics : a first course—8th ed./John E. Freund, Benjamin M. Perles.
        p. cm.
    Includes bibliographical references and index.
    ISBN 0-13-046653-0
        1. Statistics. I. Perles, Benjamin M. (date) II. Title.

QA276.12F74 2004
519.5—dc22   2003054456

Executive Acquisitions Editor: *Petra Recter*
Editor in Chief: *Sally Yagan*
Media Project Manager: *Jacquelyn M. Riotto*
Production Editor: *Barbara Mack*
Senior Managing Editor: *Linda Mihatov Behrens*
Executive Managing Editor: *Kathleen Schiaparelli*
Vice President/Director of Production and Manufacturing: *David W. Riccardi*
Assistant Managing Editor, Math Media Production: *John Matthews*
Media Production Editor: *Allyson Kloss*
Assistant Manufacturing Manager/Buyer: *Michael Bell*
Manufacturing Manager: *Trudy Pisciotti*
Marketing Manager: *Krista M. Bettino*
Marketing Assistant: *Annett Uebel*
Editorial Assistant/Print Supplements Editor: *Joanne Wendelken*
Art Director: *Maureen Eide*
Interior Designer: *Kiwi Design*
Cover Designer: *Jack Robol*
Art Editor: *Thomas Benfatti*
Creative Director: *Carole Anson*
Director of Creative Services: *Paul Belfanti*
Art Studio: *Laserwords Private Limited*
Composition: *Laserwords Private Limited*

© 2004, 1999, 1995, 1991, 1986, 1981, 1976, 1970 Pearson Education, Inc.
Pearson Prentice Hall
Pearson Education, Inc.
Upper Saddle River, New Jersey 07458

Pearson Prentice Hall® is a trademark of Pearson Education, Inc.

MINITAB is a registered trademark of MINITAB statistical software.

Printed in the United States of America

10 9 8 7 6 5 4 3 2 1

ISBN 0-13-046653-0

Pearson Education LTD., *London*
Pearson Education Australia PTY, Limited, *Sydney*
Pearson Education Singapore, Pte. Ltd
Pearson Education North Asia Ltd, *Hong Kong*
Pearson Education Canada, Ltd., *Toronto*
Pearson Educación de Mexico, S.A. de C.V.
Pearson Education—Japan, *Tokyo*
Pearson Education Malaysia, Pte. Ltd

CONTENTS

PREFACE

This book is intended primarily for use in a first course in statistics. There is a systematic academic approach in *Statistics: A First Course*, 8th edition. Its emphasis is on instruction in meaningful, well-established statistical techniques. The future would-be medical doctor, business executive, scientist, teacher, or other professional specialist must comprehend and be skillful in the application of basic statistical tools and methodology. The student's knowledge of statistics is greatly enhanced by repeated exposure to statistical exercises. The book contains a wide selection of nearly 1,000 exercises, presented throughout the text. The book also offers a large number of detailed illustrations and examples, in each chapter, designed to provide guidance to readers.

Important formulas, definitions, and rules are highlighted within colored boxes. The frequently used tables for normal and *t* distributions are repeated on the inside back cover, and the inside front cover contains an indexed glossary of statistical symbols.

As in previous editions, controversy has not been avoided. The reader is exposed to the weaknesses of statistical techniques, as well as their strengths. It is hoped that this honest approach will provide a stimulus as well as a challenge.

For instructors who have used previous editions of this textbook, it should be noted that the problem sets have been extensively revised. Many additional real data–based problems have been added. Internal referencing within the problem sets has been minimized. Where necessary, pertinent data has been repeated to free students from the unproductive "busy work" of flipping back to previous problems.

USE OF TECHNOLOGY

The book makes extensive use of technology such as the computer, graphing calculator, and MINITAB, but their importance should not be overstated. The entire text is built around the authors' goal of presenting statistics at a level that is appropriate and interesting for beginning students. *In spite of their obvious value, computers and graphing calculators can safely be eliminated by readers of this textbook without loss of continuity.*

This edition uses, but does not require, MINITAB. MINITAB printouts, showing both output and session commands, are set off in boxes and are referenced in the examples. MINITAB users will be pleased by the new opportunity for students to purchase this textbook packaged with a CD-ROM containing the data files for this book.

Although it is not required, this edition also uses the graphing calculator and acquaints the reader with its numerous benefits. *Using Technology: The Graphing*

Calculator boxes appear throughout the book. These boxes provide keystroke-level instruction for using the TI-83 to find solutions to selected examples in the book.

Appendix A, *TI-83 Tips*, written by Dorothy Wakefield of the University of Connecticut and Kathleen McLaughlin of Manchester Community College, provides additional help in using the graphing calculator.

Many problems are designed to be solved with the aid of a computer or a graphing calculator and are identified with the technology 💻 icon.

OTHER FEATURES

Review sections are presented at two- or three-chapter intervals throughout the book. Five review sections reinforce connections outside the context of a particular section or chapter. Identified with a pale blue screen, they include

- A list of *Achievements* that students should have mastered after studying the group of chapters;
- A *Checklist of Key Terms* annotated with specific page references;
- A cumulative section of *Review Exercises* that tests concepts from the immediately preceding chapters.

Representative problems within each exercise set are highlighted with a check ✓ to identify them as *Practice Exercises*. Fully worked out solutions to these exercises are presented at the end of each chapter in the *Solutions to Practice Exercises* section.

SUPPLEMENTS FOR THE INSTRUCTOR

Instructor's Solutions Manual (by Benjamin M. Perles and John E. Freund) (ISBN: 0-13-046654-9) Solutions to all of the exercises are given in this manual. Careful attention has been paid to ensure that all methods of solution and notation are consistent with those used in the core text.

Test Item File (by Singh Kelly, University of West Florida) (ISBN: 0-13-046656-5) Thoroughly revised, the Test includes Multiple Choice, Fill-in, and True/False questions that correlate to problems presented in the text.

SUPPLEMENTS AVAILABLE FOR STUDENTS

Student's Solutions Manual (by Benjamin M. Perles and John E. Freund) (ISBN: 0-13-046655-7) Fully worked out solutions to all of the odd-numbered exercises are provided in this manual. Careful attention has been paid to ensure that all methods of solution and notation are consistent with those used in the core text.

A MINITAB Guide to Statistics, 2nd ed. (by Ruth Meyer and David Krueger) (ISBN: 0-13-014156-9) This manual assumes no prior knowledge of MINITAB. Organized to correspond to the table of contents of most statistics texts, this manual provides step-by-step instruction for using MINITAB in statistical analysis.

TI-83 Graphing Calculator Manual for Statistics (by Stephen Kelly)
(ISBN: 0-13-020911-2) This brief, spiral bound manual provides simple keystroke instructions for using the TI-83 graphing calculator in statistics. Class tested for many years, this manual is the perfect answer for frustrated students and professors.

Student Versions of SPSS and SYSTAT Student versions of SPSS, the award-winning and market-leading commercial data analysis package, and SYSTAT are available for student purchase. Details on all current products, including integrated packages, are available from the publisher or via the SPSS Web site at http://www.spss.com.

ACKNOWLEDGMENTS

The authors are indebted to Professor E. S. Pearson and the Biometrika trustees for permission to reproduce parts of Tables 8 and 18 from their *Biometrika Tables for Statisticians*; to Prentice Hall for permission to reproduce part of Table 2 from R. A. Johnson and D. W. Wichern, *Applied Multivariate Statistical Analysis*; to the Addison-Wesley Publishing Company for permission to base Table VII on Table 11.4 of D. B. Owen, *Handbook of Statistical Tables*; and to the editor of the *Annals of Mathematical Statistics* for permission to reproduce the material in Table VIII. The changes in this edition are the authors' responsibility.

We express appreciation to Dorothy Wakefield and Kathleen McLaughlin for writing Appendix A, *TI-83 Tips*, and for updating the MINITAB printouts and the *Using Technology* boxes; to Sarah Streett for accuracy checking the text and solutions manuals; and to the following people at Prentice Hall: Petra Recter, Executive Editor; Joanne Wendelken, Supplements Editor; Jacquelyn Riotto, Media Project Manager; Krista Bettino, Marketing Manager; Annett Uebel, Marketing Assistant; Barbara Mack, Production Editor; Linda Behrens, Senior Managing Editor; Michael Bell, Assistant Manufacturing Manager/Buyer; Maureen Eide, Art Director; Thomas Benfatti, Art Editor; John Matthews, Assistant Managing Editor, Math Media Production; and Allyson Kloss, Media Production Editor.

Our gratitude is also extended to the many students and colleagues whose suggestions have contributed greatly to this and to previous editions, and in particular to E. Ray Bobo, Georgetown University; Tom Brieske, Georgia State University; Neil S. Dickson, Weber State University; Dale O. Everson, University of Idaho; Elizabeth Farber, Bucks County Community College; David A. Ford, Emory University; Ronald Friesen, Bluffton College; Frank Gunnip, Oakland Community College; Steven L. Harvey, Seminole Junior College; John Heinen, Regis College; Tom Hogenkamp, Erie Community College; John Van Iwaarden, Hope College; Elizabeth S. Low, University of Vermont; Raymond Mueller, DeVry Institute of Technology; Iosif Pinelis, Michigan Technological University; Mark Sherrin, Flagler College; James Siefert, ICM School of Business; Lloyd B. Smith, Jr., Lenoir-Rhyne College; Jake Uhrich, South Puget Sound Community College; Ian Walton, Mission College; and Arthur Larry Wright, University of Arizona.

JOHN E. FREUND
BENJAMIN M. PERLES

1

INTRODUCTION

Statistics deals with all aspects of the collection, processing, presentation, and interpretation of measurements or observations, that is, with all aspects of the handling of **data**. Thus, data constitutes the raw material we deal with in statistics, and its collection is of major concern in any statistical investigation. Data is obtained by taking measurements, by counting, by asking questions, or by referring to data made available in published form. Note that we said "data constitutes" and "data is," even though *data* is the plural form of *datum*, a term that in actual practice is rarely used. We do this because all the data is usually regarded as one unit of information and, hence, used with a singular verb.[†]

The origin of modern statistics can be traced to two areas which, on the surface, have very little in common: *government* (political science) and *games of chance*.

Governments have long used *censuses* to count persons and property. The ancient Romans used this technique to assist in the taxation of their subjects; indeed, the Bible tells how Mary and Joseph, subjects of Rome, went to Bethlehem to have their names listed in a census. Another famous census is reported in

---

[†] In more formal writing the word *data* is plural, and we would write "Thus, these data constitute the raw material...."

the *Domesday Book* of William of Normandy, completed in the year 1086. This census covered most of England, listing its economic resources, including property owners and the land which they owned. The U.S. census of 1790 was the first "modern" census, but government agents merely counted the population. More recent U.S. censuses have become much wider in scope, providing a wealth of information about the population and the economy. They are conducted every ten years (in the years that end in zero such as 1980 and 1990). The most recent one, the 22nd decennial census, was conducted in the year 2000.

The problem of describing summarizing, and analyzing census data led to the development of methods which, until recently, constituted almost all that there was to the subject of statistics. These methods, which originally consisted mainly of presenting the most important features of data by means of tables and charts, constitute what is now referred to as **descriptive statistics**. To be more specific, this term applies to *anything done to data that does not infer anything which goes (generalizes) beyond the data itself.* Thus, when the government reported on the basis of census counts that the population of the United States was 248,718,301 people in 1990 and 281,471,906 in 2000, this belonged to the field of descriptive statistics. This would also be the case if we calculated the corresponding percentage growth which, as can easily be verified, was 13.2%, but not if we had used the 1990 data to predict, say, the population of the United States in the year 2020. Such a prediction goes beyond the available information, and we must shift our emphasis from descriptive statistics to *statistical inference* or *inductive statistics.*

To mention a few other examples, generalizations (that is, methods of statistical inference) are needed to estimate the number of long-distance calls which will be made in the United States ten years hence (on the basis of business trends and population projections); to determine the effect of a new trading stamp program on the sales of supermarkets (on the basis of experiments conducted at a few markets); to evaluate conflicting and uncertain legal evidence; to predict the extent to which elevators will be used in a large office building which has not yet been built; to estimate the demand for a new computer which is being developed; or to rate the efficiency of salespeople (on the basis of fragmentary information about their performance). In each of these examples there are uncertainties, or only partial or incomplete information.

> *Statistics has grown from the art of constructing charts and tables to the science of basing decisions on numerical data, or even more generally the science of decision making in the face of uncertainty. It is here that we must use statistical methods which find their origin in games of chance.*

Games of chance date back thousands of years, as is evidenced, for example, by the use of *astragali* (the forerunners of dice) in Egypt about 3500 B.C., but the mathematical study of such games began less than four centuries ago. The study of probability as a mathematical science began in the year 1654, when Blaise Pascal (a mathematician) wrote to Pierre de Fermat (another mathematician) with regard to a gambling problem. They solved the problem independently, using different mathematical methods. It may seem surprising that it took so long, but until then chance was looked on as an expression of divine intent, and it would have been impious, or even sacrilegious, to analyze the "mechanics" of the supernatural through mathematics.

Although the mathematical study of games of chance, called **probability theory**, dates back to the seventeenth century, it was not until the early part of the nineteenth century that the theory developed for "heads or tails," for example, or "red or black" or "even or odd," was applied also to real-life nongambling situations where the outcomes were "boy or girl," "life or death," "pass or fail," and so forth. Thus probability theory was applied to many problems in the social as well as the physical sciences, and nowadays it provides an important tool for the analysis of any situation in business, in science, or in everyday life where there is an element of uncertainty or risk. In particular, it provides the basis for methods which we use when we generalize from observed data, namely, when we use the methods of statistical inference.

In Sections 1.5, 1.6, and 1.7 we shall discuss some general questions about statistics, its past and present, its study, and its future development.

1.1 NUMERICAL DATA AND CATEGORICAL DATA

Fundamentally, there are two types of data: numerical and categorical. **Numerical data** contains numbers that can be treated by ordinary arithmetical methods. For instance, if we counted the number of passengers on three buses, we might get 32, 41, and 28 passengers. To obtain the total number of passengers, we simply add the values 32, 41, and 28 and obtain 101. If necessary, we can multiply, divide, subtract, raise to powers, and extract roots of these values.

Categorical data results from data being sorted into nonnumerical categories. For instance, if an interviewer determines whether a person is single (never married), married, a widow or widower, or divorced, this information has been categorized. Questions that result in a choice of answers such as yes or no; agree or disagree; true or false; or poorly, acceptably; or superior tasting; result in categorical data. For ease in manipulating categorical data, it is sometimes coded. In the marital status illustration, the categories of single (never married), married, widow/widower, or divorced could be assigned code numbers of 1, 2, 3, and 4, respectively. Exercise 1.3 gives a simple example of how coding may be used.

1.2 NOMINAL, ORDINAL, INTERVAL, AND RATIO DATA

Data may also be classified as nominal, ordinal, interval, or ratio. The coded numbers of the previously described marital data are referred to as **nominal data** and are incapable of being manipulated arithmetically. Other examples of nominal data include the numbers on the jerseys of football players, and the identification of undergraduate college students as freshmen, sophomores, juniors, or seniors. Nominal data is the weakest level of measurement. For the marital data, we cannot add the coded numbers 1 and 2 to obtain 3, since this would result in the nonsensical statement that single + married = widow/widower. Nominal data, however, focuses on the frequency in a category and shows clearly the number of respondents who fall into each marital class.

Sometimes data is recorded in numerical form, but restated in categorical form. A meteorologist might record air temperatures in degrees Fahrenheit, but the resulting weather report might describe the data categorically as hot, cold, or normal for this time of year.

Ordinal data can be rank ordered. For instance, the five teams constituting a baseball league might be individually ranked as being in first place, in second place, in third place, in fourth place, and in last place. In connection with ordinal data, **inequality signs** can be used, but $<$ and $>$ have a broader meaning than the usual "less than" and "greater than." The sign $>$ can mean happier than, more painful than, has a better flavor than, louder than, or harder than, while $<$ would have the opposite meaning. In the army the rank of general $>$ colonel $>$ major $>$ captain $>$ first lieutenant $>$ second lieutenant. In finance, according to a recent issue of *Forbes* magazine, the six largest corporations, ranked by sales, are Wal-Mart Stores $>$ ExxonMobil $>$ General Motors $>$ Ford Motor $>$ General Electric $>$ Citigroup.

If we can form differences, but not multiply or divide, we refer to the data as **interval data**. Suppose that a meteorological map gives us temperatures of $40°$, $60°$, $80°$, and $100°$F. We can say that $40°$F $< 60°$F, which means that $40°$F is colder than $60°$F; or we can say that $60°$F $> 40°$F, which means that $60°$F is warmer than $40°$F. Or we can say that $100°$F $- 80°$F $= 60°$F $- 40°$F, which means that the temperature difference of $20°$F is the same on both sides of the equation. But we cannot say that $80°$F is twice as hot as $40°$F, although $\frac{80.00}{40.00} = 2.00$. The falsity of this statement can be seen if we convert the two temperatures to the Celsius scale, where $80°$F corresponds to $26.67°$C, and $40°$F corresponds to $4.44°$C. If we divide the two temperatures as before, we get $\frac{26.67}{4.44} \approx 6.01$, which implies that 26.67 is 6.01 times hotter than 4.44, and not merely twice as hot. Actually, both statements are incorrect, and if you wish to pursue this matter further, you would need to know that absolute zero ($0°$A), which is the lowest theoretical temperature that a gas can reach, is $-460°$ on the Fahrenheit scale and $-273°$ on the Celsius scale. The zero values for the Fahrenheit, Celsius, and absolute scales are found at different temperatures.

For **ratio data** we can form quotients, that is, divide one quantity by another quantity of the same kind. Both the dividend and divisor must be expressed in the same units, such as inches, gallons, or dollars, and the unit must have a true zero as its origin. The resulting ratio is a pure, or abstract, number and is not expressed in units. The ratio of \$2 to \$5 is expressed as $\frac{2}{5}$ or 0.40. Note that the ratio omits the dollar sign.

According to U.S. law, the ratio of the length to the width of an official U.S. flag must be 1.9. If an official flag is 2 feet wide, it must therefore be $2 \times 1.9 = 3.8$ feet long. If it is 150 centimeters wide, it must be $150 \times 1.9 = 285$ centimeters long.

1.3   SAMPLE DATA AND POPULATIONS

In the introduction to this chapter we said that "statistics deals with all aspects of the collection, processing, presentation, and interpretation of measurements or observations, that is, with all aspects of the handling of data." This broad definition of statistics includes both populations and samples, which we distinguish as follows:

*If a set of data consists of all conceivably possible (or hypothetically possible) observations of a certain phenomenon, we call it a "**population**"; if a set of data consists of only a part of these observations, we call it a* **sample***.*

We added the phrase "hypothetically possible" to take care of such clearly hypothetical situations in which we consider the outcomes (heads or tails) of 12 flips of a coin to be a sample from the population of all possible flips of the coin, or we consider the weights of ten 30-day-old calves to be a sample of the weights of all (past, present, and future) 30-day-old calves, or we consider four determinations of the copper content of an ore to be a sample of all possible determinations of the copper content of the ore. In fact, we often consider the results of an experiment to be a sample of what we might obtain if the experiment were repeated over and over again.

The term *population* comes from the origin of statistics, which attempted to describe human populations as illustrated on page 2 of this chapter. Even though it may sound strange to refer to the heights of all the trees in a forest as a "population of heights" or to the speeds of all the cars passing a checkpoint as a "population of speeds," in statistics *population* is a technical term with a meaning of its own.

Although we are free to call any set of data a population, what we do in practice depends on the context in which the data is to be viewed. For instance, the complete figures for a recent year, giving the weekly amounts of sales tax collected by the 32 restaurants of a fast-food chain, can be viewed as either a population or a sample. If a tax collector, auditing the records of this fast-food chain, is interested only in the figures for the given year, they constitute a population; on the other hand, if the management of the chain wants to estimate future collections of sales tax or guess at the amounts of sales tax that were collected by its competitors, the original data must be viewed as a sample.

EXAMPLE 1.1   According to a recent edition of *Current Population Reports* by the U.S. Bureau of the Census, the following are the percentages of persons not covered by health insurance in the Rocky Mountain states.

State	Percent not covered
Arizona	21.2
Colorado	16.8
Idaho	19.1
Montana	18.6
Nevada	20.7
New Mexico	25.8
Utah	14.2
Wyoming	16.1

Give one example each of a problem in which this set of data would be viewed as

(a) a population;

(b) a sample.

Solution
(a) The data would constitute a population if a health insurance company used it to explore the possibility of marketing its health insurance policies exclusively in the Rocky Mountain states.
(b) The data would constitute a sample if the health insurance company used it to estimate the percentage of people not covered by health insurance in the United States. ∎

1.4 BIASED DATA

In all statistical studies that use samples, great care must be exercised to ensure that the data lends itself to valid generalizations. A key issue here is the question of **bias**. A sample is said to be biased if it is not representative of the population that it is supposed to represent. Every precaution must be taken to avoid inadvertent biases. It is, of course, unethical to introduce deliberate biases to prove particular points.

Direct information is usually gathered by personal interviews, telephone calls, mail questionnaires, or any combination of the three. There is no general answer as to which method is best, although mail questionnaires are often preferred, because they are easier to get into homes and businesses that are widely dispersed. For one reason or another, however, many people fail to return even well-constructed questionnaires unless the questions are devoted to matters in which they are interested. This may introduce a bias that makes it very difficult to arrive at valid conclusions. Whether used for personal interviews or sent through the mail, the construction of questionnaires is the subject of extensive study. Statistical investigations have often been challenged on the grounds that questions were not properly asked. The question "Why do you prefer Coca-Cola to Pepsi-Cola?" is apt to elicit a different reply than "Which do you prefer, Coca-Cola or Pepsi-Cola?"

Mail questionnaires are often cheaper and easier to distribute, but an advantage of the personal interview is that a tactful interviewer can often obtain answers to personal questions relating to age, health, and income that many persons would refuse to answer in a mail questionnaire.

Telephone calls are often used to get on-the-spot responses, for example, to television viewing and other habits. It is exposed to the criticism, however, that it does not reach homes where there are no telephones and thus eliminates a portion of persons who may be relevant to the survey. The widespread use of telephone answering machines in homes and businesses also reduces the response rate. Also, telephone calls made during normal working hours may fail to reach residents, especially in homes where there is more than one worker in the family.

EXERCISES

Exercises 1.1, 1.4, and 1.7 are practice exercises; their complete solutions are given on pages 13–14.

✓ **1.1** An intercity bus was traveling down a highway when its driver was questioned, on a cell phone, by the starter at the next terminal. Which of the driver's answers would result in numerical data, and which would result in categorical data?

(a) How many passengers are aboard the bus?

(b) How many passengers are seated, and how many passengers are standing?

(c) How many miles is the bus from the terminal?

(d) What is the length of the bus?

(e) From what city did the trip originate?

(f) How many gallons of fuel do you think the bus will need when it gets to the terminal?

1.2 Which of the following are nominal data and which are ordinal data?

(a) The identification numbers painted on the bows of Navy battleships sailing with a fleet

(b) The gold, silver, and bronze medal winners of an Olympic athletic contest where the gold medal is awarded for first place, the silver medal is awarded for second place, and the bronze medal is awarded for third place

(c) The Social Security numbers of a group of factory workers

(d) Coding of the faculty of a university, for use in a computer analysis, by rank if 1 represents professor, 2 represents associate professor, 3 represents assistant professor, and 4 represents instructor

(e) Numbering of horses at the conclusion of a race to indicate which horse crossed the finish line first, second, and third

(f) The numbers displayed on the uniforms of major league baseball players who are members of the Arizona Diamondbacks team

1.3 Referring to the marital status categories in Section 1.1, suppose that we code *single* (never married) as 1, *married* as 2, *widow or widower* as 3, and *divorced* as 4. If an interviewer obtains marital information from 24 respondents, determine how many are *single* (never married), *married, widow or widower*, and *divorced* if they are coded as follows:

$$1 \quad 3 \quad 2 \quad 2 \quad 2 \quad 1 \quad 4 \quad 2 \quad 3 \quad 4 \quad 1 \quad 4$$
$$2 \quad 1 \quad 1 \quad 2 \quad 1 \quad 4 \quad 1 \quad 3 \quad 2 \quad 1 \quad 1 \quad 4$$

✓ **1.4** Suppose that we are given complete information about the number of patients admitted and the length of their stays at a certain hospital during a certain year. Give one example each of problems in which we would consider this set of data to be

(a) a population

(b) a sample

1.5 Suppose we are given complete information about the tips received in a certain week by waitresses employed by a large restaurant. Give one example each of problems in which we would consider this data

(a) a population

(b) a sample

1.6 Suppose we are given complete information about the application for and the time elapsed before issuance of residential building permits in Chicago for the past year. Give one example each of problems in which we would consider this data

(a) a population

(b) a sample

✓ **1.7**   "Bad" statistics may well result from asking questions in the wrong way or of the wrong persons. Explain why the following may lead to useless data.

    (a) To determine public opinion concerning the zoning laws in a certain community, an interviewer asks its voters, "Are you in favor of reforming our obsolete zoning laws?"

    (b) To obtain information for a toothpaste manufacturer, an interviewer asks, "Do you brush your teeth at least once a day?"

    (c) To determine the spending patterns of the general public, a detailed questionnaire is sent to a mailing list of people residing in a retirement community.

    (d) To determine whether the courses offered by a certain college adequately prepare students for their careers, an interviewer directs his questions to incoming freshmen students.

1.8   "Bad" statistics may also result from asking questions in the wrong place or at the wrong time. Explain why the following may lead to useless data.

    (a) To determine the average number of pedestrians who walk through a downtown shopping mall, a researcher makes his counts during mid-December.

    (b) To determine the average number of motor vehicles that use a certain highway, a researcher makes his counts every weekend for several weeks.

    (c) To determine the attitudes of residents of a town toward a proposed bond issue for the construction of a high school, the researcher interviews parents as they pick up and deliver their children who attend grade school.

    (d) To determine the proportion of women who would purchase a new line of expensive, high-fashion shoes, a student employee of a research company interviews all of the women students in her statistics class.

1.9   Explain why the following may lead to useless data.

    (a) To determine the proportion of improperly filled cans of paint, an inspector at the filling machine examines every 10th can as it leaves the filling machine.

    (b) To determine how voters in a city feel about an issue, a pollster randomly selects a page in the telephone directory and calls every telephone number on the page.

    (c) The wheat in several freight cars is sampled by taking samples from the top of each car.

1.5 STATISTICS, PAST AND PRESENT

The problem of describing, summarizing, and analyzing census data led to the development of methods that, until recently, constituted almost all there was to the subject of statistics. These methods, which originally consisted mainly of presenting data in the form of tables and charts, constitute what we now call **descriptive statistics**. This includes anything done to data that is designed to summarize or describe it without going any further, that is, without trying to infer anything that goes beyond the data itself. For instance, if a newspaper reports net paid circulations of 172,316 in 1998 and 207,185 in 2002 and we perform the calculation to show that there was an increase of 20.2%, our work belongs to the field of descriptive statistics. This would not be the case, however, if we used the given data to predict the newspaper's circulation in the year 2010.

Although descriptive statistics is an important branch of statistics and continues to be widely used, statistical information usually arises from samples (from observations made on only part of a large set of items), and thus its analysis will require generalizations that go beyond the data. As a result, an important feature of the growth of statistics in this century has been the shift in emphasis from methods that merely describe to methods of **statistical inference** that serve to make generalizations.

Such methods are required, for instance, to predict the operating life span of a computer (on the basis of the performance of several such computers); to estimate the assessed value of all privately owned property in Orange County, California in the year 2025 (on the basis of business trends, population projections, and so forth); to compare the effectiveness of two reducing diets (on the basis of the weight losses of persons who have been on the diets); to determine the optimum dose of a medication (on the basis of tests performed with volunteer patients from selected hospitals); or to predict the flow of traffic on a freeway that has not yet been built (on the basis of past traffic counts on alternate routes).

In each situation described in the preceding paragraph, there are uncertainties because only partial, incomplete, or indirect information is available, and it is with the use of the methods of statistical inference that we judge the merits of the results and, perhaps, suggest a "most profitable" choice, a "most promising" prediction, or a "most reasonable" course of action.

In view of the uncertainties, we handle problems like these with statistical methods that find their origin in games of chance. Although the mathematical study of games of chance dates back many centuries, it was not until the early part of the nineteenth century that the theory developed for "heads or tails," "red or black," or "even or odd" was also applied to real-life situations in which the outcomes were "boy or girl," "life or death," or "pass or fail." Thus, **probability theory** was applied to many problems in the behavioral, natural, and social sciences, and nowadays it provides an important tool for the analysis of any situation (in science, business, or everyday life) that in some way involves an element of uncertainty or chance. In particular, it provides the basis for the methods we use when we generalize from observed data, that is, when we use the methods of statistical inference.

1.6 THE STUDY OF STATISTICS

The scope of statistics and the need to study statistics have grown enormously in the last few decades for two reasons. One is the increasingly quantitative approach employed in all the sciences, as well as in business and many other activities that directly affect our lives. This includes the use of mathematical techniques in such problems as the evaluation of antipollution controls, inventory planning, the analysis of cloud formations, the study of diet and longevity, and the evaluation of teaching techniques.

The other reason is that the amount of statistical information that is collected, processed, and disseminated to the public for one reason or another has increased almost beyond comprehension, and what part of it is "good" statistics and what part is "bad" statistics is anybody's guess. To act as watchdogs, more and more persons with some knowledge of statistics are needed to take an active part in the collection and statistical analysis of the data and, what is equally important, in all the preliminary planning. Without the latter, it is frightening to think of all the things that can go wrong. The results of costly studies can be completely useless if, for example, questions are ambiguous or asked in the wrong way, instruments are poorly adjusted, or all relevant factors are not taken into account.

This book is a general statistics book that provides preparation in the basic statistics underlying the many specialties in which a student may major. Therefore, the exercises and examples utilized throughout the text are drawn from an assortment of disciplines. This is possible because the solution of statistical problems is very much the same regardless of the data source.

To clarify this point, let us examine the following example, which comes from the field of nutrition (see Example 11.3). Suppose that a nutritionist claims that at least 75% of the preschool children in a certain country have protein-deficient diets, and that a sample survey reveals that this is true for 206 preschool children in a sample of 300. Test the claim at the 0.05 level of significance. This example is followed by a step-by-step solution in the example, which the student can use as a model for similar problems.

If this book specialized in medical statistics, the same example might have been rephrased as follows: Suppose that a medical doctor claims that at least 75% of his patients suffer from obesity, and that a sample survey conducted by a medical research team reveals that this is true for 206 similar patients in a sample of 300. Test the claim at the 0.05 level of significance.

As the reader may discern, the step-by-step solution to the nutritionist example provides a model for the solution of this medical problem, and the answer is, of course, identical.

Similarly, if this were a book which specialized in, say, business statistics, the nutritionist example might have been rephrased as follows: Suppose that the manager of a supermarket in a chain of supermarkets asserts that 75% of her customers prefer nationally advertised brands of dishwasher detergent over the supermarket's house brand of detergent. A market research team observes that 206 of 300 purchasers of dishwasher detergent bought the nationally advertised brand. Test the manager's assertion at the 0.05 level of significance. Again, the nutritionist example provides a model for the solution of this business problem, and the answer is identical.

In contrast to this textbook, which presents a general introduction to the subject of statistics, numerous books have been written on business statistics, educational statistics, medical statistics, psychological statistics, . . ., and even on statistics for historians. Although problems arising in these various disciplines sometimes require special statistical techniques, none of the basic methods described in this book is restricted to any particular field of application. In the same way in which $3 + 3 = 6$ regardless of whether we are adding dollar amounts, horses, or trees, the methods that we present provide appropriate *statistical models*, regardless of whether the data are Intelligence Quotients, tax payments, reaction times, test scores, or humidity readings.

1.7 STATISTICS: WHAT LIES AHEAD

Although the point is arguable, in the last few decades the emphasis in statistics has shifted from summarizing data by means of charts and tables to making inferences (that is, generalizations) on the basis of partial, incomplete, or indirect information. This is not meant to imply, however, that the subject of statistics has now become stable and inflexible and that it has ceased to grow. Aside from the fact that new statistical techniques are constantly being developed to meet particular needs, the whole philosophy of statistics continues to be in a state of change. For example, attempts have been made to treat all problems of statistical inference within the framework of a unified theory called **decision theory**, which, so to speak, covers everything "from cradle to grave." One main feature of this theory is that we must account for all the consequences that can arise when we base decisions on statistical data. This poses serious problems, because it is generally difficult, if not impossible, to put cash values on the consequences of one's actions. For instance, how can we put a cash value on the consequences of the decision of whether or not to market a new medication, especially if the wrong decision may well involve the loss of human lives?

However, some statisticians suggest that the emphasis has swung too far from descriptive statistics to statistical inference; rightly so, they feel that the solution of many problems requires only descriptive methods. To accommodate their needs, new descriptive techniques have recently been developed under the general heading of **exploratory data analysis**. These will be introduced in Sections 2.2 and 3.3.

We have mentioned all this mainly to impress upon the reader that statistics, like most other fields of learning, is not static. Indeed, it is difficult to picture what a beginning course in statistics will be like 20 years hence. Certain aspects will probably still be the same, which includes the role of probability theory in the foundations of statistics, as well as certain data-summarizing techniques that have been very useful in the past and will undoubtedly continue to be widely used in the future.

EXERCISES

Exercises 1.10 and 1.13 are practice exercises; their complete solutions are given on page 14.

✓ **1.10**   In the years 1997, 1998, 1999, and 2000, United States exports of manufactured goods were 550.5, 552.8, 565.5, and 624.0 billions of dollars. Which of the following conclusions can be obtained from these figures by purely descriptive methods and which require generalizations? Explain your answer.

  (a) United States exports of manufactured goods exceed $565.5 billion in only one year.

  (b) United States exports of manufactured goods increased from one year to the next.

  (c) United States exports will exceed $1000 billion within five or six years.

  (d) The difference between the largest and smallest value of exports is 73.5 billion.

1.11   In the years 1990, 1995, and 2000 the state of Hawaii had (in thousands) 16, 34, and 26 unemployed. During the same years Alaska had (in thousands) 19, 22, and 21 unemployed. Which of the following conclusions can be obtained from these figures by purely descriptive methods and which require generalizations?

(a) During two of the three specified years, Hawaii's unemployment was greater than Alaska's.

(b) Alaska's unemployment grew by 2 (thousands) from the years 1990 to 2000.

(c) Hawaii's unemployment declined by 8 (thousands) from the 5 years 1995 to 2000.

(d) Alaska's unemployment grew more slowly than Hawaii's unemployment probably because Alaska's advertising for tourists was more effective than Hawaii's advertising.

1.12 The attendance at baseball's National League regular games for the years 1995, 1996, 1997, 1998, and 1999 was (in millions of spectators) 25.1, 30.8, 31.9, 38.4, and 38.3. During the same years attendance at the American League's regular games was (in millions of spectators) 25.4, 29.7, 31.3, 31.9, and 31.8. Which of the following conclusions can be obtained by purely descriptive methods and which require generalizations?

(a) The 1995 figure for the National League was probably incorrect and should have been recorded as 30.1.

(b) The highest attendance for any League and year was 38.4 million spectators in 1998 for the National League.

(c) Attendance increased in almost every year for both the National League and the American League.

(d) Attendance was greater for the National League than the American League in every year except 1995.

✓ **1.13** According to the Current Population Survey, the number of employed civilians in the labor force during the years 1980, 1985, 1990, 1995 and 2000 was (in thousands) 167,745; 178,206; 189,164; 198,584; and 209,699. Which of the following conclusions can be obtained by purely descriptive methods and which require generalizations? Explain your answers.

(a) The number of employed civilians grew in each of the designated five-year periods.

(b) Civilian employment grew by approximately 25% from the year 1980 to 2000.

(c) It can be predicted that civilian employment will grow by approximately 25% from the years 2000 to 2020.

(d) Civilian employment must have been about 25% less in the year 1960 than it was in 1980.

(e) It can be predicted that the numbers of civilians employed will approximately double between the years 2000 and 2080.

1.14 Three 2-bedroom apartments which were rented in a large apartment building bore monthly rentals of $1,200, $700, and $800. Which of the following conclusions can be obtained from these data by purely descriptive methods and which require generalization?

(a) The average rental of these three 2-bedroom apartments is $900 per month.

(b) The average rental of a 2-bedroom apartment in this apartment building is $900 per month.

1.15 According to the secretary of the Senate of the United States, the numbers of Democratic and Republican senators in the 102nd to the 107th Congress were as follows:

Congress	Number of Democratic senators	Number of Republican senators
102nd	56	44
103rd	57	43
104th	48	52
105th	45	55
106th	45	55
107th	50	50

Which of the following conclusions can be obtained from these figures by purely descriptive methods and which require generalizations?

(a) In the 102nd Congress the number of Democratic senators exceeded the number of Republican senators.

(b) Neither the Democrats nor the Republicans will ever attain a two-thirds majority in the Senate.

(c) The maximum number of Democratic senators in the 102nd to the 107th Congress was 57, and the maximum number of Republican senators in the 102nd to the 107th Congress was 55.

(d) The smallest number of senators of either party was 43.

(e) The number of Democratic senators has steadily declined in recent years and will probably continue to decline.

(f) In the near future, the Republican senators will continue to increase their majority.

1.16 A statistically minded lawyer has his office on the third floor of a very tall office building, and whenever he leaves his office he records whether the first elevator that stops at his floor is going up or coming down. Having done this for some time, he discovers that the vast majority of the time the first elevator that stops is going down. Comment on his conclusion that fewer elevators are going up than are coming down.

✓ SOLUTIONS TO PRACTICE EXERCISES

1.1 (a)   Numerical

(b)   Categorical

(c)   Numerical

(d)   Numerical

(e)   Categorical

(f)   Numerical

1.4 (a)   The data can be viewed as a population if we are interested only in this particular hospital at this time.

(b)   It can be viewed as a sample if we use it to predict admissions and discharges at a subsequent time; or use it to make predictions about other hospitals.

1.7 (a) This is called *begging the question* because the interviewer suggests to the voters that the zoning laws are obsolete.

(b) The person being interviewed may be reluctant to answer such personal questions truthfully.

(c) The elderly people in the retirement community are not likely to have the same spending habits as the general public.

(d) Incoming freshmen are not usually well qualified to know whether the courses offered at the college adequately prepare them for their careers.

1.10 (a) Descriptive. Numbers are merely read directly from the data.

(b) Descriptive. Numbers are merely read directly from the data.

(c) Generalization. Answer cannot be read or calculated directly from the data.

(d) Descriptive. Can be calculated from the data. $624.0 - 550.5 = 73.5$.

1.13 (a) Descriptive. Growth is revealed by examination of the data.

(b) Descriptive. Can be calculated from the data.

(c) Generalization. Cannot be calculated from the data.

(d) Generalization. Cannot be calculated from the data.

(e) Generalization. Cannot be calculated from the data.

2

SUMMARIZING DATA: LISTING AND GROUPING

During recent years the collection of statistical data has grown at such a rate that it would be impossible to keep up with even part of the things that directly affect one's life unless such information were disseminated in "predigested" or summarized form. The whole matter of putting large masses of data in usable form has always been important, and it has multiplied greatly in the last few decades. This has been due partly to the development of computers, which have made it possible to accomplish in minutes what previously had to be left undone because it would have taken months or even years. It also results from the increasingly quantitative approach of the sciences, especially the social sciences, where nearly every aspect of human life is nowadays measured in one way or another.

The most common method of summarizing data is to present it in condensed form in tables or charts, and at one time this took up the better part of an elementary course in statistics. However, there is so much else to learn in statistics that very little time is now devoted to this kind of work.

In Sections 2.1 and 2.2 we present methods of **listing** (rearranging) sets of data so as to simplify their use. By *listing*, we refer to any kind of treatment in which the identity of each value (or item) remains preserved. In other words, the number 25 remains a 25 and a hospital patient remains a hospital patient,

15

although the order in which the values (or items) are presented may well be changed.

In Sections 2.3 and 2.4 we shall see how a solid grasp of a set of data can also be obtained by **grouping** the data into a number of classes (intervals or categories) and then presenting the result in the form of a table or a pictorial chart.

2.1 DOT DIAGRAMS

Once the data has been tallied (that is, we have counted how many times each value occurs), the results may be presented in the form of a dot diagram. As illustrated in Figures 2.1 and 2.2, in a dot diagram we simply indicate by means of dots how many times each value (or item) occurs.

EXAMPLE 2.1   The following data shows the number of speeding tickets per day issued by a highway police patrol during a period of 60 days.

$$
\begin{array}{cccccccccccc}
6 & 3 & 1 & 4 & 3 & 1 & 7 & 3 & 2 & 3 & 1 & 7 \\
5 & 1 & 2 & 2 & 4 & 0 & 1 & 1 & 3 & 4 & 3 & 1 \\
4 & 5 & 1 & 3 & 2 & 5 & 3 & 2 & 2 & 0 & 1 & 4 \\
6 & 2 & 3 & 5 & 6 & 2 & 4 & 4 & 2 & 5 & 3 & 2 \\
3 & 1 & 1 & 4 & 4 & 2 & 0 & 1 & 2 & 5 & 5 & 2
\end{array}
$$

Construct a dot diagram.

Solution   Counting how many times each value occurs, we get the dot diagram of Figure 2.1. It gives a clearer picture of the speeding ticket situation than the original numbers shown in the five rows of 12 numbers each.

Figure 2.1
Dot diagram of speeding tickets issued daily.

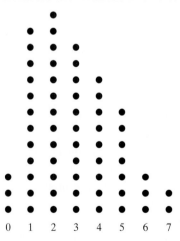

There are various ways in which dot diagrams can be modified. For instance, instead of dots we can use other symbols such as ×'s or *'s. Also, we could align the dots horizontally rather than vertically.

EXAMPLE 2.2   Thirty-six instrumentalists auditioned to fill vacancies in a symphony orchestra. The instruments that they played for the conductor were as follows:

flute	cello	viola	violin	violin	trumpet
trumpet	violin	violin	violin	cello	violin
violin	viola	violin	bass	viola	cello
violin	horn	violin	viola	horn	violin
viola	trumpet	cello	cello	bass	viola
violin	horn	horn	viola	violin	horn

Solution   Counting the number of times that each instrument appears we get the dot diagram of Figure 2.2. With the instruments thus ordered according to their frequencies, this kind of diagram is also referred to as a **Pareto diagram**.

Figure 2.2
Dot diagram of imstruments played.

2.2 STEM-AND-LEAF DISPLAYS

When we deal with large sets of data, the mere gathering of the information is no small task. Usually, much more has to be done, however, to make large sets of numbers comprehensible. Consider, for example, the following values, which represent the average travel time to work of 100 employees in a large downtown office building. The original travel times were measured to the nearest minute, and each value represents an employee's average time over five consecutive workdays.

44.0	35.4	28.4	37.0	46.0	35.4	19.4	20.4	56.4	43.2
36.2	38.4	49.2	31.8	86.4	12.6	27.4	14.0	39.4	39.4
15.8	28.8	38.0	44.0	38.4	74.0	23.0	11.4	39.8	30.2
29.2	40.6	49.6	30.4	12.2	123.8	42.0	47.0	32.4	39.2
35.2	56.4	31.0	45.0	90.2	100.0	39.0	37.0	49.4	28.2
12.6	27.0	47.8	52.6	41.0	40.0	28.0	23.6	37.6	37.8
30.0	45.8	18.0	41.0	22.6	24.2	89.6	90.4	43.0	29.8
56.2	24.8	12.6	53.6	125.4	16.2	39.0	40.8	33.6	39.4
45.6	37.4	18.0	50.6	103.4	52.4	20.2	64.6	22.2	60.0
42.2	42.0	16.2	108.2	44.0	42.6	39.4	37.6	41.4	40.4

What can be done to make this mass of information more usable? Some persons find it helpful to list such data ordered from low to high or high to low, although sorting a large set of numbers according to size can be a surprisingly difficult task. For these travel times, we get

11.4	12.2	12.6	12.6	12.6	14.0	15.8	16.2	16.2	18.0
18.0	19.4	20.2	20.4	22.2	22.6	23.0	23.6	24.2	24.8
27.0	27.4	28.0	28.2	28.4	28.8	29.2	29.8	30.0	30.2
30.4	31.0	31.8	32.4	33.6	35.2	35.4	35.4	36.2	37.0
37.0	37.4	37.6	37.6	37.8	38.0	38.4	38.4	39.0	39.0
39.2	39.4	39.4	39.4	39.4	39.8	40.0	40.4	40.6	40.8
41.0	41.0	41.4	42.0	42.0	42.2	42.6	43.0	43.2	44.0
44.0	44.0	45.0	45.6	45.8	46.0	47.0	47.8	49.2	49.4
49.6	50.6	52.4	52.6	53.6	56.2	56.4	56.4	60.0	64.6
74.0	86.4	89.6	90.2	90.4	100.0	103.4	108.2	123.8	125.4

and it can be seen, for example, that the smallest value is 11.4, the largest value is 125.4, and many of the values are in the thirties and forties.

The list of numbers, even in sorted form, is still a ponderous piece of information, and it will be helpful to have other ways of dealing with these values. A method developed in the late 1960s, the **stem-and-leaf display**, will give a good overall impression of the data.

To illustrate this technique, consider the following scores on a test of physical coordination given to 20 students who had consumed an amount of alcohol equal to 0.1% of their weight.

69	84	52	93	61	74	79	65	88	63
57	64	67	72	74	55	82	61	68	77

Now break each number into its tens and units digits, tallying together values that share the tens digit. That is, we will think of the number 69 as 6 | 9. The tens digits will then be aligned vertically with the units digits displayed to the side. For the set of 20 physical coordination scores, the picture is this:

```
5 | 2 7 5
6 | 9 1 5 3 4 7 1 8
7 | 4 9 2 4 7
8 | 4 8 2
9 | 3
```

The first row of the display, 5 | 2 7 5, tells us that the list contains the values 52, 57, and 55. The second row tells us that the list contains eight values in the sixties.

This table is referred to as a stem-and-leaf display, since each row represents a **stem** position, and each digit to the right of the vertical line can be thought of as a **leaf**. To make this stem-and-leaf display, begin with the stems only:

```
5 |
6 |
7 |
8 |
9 |
```

This step need not be perfect; after all, it is easy to add more stem positions at the top or bottom. Then mark the leaves by going through the data items sequentially. After the first three values (69, 84, 52), the stem-and-leaf display will look like this:

$$
\begin{array}{c|c}
5 & 2 \\
6 & 9 \\
7 & \\
8 & 4 \\
9 & \\
\end{array}
$$

After just one pass through the data, the complete stem-and-leaf display will be done.

The stem-and-leaf display conveys the same information as the original list, but it is much more compact. The stem-and-leaf display highlights the important aspects of the data. For instance, in this case it reveals immediately that most of the values are in the sixties.

The stem-and-leaf display accomplishes most of the task of sorting the values. Many persons like to complete the sorting by ordering the leaves on each branch as well. In this example, it gives the following:

$$
\begin{array}{c|ccccccc}
5 & 2 & 5 & 7 \\
6 & 1 & 1 & 3 & 4 & 5 & 7 & 8 & 9 \\
7 & 2 & 4 & 4 & 7 & 9 \\
8 & 2 & 4 & 8 \\
9 & 3 \\
\end{array}
$$

Not every set of values can be placed into a stem-and-leaf display this easily. Treatment of more complicated situations is a matter of taste. Let us consider the 100 average travel times noted before. These numbers were given to *tenths* of minutes. In making the stem-and-leaf display, we recommend ignoring the tenths rather than rounding the values to the nearest minute. This introduces a half-minute bias to the display, but produces with less effort nearly the same result as proper rounding. We will use the tens digit to label the stem, and this produces the following display:

$$
\begin{array}{r|l}
1 & 122224566889 \\
2 & 0022334477888899 \\
3 & 0001123555677777788899999999 \\
4 & 00001112222334445567799 \\
5 & 0223666 \\
6 & 04 \\
7 & 4 \\
8 & 69 \\
9 & 00 \\
10 & 038 \\
11 & \\
12 & 35 \\
13 & \\
\end{array}
$$

$$\textit{Note:} \quad \frac{7}{12} \Big| \frac{4}{3} \quad \begin{matrix} \text{means} \\ \text{means} \end{matrix} \quad \begin{matrix} 74 \\ 123 \end{matrix} \quad \begin{matrix} \text{minutes} \\ \text{minutes} \end{matrix}$$

This stem-and-leaf display requires looking through the original list just once. The leaves can then be sorted. Indeed, this picture conveys the message of the original list in a very clean pictorial form. It is helpful to attach a note, as we did, to help persons read the display. This is certainly helpful when the stem labels are not the tens digits.

Exercises 2.1, 2.4, and 2.10 are practice exercises; their complete solutions are given on page 42.

EXERCISES

✓ **2.1**   The number of passengers in 22 automobiles traveling on a highway is

3 4 3 5 4 2 2 7 1 2 4 3 2 5 1 2 2 4 1 2 3 6

Construct a dot diagram that shows the number of cars carrying 1, 2, 3, 4, 5, 6, or 7 passengers.

2.2   According to the U.S. National Oceanic and Atmospheric Administration, the average number of days per month when there was precipitation (of 0.01 inch or more) in Washington, D.C. for the period of 58 years through 1999 is as follows:

Month	Number of days	Month	Number of days
January	10	July	10
February	9	August	9
March	11	September	8
April	10	October	7
May	11	November	8
June	9	December	9

Construct a dot diagram that shows the number of months in which there were 7, 8, 9, 10, or 11 days of precipitation.

2.3   An automobile dealer complains to a manufacturer's representative that there are numerous defects in the paint of 25 automobiles received in a shipment. The manufacturer's representative checked each car and found the following defects:

1, 3, 0, 2, 1, 2, 2, 1, 2, 1, 1, 1, 2, 2, 3,

1, 1, 2, 1, 2, 2, 1, 3, 2, and 2

Construct a dot diagram that shows the number of automobiles that have 0, 1, 2, and 3 paint defects.

✓ **2.4**   The maintenance department of a real-estate management company received 25 maintenance requests during a week, which included

10   *electrical*, including replacement of burned-out light bulbs;

2   *carpentry*, including repairs to broken doors;

8   *plumbing*, including replacement of washers and repair of minor leaks;

4   *paint*, including painting over scratches and scrapes in lobby and in halls;

1   *elevator*, including repair of elevator control buttons.

Construct a Pareto diagram like that of Figure 2.2 with the largest value plotted on top and the remaining values plotted below in descending order.

2.5 Forty-two people attending a conference were offered their choice of nonalcoholic beverages and selected the following:

Beverage	Number
Coffee	15
Soda pop	10
Juice	7
Tea	6
None	4
Total	42

Construct a Pareto diagram like that of Figure 2.2 with the largest value plotted on top and the remaining values plotted below in descending order.

2.6 The following are the standardized reading scores of 16 high school students:

120 105 112 108 102 117 100 108 103 107 115 143 98 126 103 114

Construct a stem-and-leaf display with the stem labels 9, 10, ..., and 14.

2.7 The number of motor vehicles washed at a car-wash establishment in 24 consecutive business days is

168 195 227 193 207 189 176 216 164 199 198 203
214 191 171 200 197 195 184 202 188 173 197 181

Construct a stem-and-leaf display with the stem labels 16, 17, 18, 19, 20, 21, and 22.

2.8 We have illustrated stem-and-leaf displays with one-digit leaves, but sometimes it is more convenient to display them with, say, two-digit leaves, three-digit leaves, or more. For example, the numbers 247, 139, 223, 148, and 115 can be displayed as

$$
\begin{array}{c|ccc}
1 & 39 & 48 & 15 \\
2 & 47 & 23
\end{array}
$$

or, ordering the leaves as

$$
\begin{array}{c|ccc}
1 & 15 & 39 & 48 \\
2 & 23 & 47
\end{array}
$$

Construct a stem-and-leaf display with two-digit leaves and with stem labels 1, 2, 3, 4, and 5 for the following weekly earnings, in dollars, of 15 salespersons:

305 255 319 167 270 291 512 183 334 362 188 217 440 195 408

2.9 List the data that corresponds to the following stems of stem-and-leaf displays:

$$
\begin{array}{lll}
\text{(a)} & 1 & 0\ 1\ 1\ 2\ 5\ 7\ 8 \\
\text{(b)} & 12 & 0\ 2\ 3\ 3\ 5 \\
\text{(c)} & 0.6 & 0\ 2\ 3\ 6 \\
\text{(d)} & 1.5 & 0\ 1\ 1\ 3\ 4\ 6
\end{array}
$$

✓ **2.10** If we want to construct a stem-and-leaf display with more stems than there would be otherwise, we can divide each stem position in two. Use the first stem position to hold leaves 0, 1, 2, 3, and 4, and use the second stem position to hold leaves 5, 6, 7, 8, and 9. For the data on page 18, which is

$$69 \quad 84 \quad 52 \quad 93 \quad 61 \quad 74 \quad 79 \quad 65 \quad 88 \quad 63$$
$$57 \quad 64 \quad 67 \quad 72 \quad 74 \quad 55 \quad 82 \quad 61 \quad 68 \quad 77$$

we would thus get the **double-stem display**

```
5* | 2
5· | 5  7
6* | 1  1  3  4
6· | 5  7  8  9
7* | 2  4  4
7· | 7  9
8* | 2  4
8· | 8
9* | 3
```

where we doubled the number of stem positions by cutting in half the interval covered by each tens digit. Construct a double-stem display for the data of Exercise 2.7.

2.11   Following are the numbers of minutes that it took a newspaper delivery service to deliver the morning newspapers on 28 days:

$$78 \quad 66 \quad 54 \quad 62 \quad 67 \quad 68 \quad 62 \quad 60 \quad 71 \quad 67 \quad 80 \quad 60 \quad 56 \quad 61$$
$$63 \quad 65 \quad 52 \quad 69 \quad 59 \quad 65 \quad 73 \quad 68 \quad 64 \quad 60 \quad 71 \quad 57 \quad 56 \quad 76$$

Construct a double-stem display for these values.

2.12   According to the Federal Bureau of Investigation, the number of motor vehicle thefts in the midwestern states during a recent year, per 10,000 of population, was

State	Thefts per 10,000 of population	State	Thefts per 10,000 of population
Illinois	43	Missouri	42
Indiana	34	Nebraska	33
Iowa	18	North Dakota	16
Kansas	23	Ohio	35
Michigan	58	South Dakota	12
Minnesota	28	Wisconsin	26

Construct a stem-and-leaf display with stem labels 1, 2, 3, 4, and 5.

2.13   The number of patients who received treatment at a hospital emergency room in each of 30 consecutive days was

$$35 \quad 37 \quad 46 \quad 33 \quad 31 \quad 29 \quad 43 \quad 31 \quad 31 \quad 46$$
$$43 \quad 49 \quad 36 \quad 38 \quad 40 \quad 41 \quad 39 \quad 40 \quad 28 \quad 34$$
$$30 \quad 56 \quad 44 \quad 38 \quad 37 \quad 51 \quad 44 \quad 36 \quad 47 \quad 50$$

Construct a double-stem plot with one-digit leaves.

2.14 Following are the mechanical aptitude scores of fifty applicants for production line jobs in a factory.

$$
\begin{array}{cccccccccc}
47 & 89 & 79 & 50 & 61 & 98 & 65 & 58 & 46 & 71 \\
63 & 54 & 99 & 61 & 75 & 52 & 68 & 65 & 79 & 73 \\
80 & 58 & 71 & 65 & 70 & 57 & 88 & 71 & 64 & 67 \\
86 & 83 & 72 & 46 & 78 & 50 & 61 & 69 & 78 & 56 \\
72 & 53 & 64 & 93 & 69 & 82 & 67 & 92 & 71 & 82 \\
\end{array}
$$

(a) Construct a stem-and-leaf display with the stem labels 4, 5, 6, 7, 8, and 9.

(b) Construct a double-stem display for the same data. Use the first stem position to hold the leaves 0, 1, 2, 3, 4, and use the second stem position to hold the leaves 5, 6, 7, 8, and 9.

(c) If the minimum acceptable score to obtain a job on the production line is 75, use the double-stem display of part (b) to determine the number of acceptable applicants.

2.3 FREQUENCY DISTRIBUTIONS

The previous section confirms our experience that untreated raw data is difficult to deal with. The dot diagram and the stem-and-leaf display were introduced as techniques to organize numbers and make them more comprehensible. In this section we discuss the more traditional technique of using frequency distributions to classify, or group, data.

Consider the problem of a social scientist who wants to examine the ages of death-row inmates, that is, persons awaiting execution, in the United States. He will likely turn to the *Statistical Abstract of the United States*, which is a reliable source for information of this nature.

A good overall understanding of a large set of data can often be created by grouping the data into a number of classes (intervals or categories), and our social scientist finds that the desired information is presented as follows for the year 1999, from the 2001 edition of the *Statistical Abstract of the United States*.

Persons under sentence of death—1999	
Age	*Number of persons*
Under 20 years	16
20–24 years	251
25–34 years	1,108
35–54 years	1,958
35–years and over	194
Total	3,527

A table like this is called a **frequency distribution**, or simply a **numerical distribution**. It shows how the ages of 3,527 death-row inmates are distributed. The data are grouped according to the numeric value, age. In some instances we group information according to nonnumeric criteria, such as color, geographic region, consumer preference, or medical diagnosis.

In the preceding example, each class covered more than one possible value. For instance, the second class covered the values 20, 21, 22, 23, and 24. Each class may also cover a single value, as illustrated by the following example based on a study in which 400 persons were asked how many full-length movies they had seen on television during the preceding week.

Number of movies	Number of persons
0	72
1	106
2	153
3	40
4	18
5	7
6	3
7	0
8	1
Total	400

The following display illustrates a **categorical distribution**. These nonnumerical categories collectively constitute the values of all products and services sold by establishments in the luggage and personal leather goods industry for a recent year. The information was provided by the U.S. Department of Commerce and other sources.

	Millions of dollars
Leather and sheep-lined clothing	201
Leather gloves and mittens	155
Luggage	960
Women's handbags and gloves	269
Personal leather goods	371
	1,956

Frequency distributions present data in a relatively compact form, give a good overall picture, and contain information that is adequate for many purposes; but some things can be obtained only from the original data. For instance, the distribution given previously on death-row inmates cannot tell us how many are 23 years old or how many are over 60. Similarly, the study of the personal leather goods industry might have provided data concerning imports, exports, other products, and competing merchandise. Nevertheless, frequency distributions present data in a more usable form, and the price we pay—the loss of certain information—is usually a favorable exchange.

The construction of numerical distributions consists essentially of four steps: (1) choosing the classes, (2) sorting (or tallying) the data into these classes, (3) counting the number of items in each class, and (4) displaying the results in the form of a chart or table. Since the second and third steps are purely mechanical and the fourth step is a matter of craftsmanship and taste, we shall concentrate here on the first step, choosing suitable classifications.

This first step involves choosing the number of classes and the range of values that each class should cover, that is, from where to where each class should go. Both of these choices are arbitrary to some extent, but they depend on the nature of the data and on the purpose that the distribution is to serve. The following are some rules that are generally observed.

> *We seldom use fewer than 6 or more than 15 classes; the exact number we use in a given situation depends mainly on the number of measurements or observations that we have to group.*

Clearly, we would lose more than we gain if we group 6 observations into 12 classes with many of them empty, and we would probably give away too much information if we group 10,000 measurements into 3 classes.

> *We always make sure that each item (measurement or observation) goes into one and only one class.*

To this end, we must make sure that the smallest and largest values fall within the classification, that none of the values can fall into possible gaps between successive classes, and that the classes do not overlap, that is, successive classes have no values in common.

> *Whenever possible, we use classes covering equal ranges (or intervals) of values.*

If we can, we also use numbers for these ranges that are easy to work with, such as 5, 10, or 100, for this will facilitate constructing, reading, and using the distribution.

In connection with these rules, the age distribution given for death-row inmates is not a good example, but, presumably, the government statisticians had good reasons for choosing the classes as they did. There are only five classes, none of the classes covers equal ranges of values, and the "55 and over" class is *open*; for all we know the oldest person may be 75 or 85.

In general, we refer to classes of the "less than," "or less," "more than," and "or more"type as **open classes**, and they are used to reduce the number of classes that are required when a few of the values are much smaller (or much greater) than the rest. However, they should be avoided when possible for, as we shall see, they make it difficult or even impossible to calculate certain further descriptions that may be of interest.

Insofar as the second rule is concerned, it is important to watch whether the data is given to the nearest dollar or to the nearest cent, whether it is given to the nearest inch or to the nearest tenth of an inch, whether it is given to the nearest ounce or to the nearest hundredth of an ounce, and so forth. For instance, to group the weights of certain animals, we could use the first of the following three classifications if the weights are given to the nearest kilogram, the second if the weights are given to the nearest tenth of a kilogram, and the third if the weights are given to the nearest hundredth of a kilogram.

Weight (kilograms)	Weight (kilograms)	Weight (kilograms)
10–14	10.0–14.9	10.00–14.99
15–19	15.0–19.9	15.00–19.99
20–24	20.0–24.9	20.00–24.99
25–29	25.0–29.9	25.00–29.99
30–34	30.0–34.9	30.00–34.99

To illustrate what we have been discussing in this section, let us now go through the actual steps of grouping a given set of data into a frequency distribution.

EXAMPLE 2.3 Construct a distribution of the following data on the amount of time (in hours) that 80 college students devoted to leisure activities during a typical school week:

```
23  24  18  14  20  24  24  26  23  21
16  15  19  20  22  14  13  20  19  27
29  22  38  28  34  32  23  19  21  31
16  28  19  18  12  27  15  21  25  16
30  17  22  29  29  18  25  20  16  11
17  12  15  24  25  21  22  17  18  15
21  20  23  18  17  15  16  26  23  22
11  16  18  20  23  19  17  15  20  10
```

Solution Since the smallest value is 10 and the largest is 38, we might choose the six classes 10–14, 15–19, 20–24, 25–29, 30–34, and 35–39; or we might choose the eight classes 10–13, 14–17, 18–21, 22–25, 26–29, 30–33, 34–37, and 38–41, to mention two possibilities. Note that in each case the classes accommodate all the data, they do not overlap, and they are all of the same size.

Generally, we prefer classes that begin or end in multiples of 5 or 10; accordingly we will decide on the first of these classifications. We now tally the 80 observations and get the results shown in the following table:

Hours	Tally	Frequency
10–14	ЖҤ |||	8
15–19	ЖҤ ЖҤ ЖҤ ЖҤ ЖҤ |||	28
20–24	ЖҤ ЖҤ ЖҤ ЖҤ ЖҤ ||	27
25–29	ЖҤ ЖҤ ||	12
30–34	||||	4
35–39			1
Total		80

The numbers given in the right-hand column of this table, which show how many items fall into each class, are called the **class frequencies**. Also, the smallest and largest values that can go into any given class are referred to as its **class limits**, and in our example they are 10 and 14, 15 and 19, 20 and 24, . . . , and 35 and 39. More specifically, 10, 15, 20, . . . , and 35 are called the **lower class limits**, and 14, 19, 24, . . . , and 39 are called the **upper class limits**.

The amounts of time that we grouped in our example were all given to the nearest hour, so that the first class actually covers the interval from 9.5 hours to 14.5 hours, the second class covers the interval from 14.5 hours to 19.5 hours, and so forth. It is customary to refer to these numbers as the **class boundaries** or the **"real" class limits**. In actual practice, class limits are used much more widely than class boundaries, and we have mentioned them here mainly because they will be needed in Chapter 3 for calculating certain descriptive measures of a distribution.

Also, since the upper class boundary of any one class always equals the lower class boundary of the next class, there is some disagreement as to what one should do about values that actually equal one of the class boundaries. For instance, MINITAB always places observations that fall on class boundaries into the class to the right (except those in the class farthest to the right, which are placed in the class to the left). To avoid situations like this, which create a lack of symmetry, it is desirable always to choose classes for which none of the values that we intend to group can actually equal one of the class boundaries.

Numerical distributions also have what we call **class marks** and **class intervals**. Class marks are simply the midpoints of the classes, and they are obtained by adding the upper and lower limits of a class (or its upper and lower boundaries) and dividing by 2. A class interval is the length of a class, or the range of values that it can contain, and it is given by the difference between its class boundaries. If the classes of a distribution are all of equal length, their common class interval, which we refer to as the **class interval of the distribution**, is also given by the difference between any two successive class marks.

EXAMPLE 2.4   Find the class marks and the class interval of the distribution obtained in the preceding example.

Solution   The class marks are

$$\frac{10+14}{2} = 12, \frac{15+19}{12} = 17, \ldots, \quad \text{and} \quad \frac{35+39}{2} = 37$$

The intervals of the classes are $14.5 - 9.5 = 5$, $19.5 - 14.5 = 5$, ..., and $39.5 - 34.5 = 5$; since they are all equal, we say that the class interval of the distribution is 5. Note that class intervals are *not* given by the differences between the respective class limits, which in our example all equal 4, and not 5. ∎

There are essentially two ways in which frequency distributions can be modified to suit particular needs. One way is to convert a distribution into a **percentage distribution** by dividing each class frequency by the total number of items grouped and then multiplying by 100%.

EXAMPLE 2.5   Convert the distribution of the amounts of time the 80 college students devoted to leisure activities into a percentage distribution.

Solution   The first class contains

$$\frac{8}{80} \cdot 100\% = 10\% \text{ of the data}$$

the second class contains

$$\frac{28}{80} \cdot 100\% = 35\% \text{ of the data}$$

the third class contains

$$\frac{27}{80} \cdot 100\% = 33.75\% \text{ of the data}$$

..., and the sixth class contains

$$\frac{1}{80} \cdot 100\% = 1.25\% \text{ of the data}$$

These results are shown in the following table:

Hours	Percentage
10–14	10
15–19	35
20–24	33.75
25–29	15
30–34	5
35–39	1.25
Total	100

The other way of modifying a frequency distribution is to convert it into a "less than," "or less," "more than," or "or more" **cumulative distribution**. To construct a cumulative distribution, we simply add the class frequencies, starting either at the top or at the bottom of the distribution.

EXAMPLE 2.6   Convert the distribution for leisure activity times of college students into a cumulative "less than" distribution.

Solution   Since none of the values is less than 10, 8 are less than 15, $8 + 28 = 36$ are less than 20, $8 + 28 + 27 = 63$ are less than 25, ..., the result is shown in the following table:

Hours	Cumulative frequency
Less than 10	0
Less than 15	8
Less than 20	36
Less than 25	63
Less than 30	75
Less than 35	79
Less than 40	80

Instead of "less than 10," we could also have written "less than or equal to 9"; instead of "less than 15," we could also have written "less than or equal to 14"; and so on. When cumulative distributions are generated by means of computers, they are usually given as "less than or equal to" distributions.

In the same way, we can also convert a percentage distribution into a **cumulative percentage distribution** by adding the percentages, starting either at the top or at the bottom of the distribution. For our example, we get the following:

Hours	Cumulative percentage
Less than 10	0
Less than 15	10
Less than 20	45
Less than 25	78.75
Less than 30	93.75
Less than 35	98.75
Less than 40	100

Exercises 2.15, 2.18, 2.21, 2.25, and 2.28 are practice exercises; their complete solutions are given on pages 42 and 43.

✓ **2.15**   According to the U.S. Department of Education, the number of pupils per teacher in public schools of the fifty states and Washington, D.C. ranges from 12.3 to 21.9. Indicate the limits of five classes into which the number of pupils per teacher (to the nearest tenth) might be grouped.

2.16   The average speed of the winners of the annual Indianapolis 500 automobile race from 1957 to 2001, in a race that is usually 500 miles in length, varied from 133.791 to 185.981 miles per hour. Indicate limits of five classes into which these winning speeds might be grouped.

2.17   The circulations of the 100 leading U.S. newspapers in the year 2000, according to the *2001 Editor and Publisher International Yearbook*, varied from 100,356 to 1,762,751. Indicate the limits of eight classes into which these values (numbers of copies) might be grouped.

✓ **2.18**   The numbers of empty seats on flights from Dallas to New Orleans are grouped into a table with the classes 0–9, 10–19, 20–29, 30–39, and 40 or more. Will it be possible to determine exactly from this table the number of flights on which there were

  (a) at least 20 empty seats;

  (b) more than 20 empty seats;

  (c) more than 19 empty seats;

  (d) at least 19 empty seats;

  (e) exactly 19 empty seats?

2.19   The declared values of packages mailed from a foreign country are grouped into a distribution with the following classes: $0.00–$49.99, $50.00–$99.99, $100.00–$149.99, $150.00–$199.99, and $200.00 or more. Will it be possible to determine exactly from the distribution the number of packages valued at

  (a) more than $100.00;

  (b) at least $100.00;

  (c) exactly $100.00;

  (d) more than $99.99?

2.20   The following is the distribution of the weights of 125 mineral specimens collected on a field trip:

EXERCISES

Weight (grams)	Number of specimens
0– 19.9	19
20.0– 39.9	38
40.0– 59.9	35
60.0– 79.9	17
80.0– 99.9	11
100.0–119.9	3
120.0–139.9	2
Total	125

If possible, find how many of the specimens weigh

(a) at most 59.9 grams;

(b) less than 40.0 grams;

(c) more than 100.0 grams;

(d) 80.0 grams or less;

(e) exactly 20.0 grams;

(f) anywhere from 40.0 or more to less than 80.

✓ **2.21** The numbers of students who are absent each day from a certain school are grouped into a distribution having the classes 0–14, 15–29, 30–44, and 45–59. Determine

(a) the lower class limits;

(b) the upper class limits;

(c) the class marks;

(d) the class interval of the distribution.

2.22 The number of farms in Puerto Rico in 1988, by size, was approximately

Farms by size	Number
0 to 9 cuerdos	7,759
10 to 19 cuerdos	4,473
20 to 49 cuerdos	4,023
50 to 99 cuerdos	1,792
100 to 174 cuerdos	809
175 to 259 cuerdos	421
260 cuerdos or more	674

Determine

(a) the lower class limits;

(b) the upper class limits;

(c) the class marks;

(d) whether we can calculate the class interval of the distribution.

2.23 The number of congressmen absent each day during a session of Congress are grouped into a distribution having the classes 0–19, 20–39, 40–59, 60–79, 80–99, 100–119, 120–139, and 140–159. Determine

(a) the lower class limits;

(b) the upper class limits;

(c) the class marks;

(d) the class interval of the distribution.

2.24 The number of dinner reservations made at a restaurant are grouped into a distribution having the classes 50–64, 65–79, 80–94, and 95–109. Find

(a) the class limits;

(b) the class boundaries;

(c) the class marks;

(d) the class interval of the distribution.

✓ **2.25** The class marks of a distribution of the daily number of traffic accidents reported in a county are 4, 11, 18, and 25. Find

(a) the class boundaries;

(b) the class limits.

2.26 The class marks of a distribution of the daily number of calls received by a small cab company are 22, 27, 32, 37, 42, 47, and 52. Find

(a) the class boundaries;

(b) the class limits.

2.27 The class marks of a distribution of temperature readings, given to the nearest degree Fahrenheit, are 36, 45, 54, 63, 72, 81, and 90. Find

(a) the class boundaries;

(b) the class limits.

✓ **2.28** To group sales invoices ranging from $5.00 to $30.00, a clerk uses the following classification: $5.00–$9.99, $10.00–$15.99, $15.00–$19.99, $20.00–$24.90, and $25.00–$29.99. Explain where difficulties might arise.

2.29 To group data on the number of rainy days reported by a weather station for the month of May during the last 50 years, a meteorologist uses the classes 0–5, 6–10, 12–17, 18–23, and 23–30. Explain where difficulties might arise.

2.30 Following are the numbers of years of service of all 55 justices of the United States Supreme Court who were appointed from the years 1900 to 2002. The data includes the nine justices who were members of the court in 2002 and all other appointees since 1900.

29	5	23	16	16
19	8	36	16	15
3	15	9	33	27
4	16	5	5	21
5	7	1	23	16
26	16	12	31	16
5	11	6	3	14
10	15	13	4	12
10	6	7	24	11
26	34	18	17	9
22	19	7	24	8

(a) Group the years of service into a table having the classes 0–4, 5–9, 10–14, 15–19, 20–24, 25–29, 30–34, and 35–39.

(b) Convert the distribution obtained in part (a) into a cumulative "or more" distribution.

2.31 Convert the distribution obtained in part (a) of Exercise 2.30 into a

(a) percentage distribution;

(b) cumulative "less than or equal to" percentage distribution;

(c) cumulative "or more" distribution.

2.32   The following are the grades that 50 students obtained on an accounting test:

$$
\begin{array}{cccccccccc}
73 & 65 & 82 & 70 & 45 & 50 & 70 & 54 & 32 & 75 \\
75 & 67 & 65 & 60 & 75 & 87 & 83 & 40 & 72 & 64 \\
58 & 75 & 89 & 70 & 73 & 55 & 61 & 78 & 89 & 93 \\
43 & 51 & 59 & 38 & 65 & 71 & 75 & 85 & 65 & 85 \\
49 & 97 & 55 & 60 & 76 & 75 & 69 & 35 & 45 & 63
\end{array}
$$

(a)   Prepare a stem-and-leaf display of these values. Use the tens digits to form the stem.

(b)   Group these grades into a distribution having the classes 30–39, 40–49, 50–59, ..., and 90–99.

(c)   Convert the distribution obtained in part (b) into a cumulative "less than" distribution, beginning with "less than 30."

2.33   The following is a distribution of the ages of the members of a video dating service for single persons:

Age (years)	Frequency
20–24	129
25–29	221
30–34	310
35–39	163
40–44	105
45–49	62
50–54	10
Total	1,000

(a)   Convert the distribution into a cumulative "less than or equal to" distribution.

(b)   Convert the distribution into a cumulative "or more" distribution.

2.34   The following are the miles per gallon obtained with 40 tankfuls of gas:

$$
\begin{array}{cccccccccc}
24.1 & 25.0 & 24.8 & 24.3 & 24.2 & 25.3 & 24.2 & 23.6 & 24.5 & 24.4 \\
24.5 & 23.2 & 24.0 & 23.8 & 23.8 & 25.3 & 24.5 & 24.6 & 24.0 & 25.2 \\
25.2 & 24.4 & 24.7 & 24.1 & 24.6 & 24.9 & 24.1 & 25.8 & 24.2 & 24.2 \\
24.8 & 24.1 & 25.6 & 24.5 & 25.1 & 24.6 & 24.3 & 25.2 & 24.7 & 23.3
\end{array}
$$

(a)   Group these figures into a distribution having the classes 23.0–23.4, 23.5–23.9, 24.0–24.4, 24.5–24.9, 25.0–25.4, and 25.5–25.9.

(b)   Convert the distribution obtained in part (a) into a cumulative "or more" distribution, beginning with "23.0 or more" and ending with "26.0 or more."

2.35   In the sport of baseball the winner of the World Series is the first team to win four of the seven or fewer games that are played. The losing team can therefore win 0,

1, 2, or 3 games. For the last forty of the World Series played, ending with the year 2000, the losing team won only the following number of games:

$$
\begin{array}{cccccccccc}
3 & 1 & 3 & 0 & 3 & 3 & 0 & 3 & 3 & 1 \\
1 & 3 & 3 & 3 & 1 & 3 & 0 & 2 & 2 & 3 \\
2 & 2 & 3 & 1 & 1 & 3 & 3 & 3 & 1 & 0 \\
0 & 3 & 2 & 2 & 2 & 2 & 3 & 0 & 0 & 1
\end{array}
$$

Construct a distribution showing how many games, 0, 1, 2, or 3, were won by the defeated team of the World Series.

2.36 A survey made at a resort city showed that 50 tourists arrived by the following means of transportation:

car, train, plane, plane, plane, bus, train, car, car, car, plane, car, plane, train, car, car, bus, car, plane, plane, train, train, plane, plane, car, car, train, car, car, plane, car, car, plane, bus, plane, bus, car, plane, car, car, train, train, car, plane, bus, plane, car, car, train, bus

Construct a categorical distribution showing the frequencies corresponding to the different means of transportation.

2.37 Asked to rate the maneuverability of a car as excellent, very good, good, fair, poor, or very poor, 40 drivers responded as follows:

very good, good, good, fair, excellent, good, good, good, very good, poor, good, good, good, good, very good, good, fair, good, good, very poor, very good, fair, good, good, excellent, very good, good, good, good, fair, fair, very good, good, very good, excellent, very good, fair, good, good, very good

Construct a distribution showing the frequencies corresponding to the different ratings of the maneuverability of the car.

2.4 GRAPHICAL PRESENTATIONS

When frequency distributions are constructed primarily to condense large sets of data and display them in an "easy to digest" form, they are often presented graphically. The most common form of graphical presentation of a frequency distribution is the **histogram**. The data in Example 2.3 on page 26 is shown as a histogram in Figure 2.3.

A histogram is constructed by representing the measurements or observations that are grouped on a horizontal scale, the class frequencies on a vertical scale, and drawing rectangles whose bases equal the class intervals and whose heights are determined by the corresponding class frequencies. The markings on the horizontal scale can be the class limits as in Figure 2.3, the class boundaries, the class marks, or arbitrary key values. For easy readability it is usually preferable to indicate the class limits, although the rectangles actually go from one class boundary to the next. Histograms cannot be used in connection with frequency distributions having open classes, and they must be used with extreme care when

Figure 2.3
Histogram of the distribution of the amounts of time that 80 students devoted to leisure activities, during a typical school week.

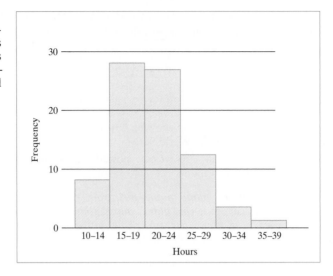

Figure 2.4
The data here is identical to the data of Figure 2.3. The histogram is reproduced from the display screen of a TI-83 graphing calculator.

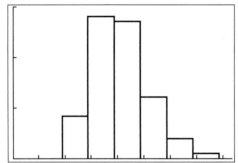

the class intervals are not all equal. In that case, it is best to represent the class frequencies by the areas of the rectangles instead of their heights.

Figure 2.4 presents the data of Figure 2.3 in similar fashion, except that the histogram is reproduced from a TI-83 graphing calculator.

Figure 2.4 is intended primarily for users of graphing calculators and can be omitted, without loss of continuity, by other readers. It is nevertheless of general interest to observe that we can obtain a histogram from a graphing calculator by entering the numerical data and following specific procedures. We do not actually draw the rectangles. The results of Figure 2.3 and Figure 2.4 are comparable.

For large sets of data, it may be convenient to construct histograms directly from raw data by using a computer package designed specially for data analysis. For instance, the MINITAB printout in Figure 2.5 shows a histogram of this distribution with the class marks 12, 17, 22, 27, 32, and 37 (as given on page 35 for the data dealing with the amounts of time that 80 students devoted to leisure activities).[†]

[†] In the printout of Figure 2.5 and in others appearing in this book, words and numbers appear that relate to the technical aspects of operating the particular computer program employed. If a computer is available, the reader should refer to the appropriate manuals for operating instructions and for a list of problems that can be solved with existing programs.

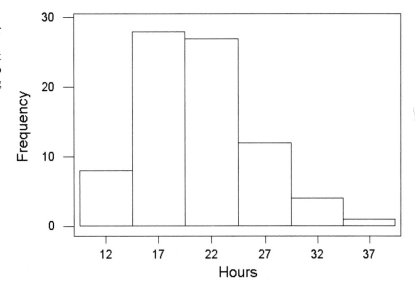

Figure 2.5
MINITAB printout for the histogram of the amounts of time that 80 students devoted to leisure activities during a typical school week.

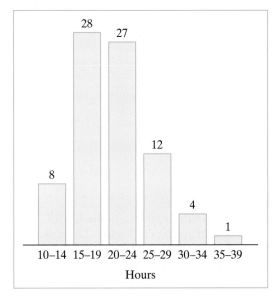

Figure 2.6
Bar chart of the distribution of the amounts of time that 80 students devoted to leisure activities during a typical school week.

Similar to histograms are **bar charts**, such as the one shown in Figure 2.6. The heights of the rectangles, or bars, again represent the class frequencies, but there is no pretense of having a continuous horizontal scale.

Another, less widely used, form of graphical presentation is the **frequency polygon** (see Figure 2.7). In this form the class frequencies are plotted at the class marks, and the successive points are connected by means of straight lines. Note that we added classes with zero frequencies at both ends of the distribution to "tie down" the graph to the horizontal scale. If we apply the same technique to a cumulative distribution, we obtain what is called an **ogive**, which rhymes with live or jive. However, the cumulative frequencies are plotted at the corresponding class limits instead of the class marks. It stands to reason that "less than or

Figure 2.7
Frequency polygon of the distribution of the amounts of time that 80 students devoted to leisure activities during a typical school week.

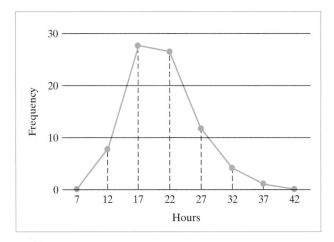

Figure 2.8
Ogive of the distribution of the amounts of time that 80 students devoted to leisure activities during a typical school week.

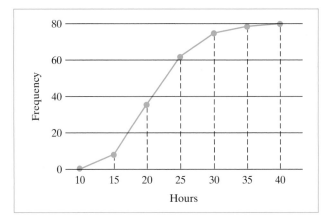

equal to 19," for example, should be plotted at 19. This value actually includes everything up to and including 19. Figure 2.8 shows an ogive of the cumulative "less than" distribution, which we constructed in Example 2.6.

Although the visual appeal of histograms, bar charts, frequency polygons, and ogives exceeds that of tables, there are various ways in which distributions can be presented more dramatically and more effectively. Two kinds of such pictorial presentations (often seen in newspapers, magazines, and reports of various sorts) are illustrated by the **pictograms** of Figures 2.9 and 2.10.

Categorical (or qualitative) distributions are often presented graphically as **pie charts**, such as the one shown in Figure 2.11, where a circle is divided into sectors (pie-shaped pieces) that are proportional in size to the corresponding frequencies or percentages. To construct a pie chart, we first convert the distribution into a percentage distribution. Then, since a complete circle corresponds to 360°, we obtain the central angles of the various sectors by multiplying the percentages by 360. For instance, the U.S. Census Bureau asserts that the marital status of 8,379 (thousands) white, female heads of household with no spouse present, year 2000, is as shown in the following table.

	Thousands
Married, but spouse absent	1,210
Widowed	1,764
Divorced	3,431
Never married	1,974
Total	8,379

The percentages are obtained by dividing each value by 8,379, and multiplying by 100, obtaining 14%, 21%, 41%, and 24%, respectively. Finally, the central angles for Figure 2.11 are obtained by multiplying each percentage by 360°, obtaining 50°, 76°, 148°, and 86°.

Figure 2.9
Electric energy production in the United States (billions of kilowatt hours).

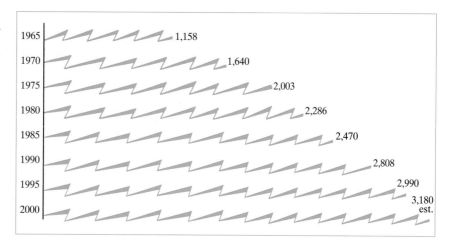

Figure 2.10
Population of the United States (1920–2000).

Each symbol = 10 million persons

Figure 2.11
Marital status of white
female heads of house-
hold with no spouse
present, 2000.

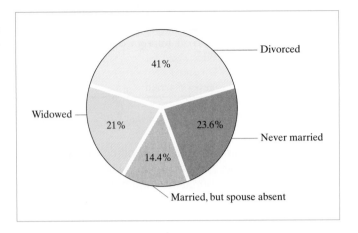

EXAMPLE 2.7   To study their attitudes toward social issues, 1,200 persons were asked (among other things) whether we are spending "too little," "about right amount," or "too much" on social welfare programs. Draw a pie chart to display the results shown in the following table:

	Number of persons
Too low	296
About right	360
Too high	544
Total	1,200

Solution   The percentages corresponding to the three categories are

$$\frac{296}{1,200} \cdot 100\% \approx 24.7\%, \quad \frac{360}{1,200} \cdot 100\% = 30.0\%$$

$$\text{and } \frac{544}{1,200} \cdot 100\% \approx 45.3\%$$

Multiplying by 360, we find that the central angles of the three sectors are

$$0.247(360) \approx 88.92°, \, 0.300(360) = 108.00°,$$

$$\text{and } 0.453(360) \approx 163.08°$$

The symbol $\approx$ means *approximately equal* (its use will be illustrated further in Section 3.1). Rounding the angles to the nearest degree and measuring the angles with a protractor, we get the pie chart shown in Figure 2.12.   ■

Many computer programs can produce pie charts directly on their display screens.

Research on the ability to read graphical displays has suggested that untrained users have trouble interpreting pie charts. Comparing information from two adjacent pie charts is particularly difficult. The comparison of bar heights in histograms is rather easy. It is recommended that pie charts be used with great care.

Figure 2.12
Responses to questions about welfare expenditures.

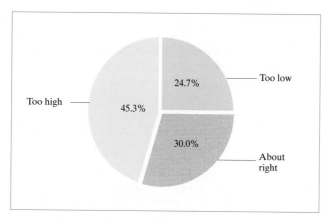

Exercises 2.38 and 2.42 are practice exercises; their complete solutions are given on pages 44 and 45.

✓ **2.38**   The following is the distribution of the total finance charges that 200 customers paid on their budget accounts at a department store:

Amount (dollars)	Frequency
1– 20	18
21– 40	62
41– 60	63
61– 80	43
81–100	14
Total	200

(a) Draw a histogram of this distribution.

(b) Draw a bar chart of this distribution.

2.39   Convert the distribution of Exercise 2.38 into a cumulative "less than" distribution and draw an ogive.

2.40   Following is the distribution of salaries of the fifty governors of the United States for the year 2001.

Salaries (in thousands of dollars)	Frequency
50– 74	3
75– 99	21
100–124	19
125–149	3
150–174	1
175–199	3
Total	50

(a) Draw a histogram of this distribution.

(b) Draw a frequency polygon of this distribution. Class marks for the frequency polygon are 62, 87, ..., 187.

2.41 Convert the distribution of Exercise 2.40 into a cumulative "less than" distribution and draw an ogive.

✓ **2.42** The following table shows how workers in Phoenix, Arizona, get to work:

Means of transportation	Percentage
Ride alone	81
Car pool	14
Ride bus	3
Varies or work at home	2
Total	100

Construct a pie chart of this percentage distribution.

2.43 Data from the Institute of International Education shows that foreign students in U.S. colleges originate from the following regions:

Region	Enrollment (in thousands)
Africa	21
Asia	290
Europe	67
Latin America	47
North America	24
Oceania	4

Construct a pie chart of this categorical distribution.

2.44 Data from the U.S. Bureau of the Census shows that the resident populations of the East North Central states are as follows:

State	Population (in thousands)
Indiana	6,080
Illinois	12,419
Michigan	9,438
Wisconsin	5,364
Total	33,301

Construct a pie chart of this categorical distribution.

2.45   According to the U.S. National Center for Educational Statistics, earned degrees conferred by level in 1998 were as follows (in thousands):

Degrees	Number of degrees
Associate's	556
Bachelor's	1,183
Master's	429
Doctor's	46
Total	2,214

Construct a pie chart of this categorical distribution.

2.46   Here, again, are the grades from Exercise 2.32 on page 32 that 50 students obtained on an accounting test:

73	65	82	70	45	50	70	54	32	75
75	67	65	60	75	87	83	40	72	64
58	75	89	70	73	55	61	78	89	93
43	51	59	38	65	71	75	85	65	85
49	97	55	60	76	75	69	35	45	63

Use a computer package to construct a histogram with the classes 30–39, 40–49, 50–59, ..., and 90–99. Construct also a histogram with the classes 20–39, 40–59, 60–79, and 80–99. Which histogram do you prefer?

2.47   Following are the body weights of 50 immature rats used in a study of vitamin deficiencies. Use a computer package to construct a histogram of these weights with the classes 80–89, 90–99, 100–109, ..., and 140–149.

136	92	115	118	121	137	132	120	104	125
119	115	101	129	87	108	110	133	135	126
127	103	110	126	118	82	104	137	120	95
146	126	119	119	105	132	126	118	100	113
106	125	117	102	146	129	124	113	95	148

2.48   Here, again, are the mileages of Exercise 2.34 on page 32, which are the miles per gallon obtained with 40 tankfuls of gas. Use a computer package to construct a histogram with the classes 23.0–23.4, 23.5–23.9, 24.0–24.4, ..., and 25.5–25.9.

24.1	25.0	24.8	24.3	24.2	25.3	24.2	23.6	24.5	24.4
24.5	23.2	24.0	23.8	23.8	25.3	24.5	24.6	24.0	25.2
25.2	24.4	24.7	24.1	24.6	24.9	24.1	25.8	24.2	24.2
24.8	24.1	25.6	24.5	25.1	24.6	24.3	25.2	24.7	23.3

✓ SOLUTIONS TO PRACTICE EXERCISES

2.1

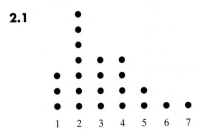

```
  1   2   3   4   5   6   7
```

2.4   Electrical   ● ● ● ● ● ● ● ● ● ● 10

   Plumbing   ● ● ● ● ● ● ● ● 8

   Painting   ● ● ● ● 4

   Carpentry   ● ● 2

   Elevator   ● 1

2.10
```
16* | 4
16· | 8
17* | 1  3
17· | 6
18* | 4  1
18· | 9  8
19* | 3  1
19· | 5  9  8  7  5  7
20* | 3  0  2
20· | 7
21* | 4
21· | 6
22* |
22· | 7
```

2.15 The number of students per teacher covers a range of 12.3 to 21.9 = 9.6. It is reasonable that we use the following five classes:

Number of students per teacher
12.0–13.9
14.0–15.9
16.0–17.9
18.0–19.9
20.0–21.9

2.18 (a) Yes;

(b) no;

(c) yes;

(d) no;

(e) no.

2.21 (a) The lower class limits are 0, 15, 30, and 45.

(b) The upper class limits are 14, 29, 44, and 59.

(c) The class marks are

$$\frac{0+14}{2} = 7, \frac{15+29}{2} = 22, \frac{30+44}{2} = 37, \text{ and } \frac{45+59}{2} = 52$$

(d) The class interval of the distribution is $22 - 7 = 15$ or $29.5 - 14.5 = 15$.

2.25 (a) The boundary between the first two classes is

$$\frac{4+11}{2} = 7.5$$

the boundary between the second and third classes is

$$\frac{11+18}{2} = 14.5$$

and the boundary between the third and fourth classes is

$$\frac{18+25}{2} = 21.5$$

Since the differences between successive class boundaries are $14.5 - 7.5 = 7$, the lower boundary of the first class is $7.5 - 7 = 0.5$, and the upper boundary of the fourth class is $21.5 + 7 = 28.5$.

(b) The limits of the first class are the smallest and largest integers falling between 0.5 and 7.5, that is, 1 and 7; the limits of the second class are the smallest and largest integers falling between 7.5 and 14.5, that is, 8 and 14; similarly, the limits of the third class are 15 and 21, and the limits of the fourth class are 22 and 28.

2.28 There is an ambiguity because values from $15.00 to $15.99 can be put into the second or third class; also, there is no place to put values from $24.91 to $24.99 and no place to put $30.00.

2.38

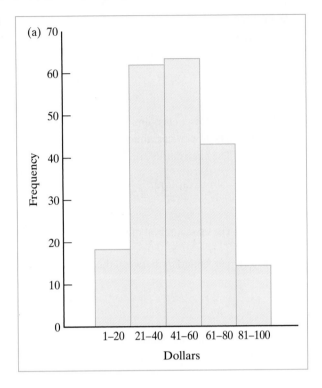

Figure 2.13   Histogram of the distribution.

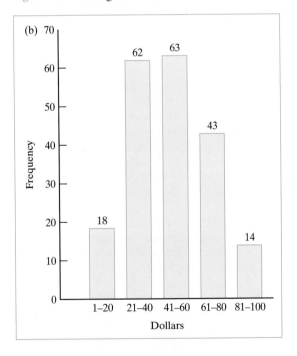

Figure 2.14   Bar chart of the distribution.

2.42 Multiplying 81%, 14%, 3%, and 2% by 360, we get the central angles of 291.6°, 50.4°, 10.8°, and 7.2°.

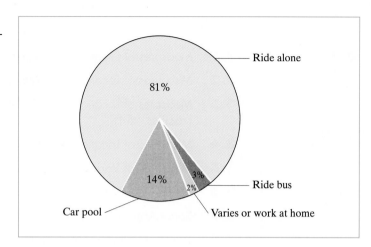

3

SUMMARIZING DATA: STATISTICAL DESCRIPTIONS

The introduction to Chapter 1 stated that statistics is the subject that deals with all aspects of the handling of data. Actually, *statistics* has several other meanings. It is also used to refer to the data itself—for instance, when we refer to birth and death records as **vital statistics**. Furthermore, yet another connotation of *statistics* is the plural of *statistic*, that is, a quantity (such as an average) that is computed from sample data. In this sense, a **statistic** is also referred to as a **statistical measure** or simply as a **statistical description**.

It is customary to classify statistical measures according to the particular features of a set of data that they are intended to describe. In this chapter we shall be concerned mainly with **measures of location**, especially **measures of central location**, and **measures of variation**. The former may be described crudely as *averages*, in the sense that they are generally indicative of the "center," "middle," or the "most typical" of a set of data; the latter are indicative of the variability, spread, or dispersion of a set of data. Other statistical measures will be taken up later, as needed.

For most statistical descriptions, the calculations are fairly simple. However, for large sets of data we may well want to use a computer, a graphing calculator, a statistical calculator, or other type of calculating device.

Statistical descriptions can be simple or complex, depending on our purpose in collecting them. Sometimes it is desirable to present the data in its raw form. In other situations we may group the data, presenting it in the form of tables or graphs.

Measures of location are discussed in Sections 3.1 through 3.4, and measures of variation in Sections 3.5 through 3.7. The description of grouped data is treated in Section 3.8, and some further kinds of descriptions are presented in Section 3.9.

Since the term *statistic* was introduced in connection with sample data, let us add that there is also a name for statistical descriptions of populations—we call them **parameters**. As we shall see, the distinction between statistics and parameters will serve to simplify and clarify our language. Indeed, we shall even use different symbols for statistical measures, depending on whether they are used to describe samples or populations.

3.1 MEASURES OF LOCATION: THE MEAN

The most popular measure of central location is what the layperson calls an "average" and what the statistician calls a **mean**.[†] We put the word "average" in quotes because in everyday language it has all sorts of connotations—we speak of a baseball player's batting average, we talk about the average suburban family, we describe a holdup man's appearance as average, and so on. Formally, then, the mean is defined as follows:

The mean of n numbers is their sum divided by n.

EXAMPLE 3.1 On a certain day, nine students received, respectively, 1, 3, 2, 0, 1, 5, 2, 1, and 3 pieces of mail. Find the mean.

Solution The total number of pieces of mail that the nine students received is

$$1 + 3 + 2 + 0 + 1 + 5 + 2 + 1 + 3 = 18$$

Since $\dfrac{18}{9} = 2$, the mean number of pieces of mail per student is 2. ■

EXAMPLE 3.2 A supermarket manager, who wants to study the traffic in her store, finds that 295, 1,002, 941, 768, and 1,283 persons entered the store during the past five days. Find the mean number of persons who entered the supermarket during these five days.

Solution Altogether the number of persons who entered the supermarket during the past five days is

$$295 + 1,002 + 941 + 768 + 1,283 = 4,289$$

Since $\dfrac{4,289}{5} = 857.8$, this is the mean (or average) number of persons who entered the store per day. ■

[†] The mean is also referred to as the *arithmetic mean*, to distinguish it from the *geometric mean* and the *harmonic mean*, two other kinds of averages that are used only in very special situations and will not be discussed in this book.

Since we shall have occasion to calculate the means of many different sets of data, it will be convenient to have a formula that is always applicable. This requires that we represent the numbers to be averaged by using symbols such as x, y, or z; the number of values in a sample, the **sample size**, is usually denoted by the letter n. Choosing the letter x, we can refer to the n values in a sample as x_1 (which is read "x sub-one"), x_2 (which is read "x sub-two"), $x_3, \ldots$, and x_n, and write the **sample mean** as

$$\frac{x_1 + x_2 + x_3 + \cdots + x_n}{n}$$

This formula will take care of any set of sample data, but it can be made more compact by assigning the sample mean the symbol $\bar{x}$ (which is read "x bar") and by using the $\sum$ **notation**. The symbol $\sum$ is capital *sigma*, the Greek letter for S. In this notation, we let $\sum x$ stand for "the sum of the x's" (that is, $\sum x = x_1 + x_2 + x_3 + \cdots + x_n$), so that we can write

SAMPLE MEAN

$$\bar{x} = \frac{\sum x}{n}$$

If we refer to the measurements as y's or z's, we write their mean as $\bar{y}$ or $\bar{z}$ and substitute $\sum y$ or $\sum z$ for $\sum x$.

In the formula for $\bar{x}$, the expression $\sum x$ does not state explicitly which x's are to be added; let it be understood, therefore, that $\sum x$ always stands for the sum of all the x's under consideration in a given situation. In Section 3.10, the use of the sigma notation is discussed in some detail.

The mean of a population of N items is defined in the same way. It is the sum of the N items, $x_1 + x_2 + x_3 + \cdots + x_N$, or $\sum x$, divided by the **population size N**. Assigning the **population mean** the symbol μ (lowercase mu, the Greek letter for m), we write

POPULATION MEAN

$$\mu = \frac{\sum x}{N}$$

with the reminder that $\sum x$ is now the sum of all N values that constitute the population. To distinguish between parameters and statistics, namely, descriptions of populations and samples, it is common practice to denote the former with Greek letters. We note also that uppercase N is the common symbol for the population size, while lowercase n is used for the sample size.

The division step in finding the sample mean or population mean generally yields an answer which needs to be rounded. In our work in this book, we try to retain a reasonable number of figures after the rounding, but it must be admitted that one cannot form a dogmatic rule for dealing with rounding. We will use the *approximately equal sign* $\approx$ to indicate that a result has been rounded. As an illustration, we will write $\bar{x} = \dfrac{38}{7} \approx 5.43$. We will generally use the $\approx$ sign only at the step in which the arithmetic occurs; for example, a later reference to the same information would note $\bar{x} = 5.43$.

The popularity of the mean as a measure of the *middle* or *center* of a set of data is not accidental. Any time we use a single number to describe some aspect of a set of data, there are certain requirements, or desirable features, that should be kept in mind. Aside from the fact that the mean is a simple and familiar measure, the following are some of its noteworthy properties:

1. **The mean can be calculated for any set of numerical data.**
2. **For any set of numerical data, the mean is a unique, unambiguous value.**
3. **The mean lends itself to further statistical treatment; for example, the means of several sets of data can always be combined into the overall mean of all the data. Most other statistics do not have this property.**
4. **If each value in a sample were replaced by the mean, then $\sum x$ would remain unchanged.**
5. **The mean takes into account the value of each item in a set of data.**
6. **The mean is relatively reliable in the sense that means of many samples drawn from the same population generally do not fluctuate, or vary, as widely as other statistics used to estimate the mean of a population. (See Exercises 3.37 and 3.38.)**

This last property is of fundamental importance in statistical inference, and we shall study it in some detail in Chapter 8.

Sometimes a set of data may contain very small or very large values that are so far removed from the main body of the data that the appropriateness of including them when describing the data is questionable. Such **outliers** may be due to chance, or they may be due to gross errors in recording the data, gross errors in calculations, malfunctioning of equipment, or other identifiable sources of contamination. When such values are averaged in with the other values, they can affect the mean to such an extent that it is debatable whether it really provides a useful description of the middle of the data. In such cases, we may trim the data by deleting the upper and lower 5% and refer to the mean of the remaining data as the **trimmed mean**.

EXAMPLE 3.3   Five lightbulbs burned out after lasting, respectively, for 867, 849, 840, 852, and 822 hours of continuous use. Find the mean and also determine what the mean would have been if the second value had been recorded incorrectly as 489 instead of 849.

Solution   For the original data we get

$$\bar{x} = \frac{867 + 849 + 840 + 852 + 822}{5} = \frac{4{,}230}{5} = 846$$

and with 489 instead of 849 we get

$$\bar{x} = \frac{867 + 489 + 840 + 852 + 822}{5} = \frac{3{,}870}{5} = 774$$

This shows that a very small or very large value, here due to a careless error in recording the data, can have a pronounced effect on the mean.

To avoid the possibility of being misled by very small or very large values, we sometimes describe the middle or center of a set of data with other kinds of statistical measures, for instance, those given in Sections 3.3 and 3.4.

3.2   MEASURES OF LOCATION: THE WEIGHTED MEAN

When averaging quantities, it is often necessary to account for the fact that not all of them are equally important in the phenomenon being described. For instance, in 2000 the production of milk, in thousands of pounds per year per cow, was 19.0 in Michigan, 21.2 in Wisconsin, and 17.8 in Minnesota. The mean of these three milk production numbers is

$$\frac{19.0 + 21.2 + 17.8}{3} = \frac{58.0}{3} \approx 19.3$$

but we cannot very well say that this is the average milk production for the three states. The three figures do not carry equal weight because there are not equally many cows in the three states (see Exercise 3.19).

To give quantities being averaged their proper degree of importance, it is necessary to assign them **weights** (measures of their relative importance) and then calculate a **weighted mean**. In general, the weighted mean $\bar{x}_w$ of a set of numbers $x_1, x_2, x_3, \ldots,$ and x_n, whose relative importance is expressed numerically by a corresponding set of numbers $w_1, w_2, w_3, \ldots,$ and w_n, is given by

WEIGHTED MEAN

$$\bar{x}_w = \frac{w_1 x_1 + w_2 x_2 + \cdots + w_n x_n}{w_1 + w_2 + \cdots + w_n} = \frac{\sum w \cdot x}{\sum w}$$

Here $\sum w \cdot x$ stands for the sum of the products obtained by multiplying each x by the corresponding weight, and $\sum w$ is simply the sum of the weights. If the weights are all equal, the formula for the weighted mean becomes

$$\bar{x}_w = \frac{w x_1 + w x_2 + \cdots + w x_n}{w + w + \cdots + w}$$

$$= \frac{w(x_1 + x_2 + \cdots + x_n)}{n \cdot w}$$

$$= \frac{\sum x}{n}$$

and thus reduces to that of the (ordinary) mean.

EXAMPLE 3.4   Three of the grains produced by farmers in the United States in the year 2000 brought per-bushel prices of $2.65 for wheat, $1.85 for corn (for grain), and $1.05

for oats. Production in billions of bushels was 2.22, 9.97, and 0.15, respectively. What is the overall average price that farmers received per bushel for these three grains?

Solution   Substituting $x_1 = 2.65$, $x_2 = 1.85$, and $x_3 = 1.05$, $w_1 = 2.22$, $w_2 = 9.97$, and $w_3 = 0.15$ into the formula for $\bar{x}_w$, we get

$$\bar{x}_w = \frac{(2.22)(2.65) + (9.97)(1.85) + (0.15)(1.05)}{2.22 + 9.97 + 0.15}$$

$$= \frac{24.485}{12.34} \approx \$1.98$$

Note that the figure in the denominator, 12.34, is the total production of these grains in billions of bushels and the figure in the numerator, 24.485, is the total value of these grains in billions of dollars. Also, if we had averaged 2.65, 1.85, and 1.05, we would have obtained

$$\frac{2.65 + 1.85 + 1.05}{3} \approx \$1.85$$

which is thirteen cents less than the actual average price of $1.98.

A special application of the formula for the weighted mean arises when we must find the overall mean, or **grand mean**, $\bar{\bar{x}}$, of k sets of data having the means $\bar{x}_1, \bar{x}_2, \bar{x}_3, \ldots$, and $\bar{x}_k$, consisting, respectively, of $n_1, n_2, n_3, \ldots$, and n_k measurements or observations. The result is given by

GRAND MEAN OF COMBINED DATA

$$\bar{\bar{x}} = \frac{n_1\bar{x}_1 + n_2\bar{x}_2 + \cdots + n_k\bar{x}_k}{n_1 + n_2 + \cdots + n_k} = \frac{\sum n \cdot \bar{x}}{\sum n}$$

where the weights are the sizes of the respective samples, the numerator is the total of all the measurements or observations, and the denominator is the sample size of the combined data.

EXAMPLE 3.5   In a biology class there are 20 freshmen, 18 sophomores, and 12 juniors. If the freshmen averaged 68 on an examination, the sophomores averaged 75, and the juniors averaged 86, find the mean grade for the entire class.

Solution   Substituting $n_1 = 20$, $n_2 = 18$, $n_3 = 12$, $\bar{x}_1 = 68$, $\bar{x}_2 = 75$, and $\bar{x}_3 = 86$ into the formula for the grand mean of combined data, we get

$$\bar{\bar{x}} = \frac{20 \cdot 68 + 18 \cdot 75 + 12 \cdot 86}{20 + 18 + 12}$$

$$= \frac{3{,}742}{50} = 74.84 \approx 75$$

Exercises 3.1, 3.4, 3.10, 3.17, and 3.18 are practice exercises; their complete solutions are given on page 96.

✓ **3.1**   Which of the following are statistics, and which are parameters?

   (a) A statistical measure that is calculated from or used to describe a population

   (b) A statistical measure that is calculated from or used to describe a sample

   (c) A statistical measure that is calculated from or used to describe a large sample

   (d) A statistical measure that is calculated from or used to describe a small population

3.2   Which of the following are statistics, and which are parameters?

   (a) The mean height of a sample of basketball players in the National Basketball Association

   (b) The mean weight of the population of football players on the Washington Redskins team

   (c) The mean blood pressure of the population of patients in the cardiac ward of a hospital

   (d) The mean grade of students in a certain elementary algebra section at a college, which is used to estimate the mean grade of all students in all sections of elementary algebra at the college

3.3   Suppose that the research department of a pharmaceutical company has complete information about the effects of a weight reducing pill on a group of 100 voluntary subjects. Give one illustration each of a situation where the mean weight reduction of the voluntary subjects in the study would be looked upon as

   (a) a parameter;

   (b) a statistic.

✓ **3.4**   The one-year survival rate for patients receiving organ transplants of heart, heart-lung, lung, liver, kidney, kidney-pancreas, pancreas, and intestine are 82.8, 53.3, 72.4, 78.2, 90.0, 92.2, 80.3, and 50.9%, respectively. Find the mean overall percentage of patients who survive their organ transplants by at least one year.

3.5   According to the U.S. Census Bureau, the expectation of life at birth (in years) in the year 2001 for selected nations was as follows:

Country	Years	Country	Years
Argentina	75.3	India	66.1
Australia	79.9	Italy	80.2
Belgium	78.0	Japan	81.7
Brazil	63.2	Mexico	74.1
Canada	79.6	Netherlands	79.8
Chile	75.9	Russia	68.8
China	71.6	Spain	80.2
Egypt	63.7	Ukraine	66.2
France	78.9	United Kingdom	77.8
Germany	77.6	United States	77.3
Greece	78.6		

   (a) Find their mean expectation of life at birth.

   (b) Explain how part (a) would change if the years were weighted by the size of the human population in these nations.

3.6   According to U.S. Energy Information, the state gasoline tax in the Mountain States in cents per gallon in March 1999 was as follows:

State	Cents per gallon
Montana	28.0
Idaho	25.0
Wyoming	14.0
Colorado	22.0
New Mexico	19.0
Arizona	18.0
Utah	25.0
Nevada	24.0

Find their mean state tax on gasoline.

3.7   According to the National Association of Insurance Commissioners, the average expenditure for automobile insurance per insured vehicle in the states listed in Exercise 3.6 was $511, $493, $491, $744, $664, $789, $615, and $821. Find the overall mean expenditure per insured vehicle.

3.8   According to the Bureau of Labor Statistics, the average pay (in thousands of dollars) paid in the top 15 metropolitan areas in the United States in 1999 was

> 51.4   50.4   45.6   44.8   44.6
> 42.9   41.6   40.6   40.1   40.1
> 39.8   39.5   38.6   38.5   38.3

Find their mean annual pay in thousands of dollars.

3.9   The records of 15 persons convicted of various crimes showed that, respectively, 4, 3, 0, 0, 2, 4, 4, 3, 1, 0, 2, 0, 2, 1, and 4 of their grandparents were foreign born. Find the mean and discuss whether it can be used to support the contention that the "average criminal" has two foreign-born grandparents.

✓ **3.10**   An elevator in a hotel is designed to carry a maximum load of 2,000 pounds. Is it overloaded if at one time it carries eight women whose mean weight is 124 pounds and six men whose mean weight is 165 pounds?

3.11   An airplane cargo bay is designed to carry a maximum load of 12,000 pounds. If at a given moment it is loaded with 168 crates having a mean weight of 64 pounds, is there any real danger that the cargo bay is overloaded?

3.12   By mistake, an instructor has erased the grade that one of ten students in her class received in a final examination. However, she knows that the students averaged (had a mean grade of) 71 on the examination, and that the other nine students received grades of 96, 44, 82, 70, 47, 74, 94, 78, and 56. What must have been the grade that the teacher erased?

3.13   The hours of sleep that a student has had during each of the last 10 nights are 6, 8, 2, 13, 9, 7, 6, 5, 10, and 7. Find the mean and discuss its usefulness in describing the middle of the data.

3.14   A bill was introduced in a state legislature to repeal the sales tax on prescription drugs. Comment on the argument of the state finance director that the average (mean) per capita prescription bill for the past four years was a trifling $7.20, which is not really a burden to anyone.

3.15   Records show that in Phoenix, Arizona, the normal daily maximum temperature for each month is 65, 69, 74, 84, 93, 102, 105, 102, 98, 88, 74, and 66 degrees Fahrenheit. Verify that the mean of these figures is 85 and comment on the claim that, in Phoenix, the average daily maximum temperature is a very comfortable 85 degrees.

3.16 Careful measurements show that the actual coffee content of six jars of instant coffee is 6.03, 5.98, 6.04, 6.00, 5.99, and 6.02 ounces of coffee. What would have been the error in the mean coffee content of the six jars if the third value had been recorded incorrectly as 6.40?

✓ **3.17** If somebody invests $1,000 at 7%, $3,000 at 7.5%, and $16,000 at 8%, what is the overall percentage yield of these investments?

✓ **3.18** In the year 2000, the average salaries of elementary school teachers in Oregon, Washington, and Alaska were (in thousands of dollars) 40.9, 41.1, and 47.3. Given that there were 19.7, 28.1, and 5.3 thousand teachers in these states, find the average salary of all the elementary school teachers in the three states.

3.19 Previously, we stated that in 2000 the production of milk, in thousands of pounds per year per cow, was 19.0 in Michigan, 21.2 in Wisconsin, and 17.8 in Minnesota. Given that there were 300,000 cows in Michigan, 1,340,000 in Wisconsin, and 534,000 cows in Minnesota, what is the average annual milk production for the three states combined?

3.20 An instructor counts the final examination in a course three times as much as each of the three one-hour examinations. What is the average grade of a student who received grades of 75, 77, and 58 on the three one-hour examinations and 82 on the final examination?

3.21 In a recent season, a baseball team's five best hitters had batting averages of 0.307, 0.299, 0.297, 0.291, and 0.283. What is their combined batting average if they had, respectively, 488, 137, 646, 533, and 502 at bats?

3.22 A wholesaler sold 575, 410, and 520 microwave ovens at prices of $75, $125, and $100, respectively. What is the mean price of the ovens sold?

3.23 In a study of home replacement costs, it was found that 12 houses in one city were underinsured on the average by $5,000, 24 homes in another city were underinsured on the average by $6,200, and 14 homes in a third city were underinsured on the average by $6,900. On the average, by how much were these 50 homes underinsured?

3.24 The following ordered data relates to the average travel time to work, in minutes, of 100 employees in a large downtown office building.

11.4	12.2	12.6	12.6	12.6	14.0	15.8	16.2	16.2	18.0
18.0	19.4	20.2	20.4	22.2	22.6	23.0	23.6	24.2	24.8
27.0	27.4	28.0	28.2	28.4	28.8	29.2	29.8	30.0	30.2
30.4	31.0	31.8	32.4	33.6	35.2	35.4	35.4	36.2	37.0
37.0	37.4	37.6	37.6	37.8	38.0	38.4	38.4	39.0	39.0
39.2	39.4	39.4	39.4	39.4	39.8	40.0	40.4	40.6	40.8
41.0	41.0	41.4	42.0	42.0	42.2	42.6	43.0	43.2	44.0
44.0	44.0	45.0	45.6	45.8	46.0	47.0	47.8	49.2	49.4
49.6	50.6	52.4	52.6	53.6	56.2	56.4	56.4	60.0	64.6
74.0	86.4	89.6	90.2	90.4	100.0	103.4	108.2	123.8	125.4

Find the trimmed mean for the average travel time of these 100 employees.

3.3 MEASURES OF LOCATION: THE MEDIAN AND OTHER FRACTILES

In the preceding section we introduced the mean as a measure of location which describes the middle, or center, of a set of data. It is by far the most widely used measure of central location, but as we pointed out, it has the shortcoming that it can be affected by a few very small or very large values. Such outliers may be due to chance factors, or they may be due to careless mistakes. Another shortcoming of the mean is that, for grouped data, it cannot be determined when there is an open class; not even with the assumption that the values in each class are equal to the corresponding class mark. For open classes, the class marks are not defined. A measure of central location without these shortcomings is the median, which will be treated here. Next to the mean, the median is one of the most widely used measures of location. The symbol $\tilde{x}$ is often used to denote the median of a sample, and $\tilde{\mu}$ to denote the median of a population. To avoid the possibility of being misled by a few very small or very large values, as demonstrated previously, it is sometimes preferable to describe the *middle* or *center* of a set of data with statistical measures other than the mean.

The definition of one of these, the **median of *n* values**, requires that we arrange the data according to size. Then

When n is odd, the median is the value of the item that is in the middle.

When n is even, the median is the mean of the two items that are nearest to the middle.

EXAMPLE 3.6 In a recent month, a state's Game and Fish Department reported 53, 31, 67, 53, and 36 hunting or fishing violations for five different regions. Find the median number of violations for these regions.

Solution The median is not 67, the third (or middle) item, because the figures must first be ordered according to size. Thus, we get

$$31\ 36\ 53\ 53\ 67$$

and it can be seen that the third item in this ordered set of five items is 53. The fact that our set contains two 53s does not affect our answer. ■

EXAMPLE 3.7 According to the U.S. Department of Agriculture, the leading wheat producing states in the year 2000 and their production (in millions of bushels) were

States	Wheat (millions of bushels)
Idaho	108
Kansas	348
Minnesota	97
Montana	135
North Dakota	314
Oklahoma	143
South Dakota	114
Washington	165

Find the median.

Solution Arranging these figures according to size, we get

$$97\quad 108\quad 114\quad 135\quad 143\quad 165\quad 314\quad 348$$

and it can be seen that the median is

$$\frac{135 + 143}{2} = 139$$

that is, the mean of the two items nearest the middle.

We cannot give a general formula for the value of the median, but we can give one for the **median position**. With data ranked from low to high or high to low and counting from either end,

The median is the value of the $\dfrac{n+1}{2}$ *th item.*

When n is odd, $\dfrac{n+1}{2}$ is an integer and it gives the position of the median; when n is even, $\dfrac{n+1}{2}$ is midway between two integers and the median is the mean of the values of the corresponding items.

EXAMPLE 3.8   Find the median position for (a) $n = 11$, (b) $n = 25$, and (c) $n = 75$.

Solution   With the data arranged according to size and counting from either end,

(a) $\dfrac{n+1}{2} = \dfrac{11+1}{2} = 6$, so that for $n = 11$ the median is the value of the 6th item;

(b) $\dfrac{n+1}{2} = \dfrac{25+1}{2} = 13$, so that for $n = 25$ the median is the value of the 13th item;

(c) $\dfrac{n+1}{2} = \dfrac{75+1}{2} = 38$, so that for $n = 75$ the median is the value of the 38th item.

EXAMPLE 3.9   Find the median position for (a) $n = 8$, (b) $n = 20$, and (c) $n = 100$.

Solution   With the data arranged according to size and counting from either end,

(a) $\dfrac{n+1}{2} = \dfrac{8+1}{2} = 4.5$, so that for $n = 8$ the median is the mean of the values of the 4th and 5th items;

(b) $\dfrac{n+1}{2} = \dfrac{20+1}{2} = 10.5$, so that for $n = 20$ the median is the mean of the values of the 10th and 11th items;

(c) $\dfrac{n+1}{2} = \dfrac{100+1}{2} = 50.5$, so that for $n = 100$ the median is the mean of the values of the 50th and 51st items.

It is important to remember that $\dfrac{n+1}{2}$ is a formula for the median position, and not a formula for the value of the median. To simplify the determination of a median, it sometimes helps to utilize the grouping provided by a stem-and-leaf display.

EXAMPLE 3.10   There have been 43 Presidents of the United States. Thirty-seven of them, beginning with George Washington, died at ages 67, 90, 83, 85, 73, 80, 78, 79, 68,

71, 53, 65, 74, 64, 77, 56, 66, 63, 70, 49, 56, 71, 67, 71, 58, 60, 72, 67, 57, 60, 90, 63, 88, 78, 46, 64, and 81. Construct a stem-and-leaf display with one-digit leaves and use it to find the median.

Solution First, constructing the stem-and-leaf display and sorting the leaves, we get

$$
\begin{array}{c|ccccccccccc}
4 & 6 & 9 \\
5 & 3 & 6 & 6 & 7 & 8 \\
6 & 0 & 0 & 3 & 3 & 4 & 4 & 5 & 6 & 7 & 7 & 7 & 8 \\
7 & 0 & 1 & 1 & 1 & 2 & 3 & 4 & 7 & 8 & 8 & 9 \\
8 & 0 & 1 & 3 & 5 & 8 \\
9 & 0 & 0 \\
\end{array}
$$

Since the median position is $\dfrac{37+1}{2} = 19$ and 7 values are found on the first two stems, we must find the twelfth value on the third stem. As can be seen by inspection, it is 68.

If we had calculated the mean in the example on page 55 that dealt with the hunting and fishing violations, we would have obtained

$$
\frac{53 + 31 + 67 + 53 + 36}{5} = \frac{240}{5} = 48
$$

and it should not come as a surprise that it differs from the median, which was 53. Each of these averages describes the middle or center of the data in its own way. The median is average in the sense that it splits the data into two parts so that as many of the values are to the left of the median position as there are to its right. If the values in a set of data are all unequal, we can substitute "below" for "to the left of" and "above" for "to the right of." When some of the values are equal, this may or may not be done. For instance, it cannot be done for 1, 2, 3, 3, 3, 3, and 5, where the median is 3 and there are two values below the median and only one value above it.

The mean, on the other hand, is typical in the sense that if each value in a set of data is replaced by the same number while the total remains unchanged, this number will have to be the mean. This is the fourth of the six properties listed in Section 3.1, and it follows from the fact that

$$
n \cdot \overline{x} = n \cdot \frac{\sum x}{n} = \sum x
$$

The median shares the first two properties of the mean as listed in Section 3.1; that is, it can be determined for any set of numerical data and it is always unique.[†] Also like the mean, the median is simple enough to find once the data has been arranged according to size, but it should be kept in mind that ordering a large set of data manually can be a very tedious job. As far as the third and sixth properties are concerned, the medians of several sets of data can generally not be combined into the overall median of all the data, and in problems of inference

---

[†] There are other definitions for the median that allow it to be nonunique when n is even. The definition used in this book leads to a unique median.

the median is generally not as reliable as the mean: that is, it is subject to greater chance fluctuations, as is illustrated by Exercises 3.37 and 3.38. When it comes to the fifth property, the median is actually preferable to the mean, because it is not so easily affected by extreme (very large or very small) values.

EXAMPLE 3.11   In Example 3.3 on page 49 we showed that the mean lifetime of five lightbulbs is $\bar{x} = 846$ hours, but if one of the values, 849, is recorded incorrectly as 489, the mean lifetime is 774, which is off by $846 - 774 = 72$ hours. What would have been the corresponding error if we had used the medians?

Solution   Since the original values, sorted, are 822, 840, 849, 852, and 867, the correct value of the median is $\tilde{x} = 849$. With 489 substituted for 849, the values, arranged according to size, are 489, 822, 840, 852, and 867, and $\tilde{x} = 840$, which is off by $849 - 840 = 9$ hours. This is much smaller than the error we made when we used the mean.   ∎

Also, unlike the mean, the median can be used to define the middle of a number of objects, properties, or attributes. It is possible, for example, to rank a number of tasks according to their difficulty and then describe the middle (or median) one as being of *average* difficulty; also, we might rank samples of chocolate fudge according to their consistency and then describe the middle (or median) one as having average consistency.

The median is but one of many different **fractiles** that divide a set of data into two or more equal parts. Also of importance in statistics are **quartiles** and **percentiles**; but since percentiles are used mainly in connection with large sets of data, we shall discuss them for grouped data in Section 3.8. Thus, let us treat here the three quartiles Q_1, Q_2, and Q_3, which are defined as follows: With the data arranged from left to right in an increasing order of magnitude,

> *The first quartile, Q_1, is the median of all the values to the left of the median position for the whole set of data.*
>
> *The second quartile, Q_2, is the median.*
>
> *The third quartile, Q_3, is the median of all the values to the right of the median position for the whole set of data.*

Generally, we find Q_3 by counting as many places as for Q_1, starting at the other end, that is, from the right rather than the left for data ordered from low to high.

Note that with this definition there are always as many values to the left of the Q_1 position as there are between the Q_1 position and the median position, between the median position and the Q_3 position, and to the right of the Q_3 position.

EXAMPLE 3.12   Verify the preceding sentence for $n = 23$.

Solution   The median position is $\dfrac{23 + 1}{2} = 12$, so there are 11 values to its left and 11 values to its right. Thus, the Q_1 position is $\dfrac{11 + 1}{2} = 6$ and, correspondingly, Q_3 is the 6th value counting from the other end; that is, the Q_3 position is 18 (see Figure 3.1). It follows that there are five values to the left of the Q_1 position, five values between the Q_1 position and the median position (between 6 and 12), five values between the median position and the Q_3 position (between 12 and 18), and five values to the right of the Q_3 position. ■

EXAMPLE 3.13 Use the following stem-and-leaf display constructed on page 57 to find Q_1 and Q_3 for the data concerning the lives of the Presidents. The median position is $\dfrac{37+1}{2} = 19$.

$$
\begin{array}{c|ccccccccccccc}
4 & 6 & 9 \\
5 & 3 & 6 & 6 & 7 & 8 \\
6 & 0 & 0 & 3 & 3 & 4 & 4 & 5 & 6 & 7 & 7 & 7 & 8 \\
7 & 0 & 1 & 1 & 1 & 2 & 3 & 4 & 7 & 8 & 8 & 9 \\
8 & 0 & 1 & 3 & 5 & 8 \\
9 & 0 & 0 \\
\end{array}
$$

Solution Since we have already shown that the median position is 19, we find that the Q_1 position is $\dfrac{18+1}{2} = 9.5$. Thus, Q_1 is the mean of the ninth and tenth values and it can be seen from the stem-and-leaf plot that it equals $\dfrac{60+63}{2} = 61.5$. To find Q_3 we count 9.5 values starting from the other end. Since the ninth value is 78 and the tenth value is 78, we get $\dfrac{78+78}{2} = 78$. ■

EXAMPLE 3.14 The following are the average prices of eggs in cents per dozen paid to farmers in the United States from the year 1991 to 2000: 67.8, 57.6, 63.4, 61.4, 62.4, 74.9, 70.3, 65.5, 62.2, and 61.8. Find the median, Q_1 and Q_3.

Solution For $n = 10$, the median position is $\dfrac{10+1}{2} = 5.5$ so that the Q_1 position is $\dfrac{5+1}{2} = 3$, and Q_3 is the third value from the other end. Since the prices arranged according to size are 57.6 61.4 61.8 62.2 62.4 63.4 65.5 67.8 70.3 74.9 it can be seen that the median is $\dfrac{62.4+63.4}{2} = 62.9$, $Q_1 = 61.8$, and $Q_3 = 67.8$. ■

The information provided by the median, the two quartiles, and also the smallest and largest values, is sometimes presented in the form of a **box-and-whisker plot**, often just called a **box plot**. Such a plot, reproduced from the display screen of a TI-83 graphing calculator, is shown in Figure 3.2 for the data of the preceding example. There is no numerical scale on the horizontal axis, but the TRACE button allows you to trace through the graph. Note that the rectangle extends from $Q_1 = 61.8$ to $Q_3 = 67.8$. The whisker on the left extends to the smallest value, which is 57.6. The whisker on the right extends to the largest value, which is 74.9. The vertical line that divides the box is at the median, which is 62.9. Box plots provide useful information for the exploratory analysis of the overall shape of distributions (see also Section 3.8). They are sometimes

Figure 3.2
Box-and-whisker plot
reproduced from the dis-
play screen of a TI-83
graphing calculator.

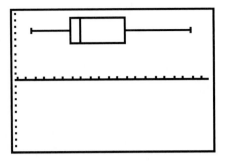

embellished with other features of the data, but the simple form given here is adequate for most purposes.

Optional directions for graphing the box plot with a TI-83 graphing calculator are provided. These are intended primarily for users of graphing calculators and can be omitted by other readers without loss of continuity.

Two other measures of central location, the midrange and the midquartile, are introduced in Exercises 3.41 and 3.48.

How to graph the box plot shown in Figure 3.2 using the TI-83.

(**Tip**: Before beginning, see Appendix A for clearing lists and entering data.)

Step 1. Clear list L1.

Step 2. Input the example data into list L1.

Step 3. Select the type of graph.
Press 2nd, Y = to choose the STAT PLOTS function.
Select PLOT 1, press ENTER.
Highlight ON.
Scroll to the box plot icon (the fifth graph icon), press ENTER.
Set Xlist: L1
Freq: 1

Step 4. View the graph.
Press ZOOM and Select 9: Zoom Stat.

USING TECHNOLOGY The Graphing Calculator

Exercises 3.25, 3.28, 3.37, 3.42, 3.45, and 3.48 are practice exercises; their complete solutions are given on pages 97–98.

✓ **3.25** Find the median position for (a) $n = 15$ and (b) $n = 40$.

3.26 Find the median position for (a) $n = 17$ and (b) $n = 30$.

3.27 Find the median position for (a) $n = 39$ and (b) $n = 150$.

✓ **3.28** According to the U.S. Department of Energy, retail gasoline prices expressed in U.S. dollars per gallon were as follows in the year 2000.

Country	Prices in dollars per gallon
Australia	1.94
Canada	1.86
Germany	3.45
Japan	3.74
Mexico	2.02
United States	1.51

(a) Find the median of this list of 6 values.

(b) Find the mean.

3.29 The following are the numbers of restaurant meals that 13 persons ate during a given week:

$$3 \quad 10 \quad 5 \quad 1 \quad 8 \quad 5 \quad 6 \quad 12 \quad 15 \quad 1 \quad 0 \quad 6 \quad 5$$

Find the median.

3.30 The melting points, in degrees Fahrenheit, of 16 selected metals are as follows:

Metals	Degrees Fahrenheit	Metals	Degrees Fahrenheit
Aluminum	1,220	Pewter	563
Chromium	3,430	Platinum	3,224
Copper	1,981	Plutonium	1,384
Gold	1,945	Silver	1,761
Iron	2,802	Tin	449
Lead	621	Titanium	3,300
Magnesium	1,202	Tungsten	6,170
Nickel	2,651	Zinc	787

Find

(a) the mean;

(b) the median.

3.31 The following list gives the duration in minutes of 24 power failures:

18	125	44	96	31	53
26	80	49	125	63	58
45	33	89	12	103	127
75	40	80	61	28	129

Find the median.

3.32 Unemployment rates in selected countries, as a percentage of civilian working age population, in the year 2000, were as follows:

Country	Unemployment rates (%)
United States	4.0
Australia	6.5
Canada	5.7
France	9.2
Germany	8.1
Italy	10.1
Japan	4.6
Sweden	5.2
United Kingdom	5.5

Find the median percentage.

3.33   The numbers of full-length movies observed by a group of 400 persons were as follows:

Number of movies	Number of persons
0	72
1	106
2	153
3	40
4	18
5	7
6	3
7	0
8	1
Total	400

Find the median.

3.34   The automatic telephone answering machine of a large office transferred the following numbers of incoming calls to various personnel during 25 business days.

$$
\begin{array}{ccccc}
147 & 123 & 135 & 154 & 146 \\
182 & 132 & 128 & 144 & 177 \\
150 & 142 & 169 & 137 & 193 \\
129 & 198 & 156 & 154 & 140 \\
138 & 107 & 146 & 125 & 117 \\
\end{array}
$$

Find the median number of these telephone calls

(a) directly;

(b) by first constructing a stem-and-leaf display.

3.35   Using the following stem-and-leaf display concerning the scores obtained by 20 students on a physical coordination test (see Section 2.2),

```
5 | 2 7 5
6 | 9 1 5 3 4 7 1 8
7 | 4 9 2 4 7
8 | 4 8 2
9 | 3
```

determine the median score.

3.36   According to the U.S. Bureau of the Census, the 12 fastest-growing states in the United States from 1990 to 2000 were Arizona, Colorado, Florida, Georgia, Idaho, Nevada, New Mexico, North Carolina, Oregon, Texas, Utah, and Washington. The percentage population increase in these states was, respectively, 40.0, 30.6, 23.5, 26.4, 28.5, 66.3, 20.1, 21.4, 20.4, 22.8, 29.6, and 21.1. Find

(a) the mean;

(b) the median.

✓ **3.37**   To verify the claim that the mean is generally more reliable than the median (meaning that it is subject to smaller chance fluctuations), a student conducted an experiment consisting of 15 times drawing 3 cards from a modified deck of cards. This modified deck had 40 cards and was obtained by removing the jacks, queens, and kings from an ordinary deck of 52. The aces were assigned the numeric value 1. Here are the results:

Trial number	Cards selected			Trial number	Cards selected		
1	8	3	2	9	4	5	9
2	4	4	10	10	1	8	8
3	7	10	4	11	3	5	9
4	8	10	3	12	10	1	4
5	3	6	7	13	5	1	9
6	6	6	7	14	5	3	8
7	8	3	8	15	10	5	6
8	1	3	1				

(a) Calculate the 15 medians and 15 means.

(b) Group the medians and the means obtained in part (a) into separate distributions having the classes 0.5–2.5, 2.5–4.5, 4.5–6.5, and 6.5–8.5. (There can be no ambiguities since the medians and the means of three whole numbers cannot equal 2.5, 4.5, or 6.5.)

(c) Draw histograms of the two distributions obtained in part (b) and explain how they illustrate the claim that the mean is generally more reliable than the median.

3.38   To verify the claim that the mean is generally more reliable than the median (meaning that it is subject to smaller chance fluctuations), a student conducted an experiment consisting of 12 tosses of three dice. The results were as follows:

Toss number	Dice values			Toss number	Dice values		
1	2	4	6	7	5	5	2
2	5	2	3	8	3	3	3
3	3	1	4	9	4	5	3
4	1	6	2	10	3	2	1
5	5	3	5	11	3	3	4
6	6	1	5	12	4	5	3

(a) Calculate the 12 medians and the 12 means.

(b) Group the medians and the means obtained in part (a) into separate distributions having the classes 1.5–2.5, 2.5–3.5, 3.5–4.5, and 4.5–5.5. (There will be no ambiguities since the medians and means of three whole numbers cannot equal 2.5, 3.5, or 4.5.)

(c) Draw histograms of the two distributions obtained in part (b), and explain how they illustrate the claim that the mean is generally more reliable than the median.

3.39 Repeat Exercise 3.38 with your own data by repeatedly rolling three dice (or one die three times) and construct corresponding distributions for the 12 medians and the 12 means. (If no dice are available, simulate the experiment mentally or by drawing numbered slips of paper out of a hat.)

3.40 A consumer testing service obtained the following miles per gallon in five test runs performed with each of three compact cars:

Car A:	27.9	30.4	30.6	31.4	31.7
Car B:	31.2	28.7	31.3	28.7	31.3
Car C:	28.6	29.1	28.5	32.1	29.7

(a) If the manufacturers of car A want to advertise that their car performed best in this test, which of the averages discussed in this text could they use to substantiate their claim?

(b) If the manufacturers of car B want to advertise that their car performed best in this test, which of the averages discussed in this text could they use to substantiate their claim?

3.41 Suppose that the manufacturers of car C of Exercise 3.40 hire an unscrupulous statistician and instruct him to find some kind of "average" that will show that their car performed best in the test. Show that the **midrange**, which is defined as the mean of the smallest and largest values, will serve their purpose. The midrange is occasionally used as a measure of central location.

✓ **3.42** Find the positions of the median, Q_1, and Q_3 when $n = 21$, and verify that there are as many values to the left of the Q_1 position as there are between the Q_1 position and the median position, between the median position and the Q_3 position, and to the right of the Q_3 position.

3.43 Rework Exercise 3.42 with $n = 18$.

3.44 Rework Exercise 3.42 with $n = 24$.

✓ **3.45** In Exercise 3.29, 13 persons ate 3, 10, 5, 1, 8, 5, 6, 12, 15, 1, 0, 6, and 5 restaurant meals during a week. Calculate the values of Q_1 and Q_3 for the numbers of restaurant meals.

3.46 Selected major bodies of water in the Gulf Coast are Tampa Bay, Galveston Bay, Atchafalaya Bay, Matagorda Bay, Lake Borgne, Mobile Bay, Breton Sound, Florida Bay, Lake Pontchartrain, Laguna Madre, Mississippi Sound; and their areas in square miles are 212, 236, 245, 253, 271, 310, 511, 616, 631, 733, and 813.

(a) Calculate the values of Q_1, Q_3, and the median.

(b) Construct a box-and-whisker plot for the areas of the water bodies.

3.47 In Exercise 3.8, the Bureau of Labor Statistics disclosed the average pay (in thousands of dollars) paid in the top 15 metropolitan areas in the United States for the year 1999. The data are repeated here.

$$51.4 \quad 50.4 \quad 45.6 \quad 44.8 \quad 44.6$$
$$42.9 \quad 41.6 \quad 40.6 \quad 40.1 \quad 40.1$$
$$39.8 \quad 39.5 \quad 38.6 \quad 38.5 \quad 38.3$$

Given that the median is $40.6 thousand, that $Q_1 = 39.9$, and $Q_3 = 44.8$, construct a box-and-whisker plot.

✓ **3.48** The **midquartile** is occasionally used as a measure of central location, and its formula is $\dfrac{Q_1 + Q_3}{2}$. In Example 3.14 we gave the prices of eggs paid to farmers from the year 1991 to 2000 and we determined that Q_1 is 61.8 cents per dozen and Q_3 is 67.8 cents per dozen. The midquartile is therefore

$$\frac{Q_1 + Q_3}{2} = \frac{61.8 + 67.8}{2} = 64.8 \text{ cents per dozen}$$

which is a measure of central location.

Previously we found that the median of this data is 62.9 cents per dozen and we can readily calculate that the mean is $67.8 + 57.6 + 63.4 + 61.4 + 62.4 + 74.9 + 70.3 + 65.5 + 62.2 + 61.8 = \dfrac{647.3}{10} = 64.7$.

Finally, the midrange of the data is the mean of the smallest and largest value, which is $\dfrac{57.6 + 74.9}{2} = 66.2$.

(a) Exercise 3.29 relates to the number of restaurant meals 13 people ate during a given week. Given that $Q_1 = 2$, $Q_3 = 9$, the smallest number of meals is zero, and the largest number of meals is 15, find the value of the midquartile and that of the midrange.

(b) Exercise 3.46 relates to the areas of bodies of water in the Gulf Coast. Given that $Q_1 = 245$ square miles, $Q_3 = 631$ square miles, the smallest body of water is 212 square miles, and the largest body of water is 813 square miles, find the size of the midquartile and also that of the midrange.

3.49 In a study of the breeding habits of the phoebe, a small grayish-brown bird, an ornithologist watched 15 nests plastered to agricultural barns and found that the number of eggs laid was, respectively,

$$4 \quad 3 \quad 3 \quad 5 \quad 8 \quad 5 \quad 4 \quad 2 \quad 4 \quad 5 \quad 4 \quad 4 \quad 7 \quad 4 \quad 3$$

Find

(a) the mean;

(b) the median;

(c) Q_1 and Q_3;

(d) the midquartile;

(e) the midrange.

3.4 MEASURES OF LOCATION: THE MODE

The **mode** is another measure of location that is sometimes used to describe the middle of a set of data. It is defined as follows:

The mode is the value that occurs with the highest frequency.

In this sense it is "most typical" of a set of data; its two main advantages are that it requires no calculations, only counting, and that it can be determined for qualitative as well as quantitative data.

The mode, however, is a poor measure of central location in statistical inference. It also has the disadvantage of not existing at all, or of not being unique. For example, the mode does not exist in the distribution 20, 24, 27, 35, and 18. There are two modes, however, in the distribution 20, 24, 24, 35, 18, and 18.

EXAMPLE 3.15   U.S. Bureau of the Census data shows that the percentages of children not covered by health insurance in the 50 states and the District of Columbia, in 1999, were as follows:

11	16	22	12	18	16	9	7	17	16	13	11	20	12	9	6	12
14	25	7	10	9	10	7	16	5	19	9	22	6	10	28	12	12
10	9	17	13	8	7	19	9	9	24	11	8	14	13	14	11	14

Find the mode.

Solution   Among the numbers, the value 9 appears 7 times, which is more often than any other value occurs. Thus, 9 is the modal percentage.   ■

EXAMPLE 3.16   Asked to name the best collegiate basketball team in the country, 25 sportswriters named the following teams: Duke, Kansas, UCLA, Arizona, North Carolina, Indiana, Kansas, Kentucky, Kansas, UCLA, Kansas, Duke, Michigan, North Carolina, Kentucky, Washington, UCLA, North Carolina, Arizona, Kansas, Kansas, Duke, UCLA, Kentucky, and North Carolina. Find the mode.

Solution   Since Indiana, Michigan, and Washington are named once, Arizona is named twice, Duke and Kentucky are named three times, UCLA and North Carolina are named four times, and Kansas is named six times, Kansas is the modal choice.■

In a great many sets of data, all values are different. Certainly, the mode is useless in such a situation. A serious disadvantage of the mode is that its value is badly altered when the values in the set of data are rounded. Exercise 3.58 illustrates this point. We recommend that the mode be avoided for sets of values presented to several significant figures.

EXAMPLE 3.17   A sample from the records of a department store shows that two modes exist in a sample of 18 dresses which were sold. The dresses sold were size 10, 7, 14, 9, 9, 14, 18, 9, 11, 12, 16, 14, 9, 14, 14, 11, 9 and 20. Find the two modes.

Solution   A count of the dress sizes sold reveals that 5 size nines, and an equal number of size fourteens, were sold. All other sizes had fewer numbers sold. The fact that the data has more than one mode (in this case, bimodal) may indicate that the data is not homogeneous. The department store is patronized by younger women and their mothers (or other older women) who wear different dress sizes.   ■

There are many other measures of location besides the mean, the median, and the mode, and the question of what particular average should be used in a given situation is not always easily answered. The fact that the selection of statistical descriptions is to some extent arbitrary has led some persons to believe that the magic of statistics can be used to prove almost anything. A famous nineteenth-century British statesman is often quoted as having said that there are three kinds of lies: lies, damned lies, and statistics. Exercises 3.40 and 3.41 described a situation where this kind of criticism would well be justified.

Exercises 3.50 and 3.55 are practice exercises; their complete solutions are given on page 98.

✓ **3.50** The following data provided by the U.S. National Center for Health Statistics is the number of live births per 1,000 population, year 1999, in 17 eastern states and also in the nation's capital. These are Maine, New Hampshire, Vermont, Massachusetts, Rhode Island, Connecticut, New York, New Jersey, Pennsylvania, Delaware, Maryland, District of Columbia, Virginia, West Virginia, North Carolina, South Carolina, Georgia, and Florida. The values are, respectively,

11 12 11 13 12 13 14 14 12 14 14 14 14 12 15 14 16 13

Find the mode.

3.51 The following data provided by the Current Population Survey is the percent of total unemployment (as a percent of the civilian labor force) for the year 2000. The data refers to the eastern states and the District of Columbia that are specified in Exercise 3.50.

4 3 3 3 4 2 5 4 4 4 4 6 2 6 4 4 4 4

(a) Find the mode;
(b) find the median;
(c) find the mean.

3.52 Referring to Exercise 3.51, the actual number of unemployed (in thousands) in the 17 states and the District of Columbia (*not* the percentages) is

24 19 10 86 21 39 408 157 250 16 108 16 80 46 144 77 154 269

(a) Find the mode;
(b) find the median;
(c) find the mean.

3.53 Exercise 3.29 gives the numbers of restaurant meals that 13 persons ate during a given week as

3 10 5 1 8 5 6 12 15 1 0 6 5

Find the mode.

3.54 The receiving clerk at a warehouse recorded the following number of deliveries received during 22 business days of a month:

4 7 9 3 5 5 6 5 8 3 7 8 6 10 5 4 5 5 6 9 4 8

Find the mode.

✓ **3.55** The following are the amounts of time (in minutes) that 16 persons spent standing in line waiting to buy tickets for a concert:

8 2 9 1 16 5 7 11 9 1 14 12 9 0 8 4

Find the mode.

3.56   Asked for their favorite color, 50 persons gave the following responses:

red	blue	blue	green	yellow
blue	brown	red	blue	red
red	green	white	blue	red
green	blue	red	green	green
purple	white	yellow	blue	blue
blue	red	red	brown	orange
white	green	blue	blue	blue
red	blue	red	yellow	green
yellow	blue	blue	orange	red
green	white	purple	blue	red

What was their modal choice?

3.57   A specialty shop permits customers to pay for purchases using cash, check, credit card, or a store charge. A random sample of sales slips showed that customers used the following methods of purchases: store charge, store charge, cash, credit card, check, credit card, check, credit card, store charge, credit card, credit card, cash, store charge, credit card, check, credit card, store charge, store charge, credit card, store charge, store charge, credit card, store charge, cash, cash, store charge, credit card, check, credit card, credit card, store charge, check, cash, check, credit card, store charge, check, and cash. What is the modal choice of customers for making purchases?

3.58   The following figures are a local bakery's daily flour utilization, in pounds, for 20 consecutive weekdays:

$$452 \quad 677 \quad 481 \quad 690 \quad 707 \quad 514 \quad 671 \quad 488 \quad 483 \quad 534$$
$$611 \quad 638 \quad 572 \quad 514 \quad 623 \quad 664 \quad 631 \quad 570 \quad 484 \quad 612$$

(a) Find the mean, median, and mode for this set of values.

(b) Round the 20 values to the nearest 10 pounds. Then again give the mean, median, and mode.

(c) Round the 20 values to the nearest 100 pounds. Provide the mean, median, and mode.

(d) State a conclusion about the effect of rounding on the mean, median, and mode.

3.5 MEASURES OF VARIATION: THE RANGE

An important feature of almost any kind of data is that the values are not all alike, and the extent to which they are unalike or vary among themselves is of basic importance in statistics. The following are some examples that illustrate the importance of measuring the variability of statistical data.

The chocolate chip ice cream produced by one company averages 360 chocolate chips per quart, with all quarts containing anywhere from 340 to 380 chocolate chips. Another company's chocolate chip ice cream also averages 360 chocolate chips per quart, but some quarts contain as few as 20 chips and others as many as 740. If the ice cream produced by the two companies is the same in all other respects, it stands to reason that most persons would prefer the chocolate chip ice cream made by the first company—this should give them a better chance of getting ice cream with a "desirable" number of chocolate chips.

Suppose that in a hospital each patient's pulse rate is taken three times a day and that on a certain day the records of two patients show

Patient A:	72	76	74
Patient B:	59	92	71

The mean pulse rates are the same, as can easily be checked, but observe the difference in variability. Whereas patient A's pulse rate is quite stable, that of patient B fluctuates widely, and this should be taken into account when treatments are prescribed.

Finally, suppose that we have a coin that is slightly bent and we wonder whether it is still balanced or *fair*. Suppose that we toss the coin 200 times. If we get 100 heads and 100 tails, there would be no doubt. What if we get 130 heads, however, or only 92? Are the discrepancies due to chance or due to the coin's being bent? Clearly, we need some indication of how much variability can be attributed to chance.

These examples show the need for statistical descriptions that measure the extent to which data is dispersed, or spread out, namely, the need for measures of variation.

To introduce one of the simplest ways of measuring variability, let us refer to the second of the preceding examples, and let us observe that the pulse rate of patient A varied from 72 to 76, while that of patient B varied from 59 to 92. These extreme (smallest and largest) values tell us something about the variability of the respective sets of data, and so do their differences. Thus, we define the **range** as follows:

The range of a set of data is the largest value minus the smallest.

For patient A we get a range of $76 - 72 = 4$, and for patient B we get a range of $92 - 59 = 33$.

The range is easy to calculate and easy to understand, and there is a natural curiosity about the minimum and maximum values. Nonetheless, it is not generally a useful measure of variation. Its main shortcoming is that it does not tell us anything about the dispersion of the values that fall between the two extremes. Each of the following sets of data

Set 1:	5	20	20	20	20	20	20	20
Set 2:	5	5	5	5	20	20	20	20
Set 3:	5	7	9	12	15	17	19	20

has a range of $20 - 5 = 15$, but the dispersion is entirely different in each case. Thus, the range is used mainly as a quick-and-easy indication of variability. It is used, for instance, in industrial quality control to keep a close check on raw materials or products by observing and charting the range of small samples taken at regular intervals of time.

3.6 MEASURES OF VARIATION: THE STANDARD DEVIATION

To introduce the **standard deviation**, by far the most generally useful measure of variation, let us observe that the dispersion of a set of data is small if the values are closely bunched about their mean, and that it is large if the values are scattered widely about their mean. It would seem reasonable, therefore, to measure the variation of a set of data in terms of the amounts by which the individual values differ from their mean. If a set of numbers $x_1, x_2, x_3, \ldots$, and x_N, constituting a population, has the mean μ, the differences

$$x_1 - \mu, \quad x_2 - \mu, \quad x_3 - \mu, \quad \ldots, \quad \text{and} \quad x_N - \mu$$

are called the **deviations from the mean**. It is plausible that we might use their average (that is, their mean) as a measure of the variation in the population. Unfortunately, this will not do. Unless the x's are all equal, some of the deviations will be positive and some will be negative, and, as the reader will be asked to show in Exercise 3.120, the sum of the deviations from the mean, $\sum(x - \mu)$, and consequently also the mean of the deviations is always zero.

Since we are really interested in the magnitude of the deviations, and not in whether they are positive or negative, we might simply ignore the signs and define a measure of variation in terms of the absolute values of the deviations from the mean. Indeed, if we add the deviations from the mean as if they were all positive or zero and divide by N, we obtain the statistical measure called the **mean deviation**. This measure has intuitive appeal, but because of the absolute values it leads to theoretical difficulties in problems of inference, and it is rarely used.

An alternative approach is to work with the squares of the deviations from the mean, as this will also eliminate the effect of the signs. Squares of real numbers cannot be negative; in fact, squares of the deviations from a mean are always positive unless a value happens to coincide with the mean, in which case both $x - \mu$ and $(x - \mu)^2$ are equal to zero. Then, if we average the squared deviations from the mean and take the square root of the result (to compensate for the fact that the deviations were squared), we get the **population standard deviation**.

POPULATION STANDARD DEVIATION

$$\sigma = \sqrt{\frac{\sum(x - \mu)^2}{N}}$$

This measure of variation is denoted by σ (lowercase sigma, the Greek letter for s), and, expressing literally what we have done here mathematically, it is also called the **root-mean-square deviation**. The square of σ is called the **population variance**. The formula for the population variance is obtained by removing the square root symbol from the formula for the population standard deviation, getting

POPULATION VARIANCE

$$\sigma^2 = \frac{\sum(x - \mu)^2}{N}$$

It may seem logical to use the same formula, with n and $\bar{x}$ substituted for N and μ, for the standard deviation of a sample, but this is not quite what we do. Instead of dividing the sum of the squared deviations from the mean by n, we divide it by $n - 1$ and define the **sample standard deviation**, denoted by s, as

SAMPLE STANDARD
DEVIATION

$$s = \sqrt{\frac{\sum(x - \bar{x})^2}{n - 1}}$$

Its square, s^2, is called the **sample variance**. The formula for the sample variance is obtained by removing the square root symbol from the formula for the sample standard deviation, getting

SAMPLE VARIANCE

$$s^2 = \frac{\sum(x - \bar{x})^2}{n - 1}$$

In dividing by $n - 1$ instead of n, we are not just being arbitrary. There is a good theoretical reason for doing this; if we divided by n and used s^2 as an estimate of σ^2 (that is, used the variance of a sample to estimate the variance of the population from which it came), our result would tend to be too small, and we correct for this by dividing by $n - 1$ instead of n. If n is large, it generally will not matter much whether we divide by $n - 1$ or n, but it is common practice to define σ and s as we did.

In calculating the sample standard deviation using the formula by which it is defined, we must (1) find $\bar{x}$, (2) determine the n deviations from the mean $x - \bar{x}$, (3) square these deviations, (4) add all the squared deviations, (5) divide by $n - 1$, and (6) take the square root of the result obtained in step 5. In actual practice, this formula is rarely used, but we shall illustrate it here to emphasize what is really measured by σ and s.

EXAMPLE 3.18   On six consecutive Sundays, a tow-truck operator received 9, 7, 11, 10, 13, and 7 service calls. Calculate s.

Solution   First calculating the mean, we get

$$\bar{x} = \frac{9 + 7 + 11 + 10 + 13 + 7}{6} = \frac{57}{6} = 9.5$$

and the work required to find $\sum(x - \bar{x})^2$ may be arranged as in the following table:

x	$x - \bar{x}$	$(x - \bar{x})^2$
9	−0.5	0.25
7	−2.5	6.25
11	1.5	2.25
10	0.5	0.25
13	3.5	12.25
7	−2.5	6.25
Total 57	0.0	27.50

Then, dividing by $6 - 1 = 5$ and taking the square root, we get

$$s = \sqrt{\frac{27.50}{5}} = \sqrt{5.5} \approx 2.3$$

Note in this table that the total for the middle column is zero; since this must always be the case, it provides a check on the calculations. ■

It was easy to calculate s in this example because the data were whole numbers and the mean was exact to one decimal. Otherwise, the calculations required by the formula defining s can be quite tedious, and, unless we can get s directly with a statistical calculator or a computer, it helps to use the computing formula:

COMPUTING
FORMULA FOR
THE SAMPLE
STANDARD
DEVIATION

$$s = \sqrt{\frac{S_{xx}}{n-1}} \quad \text{where } S_{xx} = \sum x^2 - \frac{\left(\sum x\right)^2}{n}$$

The advantage of this computing formula is that we merely have to get $\sum x$ and $\sum x^2$ from the data, without having to determine the value of $\bar{x}$ and work with the deviations from the mean. In Exercises 3.64 and 3.65, the reader is asked to verify that the computing formula is, indeed, equivalent to the one defining s^2.

EXAMPLE 3.19   Use the computing formula for s to rework the preceding example.

Solution   First, we calculate $\sum x$ and $\sum x^2$, getting

	x	x^2
	9	81
	7	49
	11	121
	10	100
	13	169
	7	49
Total	57	569

Then, substituting $\sum x = 57$ and $\sum x^2 = 569$, together with $n = 6$, we find that

$$S_{xx} = 569 - \frac{(57)^2}{6} = 27.50$$

This gives us

$$s = \sqrt{\frac{27.50}{6-1}} = \sqrt{5.5} \approx 2.3$$

which agrees with the result we obtained before. ■

Since beginners often seem to confuse $\sum x^2$ with $\left(\sum x\right)^2$, let us emphasize the point that for $\sum x^2$ we first square each x and then add all the squares; for $\left(\sum x\right)^2$, on the other hand, we first add all the x's and then square their sum.

The sum $\sum x$ can be obtained by merely pressing calculator keys; it is not necessary to write down intermediate steps. On most calculators it is possible to compute $\sum x^2$ without writing down the squares of the individual numbers, though the specific techniques vary from one calculator brand to another.

The usual method for verifying a calculation is to repeat it and see if the same result is obtained. When different results arise, the calculation must be redone. This checking can be time consuming for the sample standard deviation. Some calculators have keys that compute the standard deviation directly from the list of values; even with such calculators the checking is inconvenient. Accordingly, we recommend that hand-held calculators be used only for small problems, say $n \leq 20$; computers should be used for larger problems.

EXERCISES

Exercises 3.59 and 3.64 are practice exercises; their complete solutions are given on page 98.

✓ **3.59** Following are the numbers of pupils per teacher in the public schools of the eight Mountain States:

Montana	14.3
Idaho	18.0
Wyoming	13.3
Colorado	17.4
New Mexico	16.4
Arizona	19.4
Utah	22.0
Nevada	18.7

What is their range?

3.60 Annual student enrollment (in millions of students) in two- and four-year colleges and universities in the United States for the years 1991, 1992, ..., 1999 was 14.1, 14.0, 13.9, 15.0, 14.7, 15.2, 15.4, 15.5, and 15.2. Find the range.

3.61 The following are the closing prices (in dollars) of two stocks on five consecutive Fridays:

Stock A:	18.25	17.03	18.41	17.44	18.10
Stock B:	20.31	20.27	19.55	20.60	20.40

Calculate the range for each stock and decide which is more stable, that is, less variable.

3.62 Forty flight attendants are given an intensive course in cardiopulmonary resuscitation (CPR). The following are the numbers of correct answers they gave on a test administered after the completion of the course:

21	18	19	10	12	20	14	18	9	13
18	8	15	10	22	21	15	10	18	9
20	18	11	8	5	18	23	16	17	17
21	7	15	18	16	22	13	9	18	20

Find the range of these figures.

3.63 The densities of selected gases in kilograms per cubic meter are

air	1.17	neon	0.90
carbon dioxide	1.98	nitrogen	1.25
carbon monoxide	1.25	nitrous oxide	2.00
helium	0.18	oxygen	1.43
hydrogen	0.09	propane	2.02
methane	0.72	sulfur dioxide	2.93

Find the range of the densities of the selected gases.

✓ **3.64** In five attempts, it took a person 11, 15, 12, 8, and 14 minutes to change a tire on a car.

(a) Calculate s using the formula that defines s.

(b) Rework this exercise, using the computing formula.

(c) Verify that the solutions of parts (a) and (b) are identical.

3.65 According to the Insurance Information Institute, per capita fire losses (excluding losses from forest fires and government property) in the United States from 1990 to 1999 were, respectively, 38, 45, 53, 44, 49, 45, 47, 48, 46, and 42 dollars.

(a) Calculate s using the formula that defines s.

(b) Rework this exercise, using the computing formula.

(c) Verify that the solutions of parts (a) and (b) are identical.

3.66 A filling machine in a high-production bakery is set to fill open-face pies with 16 fluid ounces of fill. A sample of four pies from a large production lot shows fills of 16.2, 15.9, 15.8, and 16.1 fluid ounces.

(a) Calculate s using the formula that defines s.

(b) Rework this exercise, using the computing formula.

3.67 The per capita government debt of the South Atlantic States in 1998 was

	Thousands of dollars
Delaware	3.7
Maryland	2.2
Virginia	11.9
West Virginia	3.7
North Carolina	8.2
South Carolina	5.1
Georgia	0.8
Florida	1.2

Use the computing formula to determine the value of s.

3.68 The following are the numbers of hours that 12 students studied for a final examination: 7, 14, 22, 19, 20, 13, 25, 28, 32, 11, 20, and 24. Use the computing formula to determine s^2 and s for these values:

$$\sum x = 235 \quad \text{and} \quad \sum x^2 = 5{,}189$$

3.69 The numbers of congressional bills vetoed by Presidents Kennedy, Johnson, Nixon, Ford, Carter, Reagan, Bush, and Clinton are, respectively, 21, 30, 43, 66, 31, 78, 44, and 30. Use the computing formula to determine s^2 and s for these values:

$$\sum x = 343 \quad \text{and} \quad \sum x^2 = 17{,}427$$

3.70 According to Ward's Communications, the best-selling passenger cars in the United States in the year 2000 were Toyota Camry, Honda Accord, Ford Taurus, Honda Civic, Ford Focus, Chevrolet Cavalier, Toyota Corolla, Pontiac Grand Am, Chevrolet Malibu, and Saturn S. Their sales in hundreds of thousands were 4.2, 4.0, 3.8, 3.2, 2.9, 2.4, 2.3, 2.1, 2.1, and 1.8, respectively.

(a) Verify the values $\left(\sum x\right)^2 = 28.8^2 = 829.44$; $\sum x^2 = 89.84$.

(b) Find s for these values, using the computing formula for the sample standard deviation.

3.71 In 1999 the electric energy generated in units of billions of kilowatt-hours in the seven west north-central states of North Dakota, South Dakota, Nebraska, Kansas, Minnesota, Iowa, and Missouri was 31.3, 10.6, 30.0, 42.0, 44.0, 37.0, and 73.5. Modify the computing formula for the standard deviation so that it applies to populations (that is, replace the $n-1$ by n and then replace each n by N) and then use it to calculate σ for the given data.

3.72 With reference to Exercise 3.62, determine s for the number of correct answers that the flight attendants got on the test.

$$n = 40, \quad \sum x = 612 \quad \text{and} \quad \sum x^2 = 10{,}282$$

3.73 It has been claimed that for samples of size $n = 4$ the range should be roughly twice as large as the standard deviation. Check this claim with reference to the following data, representing the number of times that four students were absent from a course in anthropology: 4, 7, 3, and 7.

3.74 It has been claimed that for samples of size $n = 10$ the range should be roughly three times as large as the standard deviation. Check this claim with reference to the following test scores, which 10 students obtained on an examination: 77, 90, 72, 71, 91, 84, 64, 80, 75, and 86.

3.75 The United States imported a total of $1,115,900,000 of goods and services in 1998, of which $182,945,000 was for information and communication technology (ICT). It can easily be verified that the proportion of ICT imports to total imports is approximately 16.4%. Following is the percentage of ICT imports to total imports of 28 nations in 1998.

Nations	Percent	Nations	Percent
Australia	13.4	Japan	13.5
Austria	8.2	Korea (South)	12.0
Belgium-Luxembourg	7.4	Mexico	17.7
Canada	12.4	Netherlands	16.7
Czech Republic	10.5	New Zealand	1.3
Denmark	12.7	Norway	7.8
Finland	16.1	Poland	9.6
France	11.1	Portugal	8.4
Germany	11.0	Spain	8.6
Greece	8.6	Sweden	14.2
Hungary	19.7	Switzerland	9.4
Iceland	8.1	Turkey	8.6
Ireland	33.9	United Kingdom	14.9
Italy	8.5	United States	16.4

Use a computer package to find $\bar{x}$ and s for these percentages.

3.76 The following are the ignition times of certain upholstery materials exposed to a flame (in seconds):

2.50	4.50	5.11	9.70	5.62	6.77	3.49	4.90	10.21	8.76
9.33	4.12	3.85	4.97	5.04	2.97	3.81	10.60	7.95	7.41
8.64	5.33	3.90	11.25	1.92	1.42	12.80	9.45	6.25	4.71
7.86	2.65	4.79	6.20	1.52	1.38	3.87	4.54	5.12	5.15
11.75	7.35	2.80	6.85	1.20	9.20	1.76	5.21	3.40	7.29
8.66	5.04	10.25	6.43	2.97	4.45	5.50	5.92	4.56	2.46
6.90	1.47	2.11	2.32	4.19	2.20	4.32	1.58	6.43	4.04
2.51	2.58	3.78	3.75	3.10	6.43	1.70	6.40	3.24	1.79
8.75	2.46	3.62	4.72	7.40	8.81	5.83	6.75	7.65	8.79
10.92	9.65	5.09	4.11	6.37	5.40	2.51	10.28	5.49	3.76

Use a computer package to find the mean and variance of the sample data.

3.7 SOME APPLICATIONS OF THE STANDARD DEVIATION

After this chapter, we shall use sample standard deviations primarily as estimates of population standard deviations in problems of inference. To get more of a feeling for what a standard deviation really measures, let us devote this section to some applications.

In the argument that led to the definition of the standard deviation, we observed that the dispersion of a set of data is small if the values are bunched closely about their mean, and that it is large if the values are scattered widely about their mean. Correspondingly, we can now say that, if the standard deviation of a set of data is small, the values are concentrated near the mean, and if the standard deviation is large, the values are scattered widely about the mean. This idea is expressed more formally by **Chebyshev's theorem**, named for the Russian mathematician P. L. Chebyshev (1821–1894):

CHEBYSHEV'S
THEOREM

> *For any set of data (population or sample) and any constant k greater than 1, at least $1 - 1/k^2$ of the data must lie within k standard deviations on either side of the mean.*

Thus, we can be sure that at least

$$1 - \frac{1}{2^2} = 1 - \frac{1}{4} = \frac{3}{4} \quad \text{or} \quad 75\%$$

of the values in any set of data must lie within 2 standard deviations on either side of the mean; at least

$$1 - \frac{1}{3^2} = 1 - \frac{1}{9} = \frac{8}{9} \quad \text{or about} \quad 89\%$$

must lie within 3 standard deviations on either side of the mean; and at least

$$1 - \frac{1}{10^2} = 1 - \frac{1}{100} = \frac{99}{100} \quad \text{or} \quad 99\%$$

must lie within 10 standard deviations on either side of the mean.

EXAMPLE 3.20 If all the 1-pound cans of coffee filled by a food processor have a mean weight of 16.00 ounces with a standard deviation of 0.02 ounce, at least what percentage of the cans must contain anywhere from 15.95 to 16.05 ounces of coffee?

Solution The range of values 15.95 to 16.05 represents the mean plus or minus 0.05. This 0.05 is to be interpreted as k standard deviations, so that

$$k = \frac{0.05}{0.02} = 2.5$$

Thus, at least

$$1 - \frac{1}{2.5^2} = 1 - \frac{1}{6.25} = 0.84 \quad \text{or} \quad 84\%$$

of the cans must contain anywhere from 15.95 to 16.05 ounces of coffee. ▮

Chebyshev's theorem applies to any kind of data, but it tells us only "at least what percentage" must lie between certain limits. For nearly all sets of data, the actual percentage of data lying between the limits is much greater than that specified by Chebyshev's theorem. For many distributions, we can make the following much stronger statement:

> *For distributions having the general shape of the cross section of a bell (see Figure 3.3),*
>
> **1.** *about 68% of the values will lie within 1 standard deviation of the mean, that is, between $\bar{x} - s$ and $\bar{x} + s$;*
> **2.** *about 95% of the values will lie within 2 standard deviations of the mean, that is, between $\bar{x} - 2s$ and $\bar{x} + 2s$;*
> **3.** *about 99.7% of the values will lie within 3 standard deviations of the mean, that is, between $\bar{x} - 3s$ and $\bar{x} + 3s$.*

This result is sometimes referred to as the **empirical rule**, although actually it is a theoretical result based on the normal distribution, which we shall discuss in Chapter 7.

Figure 3.3
Bell-shaped distribution.

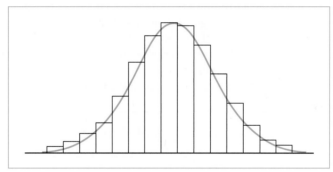

There are many ways in which knowledge of the variability of a set of data can be of importance, and some of them were previously shown in Section 3.6. Another application arises in the comparison of numbers belonging to different sets of data.

EXAMPLE 3.21

In an English class the final examination grades average 60 with a standard deviation of 16, and in a mathematics class the final examination grades average 58 with a standard deviation of 10. If a student gets a 72 on the English examination and a 68 on the mathematics examination, how many standard deviations is each of her grades above the average of the respective class? What does this tell us about her performance in the two subjects?

Solution

On the English examination her grade is $72 - 60 = 12$ points and, hence, $\dfrac{12}{16} = 0.75$ standard deviation above average; on the mathematics examination her grade is $68 - 58 = 10$ points and, hence, $\dfrac{10}{10} = 1.00$ standard deviation above average. Even though her English grade is higher than her mathematics grade, and her English grade is 12 points above average, while her mathematics grade is only 10 points above average, her performance relative to the two classes is better in mathematics than it is in English. Being 1.00 standard deviation above average in mathematics puts her in a relatively better position than being 0.75 standard deviation above average in English. ■

What we have done here consisted of converting the grades into **standard units** or **z-scores**. In general, if x is a measurement belonging to a set of data having the mean $\bar{x}$ (or μ) and the standard deviation s (or σ), then its value in standard units, denoted by z, is

FORMULA FOR CONVERTING TO STANDARD UNITS

$$z = \frac{x - \bar{x}}{s} \quad \text{or} \quad z = \frac{x - \mu}{\sigma}$$

depending on whether the data constitute a sample or a population. The z-score tells us how many standard deviations a value lies above or below the mean of the set of data to which it belongs. Standard units will be used frequently in later chapters. The standard deviation depends on the units of measurement. For instance, the weights of certain objects may have a standard deviation of 0.1 ounce or 2,835 milligrams, which is the same, but neither value really tells us whether it reflects a great deal of variation or very little variation. If the objects we are weighing are the eggs of small birds, either figure would reflect a great deal of variation, but this would not be the case if the objects that we are weighing are 100-pound bags of potatoes. What we need in a situation like this is a **measure of relative variation**, such as the **coefficient of variation**,

COEFFICIENT OF VARIATION

$$V = \frac{s}{\bar{x}} \cdot 100\% \quad \text{or} \quad V = \frac{\sigma}{\mu} \cdot 100\%$$

which expresses the standard deviation as a percentage of what is being measured, at least on the average. Thus, if V is smaller for one set of data than another, there is relatively less variability in the first set of data, and we say that it is "more consistent."

EXAMPLE 3.22 During the past few months, one runner averaged 12 miles per week with a standard deviation of 2 miles, while another runner averaged 25 miles per week with a standard deviation of 3 miles. Which of the two runners is relatively more consistent in his weekly running habits?

Solution The two coefficients of variation are, respectively,

$$\frac{2}{12} \cdot 100\% \approx 16.7\% \quad \text{and} \quad \frac{3}{25} \cdot 100\% = 12.0\%$$

Since 12.0% is less than 16.7%, the second runner is relatively more consistent in his weekly running habits. ■

Exercises 3.77, 3.80, 3.84, and 3.88 are practice exercises; their complete solutions are given on pages 98 and 99.

✓ **3.77** According to Chebyshev's theorem, what can we assert about the percentage of any set of data that must lie within k standard deviations on either side of the mean when
  (a) $k = 4$;
  (b) $k = 12$?

3.78 The records of a charitable center for the collection of used clothing and household items (for later resale) show that, on the average, they receive 100 contributions daily, with a standard deviation of 5 contributions daily. Use Chebyshev's theorem to determine at least what percentage of the days the contributions will number between
  (a) 90 and 110;
  (b) 85 and 115.

3.79 If the empirical rule was used to solve Exercise 3.78 (a) and (b), would the percentages be larger or smaller? Explain (do not solve).

✓ **3.80** An airline's records show that its flights between two cities arrive on the average 5.4 minutes late with a standard deviation of 1.4 minutes. At least what percentage of its flights between the two cities arrive anywhere between
  (a) 2.6 minutes late and 8.2 minutes late;
  (b) 1.6 minutes early and 12.4 minutes late?

3.81 According to Chebyshev's theorem, what can we assert about the percentage of any set of data that must lie within k standard deviations on either side of the mean when
  (a) $k = 5$;
  (b) $k = 8$;
  (c) $k = 10$;
  (d) $k = 20$?

3.82 With reference to Exercise 3.80, at least what percentage of the flights between the two cities must arrive anywhere between
  (a) 4.0 minutes late and 6.8 minutes late;
  (b) 2.6 minutes late and 8.2 minutes late;
  (c) 1.2 minutes late and 9.6 minutes late;
  (d) 0.5 minute late and 10.3 minutes late?

3.83 Consider again Exercise 3.80. Using the empirical rule on page 77, about what percentage of the flights between the two cities will arrive between
  (a) 4.0 minutes late and 6.8 minutes late;
  (b) 2.6 minutes late and 8.2 minutes late;
  (c) 1.2 minutes late and 9.6 minutes late?

✓ **3.84** The mean amount of time for college students to vacate a certain classroom building during a fire drill is 7 minutes, with a standard deviation of 0.5 minutes. Use Chebyshev's theorem to determine at least what percentage of the time the classroom building can be vacated during a fire drill

EXERCISES

(a)   between 6 and 8 minutes;

(b)   between 5 and 9 minutes.

3.85   In a certain department store, the mean number of items returned for a refund each day is 45, and the standard deviation is 6. Use Chebyshev's theorem to determine between what two numbers must lie

(a)   at least $\frac{3}{4}$ of the daily number of items returned for a refund;

(b)   at least $\frac{15}{16}$ of the number of items returned for a refund.

3.86   Of two persons on a reducing diet, the first belongs to an age group for which the mean weight is 146 pounds with a standard deviation of 14 pounds, and the second belongs to an age group for which the mean weight is 160 pounds with a standard deviation of 17 pounds. If their respective weights are 178 pounds and 193 pounds, which of the two is more seriously overweight for his or her age group?

3.87   The applicants to one state university have an average ACT mathematics score of 21.4 with a standard deviation of 3.1, while the applicants to another state university have an average ACT mathematics score of 22.1 with a standard deviation of 2.8. With respect to which of these two universities is a student in a relatively better position if he or she scores

(a)   26 on this test;

(b)   31 on this test?

✓ **3.88**   To compare the precision of two measuring instruments, a laboratory technician studies recent measurements made with both instruments. The first was recently used to measure the diameter of a ball bearing and the measurements had a mean of 4.92 mm with a standard deviation of 0.018 mm; the second was recently used to measure the unstretched length of a spring, and the measurements had a mean of 2.54 in. with a standard deviation of 0.012 in. Which of the two measuring instruments is relatively more precise?

3.89   A sample of the ages of five women in an aerobics class revealed the following ages: 22, 18, 26, 20, and 24. Their weights were 115, 159, 141, 137, and 130 pounds.

(a)   Find the mean and standard deviation of the ages and use these values to determine the coefficient of variation.

(b)   Find the mean and standard deviation of the weights and use these values to determine the coefficient of variation.

(c)   Using the coefficients of variation of parts (a) and (b), determine which of the two sets of data are more variable.

3.90   If we add or subtract an arbitrary value to all the original values of a distribution, the value of s remains unchanged.

(a)   In Example 3.19 we obtained $s \approx 2.3$ for the values 9, 7, 11, 10, 13, and 7. Subtract, say, 7 from each of these values and we get 2, 0, 4, 3, 6, and 0. Recalculate s for the new values, and verify that s remains approximately 2.3.

*3.8   THE DESCRIPTION OF GROUPED DATA

In the past, considerable attention was paid to the description of grouped data, because it was generally advantageous to group data before calculating the appropriate descriptions. This was true, particularly, in connection with large sets of data, where the manual determination of quantities such as $\sum x$ and $\sum x^2$ entailed a considerable amount of work. Today, this is no longer the case, since such quantities can be determined quickly with a computer. Nevertheless, we shall devote this section to the description of grouped data, since some data (published government figures, for example) is available only in the form of frequency distributions. As we saw in Chapter 2, the grouping of data entails some loss of information. Each item loses its exact identity by being placed into a class. We only know the number of items in each class, so we must be satisfied with approximations. In the case of the mean and the standard deviation, we can get excellent approximations by proceeding as follows:

We assign to each item falling into a class the value of the class mark.

For instance, restating the distribution that pertains to the amount of time that 80 students devoted to leisure activities, we have the following:

Hours	Frequency
10–14	8
15–19	28
20–24	27
25–29	12
30–34	4
35–39	1
Total	80

To calculate the mean or the standard deviation of the distribution, we treat the 8 values falling into the class 10–14 as if they were all equal to 12, the 28 values falling into the class 15–19 as if they were all equal to 17,..., and the value falling into the class 35–39 as if it were equal to 37. This procedure is usually quite satisfactory, since the errors that are thus introduced into the calculations will tend to average out.

To give general formulas for the mean and the standard deviation of a distribution with k classes, let us denote the successive class marks by $x_1, x_2, x_3, \ldots,$ and x_k, and the corresponding class frequencies by $f_1, f_2, f_3, \ldots,$ and f_k. The number of values is

$$f_1 + f_2 + f_3 + \cdots + f_k = \sum f = n$$

The sum of all the measurements or observations is represented by

$$x_1 f_1 + x_2 f_2 + x_3 f_3 + \cdots + x_k f_k = \sum x \cdot f$$

The sum of their squares is represented by

$$x_1^2 f_1 + x_2^2 f_2 + x_3^2 f_3 + \cdots + x_k^2 f_k = \sum x^2 \cdot f$$

The formula for $\bar{x}$ and the computing formula for s can be written as

$$\bar{x} = \frac{\sum x \cdot f}{n} \quad \text{and} \quad s = \sqrt{\frac{S_{xx}}{n-1}}$$

where

$$S_{xx} = \sum x^2 \cdot f - \frac{\left(\sum x \cdot f\right)^2}{n}$$

To get the corresponding formulas for a population, we replace n by N, and in the formula for the standard deviation we replace $n - 1$ by N. That is, we write

$$\mu = \frac{\sum xf}{N} \quad \text{and} \quad \sigma = \sqrt{\frac{S_{xx}}{N}}.$$

EXAMPLE 3.23   Find the mean and the standard deviation of the previous distribution pertaining to the amounts of time that 80 college students devoted to leisure activities.

Solution   To obtain $\sum x \cdot f$ and $\sum x^2 \cdot f$, we perform the calculations shown in the following table:

Class Mark		Frequency		
x	x^2	f	$x \cdot f$	$x^2 \cdot f$
12	144	8	96	1,152
17	289	28	476	8,092
22	484	27	594	13,068
27	729	12	324	8,748
32	1,024	4	128	4,096
37	1,369	1	37	1,369
	Total	80	1,655	36,525

Then, substitution into the formulas yields

$$\bar{x} = \frac{1,655}{80} = 20.6875 \approx 20.69$$

and

$$S_{xx} = 36,525 - \frac{(1,655)^2}{80} \approx 2,287.19$$

and then

$$s = \sqrt{\frac{2,287.19}{79}} \approx 5.381$$

To check the grouping error for this set of data, that is, the error introduced by first grouping the data, let us refer to the MINITAB printout of Figure 3.4. It shows that the mean of this ungrouped data is 20.613 rounded to three decimals and that the standard deviation is 5.563 rounded to three decimals. Thus, both errors are fairly small. ■

Once a set of data has been grouped, we can still determine (at least approximately) most other statistical measures besides the mean and the standard deviation, but we must modify their definitions. Although we can no longer determine the exact value of the median, we can find an approximate value as follows:

Figure 3.4
MINITAB printout for the mean and the standard deviation of the amounts of time that 80 students devoted to leisure activities.

For grouped data, the median is such that half the total area of the rectangles of a histogram of their distribution lies to its left and the other half lies to its right (see Figure 3.5).

Figure 3.5
The median of the distribution of the amounts of time that 80 students devoted to leisure activities.

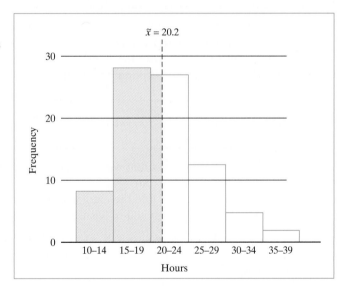

This definition is equivalent to the assumption that the values in the class containing the median are distributed evenly (that is, spread out evenly) throughout that class. To find the dividing line between the two halves of a histogram (each of which represents $\frac{n}{2}$ of the items grouped), we must somehow count $\frac{n}{2}$ of the items, starting at either end of the distribution. How this is done is illustrated by the following example.

EXAMPLE 3.24   Find the median of the distribution of the amounts of time that the students devoted to leisure activities.

Solution   Since $\frac{n}{2} = \frac{80}{2} = 40$, we must count 40 of the items, starting at either end. Starting at the bottom of the distribution (beginning with the smallest values), we find that

$8 + 28 = 36$ of the values fall into the first two classes and that $8 + 28 + 27 = 63$ of the values fall into the first three classes. Therefore, we must count $40 - 36 = 4$ of the values beyond the 36 that fall into the first two classes. On the assumption that the 27 values in the third class are spread evenly throughout that class, we add $\dfrac{4}{27}$ of the class interval of 5 to the lower boundary of the third class. This gives us

$$\tilde{x} = 19.5 + \frac{4}{27} \cdot 5 \approx 20.2$$

for the median of the distribution. ■

In general, if L is the lower boundary of the class into which the median must fall, f is its frequency, c is its interval, and j is the number of items we still lack when we reach L, then the median of the distribution is given by

MEDIAN OF GROUPED DATA

$$\tilde{x} = L + \frac{j}{f} \cdot c$$

If we prefer, we can find the median of a distribution by starting to count at the other end (beginning with the largest values) and subtracting an appropriate fraction of the class interval from the upper boundary U of the class into which the median must fall. The corresponding formula is

ALTERNATIVE FORMULA FOR THE MEDIAN OF GROUPED DATA

$$\tilde{x} = U - \frac{j'}{f} \cdot c$$

where j' is the number of items that we still lack when we reach U.

EXAMPLE 3.25   Use the alternative formula to find the median of the distribution of the amounts of time that the 80 students devoted to leisure activities.

Solution   Since $1 + 4 + 12 = 17$ of the values fall into the three classes at the top of the distribution, we need $40 - 17 = 23$ of the 27 values in the next class to reach the median, and we write

$$\tilde{x} = 24.5 - \frac{23}{27} \cdot 5 \approx 20.2$$

The result is, of course, the same. ■

Note that the median of a distribution can be found by this method regardless of whether the class intervals are all equal; in fact, it can usually be found even when either or both of the classes at the top and at the bottom of a distribution are open.

The method by which we found the median of a distribution can also be used to determine other fractiles. For instance, the quartiles Q_1 and Q_3 of a distribution are defined so that 25% of the total area of the rectangles of the histogram lies to the left of Q_1 and 25% to the right of Q_3. To find them, we use either of the two formulas for the median of grouped data.

EXAMPLE 3.26 Find Q_1 and Q_3 for the distribution of the amounts of time that the 80 students devoted to leisure activities.

Solution Since $\dfrac{n}{4} = \dfrac{80}{4} = 20$, we must count 20 of the items, starting at the bottom of the distribution, to find Q_1. Since there are eight values in the first class, we must count $20 - 8 = 12$ of the 28 values in the second class to reach Q_1, and we write

$$Q_1 = 14.5 + \frac{12}{28} \cdot 5 \approx 16.6$$

To find Q_3 we must count 20 of the items starting at the other end of the distribution. Since $1 + 4 + 12 = 17$ of the values fall into the three classes at the top of the distribution, we must count $20 - 17 = 3$ of the 27 values in the next class to reach Q_3, and we write

$$Q_3 = 24.5 - \frac{3}{27} \cdot 5 \approx 23.9$$ ■

The results that we have obtained here for the median, Q_1, and Q_3, together with the fact that the smallest value is 10 and the largest value is 38 (as shown in the ungrouped data on page 26), are summarized in the box-and-whisker plot of Figure 3.6.

Another set of useful fractiles consists of the **percentiles**. The first percentile is defined so that 1% of the total area of the rectangles of the histogram lies to its left. The second percentile is defined so that 2% of the total area of the rectangles of the histogram lies to its left, and the other percentiles are defined in an analogous manner. They are determined by using either of the two formulas on page 84.

EXAMPLE 3.27 Find P_{33} and P_{85} for the distribution of the amounts of time that the 80 students devoted to leisure activities.

Solution To find P_{33}, we must count $0.33(80) = 26.4$ of the items starting at the bottom of the distribution. Since there are 8 values in the first class, we must count $26.4 - 8 = 18.4$ of the 28 values in the second class to reach P_{33}, and we write

$$P_{33} = 14.5 + \frac{18.4}{28} \cdot 5 \approx 17.8$$

Figure 3.6
Box-and-whisker plot for the amounts of time that 80 students devoted to leisure activities.

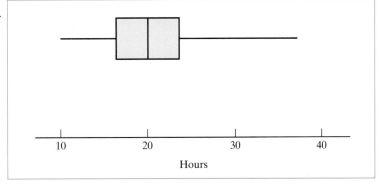

To find P_{85}, we must count $0.15(80) = 12$ of the items starting at the top of the distribution. Since $1 + 4 = 5$ of the values fall into the two classes at the top of the distribution, we must count $12 - 5 = 7$ of the 12 values in the next class to reach P_{85}, and we get

$$P_{85} = 29.5 - \frac{7}{12} \cdot 5 \approx 26.6$$

*3.9   SOME FURTHER DESCRIPTIONS

So far we have discussed statistical descriptions that come under the general heading of *measures of location* and *measures of variation*. Actually, there is no limit to the number of ways in which statistical data can be described, and statisticians continually develop new methods of describing characteristics of numerical data that are of interest in particular problems. In this section we shall consider briefly the problem of describing the overall shape of a distribution.

Although frequency distributions can take on almost any shape or form, most of the distributions that we meet in practice can be described fairly well by one or another of a few standard types. Among these, foremost in importance is the aptly described **symmetrical bell-shaped distribution** shown at the top of Figure 3.7. Indeed, there are theoretical reasons why, in many instances, distributions of actual data can be expected to follow this general pattern. The other two distributions of Figure 3.7 can still, by a stretch of the imagination, be called **bell shaped**, but they are not symmetrical. Distributions like these, having a *tail* on one side or the other, are said to be **skewed**; if the tail is on

Figure 3.7
Bell-shaped
distributions.

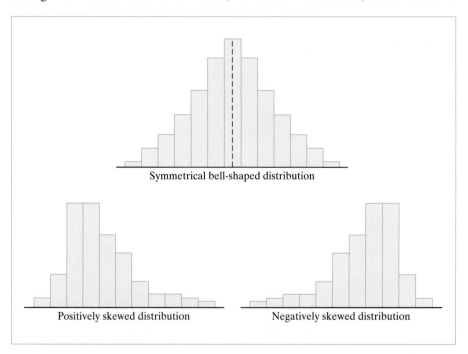

Symmetrical bell-shaped distribution

Positively skewed distribution   Negatively skewed distribution the left, we say that they are **negatively skewed**, or skewed to the left, and if the tail is on the right, we say that they are **positively skewed** or skewed to the right. In general, we say that a distribution is skewed in the direction of the tail. Distributions of incomes or wages are often positively skewed because of the presence of some relatively high values that are not offset by correspondingly low values.

The symmetry or skewness of a distribution can also be judged visually by inspection of a box-and-whisker plot. For a symmetrical distribution, the median line divides the box into equal halves. It is moved to the left of center when a distribution is positively skewed and to the right of center when a distribution is negatively skewed. In Figure 3.6, the median line is slightly to the left of center, and this reflects the mild positive skewness, which is apparent from the histogram of Figure 3.5. Note also the long whisker extending from Q_3 to the largest value, which was 38.

There are several ways of actually measuring the extent to which a distribution is skewed. A relatively easy one is based on the fact that, for a perfectly symmetrical bell-shaped distribution such as the one at the top of Figure 3.7, the values of the median and the mean coincide. The presence of some relatively high values that are not offset by correspondingly low values will tend to make the mean greater than the median. Similarly, the presence of some relatively low values that are not offset by correspondingly high values will tend to make the mean less than the median. We can use this relationship between the mean and the median to define a relatively simple measure of the extent to which a distribution is skewed. It is called the **Pearsonian coefficient of skewness** and is given by

PEARSONIAN COEFFICIENT OF SKEWNESS

$$SK = \frac{3(\text{mean} - \text{median})}{\text{standard deviation}}$$

For a perfectly symmetrical distribution, the value of SK is 0, and in general its values must fall between -3 and 3.

EXAMPLE 3.28   Find the Pearsonian coefficient of skewness for the distribution of the amounts of time the 80 students devoted to leisure activities.

Solution   Substituting into the formula the values obtained earlier for the mean, the median, and the standard deviation, $\bar{x} = 20.7$, $\tilde{x} = 20.2$, and $s = 5.4$, we get

$$SK = \frac{3(20.7 - 20.2)}{5.4} \approx 0.3$$

This reflects the fact that the distribution has a slight positive skewness.   ■

Two other kinds of distributions that sometimes arise in practice are the **reverse J-shaped** and **U-shaped distributions** shown in Figure 3.8. As can be seen from the diagram, the names of such distributions quite literally describe their shape. Examples of such distributions may be found in Exercises 3.110 and 3.111.

Figure 3.8
Reverse J-shaped and
U-shaped distributions.

EXERCISES

Exercises 3.91, 3.95, 3.99, 3.104, and 3.107 are practice exercises; their complete solutions are given on pages 99–100.

✓ ***3.91**   Find $\bar{x}$ and s for the following distribution of the weekly earnings of 125 wage earners. Observe that the class marks are $124.995, $134.995, $144.995, and so on. The computations will be somewhat easier if you raise these by $0.005 to $125, $135, $145, and so on. The effect of this on the mean will be negligible, and it will not affect the standard deviation.

Weekly earnings (dollars)	Frequency
120.00–129.99	9
130.00–139.99	20
140.00–149.99	36
150.00–159.99	30
160.00–169.99	15
170.00–179.99	11
180.00–189.99	4
Total	125

*3.92   Following is a summary of the gasoline tax rates that were in effect in each of the states at the end of the year 1999:

Cents per gallon	Number of states
5–9	2
10–14	3
15–19	15
20–24	23
25–29	7
Total	50

Find the mean, $\bar{x}$, and the standard deviation, s, of this distribution.

*3.93   According to the National Climatic Data Center, the highest temperatures recorded in the 50 states of the United States (through 1999) in degrees Fahrenheit were as follows:

Degrees fahrenheit	Number of states
100–104.9	2
105–109.9	9
110–114.9	19
115–119.9	10
120–124.9	7
125–129.9	2
130–134.9	1
Total	50

Find the mean, $\bar{x}$, and the standard deviation, s, of the distribution. The lowest of the high temperatures were in Alaska and Hawaii, both with 100°F. The highest was in California, with 134°F.

*3.94   According to the Organization for Economic Cooperation and Development (OECD), the following are the percentages of persons not in education or at work in 19 countries, year 1999. Find the median.

Percent	Number of countries
5–9	4
10–14	5
15–19	5
20–24	2
25–29	3
Total	19

✓ *3.95   Find the median of the distribution of Exercise 3.91.

*3.96   Find the mean and standard deviation of the small mortgage loans made by a private loan company:

Thousands of dollars	Number of loans
0–9	3
10–19	4
20–29	8
30–39	6
40–49	5
50–59	2
60–69	2
Total	30

*3.97   The state of North Carolina has 100 counties, and their land areas, in square miles, are as follows:

Number of square miles	Number of counties
100–199	2
200–299	17
300–399	17
400–499	20
500–599	19
600–699	10
700–799	5
800–899	6
900–999	4
Total	100

(a) Find $\bar{x}$, the mean;

(b) find $\tilde{x}$, the median.

*3.98 U.S. Bureau of the Census data reveals that in 2002, the population of the United States, by age, in millions of persons, is distributed as follows:

Age	Number of persons (in millions)
0–9	38.4
10–19	40.3
20–29	36.6
30–39	41.5
40–49	43.4
50–59	31.8
60–69	20.4
70–79	16.1
Over 80	9.5
Total	278.0

(a) Find the median;

(b) find quartile 1;

(c) find percentile 75.

✓ ***3.99** The lengths of business cycle expansions and contractions are measured from peak to peak by the National Bureau of Economic Research. Fifteen continuous cycles are measured, in months. The current business cycle is not included. Measurements for this long-term study begin in the year 1919.

Length in months	Number of cycles
10–29	2
30–49	7
50–69	1
70–89	1
90–109	3
110–129	1
Total	15

(a) Find the median, $\tilde{x}$;

(b) find the quartiles, Q_1 and Q_3.

*3.100 Using the data provided for Exercise 3.99,

(a) Find the mean;

(b) find the standard deviation.

*3.101 If a share of common stock has a market price of $30.00, and its earnings per share is $5.00, then its price-earnings (PE) ratio is $\dfrac{30.00}{5.00} = 6.0$. If the earnings of this common stock had been $2.00, $3.00, or $4.00, then the PE ratio would have been 15.0, 10.0, and 7.5, respectively. Suppose an investor's stock portfolio consisted of 40 stocks with the following PE ratios:

PE ratio	Number of stocks
0–4	3
5–9	10
10–14	12
15–19	9
20–24	4
25–29	2
Total	40

(a) Find the mean;

(b) find the standard deviation;

(c) find the median;

(d) find quartile 1 and quartile 3;

(e) find the eightieth percentile.

*3.102 A geologist weighs his rock specimens and determines that the heaviest weighs 17.2 pounds, the lightest weighs 0.5 pound, and the value of the median, Q_1, and Q_3 are 9.27, 5.95, and 13.00 pounds, respectively. Use this information to draw a box-and-whisker plot.

3.103 Determine, by inspection, whether the box plots of Figures 3.9, 3.10, and 3.11 are positively skewed, negatively skewed, or symmetrical. Justify your answers.

(a)

Figure 3.9

Figure 3.10

(b)

Figure 3.11

(c)

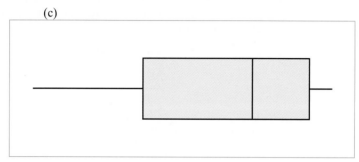

✓ *3.104 Find P_{40}, P_{60}, and P_{80} for the distribution of the world's largest companies. They are ranked by market value at the beginning of the year 2000, in billions of dollars. The largest company in this international group is General Electric (U.S.A.) with a market value of $245,000,000,000.

Billions of dollars	Number of companies
0–39	32
40–79	41
80–119	17
120–159	4
160–199	2
200–239	3
240–279	1
Total	100

*3.105 Find P_{20} and P_{80} for the following distribution of grades for 180 students enrolled in a physics class.

Grades in percent	Number of students
40–49	8
50–59	16
60–69	26
70–79	64
80–89	45
90 and above	21
Total	180

*3.106 **Deciles** are calculated in the same manner as quartiles and percentiles. Nine deciles divide a set of data into 10 parts, and $D_1, D_2, D_3, \ldots,$ and D_9 correspond to $P_{10}, P_{20}, P_{30}, \ldots, P_{90}$. Find D_2 and D_8 for the distribution of grades in Exercise 3.105 in this set of data.

✓ *3.107 In a study of the weekly earnings of a large number of wage earners, we find that the mean, $\bar{x}$, is \$210.54, the median, $\tilde{x}$, is \$204.38, and the standard deviation, s, is \$9.43. Calculate the Pearsonian coefficient of skewness.

*3.108 Find the Pearsonian coefficient of skewness for the distribution of grades in a quiz given to students in a mathematics course. The mean grade, $\bar{x}$, is 79.2, the median grade, $\tilde{x}$, is 80.1, and the standard deviation, s, is 16.5. After determining the value of SK, explain whether the skewness is high, or whether the distribution is fairly symmetrical.

*3.109 In a tree nursery, a nursery attendant measures the heights of all the blue spruce trees to measure their growth in a month and finds that the mean, $\bar{x}$, growth is 2.25 inches, their median, $\tilde{x}$, growth is 1.96 inches, and their variance, s^2, is 0.23 inch. Find the Pearsonian coefficient of skewness.

*3.110 Roll a pair of dice 120 times and construct a distribution showing how many times there were 0 sixes, how many times there was 1 six, and how many times there were 2 sixes. Draw a histogram of this distribution and describe its shape.

*3.111 If a coin is flipped five times, the result may be represented by means of a sequence of H's and T's (for example, HHTTH), where H stands for *heads* and T for *tails*. Having obtained such a sequence of H's and T's, we can then check after each successive flip whether the number of heads exceeds the number of tails. For example, for the sequence HHTTH, heads is ahead after the first flip, after the second flip, after the third flip, not after the fourth flip, but again after the fifth flip; altogether it is ahead four times. Repeat this experiment 50 times, and construct a histogram showing in how many cases heads was ahead altogether 0 times, 1 time, 2 times, $\ldots$, and 5 times. Explain why the resulting distribution should be U shaped.

*3.112 The following are the numbers of minutes that a doctor kept 60 patients waiting beyond their appointment times:

12.1	9.8	10.5	5.6	8.2	0.5	6.8	10.1	17.2	4.2
8.3	1.3	7.9	11.3	6.3	7.2	9.3	9.9	7.2	12.7
1.2	4.6	10.3	8.5	10.0	12.8	9.6	13.5	10.8	5.1
12.7	11.5	3.8	12.9	13.0	3.9	7.5	16.1	11.1	8.3
9.6	6.4	15.7	5.8	9.7	11.9	2.4	5.2	8.4	16.7
2.5	13.0	4.8	10.7	11.4	9.3	4.7	6.0	9.5	14.6

If the data is grouped into a distribution with the classes 0.0–2.9, 3.0–5.9, 6.0–8.9, 9.0–11.9, 12.0–14.9, and 15.0–17.9, and if we solve for the mean, $\bar{x}$, and standard deviation, s, we get 8.90 minutes and 4.04 minutes, respectively. Use a computer package to find the mean and standard deviation of the raw data (ungrouped data). Determine the respective grouping errors by comparing these results with those obtained from the grouped data.

3.10 TECHNICAL NOTE: SUMMATIONS

In the abbreviated notation introduced in Section 3.1, the expression $\sum x$ does not make it clear which, or how many, values of x we have to add. This is taken care of by the more explicit notation

$$\sum_{i=1}^{n} x_i = x_1 + x_2 + \cdots + x_n$$

where it is made clear that we are adding the x's whose subscripts i are 1, 2, ..., and n. Generally, we shall not use the more explicit notation in this text to simplify the overall appearance of the formulas, assuming that it is clear in each case what x's we are referring to and how many there are.

Using the $\sum$ notation, we shall also have occasion to write such expressions as $\sum x^2, \sum xy, \sum x^2 f, \ldots$, which (more explicitly) represent the sums

$$\sum_{i=1}^{n} x_i^2 = x_1^2 + x_2^2 + \cdots + x_n^2$$

$$\sum_{j=1}^{m} x_j y_j = x_1 y_1 + x_2 y_2 + \cdots + x_m y_m$$

$$\sum_{i=1}^{n} x_i^2 f_i = x_1^2 f_1 + x_2^2 f_2 + \cdots + x_n^2 f_n$$

Working with two subscripts, we shall also have occasion to evaluate **double summations** such as

$$\sum_{j=1}^{3} \sum_{i=1}^{4} x_{ij} = \sum_{j=1}^{3} (x_{1j} + x_{2j} + x_{3j} + x_{4j})$$

$$= x_{11} + x_{21} + x_{31} + x_{41}$$

$$+ x_{12} + x_{22} + x_{32} + x_{42}$$

$$+ x_{13} + x_{23} + x_{33} + x_{43}$$

To verify some of the formulas involving summations that are stated but not proved in the text, the reader will find it helpful to use the following rules:

RULES FOR SUMMATIONS

$$Rule\,A: \sum_{i=1}^{n} (x_i \pm y_i) = \sum_{i=1}^{n} x_i \pm \sum_{i=1}^{n} y_i$$

$$Rule\,B: \sum_{i=1}^{n} (k \cdot x_i) = k \cdot \sum_{i=1}^{n} x_i$$

$$Rule\,C: \sum_{i=1}^{n} k = k \cdot n$$

The first of these rules states that the summation of the sum (or difference) of two terms equals the sum (or difference) of the individual summations, and it can be extended to the sum or difference of more than two terms. The second rule states that we can, so to speak, factor a constant out of a summation. The third rule states that the summation of a constant is simply n times that constant. All these rules can be proved by actually writing out in full what each of the summations represents. For instance, for Rule B we can write

$$\sum_{i=1}^{n} kx_i = kx_1 + kx_2 + \cdots + kx_n$$

$$= k(x_1 + x_2 + \cdots + x_n)$$

$$= k \cdot \sum_{i=1}^{n} x_i$$

Exercise 3.115 is a practice exercise; its complete solution is given on page 100.

3.113 Write each of the following in full, that is, without summation signs:

(a) $\sum_{i=1}^{6} x_i^2$;   (c) $\sum_{i=1}^{3} x_i y_i$;   (e) $\sum_{i=3}^{7} x_i^2$;

(b) $\sum_{i=1}^{5} y_i$;   (d) $\sum_{j=1}^{8} x_j f_j$;   (f) $\sum_{j=1}^{4} (x_j + y_j)$.

3.114 Write each of the following as summations:

(a) $z_1 + z_2 + z_3 + z_4 + z_5$;

(b) $x_5 + x_6 + x_7 + x_8 + x_9 + x_{10} + x_{11} + x_{12}$;

(c) $x_1 f_1 + x_2 f_2 + x_3 f_3 + x_4 f_4 + x_5 f_5 + x_6 f_6$;

(d) $y_1^2 + y_2^2 + y_3^2$.

✓ **3.115** Given $x_1 = 1, x_2 = 3, x_3 = 5, x_4 = 7, x_5 = 9, f_1 = 1, f_2 = 5, f_3 = 10, f_4 = 3, f_5 = 2$, find

(a) $\sum_{i=1}^{5} x_i$;   (c) $\sum_{i=1}^{5} x_i \cdot f_i$;

(b) $\sum_{i=1}^{5} f_i$;   (d) $\sum_{i=1}^{5} x_i^2 \cdot f_i$.

3.116 Given $x_1 = -2, x_2 = 3, x_3 = 4$, and $x_4 = 4$, find

(a) $\sum_{i=1}^{4} x_i$;   (b) $\sum_{i=1}^{4} x_i^2$.

3.117 Given $x_1 = 2, x_2 = 3, x_3 = 4, x_4 = 5, x_5 = 6, x_6 = 7, f_1 = 3, f_2 = 12, f_3 = 10, f_4 = 6, f_5 = 3, f_6 = 1$, find

(a) $\sum_{i=1}^{6} x_i$;   (c) $\sum_{i=1}^{6} x_i \cdot f_i$;

(b) $\sum_{i=1}^{6} f_i$;   (d) $\sum_{i=1}^{6} x_i^2 \cdot f_i$.

3.118 Given $x_1 = 2, x_2 = 1, x_3 = 5, x_4 = 3$ $y_1 = 1, y_2 = 3, y_3 = 2, y_4 = 1$, find

(a) $\sum_{i=1}^{4} x_i$;   (d) $\sum_{i=1}^{4} y_i^2$;

(b) $\sum_{i=1}^{4} y_i$;   (e) $\sum_{i=1}^{4} x_i \cdot y_i$.

(c) $\sum_{i=1}^{4} x_i^2$;

EXERCISES

3.119 Given
$$x_{11} = 2, \quad x_{12} = 3, \quad x_{13} = -1,$$
$$x_{21} = 1, \quad x_{22} = 2, \quad x_{23} = 2,$$
$$x_{31} = 2, \quad x_{32} = -2, \quad x_{33} = 2,$$
$$x_{41} = 3, \quad x_{42} = -4, \quad x_{43} = -3,$$
find

(a) $\sum_{i=1}^{4} x_{ij}$ for $j = 1, 2,$ and 3;

(b) $\sum_{j=1}^{3} x_{ij}$ for $i = 1, 2, 3,$ and 4;

(c) $\sum_{i=1}^{4} \sum_{j=1}^{3} x_{ij}$.

3.120 Show that $\sum_{i=1}^{n} (x_i - \bar{x}) = 0$ for any set of x's whose mean is $\bar{x}$.

3.121 Is it true in general that $\left(\sum_{i=1}^{n} x_i \right)^2 = \sum_{i=1}^{n} x_i^2$? (*Hint:* Check whether the equation holds for $n = 2$.)

✓ SOLUTIONS TO PRACTICE EXERCISES

3.1   (a) parameter; (b) statistic; (c) statistic; (d) parameter.

3.4   The total of the 8 values is $82.8 + 53.3 + 72.4 + 78.2 + 90.0 + 92.2 + 80.3 + 50.9 = 600.1$,

$$\text{so that } \bar{x} = \frac{600.1}{8} \approx 75.0 \text{ percent.}$$

3.10 The total weight of the women is $8(124) = 992$, and the total weight of the men is $6(165) = 990$, so that their combined weight is $992 + 990 = 1,982$ pounds. Since this does not exceed 2,000 pounds, the elevator is not overloaded (but the numbers are close enough to make one nervous).

3.17 Substituting $w_1 = 1,000$, $w_2 = 3,000$, $w_3 = 16,000$, $x_1 = 7, x_2 = 7.5$, and $x_3 = 8$ into the formula for the weighted mean, we get

$$\bar{x}_w = \frac{1,000(7) + 3,000(7.5) + 16,000(8)}{1,000 + 3,000 + 16,000}$$

$$= \frac{157,500}{20,000} = 7.875\%$$

3.18 Substituting $x_1 = 40.9$, $x_2 = 41.1$, and $x_3 = 47.3$, $n_1 = 19.7$, $n_2 = 28.1$, and $n_3 = 5.3$ into the formula for the grand mean of the combined data, we get

$$\bar{x} = \frac{(19.7)(40.9) + (28.1)(41.1) + (5.3)(47.3)}{19.7 + 28.1 + 5.3}$$

$$= \frac{2,211.33}{53.1}$$

$$\approx \$41.6 \text{ thousand}$$

3.25 (a) $\dfrac{15+1}{2} = 8$, so that the median is the 8th value.

(b) $\dfrac{40+1}{2} = 20.5$, so that the median is the mean of the 20th and 21st values.

3.28 (a) Arranged according to size, the data is

$$1.51 \quad 1.86 \quad 1.94 \quad 2.02 \quad 3.45 \quad 3.74$$

and it can be seen that the median is $\dfrac{1.94 + 2.02}{2} = \$1.98.$

(b) The mean is $\dfrac{1.94 + 1.86 + 3.45 + 3.74 + 2.02 + 1.51}{6} = \$2.42.$

3.37 (a) The medians are

$$3 \quad 4 \quad 7 \quad 8 \quad 6 \quad 6 \quad 8 \quad 1 \quad 5 \quad 8 \quad 5 \quad 4 \quad 5 \quad 5 \quad 6$$

The means are

$$4\tfrac{1}{3} \quad 6 \quad 7 \quad 7 \quad 5\tfrac{1}{3} \quad 6\tfrac{1}{3} \quad 6\tfrac{1}{3} \quad 1\tfrac{2}{3} \quad 6 \quad 5\tfrac{2}{3} \quad 5\tfrac{2}{3} \quad 5 \quad 5 \quad 5\tfrac{1}{3} \quad 7$$

(b) Grouping the medians and the means into the indicated classes, we get the following:

Medians	Frequency	Means	Frequency
0.5–2.5	1	0.5–2.5	1
2.5–4.5	3	2.5–4.5	1
4.5–6.5	7	4.5–6.5	10
6.5–8.5	4	6.5–8.5	3

(c) The histograms of these distributions are shown in Figure 3.12, and it can be seen that there is less variability among the means than among the medians. In other words, the means are concentrated more closely about $\mu = 5.5$, the actual mean of the numbers on all the cards.

Figure 3.12
Histograms for part (c) of Exercise 3.37.

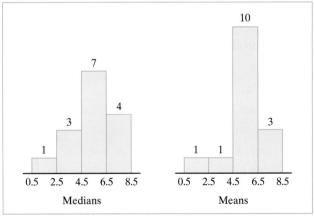

Medians   Means

3.42 The median position is $\dfrac{21+1}{2} = 11$, the Q_1 position is $\dfrac{10+1}{2} = 5.5$, and the Q_3 position is the 5.5th from the other end, or 16.5. Thus, there are five values to the left of the Q_1 position, five values between the Q_1 and median positions, five values between the median and Q_3 positions, and five values to the right of the Q_3 position.

3.45 The median position is $\dfrac{13+1}{2} = 7$, the Q_1 position is $\dfrac{6+1}{2} = 3.5$, and the Q_3 position is 3.5 from the other end, or 10.5. Arranged according to size, the data is

$$0 \quad 1 \quad 1 \quad 3 \quad 5 \quad 5 \quad 5 \quad 6 \quad 6 \quad 8 \quad 10 \quad 12 \quad 15$$

so that

$$Q_1 = \frac{1+3}{2} = 2 \quad \text{and} \quad Q_3 = \frac{8+10}{2} = 9$$

3.48 (a) Midquartile is $\dfrac{2+9}{2} = 5.5$ restaurant meals.

Midrange is $\dfrac{0+15}{2} = 7.5$ restaurant meals.

(b) Midquartile is $\dfrac{245+631}{2} = 438$ sq. miles.

Midrange is $\dfrac{813+212}{2} = 512.5$ sq. miles.

3.50 Since 11 occurs twice, 12 occurs four times. 13 occurs three times, 14 occurs 7 times, 15 occurs once, and 16 occurs once, the mode is 14 live births per thousand population.

3.55 Most of the numbers occur only once, 1 and 8 occur two times, and 9 occurs three times. The mode is 9 minutes.

3.59 The range is $22.0 - 13.3 = 8.7$ pupils per teacher.

3.64 (a) The mean is $\bar{x} = \dfrac{11+15+12+8+14}{5} = \dfrac{60}{5} = 12$, so that

$$s = \sqrt{\frac{(11-12)^2 + (15-12)^2 + (12-12)^2 + (8-12)^2 + (14-12)^2}{5-1}}$$

$$= \sqrt{\frac{1+9+0+16+4}{4}} = \sqrt{7.5} \approx 2.7 \text{ minutes}$$

(b) Substituting $n = 5$, $\sum x = 60$, and $\sum x^2 = 11^2 + 15^2 + 12^2 + 8^2 + 14^2 = 750$ into the computing formula for S_{xx}, we get

$$S_{xx} = 750 - \frac{(60)^2}{5} = 30$$

Then

$$s = \sqrt{\frac{30}{4}} \approx 2.7 \text{ minutes}$$

(c) $2.7 = 2.7$.

3.77 (a) At least $1 - \dfrac{1}{4^2} = \dfrac{15}{16} = 0.9375$, or 93.75% of the data must lie within $k = 4$ standard deviations on either side of the mean.

(b) At least $1 - \dfrac{1}{12^2} = \dfrac{143}{144} \approx 0.9931$, or 99.31%, of the data must lie within $k = 12$ standard deviations on either side of the mean.

3.80 (a) $\dfrac{8.2 - 5.4}{1.4} = 2$; $\dfrac{5.4 - 2.6}{1.4} = 2$; at least $1 - \dfrac{1}{2^2} = 0.75$, or 75% of the planes arrive anywhere between 2.6 minutes late and 8.2 minutes late.

   (b) $\dfrac{12.4 - 5.4}{1.4} = 5$; $\dfrac{5.4 + 1.6}{1.4} = \dfrac{7}{1.4} = 5$; at least $1 - \dfrac{1}{5^2} = 0.96$, or 96% of the planes arrive anywhere between 1.6 minutes early and 12.4 minutes late.

3.84 (a) $\dfrac{6 - 7}{0.5} = -2$ and $\dfrac{8 - 7}{0.5} = 2$, so that $1 - \dfrac{1}{k^2} = 1 - \dfrac{1}{2^2} = 0.75$ or 75%

   (b) $\dfrac{5 - 7}{0.5} = -4$ and $\dfrac{9 - 7}{0.5} = 4$, so that $1 - \dfrac{1}{k^2} = 1 - \dfrac{1}{4^2} = 0.9375$ or 94%

3.88 For the first measuring instrument, the coefficient of variation is

$$\frac{0.018}{4.92} \cdot 100\% \approx 0.37\%$$

and for the second measuring instrument, it is

$$\frac{0.012}{2.54} \cdot 100\% \approx 0.47\%$$

Since the first measuring instrument has the smaller coefficient of variation, it is relatively more precise.

3.91 It helps to elevate the class marks by $0.005. This gives

x	f	$x \cdot f$	$x^2 \cdot f$
125	9	1,125	140,625
135	20	2,700	364,500
145	36	5,220	756,900
155	30	4,650	720,750
165	15	2,475	408,375
175	11	1,925	336,875
185	4	740	136,900
Total	125	18,835	2,864,925

Then

$$\bar{x} = \frac{18,835}{125} = \$150.68$$

Next find

$$S_{xx} = 2,864,925 - \frac{(18,835)^2}{125} = 26,867.20$$

Then

$$s = \sqrt{\frac{26,867.20}{124}} \approx \$14.72$$

3.95 The median is

$$\bar{x} = 139.995 + \frac{33.5}{36} \cdot 10 \approx \$149.30$$

3.99

Months	Number of cycles x	Cumulative number
10–29	2	2
30–49	7	9
50–69	1	10
70–89	1	11
90–109	3	14
110–129	1	15
Total	15	

(a) Location of the median is the $\frac{15}{2} = 7.5$th value which, counting from the lowest value is located in the 30–49 class, and its value is $29.5 + \frac{5.5}{7}(20) \approx$ 45.2 months.

(b) Location of Q_1 is $\frac{15}{4} = 3.75$, which, counting from the lowest value, is located in the 30–49 class, and its value is $29.5 + \frac{1.75}{7}(20) = 34.5$ months.

Location of Q_3 is $\frac{3}{4}(15) = 11.25$, which, counting from the lowest value, is located in the 90–109 class, and it value is $89.5 + \frac{0.25}{3}(20) \approx 91.2$ months.

3.104 Location of P_{40} is the $\frac{40}{100} \cdot 100 = 40$th value, which, counting from the lowest value, is located in 40–79 class, and its value is $39.5 + \frac{8}{41} \cdot 40 \approx 47.3$ billion dollars; location of P_{60} is the $\frac{60}{100} \cdot 100 = 60$th value, which, counting from the lowest value, is located in the 40–79 class, and its value is $39.5 + \frac{28}{41} \cdot 40 \approx$ 66.8 billion dollars; location of P_{80} is the $\frac{80}{100} \cdot 80$th value, which, counting from the lowest value, is located in 80–119 class and its value is $79.5 + \frac{7}{17} \cdot 40 \approx$ 96.0 billion dollars.

3.107 $SK = \frac{3(210.54 - 204.38)}{9.43} \approx 1.96$.

3.115 (a) $1 + 3 + 5 + 7 + 9 = 25$;
(b) $1 + 5 + 10 + 3 + 2 = 21$;
(c) $1 \cdot 1 + 3 \cdot 5 + 5 \cdot 10 + 7 \cdot 3 + 9 \cdot 2 = 105$;
(d) $1^2 \cdot 1 + 3^2 \cdot 5 + 5^2 \cdot 10 + 7^2 \cdot 3 + 9^2 \cdot 2 = 605$.

1, 2, & 3

Achievements

Having read and studied these chapters and worked a good portion of the exercises, you should be able to

1. Explain the difference between descriptive statistics and statistical inference.
2. Construct frequency distributions of both numerical and categorical data.
3. Determine the class limits, class boundaries, class marks, and class interval of a distribution of numerical data.
4. Convert frequency distributions into percentage distributions.
5. Present ungrouped data in the form of dot diagrams and in other types of graphs.
6. Convert frequency distributions into cumulative distributions and percentage distributions into cumulative percentage distributions.
7. Present frequency distributions in the form of histograms, bar charts, frequency polygons, ogives, or pie charts.
8. Construct stem-and-leaf displays.
9. Explain the difference between samples and populations and the difference between statistics and parameters.
10. Determine the mean of a set of data.
11. Determine the median of a set of data.
12. Determine the mode of a set of data.
13. List some of the desirable and undesirable features of the various measures of location.
14. Calculate weighted means.
15. Determine the grand mean of combined data.
16. Determine the quartiles of a set of data.
17. Draw a box-and-whisker plot.
18. Determine the range of a set of data.
19. Determine the standard deviation (or variance) of a set of data.
20. List some of the desirable and undesirable features of the various measures of variation.
21. Explain Chebyshev's theorem.
22. Convert measurements into standard units and explain the advantage of using such units.
23. Calculate the coefficient of variation and explain its significance.
24. Determine the mean and the standard deviation of a distribution.
25. Determine the median of a distribution.
26. Determine the quartiles of a distribution.

Review

27. Determine the percentiles of a distribution.
28. Describe the shape of a distribution as symmetrical or skewed, as bell shaped, as U shaped, and so forth.
29. Determine the Pearsonian coefficient of skewness.
30. Work with the $\sum$ (summation) notation.

CHECKLIST OF KEY TERMS (with page references to their definitions)

Bar chart, 35
Bell-shaped distribution, 86
Bias, 4
Biased data, 6
Box-and-whisker plot, 59
Box plot, 59
Categorical data, 3
Categorical distribution, 22
Chebyshev's theorem, 74
Class boundary, 27
Class frequency, 26
Class interval, 27
Class interval of the distribution, 25
Class limit, 26
Class mark, 27
Coefficient of variation, 78
Cumulative distribution, 28
Cumulative percentage distribution, 29
Data, 1
Decile, 93
Decision theory, 11
Descriptive statistics, 2, 8
Deviation from mean, 7
Dot diagram, 16
Double-stem display, 22
Double summation, 94
Empirical rule, 77
Exploratory data analysis, 11
Fractile, 58
Frequency distribution, 23
Frequency polygon, 35
Grand mean of k sets of data, 51
Grouping, 16
Histogram, 33

Inequality signs, 4
Interval data, 4
J-shaped distribution, 87
Leaf, 18
Listing, 18
Lower class limit, 26
Mean, 46
Mean deviation, 70
Measures of central location, 46
Measures of location, 46
Measures of relative variation, 78
Measures of variation, 46
Median of n values, 55
Median position, 56
Midquartile, 64
Midrange, 64
Mode, 65
Negatively skewed distribution, 87
Nominal data, 3
Numerical data, 3
Numerical distribution, 23
Ogive, 35
Open class, 25
Ordinal data, 2
Parameter, 47
Pareto diagram, 17
Pearsonian coefficient of skewness, 87
Percentage distribution, 27
Percentile, 58, 86
Pictogram, 36
Pie chart, 36
Population, 5
Population mean, 48
Population size, 48

Population standard deviation, 70
Population variance, 70
Positively skewed distribution, 87
Probability theory, 3, 9
Quartile, 58
Range, 69
Ratio data, 3
"Real" class limits, 29
Reverse J-shaped distribution, 87
Root-mean-square deviation, 70
Sample, 5
Sample mean, 48
Sample size, 48
Sample standard deviation, 70
Sample variance, 71
Sigma notation ($\sum$ notation), 48
Skewed distribution, 86
Standard deviation, 69
Standard unit, 78
Statistic, 46
Statistical description, 46
Statistical inference, 9
Statistical measure, 44
Statistics, 1
Stem, 18
Stem-and-leaf display, 18
Symmetrical bell-shaped distribution, 86
Trimmed mean, 49
Upper class limit, 25
U-shaped distribution, 87
Vital statistics, 46
Weighted mean, 50
Weights, 50
z-score, 78

REVIEW EXERCISES FOR CHAPTERS 1, 2, 3

R.1 Following are the numbers of home runs hit by the baseball teams of the National League during the year 2001 season.

Team	Home runs
Colorado	213
Houston	208
St. Louis	199
Arizona	208
San Francisco	235
Florida	166
Cincinnati	176
Chicago	194
Atlanta	174
Philadelphia	164
Los Angeles	206
Montreal	131
San Diego	161
Milwaukee	209
New York	147
Pittsburgh	161
Total	2,952

Find

(a) the mean;

(b) the median.

R.2 According to Chebyshev's theorem, what can we assert about the percentage of any set of data that must lie within k standard deviations on either side of the mean when (a) $k = 1.2$; (b) $k = 6.25$; (c) $k = 11$?

R.3 On three consecutive days, a meter maid issued 28, 20, and 32 parking tickets. On the same three days, she issued 7, 4, and 5 tickets for expired inspection stickers. Which of the following conclusions can be obtained from these data by purely descriptive methods and which require generalizations? Explain your answers.

(a) The meter maid issued 80 parking tickets on the three days.

(b) On any given day, this meter maid will seldom issue more than 10 tickets for expired inspection stickers.

(c) The meter maid issued the fewest number of tickets on the second day because she took an extended lunch hour.

(d) On these three days, the meter maid issued five times as many parking tickets as tickets for expired inspection stickers.

R.4 An analysis of the outgoing telephone calls of an office reveals the lengths of time required to make 150 telephone calls:

Minutes	Frequency
1.0–2.9	9
3.0–4.9	39
5.0–6.9	45
7.0–8.9	42
9.0–10.9	11
11.0–12.9	4
Total	150

(a) Convert the data into a cumulative "less than" distribution; and

(b) a cumulative "or more" distribution.

R.5 Suppose that a Certified Public Accountant (CPA) is given complete information about the refunds the residents of New York City obtained from their federal income tax which was withheld from their paychecks in the year 2003. Give one example each of a situation in which the CPA would look upon the data as

(a) a population;

(b) a sample.

R.6 A customer in a supermarket selected 6 cartons of eggs (each containing a dozen) from a large display. The egg-filled cartons weighed 25.9, 27.8, 25.8, 26.1, 23.5, and 45.4 ounces.

(a) Find the mean weight of these cartons.

(b) Find the median weight of these cartons.

(c) Is the mean a good average in this exercise?

R.7 In the year 2000 the number of representatives to the House of Representatives of the Congress from the Pacific States of Washington, Oregon, California, Alaska, and Hawaii were 9, 5, 53, 1, and 3, respectively. Comment on the (misleading?) argument that the "average" Pacific state has 14.2 representatives.

R.8 The number of representatives that each state elects to the Congressional House of Representatives is determined by the size of the state's population, as officially determined by the Decennial Census. If the mean number of representatives in the 6 New England states is $\bar{x} = 4$, and if the number of representatives in 5 of these states is actually 2, 5, 2, 2, and 3, find the number of representatives apportioned to the sixth state (Massachusetts).

R.9 Following are the incomes of 10 families residing in a certain neighborhood: $44,000, $49,000, $55,000, $39,000, $28,000, $39,000, $180,000, $175,000, $56,000, and $64,000. Comment on the argument that the average family income in this neighborhood is $72,900.

R.10 Use the data of Exercise R.9 to perform the following calculations:

(a) Find the mean, $\bar{x}$;

(b) find the standard deviation, s.

To simplify your calculations for parts (a) and (b), round all the data to thousands so that 44,000 becomes 44, 49,000 becomes 49, ... becomes 64.

R.11 Draw a histogram and also draw a bar chart of the distribution of Exercise R.4.

R.12 The costs for sending parcels overnight by a certain firm are to be placed in a frequency distribution having the following categories:

Up to $2.00

$2.01 to $5.00

$5.00 to $10.00

$10.01 to $20.00

Over $25

Explain where difficulties might arise.

R.13 A fishery expert found the following concentrations of mercury, in parts per million, in eight fish caught in a certain stream:

$$0.058 \quad 0.062 \quad 0.070 \quad 0.063$$
$$0.068 \quad 0.061 \quad 0.068 \quad 0.060$$

(a) Find $\bar{x}$ and s.

(b) Find the median and the range.

R.14 According to the recent Martindale-Hubbell Law Directory, 192,353 of all lawyers were engaged in private practice. Of these lawyers, 116,911 practiced as individuals; 60,709 practiced as partners; and 14,733 were designated as employees by their employers. Construct a pie chart of this categorical distribution.

R.15 The following data represents the number of patients who received treatment at the emergency room of a hospital on thirty consecutive days.

$$35, \quad 32, \quad 48, \quad 54, \quad 68, \quad 35, \quad 73, \quad 35, \quad 68, \quad 37,$$
$$41, \quad 52, \quad 64, \quad 57, \quad 49, \quad 41, \quad 59, \quad 44, \quad 47, \quad 41,$$
$$49, \quad 57, \quad 44, \quad 52, \quad 41, \quad 62, \quad 43, \quad 54, \quad 59, \quad 41$$

Construct a stem-and-leaf display with the stem labels 3, 4, 5, 6, and 7.

R.16 Given $x_1 = 6$, $x_2 = 8$, $x_3 = 5$, $x_4 = -1$, $x_5 = 3$, and $x_6 = 4$, find

(a) $\sum_{i=1}^{6} x_i$;   (b) $\sum_{i=1}^{6} x_i^2$.

R.17 Explain why each of the following samples may well yield misleading information:

(a) To ascertain facts about tooth-brushing habits, a sample of the residents of a community are asked how many times they brush their teeth each day.

(b) To study the spending patterns of families in a certain income group, a sample survey is conducted during the first three weeks of December.

R.18 If a distribution has the mean 112.8, the median 96.5, and the standard deviation 24.7, find the Pearsonian coefficient of skewness.

R.19 Based on many years' information, it is known that the bus that leaves Valley Center at 8:02 A.M. takes 32 minutes, with a standard deviation of 2.25 minutes, to arrive at the downtown bus terminal. At least what percentage of the time will this bus arrive at the downtown bus terminal between 8:25 A.M. and 8:43 A.M.?

R.20 On five runs, bus A carried 15, 24, 19, 12, and 20 passengers, and bus B carried 18, 21, 16, 14, and 16 passengers.

(a) Calculate the respective means to decide which bus averaged more passengers on the five runs.

(b) Use the formula that defines the standard deviation to calculate s for each of the two sets of data.

(c) Calculate the two coefficients of variation to determine whether the number of passengers is relatively more variable for bus A or bus B.

R.21 The Federal Highway Administration and the *World Almanac* list 38 notable suspension bridges in North America (United States, Canada, and Mexico), and the length of each main span, in feet, is

4,260	2,000	1,500	1,105
4,200	1,850	1,500	1,080
3,800	1,800	1,470	1,060
3,500	1,750	1,447	1,059
2,800	1,632	1,400	1,057
2,800	1,600	1,380	1,050
2,310	1,600	1,207	1,030
2,300	1,600	1,200	1,010
2,190	1,595	1,150	
2,150	1,550	1,108	

Find (a) the median, (b) Q_1 and Q_3.

R.22 For fresh salmon sold at four stores, the prices per pound and numbers of pounds sold on a particular date were as follows:

Store	Price	Quantity
Eastside Fish Mart	$5.95	80
Murray's	$5.25	120
Thompson Specialty	$7.00	40
Bayside Fish	$6.50	60

Find the mean price of the salmon sold in the four stores.

R.23 The daily ticket sales of a movie theater are grouped into a distribution having the classes 50–99, 100–149, 150–199, 200–249, 250–299, 300–349, and 350–399. Determine

(a) the lower class limits;

(b) the upper class limits;

(c) the class boundaries;

(d) the class marks;

(e) the class interval of the distribution.

R.24 Specify the values that occur in the following stem-and-leaf display:

0.00	2
0.01	35
0.02	668
0.03	7

R.25 In the ten-year period from the year 1991 to 2000, the number of work stoppages involving 1,000 or more workers in the United States were 40, 35, 35, 45, 31, 37, 29, 34, 17, and 39, respectively. (In the year 2000, 394,000 workers were involved.) Calculate the standard deviation.

R.26 On four Saturdays, a person jogged for 46, 50, 52, and 60 minutes.

(a) Find the mean, the range, and the standard deviation of these four sample values.

(b) Subtract 50 minutes from each of the times, recalculate the mean, the range, and the standard deviation, and compare the results with those obtained in part (a).

(c)   Divide each of the original values by 2, recalculate the mean, the range, and the standard deviation, and compare the results with those obtained in part (a).

(d)   Based on these results, what effect does (1) adding or subtracting a constant and (2) multiplying or dividing by a constant have on the mean, the range, and the standard deviation of a set of data?

R.27 Is it possible to find the mean and the median of each of the following distributions? Explain your answers.

(a)

Grade	Frequency
40–49	5
50–59	18
60–69	27
70–79	15
80–89	6
Total	71

(b)

Percent of sales quota achieved	Frequency
Less than 90	3
90–99	14
100–109	22
110–119	19
More than 119	7
Total	65

(c)

Weight	Frequency
100 or less	41
101–110	13
111–120	8
121–130	3
131–140	1
Total	66

R.28 Find the median position for

(a) $n = 99$;

(b) $n = 100$.

R.29 The following is the distribution of the number of vehicles that received emission inspections during a recent month at 80 service stations in a certain county:

Number of vehicles	Frequency
0–19	4
20–39	11
40–59	28
60–79	24
80–99	13
Total	80

(a) Draw a histogram of this distribution.

(b) Convert this distribution into a cumulative "less than" distribution and draw an ogive.

R.30 With reference to Exercise R.29, find

(a) the median;

(b) the quartiles Q_1 and Q_3.

R.31 Regarding the distribution of Exercise R.29, suppose that the smallest value is 5 and the largest value is 95. Use the results of Exercise R.30 to draw a box-and-whisker plot.

R.32 Regarding the distribution of Exercise R.29, find the percentiles $P_{0.20}$ and $P_{0.80}$.

R.33 According to the American Medical Association, the mean net income of hospital based physicians in 1998 was, in $1,000s:

General/Family practice	$142.5
Internal Medicine	$182.1
Surgery	$268.2
Pediatrics	$139.6
Obstetrics/Gynecology	$214.4

(a) Calculate the mean and median of the hospital based physicians.

(b) Assuming that the 268.2 value was incorrectly printed as $862.2, recalculate the mean and median.

(c) Compare the effects of this printing error on the mean and on the median.

R.34 Using the data of Exercise R.33, find the range of the mean net income of the hospital-based physicians.

R.35 Asked whether they ever accept social invitations from their students, 40 tennis pros replied as follows: occasionally, rarely, rarely, never, rarely, frequently, occasionally, occasionally, never, rarely, rarely, never, occasionally, occasionally, frequently, occasionally, rarely, rarely, occasionally, never, occasionally, occasionally, never, never, rarely, rarely, occasionally, occasionally, frequently, occasionally, never, occasionally, rarely, rarely, never, frequently, occasionally, rarely, rarely, occasionally.

(a) Construct a categorical distribution and display it in the form of a bar chart.

(b) What is the tennis pros' modal reply?

R.36 The following are the numbers of deer observed in 60 sections of land in a wildlife count:

11	14	21	15	16	4	18	11	17	11	7	12
21	0	16	12	20	17	13	10	16	5	10	14
5	19	10	6	15	1	26	8	18	19	2	14
17	6	15	14	22	7	7	13	19	0	15	17
2	16	11	18	10	28	15	4	32	6	20	7

(a) Group these figures into a distribution having the classes 0–4, 5–9, 10–14, 15–19, 20–24, 25–29, and 30–34, and draw a bar chart.

(b) Convert the distribution obtained in part (a) into a cumulative "less than" distribution and draw an ogive.

R.37 The total number of shipments made from a warehouse for each of 25 consecutive weeks was 240, 244, 249, 224, 235, 235, 269, 216, 228, 230, 236, 229, 221, 237, 253, 233, 228, 256, 223, 222, 240, 231, 230, 245, and 230. Construct a stem-and-leaf plot for the data with one-digit leaves and with stem labels 21, 22, 23, 24, 25, and 26.

R.38 Construct a box-and-whisker plot for the data given in Exercise R.37. The largest value is 269, the smallest value is 216, the median is 233, $Q_1 = 228$, and Q_3 is 242.

R.39 Following are revenues received by each of the 50 states in the United States, stated in millions of dollars (for example, 45 is 45 million dollars) for hunting and fishing licenses: 0, 1, 2, 4, 5, 6, 7, 7, 11, 11, 11, 12, 12, 13, 13, 13, 14, 14, 14, 14, 15, 16, 16, 17, 17, 18, 18, 18, 19, 19, 20, 20, 20, 22, 22, 24, 24, 26, 26, 27, 28, 29, 29, 42, 44, 45, 60, 62, 73, and 75. For your convenience, the revenues are arranged according to value. Find

(a) the median;

(b) Quartile 1 and Quartile 3.

R.40 In Exercise R.39, the largest value is 75 and the smallest value is zero. Use these values, the values of the quartiles, and the median to construct a box-and-whisker plot.

4

POSSIBILITIES
AND
PROBABILITIES

n everyday language we use words such as *probable*, *likely*, and other terms expressing uncertainties as a matter of course. "The flight from Denver to Chicago will probably arrive on time." "She is an odds-on favorite to be elected to the school board." "The chances are that a person with malaria will fully recover." "More likely than not, the tennis player we favor will win this year's French Open Tournament."

All these statements express degrees of uncertainty, and the one thing they have in common is that they are all very vague. To make them more precise we need a yardstick for measuring uncertainties, and this leads us to the study of probability. Indeed, the subject of **probability** has been referred to as the art or science of "how to live with uncertainties." The concept of probability was first introduced in connection with games of chance.

Here we shall see how probabilities can, or may, be interpreted and how they can be used to make choices (among different courses of action) that promise to be the most profitable, or otherwise most desirable. Sections 4.1, 4.2, and 4.3 are devoted to the problem of determining *what is possible* in given situations, and in Section 4.4 we learn how to judge also *what is probable*. In Sections 4.5 and 4.6 we present the concept of a mathematical expectation and its application to problems of decision making.

4.1 COUNTING

There are essentially two ways to answer a "what is possible" question. One way consists of listing everything that can happen in a given situation. The second way determines how many different things can happen, without actually constructing a complete list. In connection with our work in this book, the second kind of answer is especially important; in most cases we will not need a complete list and hence can save ourselves a great deal of work. Although the first kind of answer, that of listing everything that can happen in a given situation, may seem easy, this is not always the case.

EXAMPLE 4.1   A woman is considering the replacement of flooring in two rooms of her apartment but has not yet decided whether to replace the flooring in two rooms, one room, or neither room. Her choices of flooring materials are carpeting, ceramic tile, or wood. The entire floor of each room must be done using one material. List the different ways in which she could replace (or not replace) the flooring.

Solution   There are clearly many possibilities. Both rooms could be carpeted; one could be carpeted and the other one tiled; one could be tiled and the other one wooden; one could be carpeted and the other not replaced; and so forth. Continuing in this manner, we may be able to complete the list, but it is likely that we will omit at least one of the possibilities.

This problem can be handled systematically by drawing a **tree diagram** such as that of Figure 4.1(a). This figure shows that at first there are 3 possibilities (3 branches) corresponding to 0, 1, or 2 of the rooms being carpeted. Then for ceramic tile flooring, there are three branches emanating from the top branch, two from the middle branch, and one from the bottom branch. Observe that there are 3 possibilities (0, 1, or 2) for the number of tiled floors when none of the floors are to be carpeted; there are 2 possibilities (0 or 1) for the number of tiled floors when one room will be carpeted; and there is only 1 possibility (0) for the number of tiled floors when two rooms will be carpeted. For the number of wooden floors, the reasoning is similar. In going from left to right in Figure 4.1(a), we find that there are altogether 10 different paths along the branches of the tree. In other words, there are 10 possibilities. From top to bottom, on the right-hand side of the diagram, we can see that the choices are the following: Do not replace either floor; replace only one floor using wood; replace two floors with wood; replace only one floor with tile; replace one floor with tile and the other with wood; replace both floors with tile; replace only one floor with carpeting; replace one floor with carpeting and the other with wood; replace one floor with carpeting and the other with tile; and, finally, replace both floors with carpeting.   ■

EXAMPLE 4.2   After completing the previous example, suppose that we learn that the woman considering changes in the flooring of two rooms in an apartment decides also to consider using other types of flooring, such as concrete, stone, terrazzo, cork, linoleum, or plastic tile. Let us combine these into a catchall category called *other flooring*. Suppose that we also learn that she has decided to go ahead with the job and will definitely replace the flooring in both rooms. Change Figure 4.1(a) to reflect this additional information.

Solution   In Figure 4.1(b), we extend each branch of Figure 4.1(a) to include the new category, other flooring. Since the flooring in both rooms will be replaced, the totals of the values on each of the 10 branches must now equal 2. On the uppermost branch of Figure 4.1(a), for instance, the sum of the choices is $0 + 0 + 0 = 0$. We must therefore insert $2 - 0 = 2$ for the new category of other flooring. The completed changes are shown in Figure 4.1(b). As in Figure 4.1(a), where we had 10 branches, Figure 4.1(b) also has 10 branches.   ■

EXAMPLE 4.3   In a medical study, patients are classified according to whether they have blood type A, B, AB, or O and also according to whether their blood pressure is low, normal, or high. In how many different ways can a patient thus be classified according to blood type and blood pressure?

Solution   As is apparent from the tree diagram of Figure 4.2, the answer is 12. Starting at the top, the first path along the branches corresponds to a patient having blood type A and low blood pressure, the second path corresponds to a patient having blood type A and normal blood pressure, ..., and the twelfth path corresponds to a patient having blood type O and high blood pressure.   ■

The answer we obtained in the preceding example is $4 \cdot 3 = 12$, that is, the product of the number of blood types and the number of blood pressure levels. Generalizing from this example, let us state the following rule:

MULTIPLICATION OF CHOICES

> *If a choice consists of two steps, of which the first can be made in m ways and for each of these the second can be made in n ways, then the whole choice can be made in m · n ways.*

To prove this rule, we have only to draw a tree diagram similar to that of Figure 4.2. First, there are m branches corresponding to the possibilities in the first step, and then there are n branches emanating from each of these branches to represent the possibilities in the second step. This leads to $m \cdot n$ paths along the branches of the tree diagram, and hence to $m \cdot n$ possibilities.

EXAMPLE 4.4   A restaurant offers nine different desserts, which it serves with coffee, decaffeinated coffee, tea, milk, or hot chocolate. In how many different ways can one order a dessert and a drink?

Solution   Since $m = 9$ and $n = 5$, the answer is $9 \cdot 5 = 45$.   ■

EXAMPLE 4.5   If an English department schedules four lecture sections and 12 discussion groups for a course in modern literature, in how many different ways can a student choose a lecture section and a discussion group?

Solution   Since $m = 4$ and $n = 12$, the answer is $4 \cdot 12 = 48$.   ■

Figure 4.1(a)
Tree diagram showing
in each case how the
flooring may be replaced.

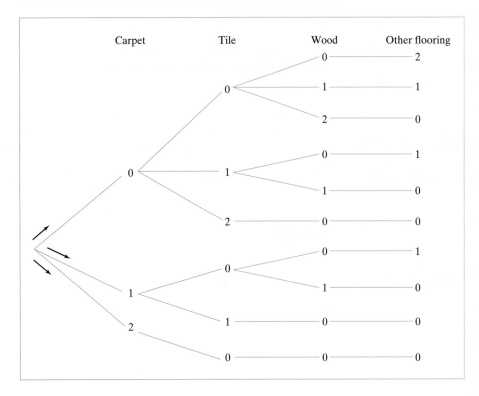

Figure 4.1(b)

Figure 4.2
Tree diagram showing combinations of blood types and blood pressures.

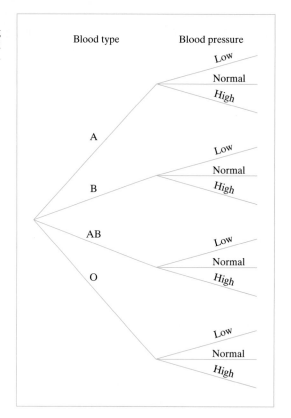

By using appropriate tree diagrams, we can readily generalize the foregoing rule for the **multiplication of choices** so that it will apply to choices involving more than two steps. For k steps, where k is a positive integer, we get the following rule:

MULTIPLICATION OF CHOICES (GENERALIZED)

> *If a choice consists of k steps, of which the first step can be made in n_1 ways, for each of these the second step can be made in n_2 ways, for each combination of choices made in the first two steps the third can be made in n_3 ways, ..., and for each combination of choices made in all the preceding steps the kth can be made in n_k ways, then the whole choice can be made in $n_1 \cdot n_2 \cdot n_3 \cdot \ldots n_k$ ways.*

We simply keep multiplying the numbers of ways in which the different steps can be made.

EXAMPLE 4.6   A new-car buyer has the choice of 4 body styles, 3 different engines, and 10 colors.

(a) In how many different ways can a person order one of these cars?
(b) If a person also has the options of ordering the car with or without air conditioning, with or without an automatic transmission, and with or without bucket seats, in how many different ways can he or she order one of these cars?

Solution   (a) Since $n_1 = 4$, $n_2 = 3$, and $n_3 = 10$, there are $4 \cdot 3 \cdot 10 = 120$ different ways in which a person can order one of the cars.

(b) Since $n_1 = 4$, $n_2 = 3$, $n_3 = 10$, $n_4 = 2$, $n_5 = 2$, and $n_6 = 2$, there are

$$4 \cdot 3 \cdot 10 \cdot 2 \cdot 2 \cdot 2 = 960$$

different ways in which a person can order one of the cars. ◼

EXAMPLE 4.7   A test consists of 12 multiple-choice questions, with each question having four possible answers. In how many different ways can a student check off one answer to each question?

Solution   Since $n_1 = n_2 = \cdots = n_{12} = 4$, there are

$$4 \cdot 4 \cdot 4 \cdot 4 \cdot 4 \cdot 4 \cdot 4 \cdot 4 \cdot 4 \cdot 4 \cdot 4 \cdot 4 = 16{,}777{,}216$$

ways in which a student can check off one answer to each question. Only in one of these cases will all the answers be correct, and in

$$3 \cdot 3 \cdot 3 \cdot 3 \cdot 3 \cdot 3 \cdot 3 \cdot 3 \cdot 3 \cdot 3 \cdot 3 \cdot 3 = 531{,}441$$

cases all the answers will be wrong. ◼

EXAMPLE 4.8   A bettor at a race track wants to place a single \$2 win bet on each of the nine races. (A win bet pays only if the selected horse finishes first.) If there are eight horses in each of the first six races and seven horses in each of the final three races, in how many ways can this bettor place his bets?

Solution   Since $n_1 = n_2 = \cdots = n_6 = 8$ and $n_7 = n_8 = n_9 = 7$, there are

$$8 \cdot 8 \cdot 8 \cdot 8 \cdot 8 \cdot 8 \cdot 7 \cdot 7 \cdot 7 = 89{,}915{,}392$$

ways in which he can place his bets. ◼

EXERCISES

Exercises 4.1, 4.7, and 4.8 are practice exercises; their complete solutions are given on page 143.

✓ **4.1**   In a baseball World Series (in which the winner is the first team to win four games), suppose that the National League champion leads the American League champion by three games to one. Construct a tree diagram to show the different ways in which these teams can win or lose the remaining game or games.

4.2   Equal numbers of vehicles turn left and turn right when entering an intersection at a certain main highway. Draw a tree diagram which shows, for the next three vehicles entering this intersection, how many could turn left and how many could turn right. In how many cases could there be

(a) exactly two vehicles that turn left;

(b) exactly one vehicle that turns left;

(c) exactly three vehicles that turn right;

(d) exactly three vehicles that turn left?

4.3   A person with \$1 in his pocket bets \$1, even money, on the flip of a coin, and he continues to bet \$1 as long as he has any money. Draw a tree diagram to show the various things that can happen during the first three flips of the coin. In how many of the cases will he be exactly $1 ahead?

4.4   (a)   Suppose that a student is allowed to take zero, one, or two elective courses per semester. Construct a tree diagram that shows that there are six ways in which a student can take exactly two electives in three semesters.

(b)   Using the tree diagram of part (a), count the number of ways a student can take exactly five electives in three semesters.

(c)   Using three diagram of part (a), count the number of ways a student can take exactly four electives in three semesters.

4.5   A quiz consists of 6 multiple choice questions, each with five choices.

(a)   In how many ways can a student mark the answers to the questions if one choice is made for each of the questions?

(b)   In how many ways can a student get a perfect score on the quiz?

(c)   In how many ways can a student mark the answers to all the questions, marking all of them incorrectly?

4.6   At the annual meeting of a social club, elections will be held for the positions of president, vice-president, treasurer, and secretary. There are 5 nominees for the position of president, 3 for the position of vice-president, 2 for the position of treasurer, and 2 for the position of secretary. In how many ways can these 4 club officers be elected from among the 12 nominees?

✓ **4.7**   There are three trails to the top of a mountain. In how many different ways can a person hike up and down the mountain if

(a)   she wants to take the same trail both ways;

(b)   she can, but need not, take the same trail both ways;

(c)   she does not want to take the same trail both ways?

✓ **4.8**   A basketball team consists of a center, two forwards, and two guards. If a basketball team has five centers, nine forwards, and eight guards, in how many ways can a most valuable player award be given to one of the centers, one of the forwards, and one of the guards?

4.9   A shopper wants to visit three of five department stores and wants to decide which stores she will visit first, second, and third. She will not visit the same store twice. In how many ways can she select the three stores?

4.10   A furniture store sells living room chairs which are available in 5 styles, 10 types of fabrics, and 8 colors. In how many ways can a customer buy a living room chair?

4.11   Twenty players of a football team are assembled in a locker room, and a competing team has 25 players meeting in another nearby locker room. A newspaper reporter wants to interview one player from each team. In how many ways can the reporter interview one player from each team?

4.12   In how many ways can three automobiles be aligned if they are to be parked in a row in a parking lot?

4.13   A producer of plastic kitchen utensils can buy the raw materials from any one of five manufacturers, process them on any one of four machines, package them in any one of three different ways, and advertise them in one of two media. In how many ways can the whole operation be performed?

4.14   Given that there are three routes by which a truck can travel from uptown to downtown, in how many different ways can a driver plan a trip from uptown to downtown and back if

(a) the driver must travel both ways by the same route;

(b) the driver can, but need not, travel both ways by the same route;

(c) the driver cannot travel both ways by the same route?

4.15 A bank identifies its charge cards by serial numbers consisting of two letters of the alphabet followed by five digits. How many identifiable charge cards can the bank issue under this plan?

4.16 A businessman's wardrobe consists of 4 suits, 6 pairs of shoes, 5 shirts, and 12 neckties. How many ways can he select a suit, a pair of shoes, a shirt, and a necktie to dress for a meeting at the office?

4.17 If a firm has 3 warehouses and supplies 20 retail stores, in how many different ways can an item be shipped from a warehouse to a retail store?

4.18 A manager wants to evaluate four trainees as outstanding, excellent, average, below average, and poor. In how many ways can the manager evaluate the four trainees using the five categories?

4.19 A housepainter has taken a job that requires the spackling and painting of a kitchen and dining room. This involves four tasks: spackle the kitchen; paint the kitchen; spackle the dining room; and paint the dining room. Spackling a room must precede painting. Construct a tree diagram to show the number of ways in which the painter can schedule the four tasks.

4.20 With reference to Exercise 4.19, suppose that the painter must also install fixtures in the kitchen after it is painted. Construct a tree diagram to show the number of ways in which the painter can schedule the five tasks.

4.2 PERMUTATIONS

In general, if r objects are selected from a set of n distinct objects, any particular arrangement (order) of these objects is called a **permutation**. For instance, the letters a, b, and c, in that order, constitute a permutation of three of the twenty-six letters of the alphabet; Montana, Nevada, and Colorado, in that order, constitute a permutation of the eight Mountain states; 6, 7, and 8, in that order, constitute a permutation of the six-digit license plate 892 637. The rule for the multiplication of choices and its generalization are often used when several selections are made from one set of objects, items, or persons, and the order in which they are selected is important.

EXAMPLE 4.9 If 16 entries are submitted to an essay contest, in how many different ways can the judges award a first prize and a second prize?

Solution Since the first prize can be awarded in 16 ways and the second prize must go to one of the other 15 entries, there are altogether $16 \cdot 15 = 240$ possibilities. ■

EXAMPLE 4.10 In how many different ways can the 42 members of a union elect a president, a vice-president, a secretary, and a treasurer?

Solution Assuming that the officers are selected in the order treasurer, secretary, vice-president, and president, we find that there are $n_1 = 42$ ways in which they can elect the treasurer, $n_2 = 41$ ways in which they can elect the secretary, $n_3 = 40$

ways in which they can elect the vice-president, and $n_4 = 39$ ways in which they can elect the president. Thus, there are altogether

$$42 \cdot 41 \cdot 40 \cdot 39 = 2,686,320$$

different possibilities. The solution would be the same for any other selection order.   ■

EXAMPLE 4.11   Determine the number of different permutations of two of the five vowels a, e, i, o, and u, and list them all.

Solution   Since $n_1 = 5$ and $n_2 = 4$, there are $5 \cdot 4 = 20$ different permutations, and they are

ae	ai	ao	au	ei	eo	eu	io	iu	ou
ea	ia	oa	ua	ie	oe	ue	oi	ui	uo

                                                                      ■

To find a formula for the total number of permutations of r objects selected from a set of n distinct objects, observe that the first selection is made from the whole set of n objects, the second selection is made from the $n - 1$ objects that remain after the first selection has been made, the third selection is made from the $n - 2$ objects that remain after the first two selections have been made, ..., and the rth and final selection is made from the

$$n - (r - 1) = n - r + 1$$

objects that remain after the first $r - 1$ selections have been made. Therefore, direct application of the generalized rule for the multiplication of choices yields the result that the total number of permutations of r objects selected from a set of n distinct objects, which we shall denote by $_nP_r$, is

$$n(n - 1)(n - 2) \cdot \ldots \cdot (n - r + 1)$$

Since products of consecutive integers arise in many problems relating to permutations and other kinds of special arrangements or selections, it is convenient to introduce here what is called the **factorial notation**. In this notation, the product of all positive integers less than or equal to the positive integer n is called n *factorial* and denoted by $n!$. Thus,

$$1! = 1$$
$$2! = 2 \cdot 1 = 2$$
$$3! = 3 \cdot 2 \cdot 1 = 6$$
$$4! = 4 \cdot 3 \cdot 2 \cdot 1 = 24$$
$$5! = 5 \cdot 4 \cdot 3 \cdot 2 \cdot 1 = 120$$
$$6! = 6 \cdot 5 \cdot 4 \cdot 3 \cdot 2 \cdot 1 = 720$$
$$\ldots\ldots\ldots\ldots\ldots\ldots\ldots\ldots\ldots$$

and, in general,

$$n! = n(n - 1)(n - 2) \cdot \ldots \cdot 3 \cdot 2 \cdot 1$$

The factorials grow rapidly; for instance, 20! exceeds $2.0 \cdot 10^{18}$ (2 quintillion). Also, to make various formulas more generally applicable, we let $0! = 1$ by definition.

To express the formula for $_nP_r$ in terms of factorials, observe that

$$12 \cdot 11 \cdot 10! = 12!$$

$$7 \cdot 6 \cdot 5 \cdot 4 \cdot 3! = 7!$$

$$35 \cdot 34 \cdot 33 \cdot 32! = 35!$$

and, similarly,

$$_nP_r \cdot (n-r)! = n(n-1)(n-2) \cdot \ldots \cdot (n-r+1) \cdot (n-r) = n!$$

so that

$$_nP_r = \frac{n!}{(n-r)!}$$

To summarize,

NUMBER OF PERMUTATIONS OF n DISTINCT OBJECTS TAKEN r AT A TIME

> *The number of permutations of r objects selected from a set of n distinct objects is*
>
> $$_nP_r = n(n-1)(n-2) \cdot \ldots \cdot (n-r+1)$$
>
> *or, in factorial notation,*
>
> $$_nP_r = \frac{n!}{(n-r)!}$$

In the first formula there are r factors; the first factor is n, and each successive factor is 1 less than the preceding factor.

EXAMPLE 4.12   Find the number of permutations of 4 objects selected from a set of 12 distinct objects (say, the number of ways in which 4 of 12 basketball teams can be ranked first, second, third, and fourth by a panel of coaches).

Solution   For $n = 12$ and $r = 4$, the first formula yields

$$_{12}P_4 = 12 \cdot 11 \cdot 10 \cdot 9 = 11{,}880$$

and the second formula yields

$$_{12}P_4 = \frac{12!}{(12-4)!} = \frac{12!}{8!} = \frac{12 \cdot 11 \cdot 10 \cdot 9 \cdot 8!}{8!} = 11{,}880$$

Essentially, the work is the same, but the second formula requires a few extra steps.

To find the formula for the number of permutations of n distinct objects taken all together, we substitute $r = n$ into either formula for $_nP_r$. Using the second formula, we get

$$_nP_n = \frac{n!}{(n-n)!} = \frac{n!}{0!} = \frac{n!}{1}$$

and hence

NUMBER OF PERMUTATIONS OF n DISTINCT OBJECTS TAKEN ALL TOGETHER

$$_nP_n = n!$$

EXAMPLE 4.13   In how many ways can 10 instructors be assigned to 10 sections of a course in economics?

Solution   Substituting $n = 10$, we get

$$_{10}P_{10} = 10! = 3,628,800$$

Throughout this discussion, it has been assumed that the n objects are all distinct. When this is not the case, the formula for $_nP_n$ can be easily modified, but the formula for $_nP_r$ with $r < n$ becomes very complicated. Exercises 4.32, 4.33, and 4.34 deal with this situation.

EXERCISES

Exercises 4.21, 4.24, and 4.25 are practice exercises; their complete solutions are given on page 143.

✓ **4.21**   Check each of the following to see whether it is true or false:
   (a) $8! = 8 \cdot 7 \cdot 6!$;
   (b) $3! \cdot 2! = 6!$;
   (c) $\dfrac{12!}{12 \cdot 11 \cdot 10} = 9!$.

4.22   Check each of the following to see whether it is true or false:
   (a) $3! + 2! = 5!$;
   (b) $14! = \dfrac{15!}{15}$;
   (c) $3! + 0! = 7$.

4.23   Check each of the following to see whether it is true or false:
   (a) $\dfrac{2}{3!} + \dfrac{1}{3!} = 1$;
   (b) $\dfrac{15!}{12!} = 15 \cdot 14 \cdot 13$;
   (c) $\dfrac{6!}{3!2!} = 0$.

✓ **4.24**   In how many different ways can a parade director line up three of seven decorative floats to lead a parade?

✓ **4.25**   In how many different ways can six football teams in a conference finish the season if we exclude the possibility of ties in the final standings?

4.26   In how many ways can a fleet commander assign five ships to the first, second, third, fourth and fifth positions in a column of ten ships?

4.27   The *Wall Street Journal* published a list of the world's largest financial companies as of the year 2000. In order of size, they are
1.   Mizuho Holdings (Japan)
2.   Citigroup (U.S.)
3.   Allianz (Germany)
4.   Sumimoto Mitsui Banking (Japan)
5.   Deutsche Bank (Germany)
6.   Fannie Mae (U.S.)
7.   Mitsubishi Tokyo Financial Group (Japan)
8.   UBS (Switzerland)
9.   BNP Paribos (France)
10.   HSBC Holdings (U.K.)

   (a)   How many permutations are there of 3 of these 10 institutions?
   (b)   How many permutations are there of two of the three largest institutions?
   (c)   List all the permutations of part (b).

4.28   Calculate the number of ways in which a research technician can designate 3 of 10 rats as first, second, and third place winners following experiments in a maze. Mazes are sometimes used for testing "skill in escape."

4.29   (a)   In how many ways can a librarian arrange six different books on a shelf?
   (b)   In how many ways can six truck drivers be assigned to six company trucks?
   (c)   In how many ways can six architectural designs for a building be ranked in order of preference?

4.30   Calculate the number of ways in which four paintings can be awarded first, second, third, and fourth prizes at an art competition in which a total of 30 paintings are displayed.

*4.31   The number of ways in which n distinct objects can be arranged in a circle is $(n - 1)!$.
   (a)   Present an argument to justify this formula.
   (b)   In how many ways can six persons be seated at a round table (if it matters only who sits on whose left and right)?
   (c)   In how many ways can a window dresser display four shirts in a circular arrangement?

*4.32   We would like to determine the number of permutations, or rearrangements, of the five letters in the word *geese*.
   (a)   Suppose that the word were rewritten as $ge_1e_2se_3$ so that the three e's become distinct. How many permutations are there?
   (b)   In fact, the three e's are not distinct. Provide an argument to show that the number of permutations in part (a) must have been overcounted by a factor of $3! = 6$.
   (c)   What is the number of permutations of the five letters in *geese*?

*4.33   Find the number of permutations of the letters of the following words:
   (a)   lullaby;
   (b)   loophole;
   (c)   paperback;
   (d)   Mississippi.

*4.34   By complete enumeration, determine the number of permutations of three letters chosen from the six letters in the word *bubble*.

4.3   COMBINATIONS

In many problems we must know the unordered number of ways in which r objects can be selected from a set of n distinct objects, but we do not care about the order in which the selection is made. For instance, we may want to know in how many ways a committee of four can be selected from the 72 staff members of a hospital, or the number of ways in which the IRS can choose 5 of 33 tax returns for a special audit. To obtain a formula that applies to problems like these, let us first examine the following 24 permutations of three of the first four letters of the alphabet:

abc	acb	bac	bca	cab	cba
abd	adb	bad	bda	dab	dba
acd	adc	cad	cda	dac	dca
bcd	bdc	cbd	cdb	dbc	dcb

If we are not concerned with the order in which the three letters are chosen from the four letters a, b, c, and d, there are only four ways in which the selection can be made: abc, abd, acd, and bcd, the values shown in the first column. Note that each row contains the $3! = 6$ different permutations of the corresponding letters shown in the first column. In general, there are $r!$ permutations of any r objects selected from a set of n distinct objects, so the $_nP_r$ permutations of r objects selected from a set of n distinct objects contain each set of r objects $r!$ times. Therefore, to write a formula for the number of ways in which r objects can be selected from a set of n distinct objects, also called the number of **combinations** of n objects taken r at a time and denoted by $_nC_r$, we must divide $_nP_r$ by $r!$, and we get

NUMBER OF COMBINATIONS OF n OBJECTS TAKEN r AT A TIME

> *The number of ways in which r objects can be selected from a set of n distinct objects is*
>
> $$_nC_r = \frac{n(n-1)(n-2)\cdot\ldots\cdot(n-r+1)}{r!}$$
>
> *or, in factorial notation,*
>
> $$_nC_r = \frac{n!}{r!(n-r)!}$$

Although the first of these formulas is generally easier to use, the one in factorial notation is easier to remember.

Instead of $_nC_r$, we often use the symbol $\binom{n}{r}$ and in this notation we refer to it as a **binomial coefficient**. The reason for this is explained in Exercise 4.46. For $n = 0$ to $n = 20$, the values of $\binom{n}{r}$ may be read from Table X at the end of the book. $\binom{n}{r}$ is not a fraction, and the division symbol (a horizontal bar) is not used. The parentheses, however, are always used.

EXAMPLE 4.14 In how many different ways can a person invite three of her eight closest friends to a party?

Solution For $n = 8$ and $r = 3$, the first formula yields

$$\binom{8}{3} = \frac{8 \cdot 7 \cdot 6}{3!} = 8 \cdot 7 = 56$$

and the second formula yields

$$\binom{8}{3} = \frac{8!}{3!\,5!} = \frac{8 \cdot 7 \cdot 6 \cdot \cancel{5!}}{3!\,\cancel{5!}} = \frac{8 \cdot 7 \cdot 6}{3 \cdot 2 \cdot 1} = 56$$

Basically, the work is the same, but the first formula requires fewer steps. ■

EXAMPLE 4.15 In how many different ways can a committee of four be selected from the 72 staff members of a hospital?

Solution For $n = 72$ and $r = 4$, we get

$$\binom{72}{4} = \frac{72 \cdot 71 \cdot 70 \cdot 69}{4!} = \frac{72 \cdot 71 \cdot 70 \cdot 69}{24} = 1{,}028{,}790$$ ■

Observe that the result of the first example, but not that of the second, can be checked in Table X.

EXAMPLE 4.16 In how many different ways can a student select two of six mathematics courses together with three of seven English courses?

Solution The student can select the two mathematics courses in $\binom{6}{2}$ ways, the three English courses in $\binom{7}{3}$ ways, and, hence, by the multiplication of choices, all five in

$$\binom{6}{2} \cdot \binom{7}{3} = 15 \cdot 35 = 525 \text{ ways}$$

The values of the binomial coefficients were obtained from Table X. ■

When we take 7 objects from a set of 10 distinct objects, then $10 - 7 = 3$ of the objects are left. Thus, there are as many ways of leaving (or selecting) 3 objects from a set of 10 distinct objects as there are ways of selecting 7 objects, and we can write

$$\binom{10}{7} = \binom{10}{3}$$

In general, when r objects are selected from a set of n distinct objects, $n - r$ of the objects are left; consequently, there are as many ways of leaving (or selecting) $n - r$ objects from a set of n distinct objects as there are ways of selecting (or leaving) r objects. Symbolically, we write

RULE FOR BINOMIAL COEFFICIENTS	$$\binom{n}{r} = \binom{n}{n-r} \quad \text{for } r = 0, 1, 2, \ldots, n$$

Sometimes this rule serves to simplify calculations and sometimes it is needed in connection with the use of Table X.

EXAMPLE 4.17   Determine the value of $\binom{75}{72}$.

Solution   To avoid having to write down the product

$$75 \cdot 74 \cdot 73 \cdot \ldots \cdot 4$$

and cancel

$$72 \cdot 71 \cdot 70 \cdot \ldots \cdot 4$$

we write directly

$$\binom{75}{72} = \binom{75}{3} = \frac{75 \cdot 74 \cdot 73}{3!} = 67,525$$

EXAMPLE 4.18   Find the value of $\binom{18}{13}$.

Solution $\binom{18}{13}$ cannot be looked up directly in Table X, but making use of the fact that

$$\binom{18}{13} = \binom{18}{18 - 13} = \binom{18}{5}$$

we look up $\binom{18}{5}$ and get 8,568.

Exercises 4.35 and 4.36 are practice exercises; their complete solutions are given on pages 143–144.

✓ **4.35**   A university catalogue lists 15 graduate courses in physics. Calculate the number of ways in which a graduate student can select
   (a) 4 courses;
   (b) 3 courses;
   (c) 2 courses;
   (d) 1 course.

✓ **4.36**   Among the 16 candidates for four positions on a city council, 9 are Democrats, 5 are Republicans, and 2 are Independents. In how many ways can the 4 councilmen be chosen so that

EXERCISES

(a) 3 are Democrats and 1 is a Republican;

(b) 2 are Democrats, 1 is a Republican, and 1 is an Independent?

4.37 Evaluate without using tables:

(a) $_{14}C_3$

(b) $_{14}C_{11}$

(c) $_{15}C_5$

(d) $_{15}C_{10}$

4.38 Express the following using factorial notation.

(a) $_{14}C_3$

(b) $_{14}C_{11}$

(c) $_{15}C_5$

(d) $_{15}C_{10}$

4.39 Symbolically, we let $\binom{n}{r}$ denote the number of combinations of r objects selected from a set of n objects. Express the following situations using the $\binom{n}{r}$ symbol. Do not do the arithmetic.

(a) A social club with 25 members wants to choose 5 members for the entertainment committee.

(b) An Army sergeant wants to pick 4 privates in a 14-man platoon for a work detail.

(c) The captain of a police detective unit wants to select 3 detectives from among his 10 detectives to form a special investigation team.

(d) A scientist wants to pick 6 of 18 guinea pigs from a cage to be subjects of an experiment.

4.40 A stock brokerage firm plans to reduce the size of its work force by dismissing two of its nine brokers. If five brokers are men and the rest are women, in how many ways can the manager dismiss

(a) any two of the male brokers;

(b) any two of the female brokers;

(c) one male and one female broker;

(d) any two brokers?

4.41 A carton of 12 transistor batteries contains 4 that are defective. In how many different ways can one choose 3 of these batteries so that

(a) none of the defective batteries is included;

(b) exactly 1 of the defective batteries is included;

(c) exactly 2 of the defective batteries are included;

(d) exactly 3 of the defective batteries are included?

4.42 A housing construction firm plans to fill three carpenter's positions from a pool of five applicants; two plumbers positions from a group of four applicants; and four electricians from six applicants. If the order does not matter, in how many ways can a foreman hire these nine craftspeople? Do not use Table X.

4.43 Use the data of Exercise 4.42, and Table X. Binomial Coefficients, to determine the number of ways a foreman can hire the carpenters, plumbers, electricians, and all nine of these craftspeople.

4.44 A firm employs six accountants in its Finance Department and four attorneys on its legal staff. In how many ways can the Chief Executive Officer of the firm consult with two of the six accountants and two of the four attorneys

(a) if the order matters;

(b) if the order does not matter?

4.45   In how many ways can a pet store owner purchase

(a) five puppies out of a litter of six puppies;

(b) three kittens out of a litter of five kittens;

(c) five puppies out of a litter of six puppies and three kittens out of a litter of five kittens?

4.46   The quantity $\binom{n}{r}$ is called a *binomial coefficient* because it is, in fact, the coefficient of $a^{n-r}b^r$ in the binomial expansion of $(a+b)^n$. Verify this for $n = 2, 3$, and 4 by expanding $(a+b)^2$, $(a+b)^3$, and $(a+b)^4$ and comparing the coefficients with the corresponding entries for $n = 2$, $n = 3$, and $n = 4$ in Table X.

4.47   A table of binomial coefficients is easy to construct by following the pattern known as **Pascal's triangle**:

$$
\begin{array}{ccccccccccc}
 & & & & & 1 & & & & & \\
 & & & & 1 & & 1 & & & & \\
 & & & 1 & & 2 & & 1 & & & \\
 & & 1 & & 3 & & 3 & & 1 & & \\
 & 1 & & 4 & & 6 & & 4 & & 1 & \\
1 & & 5 & & 10 & & 10 & & 5 & & 1 \\
 & \cdot & \cdot & \cdot & \cdot & & \cdot & \cdot & \cdot & \cdot &
\end{array}
$$

In this arrangement, each row begins with a 1 and ends with a 1, and each other entry is the sum of the nearest two values in the row immediately above. Construct the next three rows of Pascal's triangle and verify from Table X that they are, respectively, the binomial coefficients corresponding to $n = 6$, $n = 7$, and $n = 8$.

4.48   In Pascal's triangle, denote the row 1 2 1 as row 2, the row 1 3 3 1 as row 3, and in general the row beginning $1\,n\ldots$ as row n. Within row n, identify the $n + 1$ positions as $0, 1, 2, \ldots, n$. Observe that $\binom{n}{r}$ appears in position r of row n.

(a) What is row 7?

(b) Give the value of $\binom{7}{3}$.

(c) The triangle is constructed by using the relationship

$$\binom{n}{r} = \binom{n-1}{r-1} + \binom{n-1}{r}$$

By expressing the binomial coefficients in factorial form, verify that this relationship is correct.

4.4   PROBABILITY

Earlier in this chapter we considered only *what is possible* in a given situation. Now we shall also advance and judge *what is probable and what is improbable*. The most common way of measuring the uncertainties connected with events (say, the effectiveness of a heart transplant or a new genetic procedure, the commercial success of a new type of cell phone, the outcome of an election, or the total points we may roll with a pair of dice) is to assign them probabilities, or to specify the odds at which it would be fair to bet that the events will occur. As we shall see in Chapter 5, these two concepts (probabilities and odds) are very closely related. In fact, if we know one, we can always calculate the other. Historically, the oldest way of measuring uncertainty is the **classical concept of probability**. It was developed originally in connection with games of chance, and it lends itself most readily to bridging the gap between possibilities and probabilities. This concept applies only when all possible outcomes are equally likely, in which case we say that

THE CLASSICAL
PROBABILITY
CONCEPT

> *If there are n equally likely possibilities, of which one must occur and s are regarded as favorable, or as a success, then the probability of a success is $\frac{s}{n}$.*

In the application of this rule, the terms *favorable* and *success* are used rather loosely—what is favorable to one player is unfavorable to his opponent, and what is a success from one point of view is a failure from another. Thus, these terms can be applied to any kind of outcome, even if favorable means that a house gets struck by lightning, or success means that a person catches pneumonia. This usage dates back to the days when probabilities were quoted only in connection with games of chance.

EXAMPLE 4.19   What is the probability of drawing an ace from a well-shuffled deck of 52 playing cards?

Solution   There are $s = 4$ aces among the $n = 52$ cards, so that we get

$$\frac{s}{n} = \frac{4}{52} = \frac{1}{13}$$

EXAMPLE 4.20   What is the probability of rolling a 6 with a well-balanced die?

Solution   In this case, $s = 1$ and $n = 6$, so that the probability is

$$\frac{s}{n} = \frac{1}{6}$$

EXAMPLE 4.21   Find the probability that 2 cards drawn from an ordinary deck of 52 playing cards will both be black.

Solution   According to what we learned in Section 4.3, the total number of possibilities is

$$n = \binom{52}{2} = \frac{52 \cdot 51}{2} = 1,326$$

and the number of favorable possibilities is

$$s = \binom{26}{2} = \frac{26 \cdot 25}{2} = 325$$

since half of the 52 playing cards are black and the others are red. It follows that the probability of drawing two black cards is

$$\frac{s}{n} = \frac{325}{1{,}326} \approx 0.245$$

■

Although equally likely possibilities are found mostly in games of chance, the classical probability concept applies also to a great variety of situations where gambling devices are used to make random selections, say, when offices are assigned to research assistants by lot, when laboratory animals are chosen for an experiment (perhaps by the method described in Section 8.1) so that each has the same chance of being selected, when each family in a large apartment complex has the same chance of being included in a sample survey, or when machine parts are chosen for inspection so that each part has the same chance of being selected.

EXAMPLE 4.22 A manufacturer of cell phones plans to ship 12 cell phones to a customer and, as a precaution, orders 4 of the phones to be checked. The 4 phones were checked and found to be satisfactory, in spite of the fact that the 12-phone shipment contained 2 defectives.

What is the probability that the 4 phones that were checked will contain no defectives when, actually, 2 of the 12 phones are defective?

Solution There are $\binom{12}{4} = 495$ ways of choosing 4 of the 12 cell phones. Thus, $n = 495$ ways, which are assumed to be equally likely. The number of ways in which 4 of the 10 good phones can be selected is

$$s = \binom{10}{4} = 210$$

and it follows that the desired probability is

$$\frac{s}{n} = \frac{210}{495} \approx 0.424$$

The values of the two binomial coefficients, $\binom{12}{4}$ and $\binom{10}{4}$, were obtained directly from Table X, Binomial Coefficients. ■

A major shortcoming of the classical probability concept is its limited applicability, for there are many situations in which the various possibilities cannot all be regarded as equally likely (that is, as having the same probability). This would be the case, for instance, if we are concerned with the question of whether it will rain on the next day. Surely, it would be nonsensical to say that either it will rain or it will not rain, and hence the probability for rain is $\frac{1}{2}$; or that there will be no precipitation, rain, hail, or snow, and hence the probability for rain is $\frac{1}{4}$. Also, the various possibilities cannot be regarded as equally likely when we wonder whether a person will get a raise, when we want to predict the outcome of an election or the score of a football game, or when we want to judge whether food prices will go up, go down, or remain the same. Among the various probability concepts, the most widely held is the **frequency interpretation**, according to which

THE FREQUENCY INTERPRETATION OF PROBABILITY

> *The probability of an event (happening or outcome) is the proportion of the time that events of the same kind will occur in the long run.*

If we say that the probability is 0.88 that a jet from Denver to Seattle will arrive on time, we mean that such flights arrive on time 88% of the time. Also, if the Weather Service predicts that there is a 40% chance for rain (that is, the probability is 0.40 that it will rain), this is meant to imply that under the same weather conditions it will rain 40% of the time. More generally, we say that an event has a probability of, say, 0.90, in the same sense in which we might say that in cold weather our car will start 90% of the time. We cannot guarantee what will happen on any particular occasion—the car may start and then it may not—but if we kept records over a long period of time, we should find that the proportion of successes is very close to 0.90. In accordance with the frequency interpretation of probability, we estimate the probability of an event by observing what fraction of the time similar events have occurred in the past.

EXAMPLE 4.23   If data kept by a government agency shows that (over a period of time) 528 of 600 jets from Denver to Seattle arrived on time, what is the probability that any one jet from Denver to Seattle will arrive on time?

Solution   Since in the past $\frac{528}{600} = 0.88$ of the flights arrived on time, we use this figure as an estimate of the desired probability; or we say that there is an 88% chance that such a flight will arrive on time.   ■

EXAMPLE 4.24   If records show that 506 of 814 automatic dishwashers sold by a large retailer required repairs within the warranty year, what is the probability that an automatic dishwasher sold by this retailer will not require repairs within the warranty year?

Solution   Since $814 - 506 = 308$ of the dishwashers did not require repairs, we estimate the desired probability as $\frac{308}{814} \approx 0.38$.   ■

When probabilities are thus estimated, it is only reasonable to ask whether the estimates are any good. Later we shall answer this question in some detail, but for now let us refer to an important theorem called the **Law of Large Numbers**. Informally, this theorem may be stated as follows:

THE LAW OF LARGE NUMBERS

> *If a situation, trial, or experiment is repeated again and again, the proportion of successes will tend to approach the probability that any one outcome will be a success.*

We can illustrate this law by repeatedly flipping a balanced coin and recording the accumulated proportion of heads after each fifth flip. We can also use the outcomes of Figure 4.3, which show the succession of heads and tails.

Figure 4.3
Outcome of 100 flips of a balanced coin where H refers to Heads and T refers to Tails.

Rows										
1	T	T	H	H	H	H	H	T	T	H
2	H	T	T	H	T	H	H	H	T	H
3	T	T	H	T	H	H	T	H	T	T
4	H	H	T	H	T	T	H	H	H	T
5	H	T	H	T	T	T	T	H	T	T
6	H	H	T	T	T	T	T	H	T	T
7	H	H	T	T	H	H	H	T	T	T
8	H	T	H	H	T	H	H	T	T	T
9	T	T	T	H	T	T	H	T	H	H
10	H	T	H	H	H	H	T	H	T	H

Figure 4.4
Graph illustrating the Law of Large Numbers.

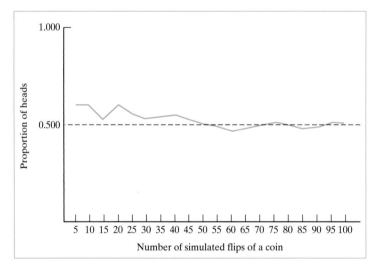

Reading across successive rows, we find that among the first 5 flips there are 3 heads, among the first 10 there are 6 heads, among the first 15 there are 8 heads, among the first 20 there are 12 heads, among the first 25 there are 14 heads, . . . , and among all 100 there are 51 heads. The corresponding proportions, plotted in Figure 4.4, are $\frac{3}{5} = 0.60$, $\frac{6}{10} = 0.60$, $\frac{8}{15} \approx 0.53$, $\frac{12}{20} = 0.60$, $\frac{14}{25} = 0.56$, . . . , and $\frac{51}{100} = 0.51$. Observe that the proportion of heads fluctuates but comes closer and closer to 0.50, the probability of heads for each flip of a coin. Theoretical support for the Law of Large Numbers will be given in Chapter 6.

In the frequency interpretation, the probability of an event is defined in terms of what happens to similar events in the long run, so let us examine briefly whether it is at all meaningful to talk about the probability of an event that can occur only once. For instance, can we assign a probability to the event that Ms. Bertha Jones will be able to leave the hospital within four days after having an appendectomy or to the event that a certain major-party candidate will win an upcoming gubernatorial election? If we put ourselves in the position of Ms. Jones's doctor, we might check medical records, discover that patients left the hospital within four days of an appendectomy in, say, 61% of hundreds of cases, and apply this figure to Ms. Jones. This may not be of much comfort to Ms. Jones, but it does provide a meaning for a probability statement about her leaving the hospital within four days—the probability is 0.61. This illustrates that when we make a probability statement about a specific (nonrepeatable) event, the frequency interpretation of probability leaves us no choice but to refer to a set of similar events. As can well be imagined, however, this can easily lead to complications, since the choice of "similar" events is generally neither obvious nor straightforward. With reference to Ms. Jones's appendectomy, we might consider as "similar" only female patients, only patients of the same age as Ms. Jones, or only cases in which the patients were of the same height and weight as Ms. Jones. Ultimately, the choice of similar events is a matter of personal judgment, often motivated by the nature of available data, and it is by no means contradictory that we can arrive at different probabilities, all valid, concerning the same event.

Regarding the question of whether a certain major-party candidate will win an upcoming gubernatorial election, suppose that we ask the persons who have conducted a poll "how sure" they are that the candidate will win. If they say they are "95 percent sure" (that is, if they assign a probability of 0.95 to the candidate's winning the election), this is not meant to imply that he would win 95% of the time if he ran for office a great number of times. Rather, it means that the prediction is based on a method that "works" 95% of the time. It is in this way that we must interpret many of the probabilities attached to statistical results.

Finally, let us mention an alternative concept in which probabilities are interpreted as **personal** or **subjective evaluations**; they measure one's belief with regard to the uncertainties that are involved. Such probabilities apply especially when there is little or no direct evidence, so that there is really no choice but to consider collateral (indirect) information, "educated guesses," and perhaps intuition and other subjective factors. Subjective probabilities can be very useful to a person who is trying to sort out the consequences of his or her point of view, but they generally do not help to resolve differences among persons with very different points of view. Subjective probabilities can sometimes be determined by putting the issue on a "put up or shut up" basis, as will be explained in Sections 4.5 and 5.4.

EXERCISES

Exercises 4.49, 4.50, and 4.61 are practice exercises; their complete solutions are given on page 144.

✓ **4.49** Using a pair of well-balanced dice, what is the probability of rolling a
  (a) two;
  (b) three;
  (c) seven;
  (d) eleven;
  (e) twelve?

✓ **4.50** A dairy farmer has 24 milk cows in a pasture, 15 of which are Holsteins and 9 of which are Jerseys. If 4 of them are randomly selected and brought into the barn for milking, what is the probability that 2 of them will be of the Holstein breed and 2 will be Jerseys?

4.51   If 1 card is drawn from a well-shuffled deck of 52 playing cards, what are the probabilities of getting

(a)   a red king;

(b)   a black card;

(c)   a 3, 4, 5, or 6;

(d)   a diamond;

(e)   not a diamond;

(f)   not an ace?

4.52   If H stands for heads and T for tails, the four possible outcomes for two flips of a coin are HH, HT, TH, and TT. If it can be assumed that these four possibilities are equally likely, what are the probabilities of getting 0, 1, or 2 heads?

4.53   If H stands for heads and T for tails, the eight possible outcomes for three flips of a coin are HHH, HHT, HTH, THH, HTT, THT, TTH, and TTT. If it can be assumed that these eight possibilities are equally likely, what are the probabilities of getting 0, 1, 2, or 3 heads?

4.54   A refrigerated case holds 24 cans of soft drinks: 6 cola, 6 root beer, 3 orange, 3 sparkling water, and 6 ginger ale. If one can is drawn at random, what is the probability of getting

(a)   a can of root beer;

(b)   a can of cola or a can of ginger ale;

(c)   a can that is neither cola, root beer, nor ginger ale?

4.55   If we roll a well-balanced die, what are the probabilities of getting either

(a)   a 1, a 3, or a 5;

(b)   an even number?

4.56   If 2 cards are drawn from a well-shuffled deck of 52 playing cards, what are the probabilities of getting

(a)   two spades;

(b)   two aces;

(c)   a king and a queen?

4.57   In how many ways can 3 good flashlights and 2 defective flashlights be chosen from a container containing 15 good and 4 defective flashlights?

4.58   A carton of 12 lightbulbs includes 3 that are defective. If 2 bulbs are chosen at random, what are the probabilities that

(a)   neither bulb will be defective;

(b)   exactly 1 bulb will be defective;

(c)   both bulbs will be defective?

4.59   A builder has 20 slabs of gray granite and 10 slabs of pink granite for use in the construction of a building. If 3 slabs of granite are randomly selected, what are the probabilities that

(a)   all will be pink;

(b)   2 will be gray and 1 will be pink?

4.60   On a tray there are six pieces of chocolate cake and four pieces of walnut cake. If a waiter randomly picks two pieces of cake from the tray and gives them to customers who ordered chocolate cake, what is the probability that he will be making a mistake?

✓ **4.61**   Data compiled by the manager of a department store shows that 1,018 of 1,956 persons who entered the store on a weekday morning made at least one purchase. Estimate the probability that a person who enters the store on a weekday morning will make at least one purchase.

4.62   According to the National Center for Health Statistics, there were 725,192 deaths from heart disease in 1999 from a total of 2,391,399 people who died of all causes. Find the probability that the cause of death of a person who died in 1999 was heart disease.

4.63   Bureau of Labor Statistics data shows that in the year 2001 there were 72,455,000 workers who were paid hourly rates; and 39,865,000 of them were paid $10.00 per hour or more. What is the probability that a randomly selected worker in this group received an hourly rate of $10.00 or more?

4.64   According to the American Medical Association, there were 797,634 physicians in the United States in the year 1999 of whom 611,028 were male and 186,606 were female. Of these physicians, 27,790 were male psychiatrists and 11,266 were female psychiatrists.

(a)   What is the probability that a randomly selected physician is male?

(b)   What is the probability that a randomly selected physician is female?

(c)   What is the probability that a randomly selected physician is a psychiatrist?

(d)   What is the probability that a randomly selected physician is not a psychiatrist?

(e)   What is the probability that a randomly selected psychiatrist is male?

4.65   In the state of Arizona, 1,035,000 out of a population of 5,176,000 (estimate) persons were not covered by health insurance in 1999. What is the probability that a person residing in Arizona during 1999 did not have health insurance?

4.66   Use a computer to simulate 100 flips of a balanced coin and plot the accumulated proportion of heads after each fifth flip (as in Figure 4.4) to illustrate the Law of Large Numbers. If a suitable computer package is not available, actually flip a coin 100 times.

4.67   Use a computer to simulate 200 flips of a balanced coin and plot the accumulated proportion of heads after each fifth flip (as in Figure 4.4) to illustrate the Law of Large Numbers. If a suitable computer package is not available, actually flip a coin 200 times.

4.68   Use a computer to simulate 150 flips of a pair of coins, recording a 1 or a 0, depending on whether or not both coins come up heads. The probability for this is 0.25 (see Exercise 4.52). Then, to illustrate the Law of Large Numbers, plot the accumulated proportion of successes (both coins come up heads) after each tenth flip. If a suitable computer package is not available, actually flip a pair of coins 150 times.

4.5 MATHEMATICAL EXPECTATION

If an insurance agent tells us that in the United States a 45-year-old woman can expect to live 33 more years, this does not mean that anyone really expects a 45-year-old woman to live until her 78th birthday and then pass away the next day. Similarly, if we read that a person living in the United States can expect to eat 276 eggs per year and 128.9 pounds of beef, or that a child in the age group from 6 to 16 can expect to visit a dentist 1.92 times a year, it must be obvious that the word *expect* is not being used in its colloquial sense. A child cannot go to the dentist 1.92 times, and it would be surprising, indeed, if we found somebody who has actually eaten 276 eggs and 128.9 pounds of beef in a given year. Insofar as 45-year-old women are concerned, some will live another 15 years, some will live another 25 years, some will live another 40 years, . . . , and the life expectancy of "33 more years" will have to be interpreted as an average or, as we shall call it here, a **mathematical expectation**.

Originally, the concept of a mathematical expectation arose in connection with games of chance, and in its simplest form it is the product of the amount a player stands to win and the probability that he or she will win.

EXAMPLE 4.25   What is our mathematical expectation if we will receive $20 if and only if a coin comes up heads?

Solution   If we assume that the coin is balanced and randomly tossed, that is, the probability of heads is $\frac{1}{2}$ our mathematical expectation is $20 \cdot \frac{1}{2} = \$10$.   ▪

EXAMPLE 4.26   What is our mathematical expectation if we buy 1 of 2,000 raffle tickets issued for a television set worth $540?

Solution   The probability that we will win is $\frac{1}{2,000}$, so our mathematical expectation is

$$540 \cdot \frac{1}{2,000} = \$0.27 \text{ or } 27 \text{ cents}$$

Thus, it would be unwise to spend more than 27 cents for the ticket, unless, of course, the proceeds of the raffle go to a worthy cause (or the difference can be credited to whatever pleasure a person might derive from placing a bet).   ▪

In both of these examples there is a single prize, but in each case there are two possible payoffs—$20 and $0 in the first example and $540 and $0 in the other. Indeed, in the second example we can argue that 1,999 of the tickets will pay $0 and one of the tickets will pay $540 (or the equivalent in merchandise). Altogether, the 2,000 tickets will thus pay $540 or, on the average,

$$\frac{540}{2,000} = \$0.27 \text{ per ticket}$$

and this is the mathematical expectation. To generalize the concept of a mathematical expectation, let us consider the following change in the raffle of the preceding example.

EXAMPLE 4.27   What is our mathematical expectation if we buy 1 of 2,000 raffle tickets issued for a first prize of a television set worth $540, a second prize of a tape recorder worth $180, and a third prize of a pocket radio worth $40?

Solution   Now we can argue that 1,997 of the raffle tickets will not pay anything at all, one ticket will pay $540 (in merchandise), another will pay $180 (in merchandise), and a third will pay $40 (in merchandise). Altogether the 2,000 tickets will thus pay $540 + 180 + 40 = \$760$ (in merchandise) or, on the average,

$$\frac{760}{2,000} = \$0.38 \text{ per ticket}$$

As before, this is the mathematical expectation for each ticket. Looking at the problem in a different way, we could argue that if the raffle were repeated many times, we would lose

$$\frac{1,997}{2,000} \cdot 100\% = 99.85\% \text{ of the time}$$

(or with probability 0.9985); and win each of the prizes

$$\frac{1}{2,000} \cdot 100\% = 0.05\% \text{ of the time}$$

(or with probability 0.0005). On average, we would thus win

$$0(0.9985) + 540(0.0005) + 180(0.0005) + 40(0.0005) = \$0.38$$

which is the sum of the products obtained by multiplying each amount by the corresponding proportion or probability.

Generalizing from this example, let us now give the following general definition:

MATHEMATICAL EXPECTATION

> *If the probabilities of obtaining the amounts $a_1, a_2, \ldots,$ or a_k are, respectively, $p_1, p_2, \ldots,$ and p_k, then the mathematical expectation is*
>
> $$E = a_1 p_1 + a_2 p_2 + \cdots + a_k p_k$$

Each amount is multiplied by the corresponding probability, and the mathematical expectation, E, is the sum of all these products. In the $\sum$ notation,

$$E = \sum a \cdot p$$

Insofar as the a's are concerned, it is important to keep in mind that they are positive when they represent profits, winnings, or gains (amounts that we receive) and that they are negative when they represent losses, penalties, or deficits (amounts that we have to pay).

EXAMPLE 4.28   What is our mathematical expectation if we win \$6 when a die comes up 1 or 2 and lose \$3 when the die comes up 3, 4, 5, or 6?

Solution   The amounts are $a_1 = 6$ and $a_2 = -3$, and if we assume that the die is balanced and randomly tossed, $p_1 = \frac{2}{6}$ and $p_2 = \frac{4}{6}$. Thus, the mathematical expectation is

$$E = 6 \cdot \frac{2}{6} + (-3)\frac{4}{6} = 0$$

This example illustrates an **equitable** or **fair game**. It is a game that does not favor either player; that is, each player's mathematical expectation is zero.

EXAMPLE 4.29   The probabilities are 0.22, 0.36, 0.28, and 0.14 that an investor will be able to sell a piece of property at a profit of $5,000, that he will be able to sell it at a profit of $2,000, that he will break even, or that he will sell it at a loss of $3,000. What is his expected profit?

Solution   If we substitute $a_1 = 5,000$, $a_2 = 2,000$, $a_3 = 0$, $a_4 = -3,000$, $p_1 = 0.22$, $p_2 = 0.36$, $p_3 = 0.28$, and $p_4 = 0.14$ into the formula for E, we get

$$E = 5,000(0.22) + 2,000(0.36) + 0(0.28) - 3,000(0.14) = \$1,400 \quad \blacksquare$$

Although we referred to the quantities $a_1, a_2, \ldots$, and a_k as *amounts*, they need not be cash winnings, losses, penalties, or rewards. When we said previously that a child in the age group from 6 to 16 can expect to visit a dentist 1.92 times a year, we referred to the result that was obtained by multiplying $0, 1, 2, 3, \ldots$, by the corresponding probabilities that a child in this age group will visit a dentist that many times a year and then adding all these products (see Exercise 4.77).

EXAMPLE 4.30   The headquarters of a fire department have determined that the probabilities of $0, 1, 2, 3, 4$, or 5 fire alarms being received from the fire alarm boxes on the streets are, respectively, 0.15, 0.29, 0.38, 0.12, 0.06, and 0.02. How many fire alarms from these fire alarm boxes can the fire department expect per day? (*Note*: Most fire alarms are called in by telephone or received from other sources.)

Solution   The expected number of alarms received from the red fire alarm boxes is $0(0.15) + 1(0.29) + 2(0.38) + 3(0.12) + 4(0.06) + 5(0.02) = 1.75$ alarms per day. $\blacksquare$

EXAMPLE 4.31   To handle a liability suit, a lawyer has to decide whether to charge a straight fee of $2,400 or a contingent fee of $9,600, which she will get only if her client wins. How does she feel about her client's chances if she prefers the straight fee of $2,400?

Solution   If she feels that the probability is p that her client will win and she decided on the contingent fee, her mathematical expectation would be $9,600p$. Since she feels that the certainty of $2,400 is preferable to the mathematical expectation of $9,600p$, we can write

$$2,400 > 9,600p$$

which yields $p < \dfrac{2,400}{9,600}$ and, hence, $p < 0.25$. $\blacksquare$

Note that this is one of the two methods of determining subjective probabilities referred to earlier. To pin it down further, we might ask the lawyer if she would still prefer the straight fee of $2,400 if the contingent fee were, say, $19,200 (see Exercise 4.80).

4.6   A DECISION PROBLEM

When we are faced with uncertainties, mathematical expectations can often be used to great advantage in making decisions. In general, if we have to choose between two or more alternatives, it is considered rational to select the one with the *most promising* mathematical expectation: the one that maximizes expected profits, minimizes expected costs, maximizes expected tax advantages, minimizes expected losses, and so on.

EXAMPLE 4.32   A clothing manufacturer must decide whether to spend a considerable sum of money to build a new factory. He knows that if the new factory is built and the clothing business has a good sales year, there will be a $451,000 profit; if the new factory is built and the clothing business has a poor sales year, there will be a deficit of $110,000; if the new factory is not built and the clothing business has a good sales year, there will be a $220,000 profit; and if the new factory is not built and the clothing business has a poor year, there will be a $22,000 profit (mostly because of lower overhead cost). If the clothing manufacturer feels that the probabilities for a good sales year or a poor sales year are, respectively, 0.40 and 0.60, would building the new factory maximize his expected profit?

Solution   In problems like this, it usually helps to present the information about profits and deficits in the following kind of table, called a **payoff table**.

	New factory built	New factory not built
Good sales year	$451,000	$220,000
Poor sales year	−$110,000	$22,000

As can be seen from this table, it will be advantageous to build the new factory only if the clothing business is going to have a good sales year, and the decision whether to build the new factory will, therefore, have to depend on the chances that this will be the case. Using the manufacturer's probabilities of 0.40 and 0.60 for a good sales year and a poor sales year, we find that if the new factory is built, his expected profit is

$$451,000(0.40) - 110,000(0.60) = \$114,400$$

If the new factory is not built, his expected profit is

$$220,000(0.40) + 22,000(0.60) = \$101,200$$

Since the first of these two figures is larger, it follows that building the new factory maximizes the manufacturer's expected profit. ∎

Decision trees are also often used in the analysis of payoff tables. The branches represent A's (actions) and E's (events). Actions emanate from small squares (■) and E's emanate from small circles (•). Figure 4.5 is an illustration of the previous payoff table in the form of a tree diagram.

The decision tree is not really necessary in this simple problem. It can be quite useful, however, when it is hard to enumerate all the decision makers'

Figure 4.5
Decision tree.

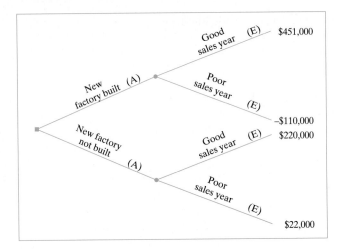

courses of action. We shall do nothing more with this decision tree beyond this point.

The way in which we have studied this problem is called a **Bayesian analysis**. In this kind of analysis, probabilities are assigned to the alternatives about which uncertainties exist (the **states of nature**, which in our example were a good sales year and a poor sales year); then we choose whichever alternative promises the greatest expected profit or the smallest expected loss. This approach to decision making has great intuitive appeal, but it is not without complications. It may be quite tricky to sort out complex situations in a logical way. In the preceding example, is it adequate to consider sales as simply "good" or "poor"? Also, it is essential that the appraisals of all relevant probabilities be close to correct and also that the values of the payoffs associated with the various possibilities be close to correct. (See Exercises 4.87 through 4.90.)

EXAMPLE 4.33   A dealer for luxury boats is concerned about the number of a very expensive line of boats to order for the April–June sales season. He would like to have available exactly the number that will be ordered. If demand for this boat exceeds supply, then the extra customers will either buy the boat elsewhere or make no purchase at all. If supply exceeds demand, then the extra boats are not likely to sell at all until the next sales season and must be returned to the manufacturer for a substantial service charge. The profit is $10,000 on each boat sold, and the loss for an unsold boat is $3,500.

The number of boats to be demanded is virtually certain to be at most 5. Therefore, the dealer must decide whether to order 0, 1, 2, 3, 4, or 5 boats for the coming season.

It is believed that the number of boats demanded will be 0 with probability 0.05, 1 with probability 0.20, 2 with probability 0.30, 3 with probability 0.25, 4 with probability 0.15, and 5 with probability 0.05.

How many boats should the dealer order to maximize his expected profit?

Solution Construct a chart showing the profit to be made under each combination of order size and number demanded. In this chart, the amounts are given in thousands of dollars.

		0	1	2	3	4	5
				Order			
	0	0	−3.5	−7	−10.5	−14	−17.5
	1	0	10	6.5	3	−0.5	−4
Number of	2	0	10	20	16.5	13	9.5
boats demanded	3	0	10	20	30	26.5	23
	4	0	10	20	30	40	36.5
	5	0	10	20	30	40	50

For example, if zero boats are ordered, the profit will certainly be zero. If one boat is ordered, the profit will be $10,000 so long as the demand is one or greater; if the demand is zero, the profit will be −$3,500. If two boats are ordered and the demand is one boat, the dealer will make $10,000 on the boat that is sold and lose $3,500 on the boat that is not sold; the amount

$$\$10,000 - \$3,500 = \$6,500$$

is presented as 6.5 in the box (Order 2, Demand 1).

If one boat is ordered, the profit will be −3.5 thousand dollars with probability 0.05 and 10 thousand dollars with probability 0.95. The expected profit is then

$$-3.5 \cdot 0.05 + 10 \cdot 0.95 = 9.325 \text{ thousand dollars}$$

If two boats are ordered, the profit will be −7 thousand dollars with probability 0.05. It will be 6.5 thousand dollars with probability 0.20, and it will be 20 thousand dollars with probability

$$0.30 + 0.25 + 0.15 + 0.05 = 0.75$$

The expected profit is then

$$-7 \cdot 0.05 + 6.5 \cdot 0.20 + 20 \cdot 0.75 = 15.95 \text{ thousand dollars}$$

Similarly, the expected profits for ordering three, four, and five boats are 18.525, 17.725, and 14.900 thousand dollars, respectively.

Ordering three boats will maximize the expected profit.

Exercises 4.69, 4.70, 4.79, and 4.85 are practice exercises; their complete solutions are given on pages 144 and 145.

✓ **4.69**   At a bazaar held to raise money for a charity, it costs 50 cents to try one's luck in drawing an ace from an ordinary deck of 52 playing cards. What is the bazaar's expected profit per card drawn by a customer if the prize is $4.00 if and only if the customer draws an ace?

✓ **4.70**   A union wage negotiator feels that the probabilities are 0.25, 0.50, 0.20, and 0.05 that the members of the union will get a $1.20 raise in their hourly wage, an 80-cent raise, a 40-cent raise, or no raise at all. What is the corresponding expected raise?

4.71   A supermarket offers its customers a free ticket to a drawing for a Thanksgiving turkey. If 300 tickets are entered in the drawing, and the turkey is worth $30.00, what is the mathematical expectation of a customer who received a free ticket?

4.72   (a)   At a football game picnic the four door prizes are a football worth $30.00, a college banner worth $25.00, and three tickets to a dinner worth $20.00 each. If students are presented with 100 free tickets from which the winning stubs are to be drawn, what is the mathematical expectation for each ticket?

(b)   Is it worthwhile to pay a parking fee of $2.00 at the picnic grounds in order to get a free ticket for the drawing?

4.73   In the game of roulette a metal ball is dropped into a spinning wheel so that it lands in one of 38 compartments. The compartments are numbered 00, 0, 1, 2, 3, ..., 36. The compartments for 00 and 0 are colored green; 18 of the remaining compartments are red and 18 are black.

(a)   You can bet $1 on red. If a red number comes up, then your bet is returned to you with another $1 so that your net winning is +$1. If the number that comes up is not red, your bet is lost, and your net winning is −$1. What is your expected net winning?

(b)   You can bet $1 on the set of numbers 1 through 12. If one of these numbers comes up, then your bet is returned to you with another $2 so that your net winning is +$2. If the number that comes up is not one of these, your bet is lost, and your net winning is −$1. What is your expected net winning?

(c)   You can bet $1 on any individual number. If this number comes up, then your bet is returned to you with another $35 so that your net winning is +$35. If your chosen number does not come up, your bet is lost, and your net winning is −$1. What is your expected net winning?

4.74   The promoter of a wrestling match offers $5,000 to the winner and $2,500 to the loser. What are the two wrestlers' mathematical expectations if

(a)   they are evenly matched;

(b)   their respective probabilities are $\frac{4}{5}$ and $\frac{1}{5}$?

4.75   The owner of a carpet cleaning service feels that the probabilities are 0.10, 0.25, 0.35, and 0.30 that the carpet cleaning business will earn profits of $100,000, $90,000, $80,000, and $70,000, respectively. According to the owner, what is the expected profitability of the carpet cleaning service?

4.76   A commodities speculator feels that the probabilities are 0.35, 0.30, 0.25, and 0.10 that the price of wheat per bushel will increase by 0.50, 0.40, 0.10, or not at all during a specified period of time. By how much does the speculator expect that the price of wheat will increase, in cents per bushel?

4.77   If the probabilities are 0.15, 0.28, 0.27, 0.17, 0.08, 0.03, and 0.02 that a child in the age group from 6 to 16 will visit a dentist 0, 1, 2, 3, 4, 5, or 6 times, respectively, a year, how many times can a child in this age group expect to visit a dentist in any given year?

4.78 The owner of an automobile repair shop knows that the probabilities of 0, 1, 2, 3, and 4 requests for towing services are 0.19, 0.38, 0.23, 0.15, and 0.05 per day. How many requests for towing services can the owner expect on a given day?

✓ **4.79** A salesperson has to choose between a straight salary of $24,000 and a salary of $20,000 plus a bonus of $8,000 if her sales exceed a certain quota. How does she assess her chances of exceeding the quota if she chooses the lower salary with the possibility of the bonus?

4.80 Suppose that a lawyer in a liability suit is asked whether she would prefer a fee of $2,400 or a contingent fee of $19,200, which she will get only if her client wins. How does she feel about her client's chances if she prefers the fee of $2,400?

4.81 An insurance company agrees to pay the promoter of a drag race $15,000 if the race has to be canceled because of rain. If the company's actuary feels that a fair net premium for this risk is $2,400, what does this tell us about his assessment of the probability that the race will have to be canceled because of rain?

4.82 A shopping-center manager finds a contractor to do a road repair job for $45,000. Another contractor offers to do the job for $50,000 with a penalty of $12,500 if the job is not finished on time. The shopping center will not experience a monetary loss if the repair is not completed on time. If the manager prefers the second offer, what does this tell us about her assessment of the probability that the second contractor will not finish the job on time?

***4.83** The records of a court clerk show that the probabilities of 0, 1, 2, 3, 4, or 5 lawsuits being filed per day are 0.25, 0.35, 0.24, 0.12, and 0.04. How many lawsuit filings can the clerk of the court expect per day?

***4.84** A dealer in luxury boats needs to place his order for his stock of Narwhal cruisers for the coming spring buying season. Each boat that he stocks and sells will generate a profit of $15,000. Each boat that he stocks but does not sell costs him $4,000 in storage and financing costs. Ideally, he would like to order exactly the number that he will be able to sell, but the number that he can sell cannot be perfectly predicted. He feels that the probabilities that he can sell 0, 1, 2, 3, or 4 of the Narwhal cruisers are 0.10, 0.25, 0.40, 0.20, and 0.05, respectively. How many Narwhal cruisers should be ordered?

✓ ***4.85** A truck driver has to deliver a load of building materials to one of two construction sites and return to the lumberyard. These sites are 16 and 20 miles from the lumberyard. He has misplaced the order telling him where the load should go, and the telephone at the lumberyard is out of order. The two construction sites are 6 miles apart. If the driver feels that the probability is $\frac{1}{5}$ that the load should go to the site that is 16 miles from the lumberyard and the probability is $\frac{4}{5}$ that the load should go to the other site, where should he go first to minimize the expected distance that he will have to drive?

***4.86** With reference to Exercise 4.85, where should the driver go first so as to minimize the expected distance that he will have to drive if he reassessed the probabilities as $\frac{1}{8}$ and $\frac{7}{8}$ instead of $\frac{1}{5}$ and $\frac{4}{5}$?

***4.87** The management of a mining company must decide whether to continue an operation at a certain location. If they continue and are successful, they will make a profit of $4,500,000; if they continue and are not successful, they will lose $2,700,000; if they do not continue but would have been successful if they had continued, they will lose $1,800,000 (for competitive reasons); and if they do not continue and would not have been successful if they had continued, they will make a profit of $450,000 (because funds allocated to the operation remain unspent). What decision would maximize the company's expected profit if it is felt that there is a 50–50 chance for success?

*4.88   Show that it does not matter what they decide to do in Exercise 4.87 if it is felt that the probabilities for and against success are, respectively, $\frac{1}{3}$ and $\frac{2}{3}$.

4.89   With reference to the following payoff table, which was previously used in Example 4.32,

	New factory built	New factory not built
Good sales year	$451,000	$220,000
Poor sales year	−$110,000	$22,000

show that building the new factory would not maximize the expected profit if the manufacturer felt that the probabilities for a good sales year and a poor sales year are, respectively, $\frac{1}{6}$ and $\frac{5}{6}$.

4.90   With reference to Exercise 4.89 and using the probabilities of 0.40 and 0.60, would the clothing manufacturer's decision not to build the new factory still maximize his expected profit if it is found that the −$110,000 figure is in error and should be a loss of $165,000?

*4.91   A realty company owns a newly constructed but unoccupied strip shopping center that has locations for four stores. The realty company needs to know the number of these stores for which they should install indoor furnishings (plumbing, counter space, display shelves, and so on). If the realty company finds more tenants than there are finished stores, the tenants will be lost. If the realty company has more finished stores than tenants, the dollar loss will be represented by the money tied up in the indoor furnishings. The value of a tenant is computed at $30,000 over the next year, while the cost of installing indoor furnishings is set at $12,000. An expert in the local real estate market believes that the following probabilities govern the number of tenants that can be found for this shopping center:

Number of tenants	Probability
0	0.1
1	0.4
2	0.3
3	0.1
4	0.1

In order to maximize the expected return, how many stores should be given indoor furnishings?

*4.92   The realty company in Exercise 4.91 consulted a second expert. The second expert gave these probabilities:

Number of tenants	Probability
0	0.1
1	0.1
2	0.1
3	0.3
4	0.4

According to the second expert, how many stores should be given indoor furnishings?

✓ SOLUTIONS TO PRACTICE EXERCISES

4.1 The tree diagram is shown in Figure 4.6, where N denotes a win by the National League champion and A denotes a win by the American League champion.

4.7 (a) There are three choices corresponding to the three trials.

(b) $m = 3, n = 3$, so there are $3 \cdot 3 = 9$ possibilities.

(c) $m = 3, n = 2$, so there are $3 \cdot 2 = 6$ possibilities.

4.8 $n_1 = 5, n_2 = 9$, and $n_3 = 8$ and there are
$$n_1 \cdot n_2 \cdot n_3 = 5 \cdot 9 \cdot 8 = 360 \text{ possibilities}$$

4.21 (a) $8! = 8 \cdot 7 \cdot (6 \cdot 5 \cdot 4 \cdot 3 \cdot 2 \cdot 1) = 8 \cdot 7 \cdot 6!$ is true; in general, $n! = n(n-1)! = n(n-1)(n-2)! = n(n-1)(n-2)(n-3)!$ and so forth.

(b) $3! = 6, 2! = 2, 6! = 720$, and $6 \cdot 2$ does not equal 720.

(c) $\dfrac{12!}{12 \cdot 11 \cdot 10} = \dfrac{12 \cdot 11 \cdot 10 \cdot 9!}{12 \cdot 11 \cdot 10} = 9!$ is true.

4.24 $7 \cdot 6 \cdot 5 = 210$ ways.

4.25 $6! = 6 \cdot 5 \cdot 4 \cdot 3 \cdot 2 \cdot 1 = 720$ ways.

4.35 (a) $\dbinom{15}{4} = \dfrac{15 \cdot 14 \cdot 13 \cdot 12}{4!} = \dfrac{32{,}760}{24} = 1{,}365$ ways.

(b) $\dbinom{15}{3} = \dfrac{15 \cdot 14 \cdot 13}{3!} = \dfrac{2{,}730}{6} = 455$ ways.

Figure 4.6
Tree diagram for Exercise 4.1.

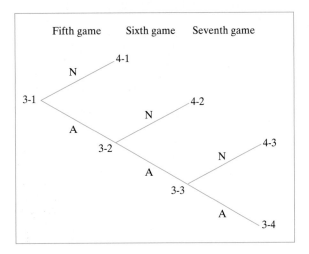

(c) $\binom{15}{2} = \frac{15 \cdot 14}{2!} = \frac{210}{2} = 105$ ways.

(d) $\binom{15}{1} = \frac{15}{1!} = 15$ ways.

4.36 (a) The three Democrats can be chosen in $\binom{9}{3} = 84$ ways, the Republican can be chosen in 5 ways, so that, by the multiplication of choices, there are $84 \cdot 5 = 420$ possibilities.

(b) The two Democrats can be chosen in $\binom{9}{2} = 36$ ways, the Republican can be chosen in 5 ways, and the Independent can be chosen in 2 ways, so that by the generalized rule for the multiplication of choices, there are $36 \cdot 5 \cdot 2 = 360$ possibilities.

4.49 (a) Altogether there are $6 \cdot 6 = 36$ possibilities, and in $s = 1$ of these (1 and 1) the total is two. Therefore, the probability is $\frac{1}{36}$.

(b) In $s = 2$ of the 36 possibilities (1 and 2, 2 and 1), the total is three. Therefore, the probability is $\frac{2}{36}$ or $\frac{1}{18}$.

(c) In $s = 6$ of the 36 possibilities (1 and 6, 6 and 1, 2 and 5, 5 and 2, 3 and 4, 4 and 3), the total is seven. Therefore, the probability is $\frac{6}{36}$ or is $\frac{1}{6}$.

(d) In $s = 2$ of the 36 possibilities (5 and 6, 6 and 5), the total is eleven and the probability is $\frac{2}{36} = \frac{1}{18}$.

(e) In $s = 1$ of the 36 possibilities (6 and 6), the total is twelve and the probability is $\frac{1}{36}$.

4.50 Altogether there are $n = \binom{24}{4} = \frac{24 \cdot 23 \cdot 22 \cdot 21}{4!} = 10{,}626$ possibilities. The 2 Holsteins can be chosen in $\binom{15}{2} = \frac{15 \cdot 14}{2!} = 105$ ways; and the 2 Jerseys can be chosen in $\binom{9}{2} = \frac{9 \cdot 8}{2!} = 36$ ways; and by the multiplication of choices $s = 105 \cdot 36 = 3780$. It follows that the probability is $\frac{s}{n} = \frac{3{,}780}{10{,}626} \approx 0.356$.

4.61 The estimate is $\frac{1{,}018}{1{,}956} \approx 0.520$.

4.69 The mathematical expectation of a person who draws a card is $4 \cdot \frac{1}{13} \approx 0.31$, so that the bazaar's approximate expected profit is $0.50 - 0.31 = \$0.19$.

4.70 Substituting $a_1 = 120$, $a_2 = 80$, $a_3 = 40$, $a_4 = 0$, $p_1 = 0.25$, $p_2 = 0.50$, $p_3 = 0.20$, and $p_4 = 0.05$ into the formula for a mathematical expectation, we get

$$120(0.25) + 80(0.50) + 40(0.20) + 0(0.05) = 78 \text{ cents}$$

4.79 If she feels that the probability of her exceeding the quota is p, the mathematical expectation corresponding to the salary plus bonus alternative is $20{,}000 + 8{,}000p$. Since this exceeds 24,000, we get $20{,}000 + 8{,}000p > 24{,}000$, and hence

$$8{,}000p > 4{,}000 \text{ and } p > \frac{4{,}000}{8{,}000} = \frac{1}{2}$$

Thus, she feels that there is a better than 50–50 chance that she will exceed the quota.

4.85 Letting I and II denote the construction sites that are, respectively, 16 and 20 miles from the lumberyard, we can show the distances corresponding to the various alternatives (depending on where the driver goes first and where he should go) as in the following table:

Driver goes first to

		I	II
Driver should go to	I	$16 + 16 = 32$	$20 + 6 + 16 = 42$
	II	$16 + 6 + 20 = 42$	$20 + 20 = 40$

Thus, if he first goes to site I, the expected distance is

$$32 \cdot \frac{1}{5} + 42 \cdot \frac{4}{5} = \frac{32 + 168}{5} = \frac{200}{5} = 40.0 \text{ miles}$$

and if he goes first to site II, the expected distance is

$$42 \cdot \frac{1}{5} + 40 \cdot \frac{4}{5} = \frac{42 + 160}{5} = \frac{202}{5} = 40.4 \text{ miles}$$

Thus, if the driver wants to minimize the expected distance that he has to drive, he should go first to site I, that is, the construction site that is 16 miles from the lumberyard.

5
SOME RULES OF PROBABILITY

n this chapter we will learn the mathematical rules that probabilities must obey. We have already studied, in Chapter 4, what we mean when we say that the probability of an event is 0.70, 0.58, or 0.05. We also learned how the numbers we call probabilities are measured in actual practice. At this point we shall study the **theory of probability** or, as it is sometimes called, the **mathematics of chance**.

The concept of probability is widely used in problems of everyday life, in business, in science and especially in statistics. The study of probability is complicated by the fact that "probability" is difficult to define, and many people associate it with nebulous, mystical ideas and with forecasting the future by looking into a crystal ball.

Probability theory was originally developed in connection with games of chance. Nowadays it has become an indispensable tool for life insurance companies who must estimate how long a policyholder will probably live; by medical researchers and other scientists who use probability theory to estimate true values from experimental data; for businesspeople who must plan inventories without knowing with certainty what products will sell; for military strategists who must commit personnel and equipment to the hazards of battle; for anyone who sends a message by mail without assurance that it will be delivered on time; and so on.

After some preliminaries in Sections 5.1 and 5.2 of this chapter, we shall study some of the basic rules in Section 5.3; the relationship between probabilities and odds in Section 5.4; addition rules in Section 5.5; conditional probabilities and related problems in Sections 5.6, 5.7, and 5.8; and Bayes's theorem in Section 5.9.

5.1 THE SAMPLE SPACE

In statistics, the word *experiment* is used in a very wide and unconventional sense. For lack of a better term, it refers to any process of observation or measurement. Thus, an **experiment** may consist of counting how many times a student has been absent; it may consist of the simple process of noting whether a light is on or off, or whether a person is single or married; or it may consist of the very complicated process of obtaining and evaluating data to predict trends in the economy, to find the source of social unrest, or to study the cause of a disease. The results that one obtains from an experiment, whether they are instrument readings, counts, yes or no answers, or values obtained through extensive calculations, are called the **outcomes** of the experiment.

For each experiment, the set of all possible outcomes is called the **sample space**, and it is usually denoted by the letter S. For instance, if a zoologist must choose 3 of his 24 guinea pigs for an experiment, the sample space consists of the

$$\binom{24}{3} = 2{,}024 \text{ ways}$$

in which the selection can be made; if the dean of a college has to assign 2 of her 84 faculty members as advisors to a political science club, the sample space consists of the

$$\binom{84}{2} = 3{,}486 \text{ ways}$$

in which this can be done. Also, if we are concerned with the number of days that it rains in Chicago during the month of January, the sample space is the set

$$S = \{0, 1, 2, 3, 4, \ldots, 30, 31\}$$

When we study the outcomes of an experiment, we usually identify the various possibilities with numbers, points, or some other kinds of symbols so that we can treat all questions about them mathematically, without having to go through long verbal descriptions of what has taken place, is taking place, or will take place. For instance, if there are eight candidates for a scholarship and we let $a, b, c, d, e, f, g,$ and h denote that it is awarded to Ms. Adams, or Mr. Bean, or Ms. Clark, and so on, then the sample space for this experiment is the set

$$S = \{a, b, c, d, e, f, g, h\}$$

The use of points rather than letters or numbers has the advantage that it makes it easier to visualize the various possibilities and perhaps discover some special features that several of the outcomes may have in common.

EXAMPLE 5.1   A used car dealer has two 2002 Mercury Sables on his lot, and we are interested in how many of them each of two salespersons will sell in a given week.

  (a) Use two coordinates so that $(0, 1)$, for example, represents the outcome that the first salesperson will sell neither of the Mercury Sables and the second will sell one, $(1, 1)$ represents the outcome that each of the two salespersons will sell one of the Mercury Sables, and $(2, 0)$ represents the outcome that the first salesperson will sell them both. Now list all possible outcomes of this experiment.

  (b) Draw a figure showing the corresponding points of the sample space.

Solution   **(a)** The six possible outcomes are $(0, 0), (1, 0), (0, 1), (2, 0), (1, 1)$, and $(0, 2)$.

  (b) The corresponding points are shown in Figure 5.1, from which it is apparent, for instance, that they sell equally as many 2002 Mercury Sables in two of the six possibilities, and that they will sell both cars in three of the six possibilities. ■

Generally, we classify sample spaces according to the number of elements, or points, that they contain. The ones we have studied so far in this section contained, respectively, 2,024, 3,486, 31, 8, and 6 elements, and we refer to them all as **finite**. In this chapter we shall consider only sample spaces that are finite, but in later chapters we shall consider also sample spaces that are **infinite**. An infinite sample space arises, for example, when we throw a dart at a target and there is a continuum of points that we may hit.

In general, if the elements of a sample space are obtained by a measuring process, then the sample space is infinite. If the elements of a sample space are obtained by counting, then the sample space may be finite (as in the preceding used-car example), or it may be infinite (as in recording the number of tails preceding the first head in repeated flips of a coin).

Figure 5.1
Sample space for two-salespersons example.

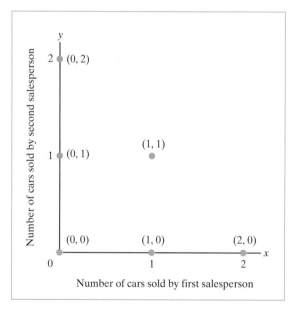

5.2 EVENTS

In statistics, any subset of a sample space is called an **event**. By *subset*, we mean any part of a set, including the set as a whole and, trivially, a set called the **empty set** and denoted by Ø, which has no elements at all. For instance, with reference to the illustration on page 147 dealing with the number of days that it rains in Chicago during the month of January, the subset

$$M = \{10, 11, 12, \ldots, 19, 20\}$$

is the event that there will be anywhere from 10 to 20 rainy days, and the subset

$$N = \{18, 19, 20, \ldots, 30, 31\}$$

is the event that there will be at least 18 rainy days.

EXAMPLE 5.2   With reference to Figure 5.1, express in words the events represented by

    (a) $C = \{(0, 0), (1, 1)\}$;

    (b) $D = \{(1, 0), (1, 1)\}$;

    (c) $E = \{(0, 2)\}$.

Solution   **(a)** C is the event that the two salespersons will sell equally as many of the 2002 Mercury Sables.

    (b) D is the event that the first salesperson will sell one and only one of the two cars.

    (c) E is the event that the second salesperson will sell both cars. ∎

As is the custom, in both of these examples we denoted events by capital letters.

In many probability problems we must deal with events that are compounded by forming **unions**, **intersections**, and **complements**.

> *The union of two events A and B, denoted by A ∪ B, is the event that consists of all the elements (outcomes) contained in event A, in event B, or in both.*

> *The intersection of two events A and B, denoted by A ∩ B, is the event that consists of all the elements (outcomes) contained in both A and B.*

> *The complement of event A, denoted by A', is the event that consists of all the elements (outcomes) of the sample space that are not contained in A.*

The symbol ∪ is commonly read as *or*. Similarly, ∩ is read as *and* while A' is read as *not A*. One sometimes encounters the symbols A^c or $\overline{A}$ for *not A*. We will use A'.

EXAMPLE 5.3   With reference to the illustration dealing with the number of days that it rains in Chicago in January, where we had

$$S = \{0, 1, 2, 3, 4, \ldots, 30, 31\}$$

$$M = \{10, 11, 12, \ldots, 19, 20\}$$

and

$$N = \{18, 19, 20, \ldots, 30, 31\}$$

express each of the following events symbolically and also in words:

(a) $M \cup N$;   (c) M';   (e) $M' \cup N'$;
(b) $M \cap N$;   (d) N';   (f) $M' \cap N'$.

Solution   (a) Since $M \cup N$ contains all the elements that are in M, in N, or in both, we find that

$$M \cup N = \{10, 11, 12, \ldots, 30, 31\}$$

and this is the event that there will be at least 10 rainy days.

(b) Since $M \cap N$ contains all the elements that are in both M and N, we find that

$$M \cap N = \{18, 19, 20\}$$

and this is the event that there will be from 18 to 20 rainy days.

(c) Since M' contains all the elements of the sample space that are not in M, we find that

$$M' = \{0, 1, 2, \ldots, 8, 9, 21, 22, \ldots, 30, 31\}$$

and this is the event that there will be fewer than 10 or more than 20 rainy days.

(d) Since N' contains all the elements of the sample space that are not in N, we find that

$$N' = \{0, 1, 2, \ldots, 16, 17\}$$

and this is the event that there will be at most 17 rainy days.

(e) Since $M' \cup N'$ contains all the elements that are not in M, not in N, or in neither, we find that

$$M' \cup N' = \{0, 1, 2, \ldots, 16, 17, 21, 22, \ldots, 30, 31\}$$

and this is the event that there will be fewer than 18 or more than 20 rainy days (or the event that there will not be 18, 19, or 20 rainy days).

(f) Since $M' \cap N'$ contains all the elements that are in both M' and N', we find that

$$M' \cap N' = \{0, 1, 2, \ldots, 8, 9\}$$

and this is the event that there will be at most 9 rainy days.   ■

Observe also that with the two salespersons of Example 5.1 on page 148 events D and E have no elements in common. Such events are called **mutually exclusive**.

Two events A and B are mutually exclusive if they cannot both occur at the same time, namely, when $A \cap B = \emptyset$.

EXAMPLE 5.4   With reference to Example 5.1, the two-salespersons example, which of the following pairs of events are mutually exclusive:

(a) C and *D*;

(b) C and *E*?

Solution   **(a)** Since events *C* and *D* both contain $(1, 1)$, they are not mutually exclusive.

(b) Since $C \cap E = \emptyset$, events *C* and *E* are mutually exclusive.   ■

Sample spaces and events, particularly relationships among events, are often pictured by means of **Venn diagrams** such as those of Figures 5.2 and 5.3. In each case, the sample space is represented by a rectangle, while events are represented by regions within the rectangle, usually by circles or parts of circles. The tinted regions of the four Venn diagrams of Figure 5.2 represent the event *X*, the complement of event *X*, the union of two events *X* and *Y*, and the intersection of two events *X* and *Y*.

EXAMPLE 5.5   If *X* is the event that a given high school student will be accepted at Duke University and *Y* is the event that he or she will be accepted at the University of Virginia, what events are represented by the tinted regions of the four Venn diagrams of Figure 5.2?

Figure 5.2
Venn diagrams.

Figure 5.3
Venn diagram.

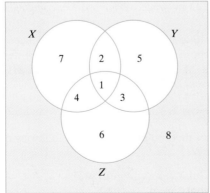

Solution   (a) The tinted region of the upper left diagram represents the event that the student will be accepted at Duke University.

(b) The tinted region of the upper right diagram represents the event that he or she will not be accepted at Duke University.

(c) The tinted region of the lower left diagram represents the event that he or she will be accepted at Duke University or at the University of Virginia, or both.

(d) The tinted region of the lower right diagram represents the event that he or she will be accepted at both Duke University and at the University of Virginia. ■

When we deal with three events, we draw the circles as in Figure 5.3. In this diagram, the circles divide the sample space into eight regions, numbered 1 through 8, and it is easy to determine whether each of the corresponding events is contained in X or in X', in Y or in Y', and in Z or in Z'.

EXAMPLE 5.6   With reference to a newly designed automobile, X represents the event that its gasoline consumption will be low, Y represents the event that its maintenance cost will be low, and Z represents the event that the dealer can sell it at a profit. Express in words what events are represented by the following regions of the Venn diagram of Figure 5.3:

(a) region 4;

(b) regions 1 and 3 together;

(c) regions 3, 5, 6, and 8 together.

Solution   (a) Since this region is contained in X and in Z but not in Y, it represents the event that the automobile's gasoline consumption will be low, that the dealer can sell it at a profit, but that its maintenance cost will not be low.

(b) Since this region is common to Y and Z, it represents the event that the automobile's maintenance cost will be low, and that the dealer can sell it at a profit.

(c) Since this is the entire region outside X, it represents the event that the automobile's gasoline consumption will not be low. ■

Exercises 5.1, 5.2, and 5.12 are practice exercises; their complete solutions are given on pages 187 and 188.

✓ **5.1**   With reference to the illustration on page 147 concerning eight candidates for a scholarship, suppose that a, b, c, d, e, f, g, and h denote that Ms. Adams, Mr. Bean, Ms. Clark, Mr. Daly, Mr. Earl, Ms. Fuentes, Ms. Gardner, or Mr. Hall will be awarded the scholarship. Suppose that

$$X = \{a, c, f, g\}, \quad Y = \{d, f, h\}, \quad \text{and} \quad Z = \{a, b, e, g\}$$

List the outcomes that comprise each of the following events, and also express the events in words:

(a) X';   (c) $X \cap Y$;   (e) $Y \cap Z$;
(b) $X \cup Y$;   (d) $X \cap Z$;   (f) $X' \cup Y$.

EXERCISES

✓ **5.2**   A small restaurant has two chefs and five waiters, and at least one chef and two waiters have to be present at all times.

(a) Using two coordinates so that $(1, 3)$, for example, represents the event that one chef and three waiters are present and $(2, 4)$ represents the event that two chefs and four waiters are present, draw a diagram similar to that of Figure 5.1, showing the eight points of the sample space.

(b) Describe in words the events represented by

$$Q = \{(1, 4), (2, 4)\}$$

$$R = \{(1, 2), (2, 3)\}$$

$$T = \{(1, 2), (1, 3), (1, 4), (1, 5)\}$$

(c) List the outcomes that comprise each of the following events and also express the events in words: $Q \cup T$, $R \cap T$, and T'.

(d) With reference to part (b), which pairs of events Q and R, Q and T, and R and T are mutually exclusive?

5.3   In an experiment we roll a balanced die and observe the total number of points, which are 1, 2, 3, 4, 5, and 6. These are the elements of the sample space. If $D = \{3, 4, 5\}$ and $E = \{2, 4, 6\}$, list the outcomes that comprise each of the following events, and also express the events in words: D', $D \cup E$, and $D \cap E$.

5.4   A laboratory technician marks six mice with the numbers 1, 2, 3, 4, 5, and 6, so that the sample space for selecting one of the mice is the set $S = \{1, 2, 3, 4, 5, 6\}$. Let

$$L = \{1, 2, 3, 4\}$$

$$A = \{3, 4, 5, 6\}$$

$$B = \{2, 3, 4, 5\}$$

If the technician selects one of the numbered mice for an experiment, list the outcomes that comprise each of the following events, and also express the events in words.

(a) $L \cap A$;

(b) $L \cup A$;

(c) $A' \cap B'$;

(d) $A \cap B'$.

5.5   If one card is drawn from an ordinary deck of 52 playing cards, the sample space may be written as

$$S = \begin{matrix} \{A\spadesuit, & 2\spadesuit, & 3\spadesuit, & \dots, & K\spadesuit, \\ A\heartsuit, & 2\heartsuit, & 3\heartsuit, & \dots, & K\heartsuit, \\ A\diamondsuit, & 2\diamondsuit, & 3\diamondsuit, & \dots, & K\diamondsuit, \\ A\clubsuit, & 2\clubsuit, & 3\clubsuit, & \dots, & K\clubsuit\} \end{matrix}$$

where $\spadesuit, \heartsuit, \diamondsuit$, and $\clubsuit$ denote the suits spades, hearts, diamonds, and clubs. If

$$M = \{Q\spadesuit, \quad K\spadesuit, \quad Q\heartsuit, \quad K\heartsuit, \quad Q\diamondsuit, \quad K\diamondsuit, \quad Q\clubsuit, \quad K\clubsuit\}$$

and

$$N = \{10\spadesuit, \quad J\spadesuit, \quad Q\spadesuit, \quad K\spadesuit\}$$

list the outcomes which comprise each of the following events, and also express the events in words:

(a) $M \cup N$;   (c) M';   (e) $M' \cup N'$;
(b) $M \cap N$;   (d) N';   (f) $M' \cap N'$.

5.6   To construct sample spaces for experiments in which we deal with categorical data, we often code the various alternatives by assigning them numbers. For instance, if persons are asked whether their favorite color is red, yellow, blue, green, brown, white, purple, or some other color, we might assign these alternatives the codes 1, 2, 3, 4, 5, 6, 7, and 8, respectively. If $A = \{3, 4\}$, $B = \{1, 2, 3, 4, 5, 6, 7\}$, and $C = \{5, 6, 7, 8\}$, list the outcomes that comprise each of the following events:

(a) B';   (c) $A \cap B$;   (e) $A' \cup C$;
(b) $A \cup B$;   (d) C';   (f) $B \cap C'$.

5.7   With reference to Exercise 5.6, which pairs of events A and B, A and C, and B and C are mutually exclusive?

5.8   A building contractor rents up to four trucks and three bulldozers by the day from an equipment rental company.

(a) Use two coordinates so that $(1, 2)$, for example, represents the fact that on a certain day the contractor rents 1 truck and 2 bulldozers, and $(3, 0)$ represents the event that the contractor rents 3 trucks and no bulldozers. Draw a diagram similar to that of Figure 5.1 showing the 20 points of the corresponding sample space.

(b) Describe in words the event which is represented by each of the following sets of points of the sample space: the event D, which consists of the points $(2, 3)$, $(3, 2)$, $(3, 3)$, $(4, 1)$, $(4, 2)$, and $(4, 3)$; the event E, which consists of the points $(0, 0)$, $(1, 1)$, $(2, 2)$, and $(3, 3)$; the event F, which consists of the points $(0, 1)$, $(0, 2)$, $(0, 3)$, $(1, 2)$, $(1, 3)$, and $(2, 3)$; and the event G, which consists of the points $(0, 3)$, $(1, 2)$, $(2, 1)$, and $(3, 0)$.

5.9   With reference to Exercise 5.8, which of the following are mutually exclusive events?

(a) E and F

(b) E and G'

(c) D and F

(d) D and G

5.10   A small marina has three fishing boats that are sometimes in dry dock for repairs.

(a) Using two coordinates so that $(2, 1)$, for example, represents the event that two of the fishing boats are in dry dock and one is rented out for the day and $(0, 2)$ represents the event that none of the boats is in dry dock and two are rented out for the day, draw a diagram similar to that of Figure 5.1 showing the 10 points of the corresponding sample space.

(b) If K is the event that at least two of the boats are rented out for the day, L is the event that more boats are in dry dock than are rented out for the day, and M is the event that all the boats that are not in dry dock are rented out for the day, list the outcomes that comprise each of these three events.

(c) With reference to part (b), list the outcomes that comprise K' and $L \cap M$, and also express these events in words.

(d) With reference to part (b), which of the three pairs of events K and L, K and M, and L and M are mutually exclusive?

5.11   Which of the following pairs of events are mutually exclusive? Explain your answers.

(a) Walking and chewing gum at the same time

(b) Drawing a queen of spades and a red card on a single draw from an ordinary deck of playing cards

(c) Being 15 years of age and voting in a national political election

(d) Wearing athletic shoes and being an athlete

(e) Driving in an automobile and calling home on a cell phone

(f) A golfer getting a hole in one in two strokes

✓ **5.12**   In Figure 5.4, D is the event that a person vacationing in Southern California visits Disneyland, and U is the event that he visits Universal Studios. Explain in words the events represented by regions 1, 2, 3, and 4.

Figure 5.4
Venn diagram for
Exercise 5.12.

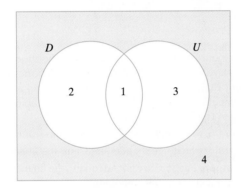

Figure 5.5
Venn diagram for
Exercise 5.14.

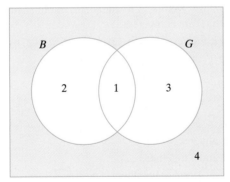

5.13   With reference to Exercise 5.12, what events are represented by

(a) regions 1 and 2 together;

(b) regions 2 and 3 together;

(c) regions 2 and 4 together?

5.14   In Figure 5.5, G is the event that a murder suspect is guilty, and B is the event that he is allowed out on bail. Explain in words the events represented by regions 1, 2, 3, and 4.

5.15   With reference to Exercise 5.14, what events are represented by

(a) regions 1 and 3 together;

(b) regions 1 and 4 together;

(c) regions 3 and 4 together?

Figure 5.6
Venn diagram for
Exercise 5.17.

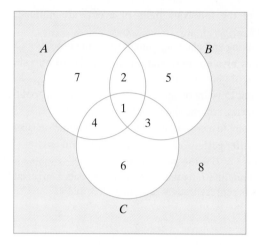

5.16 Suppose that Figure 5.6 pertains to a research laboratory established by a pharma-ceutical company where A is the event that the laboratory facilities are adequate, B is the event that the financial support is adequate, and C is the event that it is staffed by competent personnel. What regions or combinations of regions represent the following events?

(a) The laboratory facilities are adequate and are staffed by competent personnel.

(b) The laboratory facilities are adequate, and the financial support is adequate.

(c) The laboratory facilities are not adequate, the financial support is not adequate, and the personnel are not competent.

(d) The laboratory facilities are adequate, the financial support is adequate, and the personnel are competent.

5.17 Continuing with Exercise 5.16, explain in words what the following regions represent:

(a) region 1;

(b) region 3;

(c) region 6;

(d) region 8;

(e) regions 1 and 4 together;

(f) regions 3 and 5 together;

(g) regions 1, 3, 4, and 6, together;

(h) regions 2, 5, 7, and 8 together.

5.18 With reference to Exercise 5.16 and Figure 5.6, list the region, or combination of regions, that represents the following events.

(a) The financial support is adequate, but the laboratory facilities are not adequate and the personnel are not competent.

(b) The laboratory facilities are adequate, the financial support is not adequate, and the personnel are not competent.

(c) The laboratory facilities are adequate, the financial support is adequate, and the personnel are competent.

(d) The personnel are competent.

(e) The financial support is adequate, the personnel are competent, and the laboratory facilities are not adequate.

5.3 SOME BASIC RULES OF PROBABILITY

Probabilities always pertain to the occurrence or nonoccurrence of events (whether a coin will come up heads, whether a letter will arrive on time, whether it will not rain, whether a project will succeed, ...); now that we have learned how to deal with events, let us turn to the rules according to which probabilities must behave, that is, the mathematical rules that they must obey. To express these rules symbolically, we shall denote events with capital letters as in Section 5.2, and write the probability of event A as $P(A)$, the probability of event B as $P(B)$, and so forth.

The following are some of the most basic rules of probabilities, and we shall justify them here with reference to the frequency interpretation; in Exercise 5.29, the reader will be asked to justify them with reference to the classical probability concept, and in Section 5.4, we shall see to what extent they are compatible with subjective probabilities.

BASIC RULES OF PROBABILITIES

> **1.** *Probabilities are real numbers between 0 and 1, inclusive.*
>
> **2.** *If an event is certain to occur, its probability is 1, and if an event is certain not to occur, its probability is 0.*
>
> **3.** *If two events are mutually exclusive, the probability that one or the other will occur equals the sum of their probabilities.*
>
> **4.** *The sum of the probabilities that an event will occur and that it will not occur is equal to 1.*

With reference to the frequency interpretation, the first of these rules simply expresses the fact that an event cannot occur less than 0% or more than 100% of the time; that is, the proportion of successes, or the proportion of successes in the long run, cannot be negative or exceed 1. Symbolically, we write $0 \leq P(A) \leq 1$ for any event A. By the same token, the second rule expresses the fact that an event that is certain to occur will occur 100% of the time and an event that is certain not to occur will occur 0% of the time. Symbolically, we write $P(S) = 1$ for any sample space S, which expresses the certainty that one of the outcomes in any sample space S must occur, and $P(\emptyset) = 0$, which expresses the fact that an event which cannot occur (which does not include any of the outcomes) has zero probability.

To show that the third rule is satisfied by the frequency interpretation, we have only to observe that if one event occurs, say, 15% of the time, if another event occurs, say, 28% of the time, and if the two events cannot both occur at the same time (that is, they are mutually exclusive), then one or the other occurs 15% + 28% = 43% of the time. Of course, the same argument applies also to proportions. If the probabilities are, respectively, 0.45 and 0.17 that the weather will improve or that it will remain the same, then the probability is 0.45 + 0.17 = 0.62 that it will improve or remain the same. Symbolically, we write

$$P(A \cup B) = P(A) + P(B)$$

for any two mutually exclusive events A and B.

The fourth rule expresses the certainty that an event either will or will not occur. If someone is late for work 22% of the time, he is not late 78% of the time, the corresponding probabilities are 0.22 and 0.78, and their sum is $0.22 + 0.78 = 1$. Symbolically, we write

$$P(A) + P(A') = 1 \quad \text{for any event } A$$

We can also write $P(A') = 1 - P(A)$ if we want to calculate the probability that an event will not occur in terms of the probability that it will occur.

The examples that follow illustrate how these rules are put to use in actual practice.

EXAMPLE 5.7   If A is the event that a student will stay home to study and B is the event that she will instead go to a movie, $P(A) = 0.64$, and $P(B) = 0.21$, find

(a) $P(A')$;
(b) $P(A \cup B)$;
(c) $P(A \cap B)$.

Solution   (a) Using the fourth rule, we find that the probability of A', the event that the student will not stay home to study, is

$$1 - P(A) = 1 - 0.64 = 0.36$$

(b) Since A and B are mutually exclusive, we can use the third rule and write

$$P(A \cup B) = P(A) + P(B) = 0.64 + 0.21 = 0.85$$

for the probability that the student will either stay home to study or go to a movie.

(c) Since A and B are mutually exclusive, they cannot possibly both occur, and hence

$$P(A \cap B) = P(\emptyset) = 0$$

In problems like this, it often helps to draw a Venn diagram, fill in the probabilities associated with the various regions, and then read the answers directly off the diagram.

EXAMPLE 5.8   If C is the event that at 9:30 A.M. a certain doctor is in his office and D is the event that he is in the hospital, $P(C) = 0.48$ and $P(D) = 0.27$, find $P(C' \cap D')$, the probability that he is neither in his office nor in the hospital.

Solution   Drawing the Venn diagram as in Figure 5.7, we first put a 0 probability into region 1, because the events C and D are mutually exclusive. It follows that the 0.48 probability of event C must go into region 2, the 0.27 probability of event D must go into region 3, and since the probability for the entire sample space must total 1, we put $1 - (0.48 + 0.27) = 0.25$ into region 4. Since the event $C' \cap D'$ is represented by the region outside both circles, namely, region 4, we find that the answer is

$$P(C' \cap D') = 0.25$$

Figure 5.7
Venn diagram.

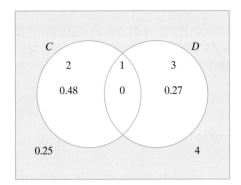

5.4 PROBABILITIES AND ODDS

If an event is twice as likely to occur as not to occur, we say that the **odds** are 2 to 1 that it will occur; if an event is three times as likely to occur as not to occur, we say that the odds are 3 to 1; if an event is ten times as likely to occur as not to occur, we say that the odds are 10 to 1; and so forth. In general,

The odds that an event will occur are given by the ratio of the probability that it will occur to the probability that it will not occur.

Symbolically, if the probability of an event is p, the odds for its occurrence are a to b, where a and b are positive values such that

$$\frac{a}{b} = \frac{p}{1-p}$$

It is customary to express odds in terms of positive integers having no common factors, if possible.

EXAMPLE 5.9   What are the odds for the occurrence of an event if its probability is

(a) $\dfrac{5}{9}$;

(b) 0.85?

Solution   (a) By definition, the odds are $\dfrac{5}{9}$ to $1 - \dfrac{5}{9} = \dfrac{4}{9}$, or 5 to 4.

(b) By definition, the odds are 0.85 to $1 - 0.85 = 0.15$, 85 to 15, or better, 17 to 3. ∎

If an event is more likely not to occur than to occur, it is customary to quote the odds that it will not occur, rather than the odds that it will occur.

EXAMPLE 5.10   What are the odds if the probability of an event is 0.20?

Solution   The odds for the occurrence of the event are 0.20 to $1 - 0.20 = 0.80$, or 1 to 4, but it is customary to say instead that the odds against the occurrence of the event are 4 to 1. ∎

In betting, the word *odds* is also used to denote the ratio of the wager of one party to that of another. For instance, if a gambler says that he will give 3 to 1 odds on the occurrence of an event, he means that he is willing to bet $3 against $1 (or perhaps $30 against $10 or $1,500 against $500) that the event *will* occur. If such **betting odds** actually equal the odds that the event will occur, we say that the betting odds are fair. If a gambler really believes that a bet is fair, then he is, at least in principle, willing to bet on either side. The gambler in this situation would also be willing to bet $1 against $3 (or $10 against $30 or $500 against $1,500) that the event *will not* occur.

EXAMPLE 5.11   During cold weather our car starts on the first try about nine-tenths of the time. If someone offers to bet $1.00 that our car will start on the first try, against our bet of $9.00 that it will not start on the first try, are these betting odds fair?

Solution   Since the probability is $1 - \dfrac{1}{10} = \dfrac{9}{10}$ that the car will start on the first try, the odds are 9 to 1. The bet is fair because we will win $1.00 about $\dfrac{9}{10}$ of the time or lose $9.00 about $\dfrac{1}{10}$ of the time. Hence we expect to break even in the long run.   ■

This discussion of odds and betting odds provides the groundwork for a way of measuring subjective probabilities. If a businessman feels that the odds for the success of a new clothing store are 3 to 2, this means that he is willing to bet (or considers it fair to bet) $300 against $200, or perhaps $3,000 against $2,000, that the new store will be a success. In this way he is expressing his belief regarding the uncertainties connected with the success of the store, and to convert it into a probability, we take the equation

$$\frac{a}{b} = \frac{p}{1 - p}$$

and solve it for p. Leaving the details to the reader in Exercise 5.37, let us merely state the result that

FORMULA RELATING PROBABILITIES TO ODDS

> *If the odds are a to b that an event will occur, the probability of its occurrence is*
>
> $$p = \frac{a}{a + b}$$

EXAMPLE 5.12   Convert the businessman's 3 to 2 odds for the success of the new clothing store into a probability.

Solution   Substituting $a = 3$ and $b = 2$ into the formula for p, we get

$$p = \frac{3}{3 + 2} = \frac{3}{5} = 0.6$$ ■

EXAMPLE 5.13   If an applicant for a managerial position feels that the odds are 7 to 4 that she will get the job, what probability is she thus assigning to her getting the job?

Solution   Substituting $a = 7$ and $b = 4$ into the formula for p, we get

$$p = \frac{7}{7+4} = \frac{7}{11} \approx 0.64$$ ∎

Let us now see whether subjective probabilities, determined in this way, behave in accordance with the four basic rules of probabilities on page 157. As far as the first and fourth rules are concerned, this is easy to see. Since a and b are positive quantities, $\frac{a}{a+b}$ is a fraction between 0 and 1; since the probabilities that an event will occur and that it will not occur are, respectively,

$$\frac{a}{a+b} \quad \text{and} \quad \frac{b}{a+b}$$

their sum is

$$\frac{a}{a+b} + \frac{b}{a+b} = \frac{a+b}{a+b} = 1$$

As for the second rule, observe that the surer we are that an event will occur, the better odds we should be willing to give, say, 100 to 1, 1,000 to 1, or perhaps even 1 million to 1.

The corresponding probabilities are

$$\frac{100}{100+1} \approx 0.9901, \quad \frac{1,000}{1,000+1} \approx 0.999001,$$

and

$$\frac{1,000,000}{1,000,000+1} \approx 0.999999$$

and it can be seen that the surer we are that an event will occur, the closer its probability will be to 1. By the same token, it can be shown that the surer we are that an event will not occur, the closer its probability will be to 0.

This leaves only the third rule,

$$P(A \cup B) = P(A) + P(B)$$

for any two mutually exclusive events A and B, and this rule is not necessarily satisfied when it comes to subjective probabilities. Indeed, proponents of the subjectivist point of view impose it as a **consistency criterion**, and this provides a means of "policing" a person's subjective probabilities.

EXAMPLE 5.14   An economist feels that the odds are 2 to 1 that the price of beef will go up during the next month, 1 to 5 that it will remain unchanged, and 8 to 3 that it will go up or remain unchanged. Are the corresponding probabilities consistent?

Solution   According to this economist, the probability that the price of beef will go up during the next month is $\frac{2}{2+1} = \frac{2}{3}$. The probability that this price will remain unchanged is $\dfrac{1}{1+5} = \dfrac{1}{6}$, and the probability that this price will go up or remain unchanged is $\dfrac{8}{8+3} = \dfrac{8}{11}$. Since $\dfrac{2}{3} + \dfrac{1}{6} = \dfrac{5}{6}$, which is not equal to $\dfrac{8}{11}$, the probabilities are not consistent. Hence, the economist's judgment must be questioned.   ■

Exercises 5.19, 5.22, 5.30, and 5.31 are practice exercises; their complete solutions are given on page 188.

✓ **5.19**   In a psychiatric study reported in the *New England Journal of Medicine* during the year 2000, psychiatrists reported on mental status evaluations. Letting A stand for the event that the patients have auditory delusions and V stand for the event that the patients have visual delusions, state in words the probabilities expressed by

(a) $P(A')$;   (d) $P(A \cap V)$;
(b) $P(V')$;   (e) $P(A' \cap V')$;
(c) $P(A \cup V)$; (f) $P(A \cap V')$.

5.20   In a study of the adequacy of fuel supplies, C stands for the event that a power plant will use coal, and E is the event that it will be able to provide enough electricity. State in words the probabilities expressed by

(a) $P(C')$;   (d) $P(C \cap E)$;
(b) $P(E')$;   (e) $P(C' \cap E)$;
(c) $P(C \cup E)$; (f) $P(C' \cap E')$.

5.21   The probabilities that a computer store will sell 0, 1, 2, 3, or at least 4 computers on a typical business day are 0.10, 0.15, 0.20, 0.25, and 0.30. What are the probabilities that, on a typical day.

(a) at most three computers will be sold;

(b) at least two computers will be sold;

(c) two or three computers will be sold?

✓ **5.22**   Explain why there must be a mistake in each of the following statements:

(a) The probability that a certain basketball player will score on a free throw is 0.67, and the probability that he will miss is 0.23.

(b) A gold-panner knows that the probability is 0.25 that a pan of gold-bearing soil will contain gold and the probability is 0.90 that it will not contain gold.

(c) The probability that a man will wear a tuxedo to a certain social event is 0.24, the probability that he will wear a business suit is 0.38, and the probability that he will wear a tuxedo or a business suit is 0.50.

5.23   Explain why there must be a mistake in each of the following statements:

(a) The probability that a new vaccine is effective is 1.09.

(b) The probability that Tom will take Mary to the senior prom is 0.65, and the probability that Bill will take the same woman to the senior prom is 0.45.

(c) The probability that an employer will hire Anne for a job is 0.30, the probability that he will hire Mildred for the same job is 0.40, and the probability that he will hire either Anne or Mildred is 0.90.

(d) Sam and George are scheduled for a boxing match in a college-sponsored tournament. The probability that Sam will win is 0.35, while the probability that George will win is 0.55. Assume that winning or losing are the only possible outcomes.

5.24 Given the *mutually exclusive* events Y and Z, for which $P(Y) = 0.28$ and $P(Z) = 0.47$, find

(a) $P(Y')$;   d) $P(Y \cup Z)$;
(b) $P(Z')$;   (e) $P(Y' \cap Z')$.
(c) $P(Y \cap Z)$;

(*Hint:* Draw a Venn diagram and fill in the probabilities associated with the various regions.)

5.25 Given the mutually exclusive events Q and R for which $P(Q) = 0.45$ and $P(R) = 0.30$, find

(a) $P(Q')$;   (d) $P(Q \cap R)$;
(b) $P(R')$;   (e) $P(Q' \cap R)$;
(c) $P(Q \cup R)$;   (f) $P(Q \cap R')$.

5.26 Events D and E are mutually exclusive. It also happens that D' and E are mutually exclusive. Find $P(E)$.

5.27 The probability that a hotel manager's reaction to a new musical tape played continuously in the lobby is 0.55 that the volume of sound is too high, and 0.10 that the volume is just right. Find the probability that the manager's reaction to the volume is that it is

(a) too low;
(b) too low or just right;
(c) too high or just right.

5.28 The probabilities that the customer service department of a department store will accept, at most, 6 merchandise returns in a day or from 7 to 10 merchandise returns are 0.65 and 0.20. Find the probability that the customer service department will accept

(a) at least 7 merchandise returns;
(b) at most 10 merchandise returns;
(c) more than 10 merchandise returns.

5.29 Show that the four basic rules of probabilities on page 157 are satisfied if we interpret probabilities in accordance with the classical probability concept, which states that if there are n equally likely possibilities, of which one must occur and s are regarded as favorable, or as a success, then the probability of a success is $\frac{s}{n}$.

✓ **5.30** Change the following odds to probabilities, and probabilities to odds.

(a) Odds are 2 to 1.   (e) Probability is $\frac{3}{4}$.

(b) Odds are 1 to 2.   (f) Probability is $\frac{1}{4}$.

(c) Odds are 3 to 2.   (g) Probability is 0.60.

(d) Odds are 2 to 3.   (h) Probability is 0.40.

✓ **5.31** A taxpayer claims that the odds are 2 to 1 that he will complete his tax return on time and 2 to 1 that it will be late. Can these odds be right? Explain.

5.32 Convert each of the following odds to probabilities:

(a) The odds that a particular horse will lose a race are 7 to 1.

(b) The odds are 3 to 5 that a sequence of four coin tosses will result in two heads and two tails.

(c) If a secretary randomly places six letters into six addressed envelopes, the odds are 1 to 719 that all letters will end up in the correct envelopes.

(d) The odds are 2 to 17 for winning a roulette bet made by placing a token at the intersection of four number boxes.

5.33   Convert each of the following probabilities to odds:

(a) The probability that the last digit of a postal zip code is 5, 6, 7, 8, 9, or 0 is $\frac{6}{10}$.

(b) The probability of randomly selecting the 8 letters alpha, beta, gamma, delta, epsilon, zeta, eta, theta from the 24 letters of the Greek alphabet is $\frac{8}{24}$.

(c) The probability of getting 2 heads in 4 flips of a coin is $\frac{6}{16}$.

(d) The probability of drawing a heart from a randomly shuffled deck of 52 playing cards is $\frac{13}{52}$.

5.34   If a student is anxious to bet $25 against $5 that she will pass a certain course, what does this tell us about the probability she assigns to her passing the course? (*Hint:* The answer should read "greater than . . . .")

5.35   A politician states that the odds are 5 to 1 that he will not run for the House of Representatives and 4 to 1 that he will not run for the Senate. Furthermore, he feels that the odds are 1 to 2 that he will run for either of these offices. Are the corresponding probabilities consistent?

5.36   An automobile salesperson feels that the odds are 9 to 1 against her selling a certain car and 4 to 1 against her being able to lease it. To be consistent, what should she consider to be fair odds that she will either sell the car or lease it?

5.37   Verify algebraically that the formula $\frac{a}{b} = \frac{p}{1-p}$, solved for p, yields $p = \frac{a}{a+b}$.

5.38   A gambler believes that the true odds favoring team Y in a coming game are 5 to 2. Which of the following bets would be attractive to him?

(a) Bet $5 on team Y to win $5.

(b) Bet $7 on team Y to win $3.

(c) Bet $3 against team Y to win $7.

(d) Bet $30 against team Y to win $70.

5.39   A gambler feels that the probability is 0.60 that Punchy Joe will win a certain boxing match. Which of the following bets would be attractive to her?

(a) Bet $6.00 on Punchy Joe to win $6.00.

(b) Bet $6.00 on Punchy Joe to win $7.00.

(c) Bet $6.00 on Punchy Joe to win $3.00.

5.5 ADDITION RULES

The third of our basic probabilities rules on page 157 is sometimes referred to as the **special addition rule**. It is special in that it applies only to mutually exclusive events, and it is an addition rule because we add the probabilities of the two events. As we shall see in this section, it can be generalized so that it will apply to more than two mutually exclusive events and also to events that need not be mutually exclusive.

Repeatedly applying the third rule on page 157 we can show that

SPECIAL ADDITION
RULE FOR TWO
OR MORE EVENTS

If k events are mutually exclusive, the probability that one of them will occur equals the sum of their respective probabilities. In symbols,

$$P(A_1 \cup A_2 \cup \cdots \cup A_k) = P(A_1) + P(A_2) + \cdots + P(A_k)$$

*for any mutually exclusive events $A_1, A_2, \ldots,$ and A_k, where, again, $\cup$ is usually read **or**.*

EXAMPLE 5.15 The probabilities that a woman shopping in New York City will buy a new dress at Macy's, Lord and Taylor, or Saks Fifth Avenue are 0.22, 0.18, and 0.35. What is the probability that she will buy the new dress at one of these stores?

Solution Since the three possibilities are mutually exclusive, the answer is

$$0.22 + 0.18 + 0.35 = 0.75$$

EXAMPLE 5.16 The probabilities that a consumer testing service will rate a new antipollution device for cars as very poor, poor, fair, good, very good, or excellent are 0.07, 0.12, 0.17, 0.32, 0.21, and 0.11. What is the probability that it will rate the device as poor, fair, good, or very good?

Solution Since the four possibilities are mutually exclusive, we get

$$0.12 + 0.17 + 0.32 + 0.21 = 0.82$$

The job of assigning probabilities to all possible events connected with a given situation can be very tedious, to say the least. If a sample space has 10 elements (outcomes), we can form more than a thousand different events, and if a sample space has 20 elements (outcomes), we can form more than a million different events.[†] Fortunately, it is seldom, if ever, necessary to find the probabilities of all possible events relating to a given situation; the following rule, which is a direct application of the special addition rule for two or more events, makes it easy to determine the probability of any event on the basis of the probabilities assigned to the individual elements (outcomes) that form the sample space.

RULE FOR
CALCULATING THE
PROBABILITY OF AN
EVENT

The probability of any event A is given by the sum of the probabilities of the individual outcomes comprising A.

[†] In general, if a sample space has n elements, we can form 2^n different events. Each element is either included or excluded for a given event, so by the multiplication of choices there are $2 \cdot 2 \cdot 2 \cdot \ldots \cdot 2 = 2^n$ possibilities. Note that

$$2^{10} = 1,024 \quad \text{and} \quad 2^{20} = 1,048,576$$

In the special case where the outcomes are all equiprobable, this rule leads to the formula

$$P(A) = \frac{s}{n}$$

which we studied earlier in connection with the classical probability concept. Here, n is the total number of outcomes and s is the number of successes, that is, the number of outcomes in event A.

EXAMPLE 5.17   In Example 5.2, dealing with two used-car salespersons trying to sell two used 2002 Mercury Sables, suppose that the six points of the sample space have the probabilities shown in Figure 5.8. Find the probabilities that

(a) the first salesperson will not sell either of the two cars;

(b) both cars will be sold;

(c) the second salesperson will sell at least one of the two cars.

Solution   (a) Adding the probabilities associated with the points $(0, 0)$, $(0, 1)$, and $(0, 2)$, we get

$$0.06 + 0.17 + 0.22 = 0.45$$

(b) Adding the probabilities associated with the points $(2, 0)$, $(1, 1)$, and $(0, 2)$, we get

$$0.22 + 0.16 + 0.22 = 0.60$$

(c) Adding the probabilities associated with the points $(0, 1)$, $(1, 1)$, and $(0, 2)$, we get

$$0.17 + 0.16 + 0.22 = 0.55$$ ■

Figure 5.8
Sample space for two-salesperson example with probabilities assigned to the outcomes.

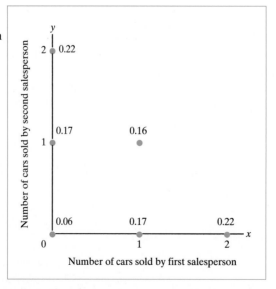

Figure 5.9
Sample space with 44
outcomes.

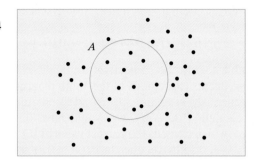

EXAMPLE 5.18   Assuming that the 44 points (outcomes) of the sample space of Figure 5.9 are all equiprobable, find $P(A)$.

Solution   Since there are $s = 10$ outcomes in A and the $n = 44$ outcomes are all equiprobable, it follows that

$$P(A) = \frac{10}{44} = \frac{5}{22}$$

Since the addition rules that we have studied so far apply only to mutually exclusive events, they cannot be used, for example, to find the probability that at least one of two friends will pass a final examination; the probability that in an automobile accident the driver will break an arm, a rib, or a leg; or the probability that a customer will buy a shirt or a sweater at a given department store. Both friends can pass the final examination; the driver of the car can break an arm and a rib, and also a leg; and the customer of the department store can buy a sweater as well as a shirt.

To find a formula for $P(A \cup B)$ that holds regardless of whether the events A and B are mutually exclusive, let us consider the Venn diagram of Figure 5.10, which concerns the election of a mayor. The letter G stands for the election of a person who is a college graduate, and the letter M stands for the election of a member of a certain minority group. It follows from the figures in the Venn diagram that

$$P(G) = 0.44 + 0.15 = 0.59, \quad P(M) = 0.15 + 0.07 = 0.22,$$

Figure 5.10
Venn diagram.

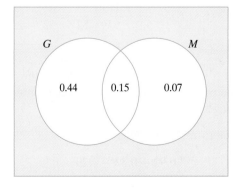

and

$$P(G \cup M) = 0.44 + 0.15 + 0.07 = 0.66$$

where we were able to add the respective probabilities because they pertain to mutually exclusive events (to regions of the Venn diagram that do not overlap).

Had we erroneously used the special addition rule, the third of the four basic rules, to calculate $P(G \cup M)$, we would have obtained $P(G) + P(M) = 0.59 + 0.22 = 0.81$, which exceeds the correct value by 0.15. This error results from including $P(G \cap M) = 0.15$ twice, once in $P(G) = 0.59$ and once in $P(M) = 0.22$, and we could correct for it by subtracting 0.15 from 0.81. Symbolically, we could thus write

$$P(G \cup M) = P(G) + P(M) - P(G \cap M)$$

$$= 0.59 + 0.22 - 0.15 = 0.66$$

and this agrees, as it should, with the result obtained before. Since the argument that we used in this example holds for any two events A and B, we can now state the following **general addition rule**, which applies regardless of whether A and B are mutually exclusive events:

GENERAL
ADDITION RULE

$$P(A \cup B) = P(A) + P(B) - P(A \cap B)$$

When A and B are mutually exclusive, note that $P(A \cap B) = 0$ and the general addition rule reduces to the special addition rule, the third of our basic rules.

EXAMPLE 5.19   If one card is drawn from an ordinary deck of 52 playing cards, what is the probability that it will be either a club or a face card (king, queen, or jack)?

Solution   If C denotes drawing a club and F denotes drawing a face card, then

$$P(C) = \frac{13}{52}, \quad P(F) = \frac{12}{52}, \quad \text{and} \quad P(C \cap F) = \frac{3}{52}$$

so that

$$P(C \cup F) = \frac{13}{52} + \frac{12}{52} - \frac{3}{52} = \frac{22}{52} = \frac{11}{26}$$

EXAMPLE 5.20   If the probabilities are, respectively, 0.92, 0.33, and 0.29 that a person vacationing in Washington, D.C., will visit the Capitol building, the Smithsonian Institution, or both, what is the probability that a person vacationing there will visit at least one of these buildings?

Solution   Substituting into the formula, we get

$$0.92 + 0.33 - 0.29 = 0.96$$

Note that if we had incorrectly used the special addition rule (for mutually exclusive events), we would have obtained the impossible answer $0.92 + 0.33 = 1.25$. ■

The general addition rule can be generalized further so that it will apply to more than two events, but we shall not go into that in this book.

EXERCISES

Exercises 5.40, 5.45, 5.49, and 5.52 are practice exercises; their complete solutions are given on pages 188–189.

✓ **5.40**   A police department needs new tires for its patrol cars, and the probabilities are, respectively, $0.22, 0.17, 0.21$, and 0.09 that it will buy Uniroyal, Goodyear, Goodrich, or Michelin tires. What is the probability that it will buy one of these four kinds of tires?

5.41   If the probabilities that a personnel manager will evaluate an employee as superior, above average, or average are, respectively, 0.08, 0.23, and 0.45, what is the probability that an employee will be evaluated as below average (assuming that this is the only remaining category)?

5.42   A professor of English with very large classes anticipates that the percentages of students who will receive grades of A, B, C, D, or F are, respectively, 6, 22, 44, 22, and 6.
(a)   What is the probability that a student will get a grade of C or higher?
(b)   What is the probability that a student will get a grade of C or lower?
(c)   What is the probability that a student will get a grade lower than A but higher than F?

5.43   The probabilities that a TV station will receive 0, 1, 2, 3, ..., 7, or at least 8 complaints after showing a controversial program are, respectively, $0.02, 0.04, 0.07, 0.12, 0.15, 0.19, 0.18, 0.14$, and 0.09. What are the probabilities that after showing such a program the station will receive
(a)   at least 5 complaints;
(b)   at most 3 complaints;
(c)   anywhere from 2 to 4 complaints?

5.44   A newspaper publisher asks its readers to evaluate the level of difficulty of its regularly published crossword puzzles, obtaining the following probabilities.

Easy	0.08
Moderately difficult	0.47
Difficult	0.32
Very difficult	0.13

Find the probabilities that the readers rate the crossword puzzles
(a)   easy or moderately difficult;
(b)   not easy, but at least moderately difficult;
(c)   difficult or very difficult.

✓ **5.45**   Figure 5.11 pertains to the number of persons who are invited to a conference and the number of persons who attend. If each of the 35 points of the sample space has the probability $\frac{1}{35}$, what are the probabilities that
(a)   all the persons who are invited will attend;
(b)   at most two persons will attend;
(c)   at least seven persons will be invited;
(d)   one or two of the persons who are invited will not attend?

Figure 5.11
Sample space for Exercise 5.45.

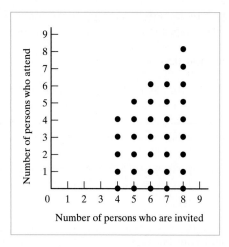

Figure 5.12
Sample space for Exercise 5.46.

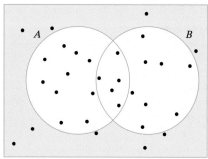

5.46 If each point of the sample space of Figure 5.12 represents an outcome having the probability $\frac{1}{32}$, find

(a) $P(A)$; (e) $P(A' \cap B)$;
(b) $P(B)$; (f) $P(A' \cap B')$;
(c) $P(A \cap B)$; (g) $P(A' \cup B)$;
(d) $P(A \cup B)$; (h) $P(A \cup B')$.

5.47 With reference to the data of Figure 5.8, find the probabilities that

(a) only one of the two cars will be sold;

(b) either salesperson will sell both cars;

(c) the two salespersons will sell equally many cars.

5.48 If H stands for heads and T for tails, the 16 possible outcomes for four flips of a coin are

HHHH   HHHT   HHTH   HTHH   THHH   HHTT
HTHT   HTTH   THHT   THTH   TTHH   HTTT
THTT   TTHT   TTTH   TTTT

Assuming that these 16 possibilities are all equally likely, what are the probabilities of getting 0, 1, 2, 3, or 4 heads in four flips of a balanced coin?

✓ **5.49** An artist who has entered a large oil painting and a small oil painting in a show feels that the probabilities are, respectively, 0.15, 0.18, and 0.11 that she will sell the large oil painting, the small one, or both. What is the probability that she will sell

(a) either or both of the two paintings;

(b) neither of the two paintings?

5.50   The probabilities that a homeowner will repair the roof, paint the house, or both are, respectively, 0.90, 0.58, and 0.50. What is the probability that the homeowner will fix the roof or paint the house?

5.51   The probabilities that a jeweler will sell a large diamond ring, a large sapphire ring, or both on a given day are, respectively, 0.06, 0.08, and 0.03. What is the probability that at least one of the rings will be sold by the jeweler on that day?

✓ **5.52**   Given $P(K) = 0.45$, $P(L) = 0.27$, and $P(K \cap L) = 0.13$, draw a Venn diagram, fill in the probabilities associated with the various regions, and thus determine

(a) $P(K \cap L')$;   (d) $P(K' \cup L)$;

(b) $P(K' \cap L)$;   (e) $P(K' \cap L')$;

(c) $P(K \cup L)$;   (f) $P(K' \cup L')$.

5.53   Among the 64 doctors on the staff of a hospital, 58 carry malpractice insurance, 33 are surgeons, and 31 of the surgeons carry malpractice insurance. If one of these doctors is chosen by lot to represent the hospital staff at an American Medical Association (AMA) convention (that is, each doctor has a probability of $\frac{1}{64}$ of being selected), what is the probability that the one chosen is not a surgeon and does not carry malpractice insurance?

5.54   Given $P(A) = 0.59$, $P(B) = 0.46$, and $P(A \cap B) = 0.28$, draw a Venn diagram, fill in the probabilities associated with the various regions, and thus determine

(a) $P(A' \cap B)$;   (c) $P(A \cup B)$;

(b) $P(A \cap B')$;   (d) $P(A' \cap B')$.

5.6   CONDITIONAL PROBABILITY

Difficulties can easily arise when we ask for the probability of an event without specifying the sample space. For instance, the probability that a person will find a bargain depends on where she shops, the chances that a person will get a traffic ticket depend on how well and how carefully he drives, and the odds that a woman will get promoted depend on how well she is trained or how hard she is willing to work.

Since the choice of the sample space (the set of all possibilities under consideration) is rarely self-evident, it helps to use the symbol $P(A|S)$ to denote the **conditional probability** of event A relative to the sample space S or, as we often say, the probability of A given S. In many cases, S will be smaller than the original sample space that we started with. The symbol $P(A|S)$ makes it explicit that we are referring to a particular sample space S. It is preferable to the abbreviated notation $P(A)$ unless the tacit choice of S is clearly understood.

To elaborate on the idea of a conditional probability, suppose that a consumer research organization has studied the service provided by the 150 appliance repair persons in a certain city. Its findings are summarized in the following table:

	Good service	Poor service	Total
Factory trained	48	16	64
Not factory trained	24	62	86
Total	72	78	150

If one of these repair persons is randomly selected (that is, each has the probability $\frac{1}{150}$ of being selected), we find that the probability of choosing one who provides good service is

$$P(G) = \frac{72}{150} = 0.48$$

and the probability of choosing one who is factory trained is

$$P(F) = \frac{64}{150} \approx 0.43$$

The probability of choosing one who provides good service and is factory trained is

$$P(G \cap F) = \frac{48}{150} = 0.32$$

All these probabilities were calculated by means of the formula $\frac{s}{n}$ for equally likely possibilities.

Since the first of these probabilities is particularly disconcerting—there is less than a 50–50 chance of choosing a repair person who provides good service—let us see what will happen if we limit the choice to those who are factory trained. Looking at the reduced sample space represented by the first row of the table, we get

$$P(G|F) = \frac{48}{64} = 0.75$$

and this is quite an improvement over $P(G) = 0.48$, as might have been expected. Note that this conditional probability, 0.75, can also be written as

$$P(G|F) = \frac{\dfrac{48}{150}}{\dfrac{64}{150}} = \frac{P(G \cap F)}{P(F)}$$

which is the ratio of the probability of choosing a repair person who provides good service and is factory trained to the probability of choosing a repair person who is factory trained.

Generalizing from this example, let us now make the following definition of conditional probability, which applies to any two events A and B belonging to a given sample space S:

DEFINITION OF
CONDITIONAL
PROBABILITY

> *If $P(B)$ is not equal to zero, then the conditional probability of A relative to B, namely, the probability of A given B, is*
>
> $$P(A|B) = \frac{P(A \cap B)}{P(B)}$$

EXAMPLE 5.21   A computer manufacturer feels that the probability is 0.75 that the computer chips (which are essential to fill an order for computers) will arrive on time, and the probability is 0.60 that the order for computers will be filled on time. If C is the event that the chips will arrive on time, and F is the event that the order for computers will be filled on time, then $P(C) = 0.75$ and $P(F \cap C) = 0.60$. What is the probability that the order will be filled on time given that the chips arrive on time?

Solution   If C and F denote the events that the chips will arrive on time and that the order will be filled on time, then $P(C) = 0.75$ and $P(F \cap C) = 0.60$. Substituting into the formula for conditional probabilities, we get

$$P(F|C) = \frac{P(F \cap C)}{P(C)} = \frac{0.60}{0.75} = 0.80 \qquad \blacksquare$$

The formula for $P(A|B)$ can be used in problems in which the possibilities are all equally likely, but this is not a requirement for its use. The only restriction is that $P(B)$ must not equal zero. Division by zero is not permissible.

EXAMPLE 5.22   The records of annual meetings of a certain corporation show that the probability is 0.30 that its stockholders will attend the annual meeting in person (instead of, say, submitting a proxy). The probability that a stockholder who attends the meeting will vote at the meeting is 0.25. What is the probability that a stockholder will attend the meeting in person and will vote at the meeting?

Solution   If A is the event that a stockholder will attend the annual meeting and V is the event that the stockholder will vote at the meeting, then $P(A) = 0.30$ and $P(A \cap V) = 0.25$. The formula for conditional probability yields

$$P(V|A) = \frac{P(A \cap V)}{P(A)} = \frac{0.25}{0.30} \approx 0.83 \qquad \blacksquare$$

EXAMPLE 5.23   A refiner of gasoline feels that the probability is 0.80 that a tanker carrying crude oil needed to fill an order for gasoline will arrive on time, and the probability is 0.60 that the crude oil will arrive on time and the order for gasoline will be filled on time. What is the probability that the order for gasoline will be filled on time given that the tanker carrying the crude oil arrives on time?

Solution   If C denotes the event that the tanker with crude oil arrives on time, and G denotes the event that the order for gasoline will be filled on time, then $P(C) = 0.80$ and $P(G \cap C) = 0.60$. Substituting into the formula for conditional probability, we get

$$P(G|C) = \frac{P(G \cap C)}{P(C)} = \frac{0.60}{0.80} = 0.75 \qquad \blacksquare$$

5.7   INDEPENDENT EVENTS

To introduce another concept that is important in the study of probability, let us consider the following problem:

EXAMPLE 5.24   The probabilities that a student will get passing grades in algebra, in literature, or in both subjects are, respectively, $P(A) = 0.75$, $P(L) = 0.84$, and $P(A \cap L) = 0.63$. What is the probability that the student will get a passing grade in algebra given that he or she gets a passing grade in literature?

Solution   Substituting into the formula that defines conditional probability, we get

$$P(A|L) = \frac{P(A \cap L)}{P(L)} = \frac{0.63}{0.84} = 0.75 \qquad \blacksquare$$

What is special and interesting about this result is that $P(A|L) = P(A) = 0.75$; that is, the probability of event A is the same regardless of whether or not event L has occurred (occurs, or will occur).

In general, if $P(A|B) = P(A)$, we say that event A is **independent** of event B; that is,

> *Event A is independent of event B if the probability of event A is not affected by the occurrence or nonoccurrence of event B.*

Since it can be shown that event B is independent of event A whenever event A is independent of event B, it is customary to say simply that *events A and B are independent* whenever one is independent of the other (see Exercise 5.73). If two events A and B are not independent, we say that they are **dependent events**.

5.8   MULTIPLICATION RULES

So far we have used the formula

$$P(A|B) = \frac{P(A \cap B)}{P(B)}$$

only to calculate conditional probabilities; but if we multiply both sides of the equation by $P(B)$, we get the following formula, called the **general multiplication rule**, which enables us to calculate the probability that two events will both occur.

GENERAL MULTIPLICATION RULE

$$P(A \cap B) = P(B) \cdot P(A|B)$$

In words, the probability that two events will both occur is the product of the probability that one event will occur and the conditional probability that the other event will occur given that the first event has occurred (occurs, or will occur). As it does not matter which event is denoted by A and which by B, the preceding formula can also be written as

GENERAL MULTIPLICATION RULE (ALTERNATIVE FORM)

$$P(A \cap B) = P(A) \cdot P(B|A)$$

EXAMPLE 5.25 A jury consists of nine persons who are native born and three persons who are foreign born. If two of the jurors are randomly picked for an interview, what is the probability that they will both be foreign born?

Solution Let A be the event that the first juror is foreign born and B the event that the second juror is foreign born. If we assume equal probabilities for each choice (which is, in fact, what we mean by the selection being random), the probability that the first juror picked will be foreign born is $P(A) = \dfrac{3}{12}$. Then, if the first juror picked is foreign born, the probability that the second juror will also be foreign born is $P(B|A) = \dfrac{2}{11}$. Hence, the probability of getting two jurors who are both foreign born is

$$P(A \cap B) = P(A) \cdot P(B|A) = \frac{3}{12} \cdot \frac{2}{11} = \frac{1}{22}$$ ■

When A and B are independent, we can substitute $P(A)$ for $P(A|B)$ in the first of the two formulas for $P(A \cap B)$, or $P(B)$ for $P(B|A)$ in the second, and we obtain the **special multiplication rule**:

SPECIAL MULTIPLICATION RULE (INDEPENDENT EVENTS)

If A and B are independent events, then

$$P(A \cap B) = P(A) \cdot P(B)$$

In words, the probability that two independent events will both occur is simply the product of their respective probabilities. This rule is sometimes used as the definition of independence; in any case, it may be used to check whether two given events are independent.

EXAMPLE 5.26 What is the probability of getting two heads in two honest flips of a balanced coin?

Solution Since *honest* and *balanced* mean that the outcomes of the two flips are independent and that the probability of getting heads is $\dfrac{1}{2}$ for each flip of the coin, the answer is $\dfrac{1}{2} \cdot \dfrac{1}{2} = \dfrac{1}{4}$. ■

EXAMPLE 5.27 If $P(C) = 0.65$, $P(D) = 0.40$, and $P(C \cap D) = 0.26$, are the events C and D independent?

Solution Since

$$P(C) \cdot P(D) = (0.65)(0.40) = 0.26 = P(C \cap D)$$

the two events are independent. ■

EXAMPLE 5.28 Select one card randomly from a standard deck of 52. Let R be the event that the card is red, and let Q be the event that the card is a queen. Are the events R and Q independent?

Solution The event $R \cap Q$ is the selection of a red queen. Since there are two red queens in the deck,

$$P(R \cap Q) = \frac{2}{52}$$

Noting also that

$$P(R) = \frac{26}{52} \text{ and } P(Q) = \frac{4}{52}$$

we find that $P(R) \cdot P(Q) = \dfrac{26}{52} \cdot \dfrac{4}{52} = \dfrac{2}{52} = P(R \cap Q)$, and therefore the two events are independent. ∎

The three previous examples illustrate common ways in which independence is asserted. In the first example, the two flips are physically unrelated, so that we simply claim independence and apply the formula. In the second example, we find $P(C \cap D) = P(C) \cdot P(D)$; in this case, the numbers allow us to say that C and D are independent. In the third example, the events R and Q are physically related (since they describe the same random selection from the deck), but the numbers still allow us to say that R and Q are independent.

EXAMPLE 5.29 What is the probability of getting two aces in a row when 2 cards are drawn from an ordinary deck of 52 playing cards, if

(a) the first card is replaced before the second card is drawn;

(b) the first card is not replaced before the second card is drawn?

Solution **(a)** Since there are four aces among the 52 cards, we get

$$\frac{4}{52} \cdot \frac{4}{52} = \frac{1}{169} \approx 0.0059$$

(b) Since there are only three aces among the 51 cards that remain if one ace has been removed from the deck, the answer is

$$\frac{4}{52} \cdot \frac{3}{51} = \frac{1}{221} \approx 0.0045$$ ∎

Observe the difference in the calculations and the results of the two parts of the preceding example. They serve to illustrate the distinction between **sampling with replacement** and **sampling without replacement**. In most statistical investigations, sampling is done without replacement.

The special multiplication rule can readily be generalized so that it applies to three or more independent events; again, we simply multiply all their probabilities.

EXAMPLE 5.30 If the probability is 0.25 that a person will name red as his or her favorite color, what is the probability that three totally unrelated persons will all name red as their favorite color?

Solution   Assuming independence, we get

$$(0.25)(0.25)(0.25) \approx 0.0156$$

When three or more events are not independent, the multiplication rule becomes more complicated: we form the product of the probability that one of the events will occur; the conditional probability that a second event will occur given that the first event has occurred; the conditional probability that a third event will occur, given that the first two events have occurred; and so on.

EXAMPLE 5.31   The great composer Ludwig Van Beethoven wrote 9 symphonies and 32 piano concertos. If an orchestra conductor randomly selects two pieces of music, without replacement, from a collection of these 41 pieces, what is the probability that

(a) the first piece selected is a symphony, and the second piece selected is a piano concerto;

(b) both pieces are symphonies;

(c) both pieces are piano concertos?

Solution   **(a)** $\dfrac{9}{41} \cdot \dfrac{32}{40} \approx 0.18$

(b) $\dfrac{9}{41} \cdot \dfrac{8}{40} \approx 0.04$

(c) $\dfrac{32}{41} \cdot \dfrac{31}{40} \approx 0.60$

Exercises 5.55, 5.56, 5.61, 5.62, and 5.80 are practice exercises; their complete solutions are given on pages 189 and 190.

✓ **5.55**   If F is the event that a student will get financial aid, J is the event that the student will find a part-time job, and G is the event that the student will graduate, express symbolically the probabilities that

(a) a student who gets financial aid will graduate;

(b) a student who gets no financial aid will find a part-time job;

(c) a student who gets no financial aid will neither find a part-time job nor graduate;

(d) a student who gets financial aid and finds a part-time job will not graduate.

✓ **5.56**   If E is the event that a female applicant for a sales position has had prior experience, C is the event that she owns a car, and G is the event that she is a college graduate, state in words what probabilities are expressed by

(a) $P(C|G)$;   (d) $P(G'|C')$;

(b) $P(E|C')$;   (e) $P(C|E \cup G)$;

(c) $P(C'|E)$;   (f) $P(E \cap C'|G)$.

5.57   If X and Y are independent events and $P(X) = 0.25$ and $P(Y) = 0.50$, find

(a) $P(X|Y)$;

(b) $P(X \cap Y)$;

(c) $P(X \cup Y)$;

(d) $P(X' \cap Y')$.

EXERCISES

5.58   If B is the event that a baseball player is well coached, and H is the event that the player is a good hitter, express symbolically the probabilities that

(a) a baseball player who is well coached will be a good hitter;

(b) a baseball player who is a good hitter is not well coached;

(c) a baseball player who is not well coached will not be a good hitter.

5.59   If $P(A) = 0.70$, $P(B) = 0.40$, and $P(A \cap B) = 0.25$, are events A and B independent?

5.60   The probability that a security guard will be hired at a shopping mall is 0.80, and the probability that the security guard will be hired and will decrease the number of thefts is 0.75. What is the probability that, if the security guard is hired, the number of thefts will decrease?

✓ **5.61**   Among the 400 inmates of a prison, some are first offenders, some are repeat offenders, some are serving sentences of less than five years, and some are serving longer sentences. The exact breakdown is

	Sentences less than five years	Longer sentences
First offenders	120	40
Repeat offenders	80	160

One of the inmates is to be selected at random to be interviewed about prison conditions. Let R be the event that he is a repeat offender, and let L be the event that he is serving a longer sentence. Determine each of the following probabilities directly from the entries and the row and column totals of the table:

(a) $P(R)$;   (d) $P(R' \cap L)$;

(b) $P(L)$;   (e) $P(L|R)$;

(c) $P(L \cap R)$;   (f) $P(R'|L)$.

✓ **5.62**   Use the results of Exercise 5.61 to verify that

(a) $P(L|R) = \dfrac{P(L \cap R)}{P(R)}$;

(b) $P(R'|L) = \dfrac{P(R' \cap L)}{P(L)}$.

5.63   In some of the delinquent charge accounts of a store, the amount owed is less than \$100, in some it is \$100 or more, some have been delinquent for less than a month, and some have been delinquent for a month or more. The exact breakdown is

	Less than a month	A month or more
Less than \$100	168	72
\$100 or more	57	43

One of these delinquent accounts is to be selected at random for a new credit check. Let L be the event that the amount owed is less than \$100, and let M be the event that the account has been delinquent for a month or more. Determine each of the following probabilities directly from the entries and the row and column totals of the table:

(a) $P(L)$;   (d) $P(M' \cap L)$;
(b) $P(M)$;   (e) $P(L|M')$;
(c) $P(L \cap M')$;   (f) $P(M'|L')$.

5.64 Use the results of Exercise 5.63 to verify that

(a) $P(L|M') = \dfrac{P(L \cap M')}{P(M')}$;

(b) $P(M'|L') = \dfrac{P(M' \cap L')}{P(L')}$.

5.65 In the following table, 60 college students are classified according to their class standing and also according to their favorite pizza topping:

	Anchovies	Onions	Mushrooms	Hamburger	
Freshman	7	6	7	3	23
Sophomore	1	9	0	9	19
Junior	3	2	5	8	18
	11	17	12	20	60

If one of these students is selected at random, if F, S, and J denote the three classes, and if A, O, M, and H denote the four pizza toppings, find

(a) $P(M \cup J)$;   (d) $P(F'|A)$;
(b) $P(H|F)$;   (e) $P(M \cup H|J')$;
(c) $P(O \cap S)$;   (f) $P(J|A \cup M)$.

5.66 With reference to Exercise 5.65, find the probabilities that the student chosen will be

(a) a freshman whose favorite pizza topping is mushrooms;

(b) an anchovy pizza eater given that he or she is a junior;

(c) a sophomore given that he or she is not a junior.

5.67 If G is the event that an antique sword is in good condition, V is the event that it is very rare, and R is the event that it is a reproduction, state in words what probability is expressed by

(a) $P(G|V)$;   (d) $P(G \cap R|V)$;
(b) $P(G|R')$;   (e) $P(V|G')$;
(c) $P(R|V')$;   (f) $P(G|V \cap R)$.

5.68 According to the mortality tables of a life insurance company, the probability that policyholder Alpha will be alive 10 years from now is 0.50, and the probability that policyholder Beta will be alive 10 years from now is 0.60. What is the probability that both policyholders, Alpha and Beta, will be alive 10 years from now? Assume that these are independent events. (If both policyholders worked together in a very hazardous occupation, their mortalities might not be independent.)

5.69 Assume that the following are all independent events, and calculate their probabilities.

(a) The probability of getting five heads in a row with a balanced coin.

(b) The probability of drawing three clubs in a row (with replacement) from an ordinary deck of 52 playing cards.

(c) The probability of drawing three clubs in a row (without replacement) from an ordinary deck of playing cards.

(d) The probability that a shooter firing at a target will, in two consecutive shots, hit the target once and then miss the target. Assume that the probability that the shooter hits the target on any one try is 0.80.

5.70 Some special cards are prepared for an unusual game. There are four varieties: cards with red letter X, cards with green letter X, cards with red letter Y, and cards with green letter Y. Deck A has 40 of these cards, and the following table gives the contents of this deck.

Deck A

	Red	Green
X	5	14
Y	5	16

If a card is selected at random from this deck, it follows that

$$P(X|\text{Red}, \text{Deck}A) = \frac{5}{10} = 0.50$$

since 5 of the 10 red cards contain the letter X. Also,

$$P(X|\text{Green}, \text{Deck}A) = \frac{14}{30} \approx 0.47$$

Deck B also has 40 of these cards, and the following table gives the makeup of Deck B.

Deck A

	Red	Green
X	5	2
Y	20	13

(a) Show that
$$P(X|\text{Red, Deck } A) > P(X|\text{Green, Deck } A)$$

(b) Show that
$$P(X|\text{Red, Deck } B) > P(X|\text{Green, Deck } B)$$

(c) Suppose that the two decks are combined to make a single deck of 80 cards. How many cards of each of the four varieties are in this deck?

(d) Suppose that a single card is to be selected randomly from this deck of 80 cards. Is it true that

$$P(X|\text{Red, Combined Deck}) > P(X|\text{Green, Combined Deck})?$$

(*Note*: Conditional probabilities behave in peculiar ways. This particular problem is an example of **Simpson's paradox**.)

5.71 Abel and Baker are quarterbacks for a football team. The coach uses both players during each game. In the first game of the season, Abel got to throw 20 passes, and

11 of them were complete. Also in the first game, Baker threw 5 passes, completing 3 of them. In the second game, Abel threw 5 passes, completing only 1. Baker threw 16 passes, completing 6.

(a) Show that Baker's proportion of pass completions was better than Abel's proportion of pass completions in the first game.

(b) Show that Baker's proportion of pass completions was better than Abel's proportion of pass completions in the second game.

(c) A newspaper combined the data from the two games to get the season-to-date data. According to the combined information, which quarterback has the higher proportion of pass completions?

5.72 If the probability that a family will buy a vacation home in Miami, Malibu, or Newport is 0.25, 0.10, and 0.35, what is the probability that the family will consummate one of these transactions?

5.73 In Example 5.24 on page 174 concerning a student's grades in algebra and literature, we had

$$P(A) = 0.75$$

$$P(L) = 0.84$$

$$P(A \cap L) = 0.63$$

$$P(A|L) = 0.75$$

Show that event L is independent of event A; that is, $P(L|A) = P(L) = 0.84$.

5.74 If $P(X) = 0.65$, $P(Y) = 0.40$, and $P(X \cap Y) = 0.26$, check whether events X and Y are independent.

5.75 (a) For two rolls of a balanced die, find the probability of getting two sixes.

(b) For a single roll of two balanced dice, find the probability of getting two sixes.

5.76 What is the probability of *not* rolling a six

(a) in a single roll of a balanced die;

(b) in two rolls of a balanced die;

(c) in three rolls of a balanced die?

5.77 A fifth-grade class consists of 16 boys and 14 girls. If one pupil is chosen each week by lot to assist the teacher, what are the probabilities that a boy will be chosen two weeks in a row if

(a) the same pupil can serve two weeks in a row;

(b) the same pupil cannot serve two weeks in a row?

5.78 If the probabilities are 0.10, 0.10, 0.20, 0.30, 0.20, and 0.10 that a consumer testing service will rate a new television set as very poor, poor, fair, good, very good, or excellent, what are the probabilities that it will rate the new television set

(a) very poor or poor;

(b) good, very good, or excellent;

(c) poor, fair, good, or very good?

5.79 If the probabilities are 0.35, 0.25, and 0.15 that a tourist in Nevada will visit Reno, Las Vegas, or both, what is the probability that the tourist will visit at least one of these cities?

✓ **5.80** If 5 of a company's 15 delivery trucks do not meet emission standards, and 4 of them are randomly chosen for inspection, what is the probability that all of them will meet emission standards?

5.81   A shipping carton contains 24 boxes of cookies, of which 10 boxes contain less than the weight specified on the label, and the remaining boxes meet the specified weight. If an inspector checks 3 of the boxes, what is the probability that
   (a) all three are less than the specified weight;
   (b) none are less than the specified weight?

5.82   A jar contains 15 numbered balls that are identical except for the numbers. These numbers are 1 through 15. Suppose that you remove 4 balls randomly and without replacement.
   (a) What is the probability that all 4 balls will have even numbers?
   (b) What is the probability that the first 2 balls will have even numbers and the last 2 balls will have odd numbers?
   (c) What is the probability that the first ball will have an even number and the last 3 balls will have odd numbers?

5.83   In a certain city the probability that it will snow on a December day is 0.40, the probability that a December day on which it snows will be followed by a day on which it snows is 0.70, and the probability that a December day on which it does not snow will be followed by a day on which it snows is 0.20. What are the probabilities that on three consecutive December days it will
   (a) snow on each day;
   (b) not snow on the first day and then snow on the next two days?

5.84   With reference to Exercise 5.83, what is the probability that it will snow, snow, not snow, and snow on four consecutive December days?

5.9   BAYES'S THEOREM

Although the symbols $P(A|B)$ and $P(B|A)$ may look very much alike, there is a great difference between the probabilities that they represent. For instance, we previously calculated the probability $P(G|F)$ that a factory-trained appliance repair person will provide good service, but what do we mean when we write $P(F|G)$? This is the probability that a repair person who provides good service will have been factory trained. Thus, we turned things around—cause becomes effect and effect becomes cause. To give another example, suppose that a person applying for a loan has his credit checked, and R denotes the event that the person is a bad risk, while J denotes the event that the person is judged a bad risk. Then $P(J|R)$ is the probability that a person who is a bad risk will be judged a bad risk, and $P(R|J)$ is the probability that a person who is judged a bad risk actually is a bad risk.

Since many problems involve such pairs of conditional probabilities, let us try to find a formula that expresses $P(B|A)$ in terms of $P(A|B)$ for any two events A and B. Fortunately, we do not have to look very far; all we have to do is equate the expressions for $P(A \cap B)$ in the two forms of the general multiplication rule on pages 174 and 175, and we get

$$P(A) \cdot P(B|A) = P(B) \cdot P(A|B)$$

and hence

$$P(B|A) = \frac{P(B) \cdot P(A|B)}{P(A)}$$

after dividing by $P(A)$.

EXAMPLE 5.32   In a state where cars have to be tested for the emission of pollutants, 25% of all cars emit excessive amounts of pollutants. When tested, 99% of all cars that emit excessive amounts of pollutants will fail, but 17% of the cars that do not emit excessive amounts of pollutants will also fail. What is the probability that a car that fails the test actually emits excessive amounts of pollutants?

Solution   Letting A denote the event that a car fails the test and B the event that it emits excessive amounts of pollutants, we can translate the given percentages into probabilities and write

$$P(B) = 0.25, \ P(A|B) = 0.99, \ \text{and} \ P(A|B') = 0.17$$

Before we can calculate $P(B|A)$ by means of the preceding formula, we will first have to determine $P(A)$, and to this end let us look at the tree diagram of Figure 5.13. Here A is reached either along the branch that passes through B or along the branch that passes through B', and the probabilities of this happening are, respectively,

$$(0.25)(0.99) = 0.2475 \ \text{and} \ (0.75)(0.17) = 0.1275$$

Since the alternatives represented by the two branches are mutually exclusive, we find that

$$P(A) = 0.2475 + 0.1275 = 0.3750$$

and substitution into the formula for $P(B|A)$ given previously yields

$$P(B|A) = \frac{P(B) \cdot P(A|B)}{P(A)} = \frac{(0.25)(0.99)}{0.3750} = 0.66$$

This is the probability that a car that fails the test actually emits excessive amounts of pollutants. ■

With reference to the tree diagram of Figure 5.13, we can say that 0.2475 is the probability that event A was reached via the upper branch of the tree, and we showed that $P(B|A)$ is given by the ratio of the probability associated with that branch to the sum of the probabilities associated with both branches of the tree. This argument can be generalized to the case where there are more than two possible causes, namely, more than two branches leading to an event A. With reference to Figure 5.14, we can say that $P(B_i|A)$ is the probability that event

Figure 5.13
Tree diagram for emission testing example.

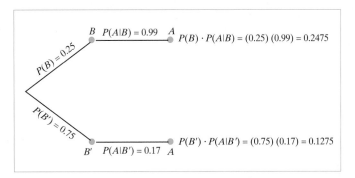

Figure 5.14
Tree diagram for Bayes's
theorem.

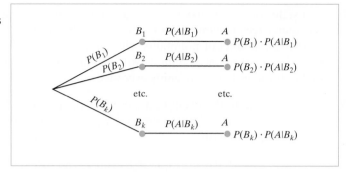

A was reached via the ith branch of the tree (for $i = 1, 2, \ldots$, or k), and it can be shown that its value is given by the ratio of the probability associated with the ith branch to the sum of the probabilities associated with all the branches. Symbolically, this result, called **Bayes's theorem**, is given by

BAYES'S
THEOREM

$$P(B_i|A) = \frac{P(B_i) \cdot P(A|B_i)}{P(B_1) \cdot P(A|B_1) + P(B_2) \cdot P(A|B_2) + \cdots + P(B_k) \cdot P(A|B_k)}$$

for $i = 1, 2, \ldots$, or k.

EXAMPLE 5.33   In a cannery, assembly lines I, II, and III account, respectively, for 37%, 42%, and 21% of the total output. If 0.6% of the cans from assembly line I are improperly sealed, while the corresponding percentages for assembly lines II and III are 0.4% and 1.2%, what is the probability that an improperly sealed can (discovered at the final inspection of outgoing products) came from assembly line III?

Solution   If we let A denote the event that a can is improperly sealed and B_1, B_2, and B_3 denote the events that a can comes from assembly lines I, II, or III, we can translate the given percentages into probabilities and write

$$P(B_1) = 0.37,\ P(B_2) = 0.42,\ P(B_3) = 0.21$$

$$P(A|B_1) = 0.006$$

Figure 5.15
Tree diagram for can-
nery example.

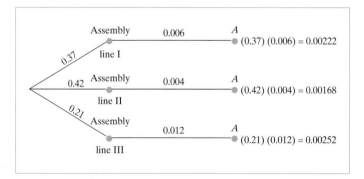

$$P(A|B_2) = 0.004$$

$$P(A|B_3) = 0.012$$

Then, with reference to the tree diagram of Figure 5.15, we find that the probabilities associated with the three branches are, respectively,

$$(0.37)(0.006) = 0.00222, \ (0.42)(0.004) = 0.00168,$$

and

$$(0.21)(0.012) = 0.00252$$

Thus, the required probability is

$$P(B_3|A) = \frac{0.00252}{0.00222 + 0.00168 + 0.00252} \approx 0.393$$

If we had wanted to work this problem without referring to a tree diagram, direct substitution into the formula for Bayes's theorem would have yielded

$$P(B_3|A) = \frac{(0.21)(0.012)}{(0.37)(0.006) + (0.42)(0.004) + (0.21)(0.012)}$$

$$\approx 0.393$$

and all the details of the calculations are, of course, the same.   ■

As can be seen from our two examples, Bayes's theorem is a relatively simple mathematical rule. There can be no question about its validity, but criticism has frequently been raised about its applicability. This is because it involves a *backward* or *inverse* sort of reasoning, that is, reasoning from effect to cause. In our examples we used Bayes's theorem to determine the probability that a car that fails the emission test actually emits excessive amounts of pollutants, and the probability that an improperly sealed can was "caused" by assembly line III. It is precisely this aspect of Bayes's theorem that makes it play an important role in statistical inference, where our reasoning goes from sample data that is observed to the populations from which it came. Discussion of such inferences, appropriately called **Bayesian inferences**, may be found in more advanced texts.

EXERCISES

Exercise 5.85 is a practice exercise; its complete solution is given on page 190.

✓ **5.85**   Suppose that 20% of the residents of an equatorial city have a certain tropical disease. Among afflicted individuals, failure in a medical diagnostic test accurately discloses the presence of the disease 95% of the time; but among persons who do not have this disease, the test falsely indicates its presence 10% of the time. Let A denote the event that a person fails the test, and let B denote the event that a person has the tropical disease; thus $P(A|B)$ denotes the probability that a person who has the disease will fail the test. Calculate

$$P(B|A) = \frac{P(B) \cdot P(A|B)}{P(A)}$$

to determine the probability that a person who fails the test actually has the disease.

5.86   Alice, Bill, and Charley (who can be labeled A, B, and C) are employees of a fast-food restaurant, taking orders from customers at a counter. From prior observation, we know that A makes mistakes 1 time in 100, B makes mistakes 3 times in 100, and C makes mistakes 5 times in 100. A, B, and C take 50%, 30%, and 20% of the orders. If the manager finds that a mistake, M, has been made in a certain order, what are the probabilities that the order was filled by A, B, or C?

5.87   In the billing department of a local city, Arthur, Beatrice, and Carla prepare and mail out 50%, 30%, and 20% of the real-estate tax bills, respectively, to property owners. The bills are verified by the treasurer before mailing. If 0.5% of the tax bills prepared by Arthur, 0.3% of the tax bills prepared by Beatrice, and 1.0% of the tax bills prepared by Carla are inaccurate, what are the probabilities that an inaccurate statement detected by the treasurer at final verification was prepared by

(a) Arthur;

(b) Beatrice;

(c) Carla?

5.88   An analysis of the causes of employee lateness in arrival at a large office shows that the probability that an employee will be late because of an automobile breakdown is 0.03. The probability that the office manager correctly attributes a late arrival to an automobile breakdown is 0.80, and the probability that the office manager incorrectly attributes a late arrival to an automobile breakdown is 0.05. What is the probability that a late arrival attributed to an automobile breakdown by the office manager is actually due to an automobile breakdown?

5.89   A driver's license examiner knows that 75% of all applicants have attended a driving school. If an applicant has attended a driving school, the probability is 0.85 that he or she will pass the license examination, and if an applicant has not attended a driving school, the probability is 0.60 that he or she will pass the examination. If an applicant passes the license examination, what is the probability that he or she has attended a driving school?

5.90   In Example 5.33 on pages 184–185, we used Bayes's theorem to show that the probability that an improperly sealed can came from assembly line III was

$$P(B_3|A) = \frac{0.00252}{0.00222 + 0.00168 + 0.00252} = \frac{0.00252}{0.00642} \approx 0.393$$

Here A denotes the event that a can is improperly sealed, and B_3 denotes the event that it came from assembly line III. Making use of the results that $P(B_1) \cdot P(A|B_1) = 0.00222$, $P(B_2) \cdot P(A|B_2) = 0.00168$, and $P(B_3) \cdot P(A|B_3) = 0.00252$, determine the probabilities that

(a) an improperly sealed can came from assembly line I;

(b) an improperly sealed can came from assembly line II.

(c) Verify that the probability that an improperly sealed can came from assembly line I, II, or III is, in fact, equal to 1.

5.91   A hotel gets cars for its guests from three rental agencies: 25% from agency X, 25% from agency Y, and 50% from agency Z. If 8% of the cars from X, 6% of the cars from Y, and 15% of the cars from Z need tune-ups, what is the probability that a car needing a tune-up that is delivered to a guest of the hotel came from rental agency Y?

5.92   A retailer of automobile parts has four employees, K, L, M, and N, who make mistakes in filling an order one time in 100, four times in 100, two times in 100, and six times in 100. Of all the orders filled, K, L, M, and N fill, respectively, 20%, 40%, 30%, and 10%. If a mistake is found in a particular order, what are the probabilities that it was filled by K, L, M, or N?

✓ SOLUTIONS TO PRACTICE EXERCISES

5.1 (a) $X' = \{b, d, e, h\}$ is the event that Mr. Bean, Ms. Daly, Mr. Earl, or Mr. Hall will be awarded the scholarship.

(b) $X \cup Y = \{a, c, d, f, g, h\}$ is the event that Ms. Adams, Ms. Clark, Ms. Daly, Ms. Fuentes, Ms. Gardner, or Mr. Hall will be awarded the scholarship.

(c) $X \cap Y = \{f\}$ is the event that Ms. Fuentes will be awarded the scholarship.

(d) $X \cap Z = \{a, g\}$ is the event that Ms. Adams or Ms. Gardner will be awarded the scholarship.

(e) $Y \cap Z$ is the empty set.

(f) $X' \cup Y = \{b, d, e, f, h\}$ is the event that Mr. Bean, Ms. Daly, Mr. Earl, Ms. Fuentes, or Mr. Hall will be awarded the scholarship.

5.2 (a) The sample space is shown in Figure 5.16.

(b) Q is the event that four waiters are present, R is the event that the number of waiters exceeds the number of chefs by 1, and T is the event that only one chef is present.

(c) $Q \cup T = \{(1, 4), (2, 4), (1, 2), (1, 3), (1, 5)\}$ is the event that four waiters are present or one chef is present, $R \cap T = \{(1, 2)\}$ is the event that one chef and two waiters are present, and $T' = \{(2, 2), (2, 3), (2, 4), (2, 5)\}$ is the event that both chefs are present.

(d) Q and R are mutually exclusive, Q and T are not mutually exclusive, and R and T are not mutually exclusive.

5.12 Region 1 represents the event that a person vacationing in Southern California visits Disneyland and Universal Studios.

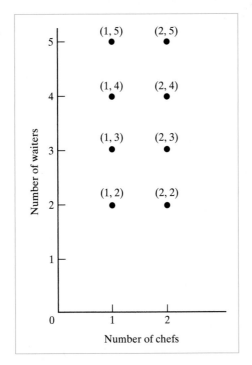

Figure 5.16
Diagram for Exercise 5.2.

Region 2 represents the event that a person vacationing in Southern California visits Disneyland, but not Universal Studios.

Region 3 represents the event that a person vacationing in Southern California visits Universal Studios, but not Disneyland.

Region 4 represents the event that a person vacationing in Southern California visits neither Disneyland nor Universal Studios.

5.19 (a) The probability that the patients do not have auditory delusions;

(b) the probability that the patients do not have visual delusions;

(c) the probability that the patients have auditory delusions, visual delusions, or both;

(d) the probability that the patients have auditory delusions and visual delusions;

(e) the probability that the patients do not have auditory delusions or visual delusions;

(f) the probability that the patients have auditory delusions but not visual delusions.

5.22 (a) The sum of the two probabilities should equal 1, but they only equal 0.90.

(b) The sum of the two probabilities cannot exceed 1, and it is 1.15.

(c) Since the two events are mutually exclusive, the sum of the first two probabilities should equal the third.

5.30 (a) $\dfrac{2}{2+1} = \dfrac{2}{3} = 0.67$

(b) $\dfrac{1}{1+2} = \dfrac{1}{3} = 0.33$

(c) $\dfrac{3}{3+2} = \dfrac{3}{5} = 0.6$

(d) $\dfrac{2}{2+3} = \dfrac{2}{5} = 0.4$

(e) $\dfrac{3}{4}$ to $1 - \dfrac{3}{4}$ or $\dfrac{3}{4}$ to $\dfrac{1}{4}$ or 3 to 1

(f) $\dfrac{1}{4}$ to $1 - \dfrac{1}{4}$ or $\dfrac{1}{4}$ to $\dfrac{3}{4}$ or 1 to 3

(g) 0.60 to $1 - 0.60$ or .60 to 0.40 or 3 to 2

(h) 0.40 to $1 - 0.40$ or .40 to 0.60 or 2 to 3

5.31 The corresponding probabilities are $\dfrac{2}{2+1} = 0.67$ and $\dfrac{2}{2+1} = 0.67$, and since the events are mutually exclusive, we can write $0.67 + 0.67 = 1.34$. Since a probability cannot exceed 1, this is an impossible value, and the original odds must have been wrong.

5.40 $0.22 + 0.17 + 0.21 + 0.09 = 0.69$.

5.45 (a) The event consists of $(4, 4), (5, 5), (6, 6), (7, 7),$ and $(8, 8)$, so that the probability is $\dfrac{5}{35} = \dfrac{1}{7}$.

(b) The event consists of $(4, 0), (4, 1), (4, 2), (5, 0), (5, 1), (5, 2), (6, 0), (6, 1), (6, 2), (7, 0), (7, 1), (7, 2), (8, 0), (8, 1),$ and $(8, 2)$, and the probability is $\dfrac{15}{35} = \dfrac{3}{7}$.

(c) The event consists of (7, 0), (7, 1), (7, 2), (7, 3), (7, 4), (7, 5), (7, 6), (7, 7), (8, 0), (8, 1), (8, 2), (8, 3), (8, 4), (8, 5), (8, 6), (8, 7), and (8, 8), and the probability is $\dfrac{17}{35}$.

(d) The event consists of (4, 2), (4, 3), (5, 3), (5, 4), (6, 4), (6, 5), (7, 5), (7, 6), (8, 6), and (8, 7), and the probability is $\dfrac{10}{35} = \dfrac{2}{7}$.

5.49 (a) $0.15 + 0.18 - 0.11 = 0.22$;

(b) $1 - 0.22 = 0.78$.

5.52 The probabilities associated with the various regions are shown in the Venn diagram of Figure 5.17.

(a) 0.32;

(b) 0.14;

(c) $0.32 + 0.13 + 0.14 = 0.59$;

(d) $0.13 + 0.14 + 0.41 = 0.68$;

(e) 0.41;

(f) $1 - 0.13 = 0.87$.

Figure 5.17
Venn diagram for
Exercise 5.52.

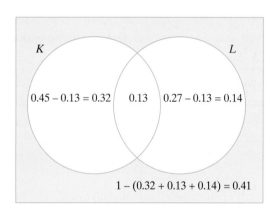

5.55 (a) $P(G|F)$;

(b) $P(J|F')$;

(c) $P(J' \cap G'|F')$;

(d) $P(G'|F \cap J)$.

5.56 (a) The probability that a female applicant who is a college graduate will own a car.

(b) The probability that a female applicant who does not own a car will have had prior experience.

(c) The probability that a female applicant with prior experience will not own a car.

(d) The probability that a female applicant who does not own a car will not be a college graduate.

(e) The probability that a female applicant who has had prior experience or is a college graduate will own a car.

(f) The probability that a female applicant who is a college graduate will have had prior experience, but not own a car.

5.61 (a) $\dfrac{80+160}{400} = \dfrac{240}{400} = 0.60$;

(b) $\dfrac{40+160}{400} = \dfrac{200}{400} = 0.50$;

(c) $\dfrac{160}{400} = 0.40$;

(d) $\dfrac{40}{400} = 0.10$;

(e) $\dfrac{160}{80+160} = \dfrac{2}{3} \approx= 0.67$;

(f) $\dfrac{40}{40+160} = 0.20$.

5.62 (a) $\dfrac{2}{3} = \dfrac{0.40}{0.60} \approx 0.67$, which checks;

(b) $0.20 = \dfrac{0.10}{0.50} = 0.20$, which checks.

5.80 The probability that the first truck chosen will meet emission standards is $\dfrac{10}{15}$; the probability that the second will meet emission standards, given that the first met emission standards, is $\dfrac{9}{14}$; the probability that the third will meet emission stan-dards, given that the first two met emission standards, is $\dfrac{8}{13}$; and the probability that the fourth will meet emission standards, given that the first three met emission stan-dards, is $\dfrac{7}{12}$. Thus, the probability that all of them will meet emission standards is

$$\dfrac{10}{15} \cdot \dfrac{9}{14} \cdot \dfrac{8}{13} \cdot \dfrac{7}{12} = \dfrac{2}{13}$$

5.85 Translate the percentages into probabilities and write $P(B) = 0.20$, $P(A|B) = 0.95$, and $P(A|B') = 0.10$. To determine $P(A)$, we must add the results that $P(B) \cdot P(A|B) = (0.20)(0.95) = 0.19$ and $P(B') \cdot P(A|B') = (0.80)(0.10) = 0.08$, getting $P(A) = 0.19 + 0.08 = 0.27$. Finally, substituting these values into the formula for $P(B|A)$, we get

$$P(B|A) = \dfrac{P(B) \cdot P(A|B)}{P(A)} = \dfrac{(0.20)(0.95)}{0.27} \approx 0.704$$

In other words, the probability that a person has the tropical disease given that he or she fails the test is approximately 0.704. The test is not particularly convincing.

4 & 5

ACHIEVEMENTS

Having read and studied these chapters and having worked a good portion of the exercises, you should be able to

1. Draw tree diagrams to determine all the alternatives that are possible in given situations.
2. Apply the formula for the multiplication of choices and its generalization.
3. Work with the factorial notation.
4. Determine the number of permutations of n distinct objects taken r at a time.
5. Determine the number of combinations of n distinct objects taken r at a time.
6. Use the table of binomial coefficients.
7. Explain the classical probability concept.
8. Explain the frequency interpretation of probability.
9. Calculate mathematical expectations.
*10. Use mathematical expectations as the basis for making rational decisions.
11. Explain what is meant by *experiment, outcome, sample space, event*, and *mutually exclusive events*.
12. Construct compound events by forming unions, intersections, and complements.
13. Picture events with the use of Venn diagrams.
14. List, justify, and apply the basic rules of probability.
15. Convert probabilities to odds and odds to probabilities.
16. Use betting odds to measure subjective probabilities.
17. State and apply the special addition rule and its generalization.
18. State and apply the general addition rule for events that need not be mutually exclusive.
19. Define *conditional probability* and apply the formula.
20. Explain what is meant by *independent events*, and check whether two events are independent or dependent.
21. State and apply the general multiplication rule.
22. State and apply the special multiplication rule.
23. State and apply generalizations of the multiplication rules.
*24. Use Bayes's theorem.

CHECKLIST OF KEY TERMS (with page references to their definitions)

R.41 A college has five professors, eight associate professors, and seven assistant professors. The dean wants to form a committee with three faculty members from each professional category. In how many ways can the nine members of the committee be selected?

R.42 Five members of a basketball team, standing on the court, are to be introduced to the fans in the stands. In how many ways can the announcer introduce the five members of the team?

R.43 A small real estate office has four part-time salespersons. Using two coordinates so that (3, 1), for example, represents the event that three of the salespersons are at work and one of them is busy with a customer, and (2, 0) represents the event that two of the salespersons are at work but none of them is busy with a customer, draw a diagram similar to that of Figure 5.1, showing the 15 points of the corresponding sample space.

R.44 With reference to Exercise R.43, if each of the 15 points of the sample space has the probability $\frac{1}{15}$, find the probabilities that

(a) all the salespersons that are at work are busy with customers;

(b) at least three of the salespersons are at work;

(c) at least three salespersons are busy with a customer;

(d) none of the salespersons who are at work are busy with customers.

R.45 The probability that a fire department will respond to 0, 1, 2, or 3 fire alarms in a day is 0.10, 0.43, 0.29 and 0.18, respectively. What is the expected number of fire alarms to which this fire department will respond during a day?

R.46 If $P(G) = 0.50$, $P(H) = 0.24$, and $P(G \cup H) = 0.62$, are events G and H dependent or independent?

R.47 Among six applicants for an executive job, A is a college graduate, is foreign born, and is single; B is not a college graduate, is foreign born, and is married; C is a college graduate, is native born, and is married; D is not a college graduate, is native born, and is single; E is a college graduate, is native born, and is married; and F is not a college graduate, is native born, and is married. One of these applicants is to get the job, and the event that the job is given to a college graduate, for example, is denoted $\{A, C, E\}$. State in a similar manner the event that the job is given to

(a) a single person;

(b) a native-born college graduate;

(c) a married person who is foreign born.

R.48 If $P(X) = 0.50$, $P(Y) = 0.60$, and $P(X \cap Y) = 0$, are events X and Y mutually exclusive or not mutually exclusive?

R.49 What is our mathematical expectation if we pay $1.00 to buy one of 1,000 raffle tickets for a first prize worth $500, a second prize worth $100, and a third prize worth $50? Is the raffle ticket a wise purchase?

R.50 If an applicant to a certain college feels that the probability he will not get accepted is $\dfrac{8}{8+2} = \dfrac{8}{10}$, what is his personal probability that he will get accepted?

R.51 Suppose that we randomly select two ball-point pens from a carton containing ten ball-point pens, three of which are defective.

(a) What is the probability that both of them will be defective?

(b) What is the probabity that neither of them will be defective?

*R.52 During a time of national emergency, a country uses lie detectors to uncover security risks. Since lie detectors are not infallible, let us suppose that the probabilities are 0.10 and 0.04 that a lie detector will fail to detect a security risk or will incorrectly label a person a security risk. If 2% of the persons who are given lie detector tests are security risks, what is the probability that a person labeled a security risk by a lie detector actually is a security risk?

R.53 A large firm has four employees named Davis: Arthur Davis, Jennifer Davis, Paul Davis, and Wendy Davis.

(a) In how many ways can the payroll department distribute their checks so that they each receive a check made out to one of the Davises?

(b) In how many of the possibilities will none of them get the correct paycheck?

(c) In how many of the possibilities will exactly one of them get the correct paycheck?

(d) In how many of the possibilities will exactly two of them get the correct paycheck?

(e) In how many of the possibilities will exactly three of them get the correct paycheck?

(f) In how many of the possibilities will all four of them get the correct paycheck?

R.54 An automobile rental agency has 10 cars in its parking lot of which 4 are white and the remainder are various other colors. If 3 cars are randomly rented to customers, what is the probability that they will all be white?

R.55 A psychologist preparing three-letter nonsense words for use in a memory test chooses the first letter from among the consonants q, w, x, and z; the second letter from among the vowels e, i, and u; and the third letter from among the consonants c, f, p, and v.

(a) How many different three-letter nonsense words can he construct?

(b) How many of these nonsense words will begin with the letter w?

(c) How many of these nonsense words will end either with the letter f or the letter p?

R.56 If an army officer thinks that 5 to 4 are fair odds that he can capture an enemy position, what subjective probability does he assign to this event?

R.57 T and W are mutually exclusive events that a factory will sell clothing to Target and Wal-Mart stores: $P(T) = 0.40$ and $P(W) = 0.25$. Find the probability that the factory will

(a) not sell to Target stores;

(b) sell to Target stores or Wal-Mart stores;

(c) not sell to Target stores or to Wal-Mart stores.

R.58 The probabilities are 0.20 that a languid student will be late to class, 0.25 that the student will not understand the lecture, and 0.15 that the student will be late to class and not understand the lecture. What is the probability that the student will be late to class or will not understand the lecture?

R.59 If there are seven members in a club, in how many ways can they be selected president, vice-president, and treasurer?

R.60 The mutually exclusive probabilities are 0.25, 0.40, and 0.15 that a certain visitor to a library will utilize an encyclopedia, a dictionary, or an atlas. What is the probability that the library patron will utilize one of these reference books?

R.61 A commuter drives to work within 30 minutes in 216 of 324 days. Estimate the probability that on any one day the commuter will make the trip within 30 minutes.

R.62 An appliance wholesaler distributes washing machines made in factories in Akron and Wheeling. Some of these washing machines receive, at customer request, an in-home warranted repair, with the exact breakdown being:

	Warranted repair	No warranted repair
Akron	20	160
Wheeling	50	150

If one of the washing machine customers is randomly selected to check on consumer satisfaction, and A is the event that the machine was made in Akron, and R is the event that it had an in-home warranted repair, determine each of the following probabilities:

(a) $P(A)$; (d) $P(A|R)$;
(b) $P(R')$; (e) $P(A' \cup R)$;
(c) $P(A \cap R)$; (f) $P(R'|A)$.

R.63 Regarding Exercise R.62, verify that

$$P(A|R') = \frac{P(A \cap R')}{P(R')}$$

R.64 If the probability is 0.90 that no students are late to class on a certain day, what is the probability that no students will be late to class for

(a) 2 days;

(b) 3 days?

R.65 An applicant feels that the odds are 5 to 2 that he will not get a certain job.

(a) What is his personal probability that he will not get the job?

(b) What are the odds that he will get the job?

(c) What is his personal probability that he will get the job?

(d) Sum the personal probabilities obtained in parts (a) and (c). The sum will equal 1 if your answers are correct.

*R.66 The probability that a theater will be sold out is 0.20.

(a) What are the odds that the theater will be sold out?

(b) What are the odds that the theater will not be sold out?

R.67 If A is the event that a school's football team is rated among the top 20 by AP and U is the event that it is rated among the top 20 by UPI, what events are represented by the four regions of the Venn diagram of Figure R.1?

R.68 A guidance department gives students various kinds of tests. If I is the event that a student scores high in intelligence, A is the event that a student rates high on a social adjustment scale, and N is the event that a student displays neurotic tendencies, express each of the following probabilities in symbolic form:

(a) The probability that a student who scores high in intelligence will display neurotic tendencies.

(b) The probability that a student who does not rate high on the social adjustment scale will not score high in intelligence.

(c) The probability that a student who displays neurotic tendencies will neither score high in intelligence nor rate high on the social adjustment scale.

(d) The probability that a student who scores high in intelligence and rates high on the social adjustment scale will not display any neurotic tendencies.

Figure R.1
Venn diagram for
Exercise R.67.

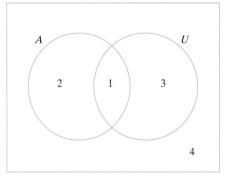

R.69 What is the probability of getting four diamonds in four random draws from an ordinary deck of playing cards

(a) if each card is replaced before the next one is drawn;

(b) if no cards are replaced before the next is drawn?

R.70 Following are the probabilities that, in a given city, 0 through 5 burglaries will be reported on any given day.

Number of burglaries	0	1	2	3	4	5
Probabilities	0.10	0.28	0.42	0.15	0.05	0.00

Find the probabilities that on any given day the number of burglaries reported will be

(a) at least 3;

(b) more than 3;

(c) no more than 3.

R.71 Six white mice and six black mice are available for a medical experiment which requires only four mice. In how many ways can the experimenter select

(a) 4 of the 6 white mice;

(b) 2 of the 6 black mice, and 2 of the 6 white mice;

(c) any 4 of the 12 mice?

R.72 In Figure R.2, E, T, and N are the events that a car brought to a garage needs an engine overhaul, transmission repairs, or new tires. Express in words the events represented by

(a) region 1;

(b) region 3;

(c) region 7;

(d) regions 1 and 4 together;

(e) regions 2 and 5 together;

(f) regions 3, 5, 6, and 8 together.

Figure R.2
Venn diagram for
Exercise R.72.

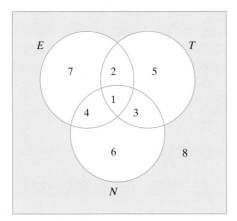

R.73 With reference to Exercise R.72 and the Venn diagram of Figure R.2, list the regions or combinations of regions that represent the events that a car brought to the garage needs

    (a) transmission repairs, but neither an engine overhaul nor new tires;

    (b) an engine overhaul and transmission repairs;

    (c) transmission repairs or new tires, but not an engine overhaul;

    (d) new tires.

R.74 Explain why there must be a mistake in the following statements.

    (a) $P(A|B) = P(B|A)$.

    (b) The probability that a policeman will issue 0, 1, 2, 3, 4, or more than 4 traffic tickets in a day is 0.02, 0.04, 0.08, 0.20, 0.30, and 0.00.

    (c) If A and B are not independent events, then $P(A \cap B) = P(A) \cdot P(B)$.

    (d) If $P(B)$ is equal to zero, then the conditional probability of A, relative to B, namely, the probability of A given B, is $P(A|B) = \dfrac{P(A \cap B)}{P(B)}$.

R.75 What is the probability of an event if the odds are

    (a) 10 to 1;

    (b) 5 to 3;

    (c) 1 to 1?

R.76 The probability that a rug will be sold during a sale is 0.70, and the probability that it will be returned or exchanged after it is sold is 0.10. What is the probability that it will be sold during the sale, and then returned or sold?

R.77 A hospital wants to buy new magazines for several of its waiting rooms, and the mutually exclusive probabilities are 0.10, 0.20, 0.30, 0.13, 0.15, and 0.12 that it will buy *Better Homes and Gardens, National Geographic, Reader's Digest, People, Time,* or *Newsweek*, respectively. If these are their only choices, find the probabilities that the hospital will buy

    (a) *National Geographic* or *Time*;

    (b) *Better Homes and Gardens, People,* or *National Geographic*;

    (c) *Reader's Digest* or *Newsweek*;

    (d) *National Geographic, People,* or *Newsweek*.

R.78 Given $P(A) = 0.33$, $P(B) = 0.59$, and $P(A \cap B) = 0$, find

    (a) $P(A')$;

    (b) $P(B')$;

    (c) $P(A \cup B)$;

    (d) $(P(A' \cap B')$.

R.79 If a student answers four questions on a true-false quiz by flipping a balanced coin, what is the probability that the student will answer all of the questions correctly?

R.80 If S is the event that a person will buy a pair of running shoes and R is the event that the person is a runner, state in words what probabilities are expressed by

    (a) $P(S|R)$;

    (b) $P(R'|S)$;

    (c) $P(R|S')$;

    (d) $P(S'|R)$.

R.81 Given $P(A) = 0.65$, $P(B) = 0.20$, and $P(A \cap B) = 0.13$, verify that

    (a) $P(A|B) = P(A)$;

    (b) $P(A|B') = P(A)$;

    (c) $P(B|A) = P(B)$;

    (d) $P(B|A') = P(B)$.

PROBABILITY DISTRIBUTIONS

Uncertainty and doubt are very much a part of everyday life. We can never be certain about tomorrow's prices of General Electric or Microsoft common stock, nor can we be certain about the number of injuries or fatalities that will result from automobile accidents today; we cannot be sure about the demand for cheese in New York, the date of the first snowstorm of the season in Vermont, or the number of misprints in today's local newspaper. In each of these, and in a profusion of other cases, we are concerned with a figure that is associated with an element of chance, namely, with the value of a **random variable**.

A random variable is a quantity that can take on the different values of a given set with specified probabilities. This definition may not be as rigorous as it could be since, strictly speaking, a random variable is a function. Most beginning students, however, find it easiest to think of random variables simply as quantities that can take on different values depending on chance. In addition to the values cited, random values can include such things as the number of speeding tickets issued each day on a freeway between two cities, the annual production of coffee in Brazil, the wind velocity at Kennedy Airport, the size of the attendance at a baseball game, and the number of mistakes a person makes computing his or her income tax.

Random values are usually classified according to the number of values that they can assume. In this chapter we shall limit our discussion to **discrete random variables**, which can take on only a finite number of values, or a countable infinity of values (as many as there are whole numbers). For instance, the number of points that we roll with a pair of dice is a discrete random variable that can take on only the finite, or fixed, number of values 2, 3, 4, 5, 6, 7, 8, 9, 10, 11, or 12. In contrast, the number of the flip on which a coin comes up heads for the first time is a discrete random variable, which can take on the countable infinity of values 1, 2, 3, 4, 5, ... It is possible, though highly unlikely, that a coin will come up tails a million times, a billion times, or even more, before it finally comes up heads.

In the study of discrete random variables, we are usually interested primarily in the probabilities associated with all their values, that is, in their **probability distributions**. A general introduction to probability distributions in Section 6.1 is followed by discussions of various special probability distributions in Sections 6.2 through 6.5, and a description of their most important properties is given in Sections 6.6 through 6.8.

There are also **continuous random variables**, whose values are determined by means of continuous measuring scales (yardsticks, thermometers, stopwatches, and so on). These will be taken up later in Chapter 7.

6.1 PROBABILITY DISTRIBUTIONS

The tables in the two examples that follow illustrate what we mean by the probability distribution of a random variable—it is a correspondence that assigns probabilities to its values.

EXAMPLE 6.1   Construct a table showing the probabilities of rolling a 1, 2, 3, 4, 5, or 6 with a balanced die.

Solution   Since *balanced* means that the outcomes are all equiprobable, each value has the probability $\frac{1}{6}$, and we get

Number of points rolled with a die	Probability
1	$\frac{1}{6}$
2	$\frac{1}{6}$
3	$\frac{1}{6}$
4	$\frac{1}{6}$
5	$\frac{1}{6}$
6	$\frac{1}{6}$

EXAMPLE 6.2   Construct a table showing the probabilities of getting 0, 1, 2, or 3 heads in three flips of a balanced coin.

Solution   The eight equally likely possibilities are HHH, HHT, HTH, THH, HTT, THT, TTH, and TTT, where H stands for heads and T for tails. Counting the number of heads in each case and using the formula $\frac{s}{n}$ for equiprobable outcomes, we obtain the following probability distribution for the total number of heads:

Number of of heads	Probability
0	$\frac{1}{8}$
1	$\frac{3}{8}$
2	$\frac{3}{8}$
3	$\frac{1}{8}$

Whenever possible, we try to express probability distributions by means of mathematical formulas that enable us to calculate the probabilities associated with the various values of a random variable.[†] For instance, for the number of points we roll with a balanced die, we can write

$$f(x) = \frac{1}{6} \quad \text{for } x = 1, 2, 3, 4, 5, \text{ and } 6$$

where $f(1)$ represents the probability of rolling a 1, $f(2)$ represents the probability of rolling a 2, and so on, in the usual functional notation. Most of the time we write $f(x)$ for the probability that a random variable takes on the value x, but we could as well write

$$g(x), \quad h(x), \quad p(x), \ldots$$

EXAMPLE 6.3   Verify that the probability distribution of the number of heads obtained in three flips of a balanced coin is given by

$$f(x) = \frac{\binom{3}{x}}{8} \quad \text{for } x = 0, 1, 2, \text{ and } 3$$

Solution   Looking up the binomial coefficients in Table X of the statistical tables in the back of the book, we find that

$$\binom{3}{0} = 1, \quad \binom{3}{1} = 3, \quad \binom{3}{2} = 3, \quad \text{and} \quad \binom{3}{3} = 1$$

Thus, the probabilities for $x = 0, 1, 2,$ and 3 are $\frac{1}{8}, \frac{3}{8}, \frac{3}{8},$ and $\frac{1}{8}$, and this agrees with the results obtained previously.

[†] It would be necessary, for example, when a random variable can take on infinitely many different values, and we cannot possibly write down all the probabilities.

Note that each probability in these examples is between 0 and 1, inclusive, and that the sum of the probabilities is in each case equal to 1. These facts suggest the following two general rules, which apply to any probability distribution:

1. Since the values of a probability distribution are probabilities, they must be numbers between 0 and 1 inclusive.
2. Since a random variable has to take on one of its values, the sum of all the values of a probability distribution must be equal to 1.

EXAMPLE 6.4   Check whether the following function can serve as the probability distribution of an appropriate random variable:

$$f(x) = \frac{x+2}{12} \quad \text{for } x = 1, 2, \text{ and } 3$$

Solution   Substituting $x = 1, 2,$ and 3, we get

$$f(1) = \frac{3}{12}, \quad f(2) = \frac{4}{12}, \quad \text{and } f(3) = \frac{5}{12}$$

Since none of these values is negative or greater than 1, and since their sum is $\frac{3}{12} + \frac{4}{12} + \frac{5}{12} = 1$, the given function can serve as the probability distribution of a random variable.   ■

6.2   THE BINOMIAL DISTRIBUTION

In many applied problems we are interested in the probability that an event will occur x times out of n. For instance, we may be interested in the probability of getting 75 responses to 200 mail questionnaires sent out as part of a sociological survey, or the probability that 12 of 50 tagged wild turkeys will be recaptured, or the probability that 45 of 300 drivers stopped at a road block will be wearing their seatbelts, or the probability that 66 of 200 television viewers (interviewed by a market research organization) will recall the products advertised on a given program. To borrow from the language of games of chance, we could say that in each of these examples we are interested in the probability of getting "x successes in n trials" or, in other words, "x successes and $n - x$ failures in n attempts." In the problems that we study in this section, we shall always make the following assumptions:

1. There is a fixed number of trials, each resulting in success or failure.
2. The probability of a success is the same for each trial.
3. The trials are all independent.

This means that the theory that we develop will not apply, for example, if we are interested in the number of dresses a woman may try on before she buys one (where the number of trials is not fixed), or if we check every hour whether traffic is congested at a certain intersection (where the probability of "success" is not constant), or if we are interested in the number of times that a person voted for the Republican candidate in the last five presidential elections (where the trials are not independent). As an example of a situation in which the three conditions apply, consider the plight of a student who is reduced to pure guessing on eight questions of a multiple-choice test. Suppose that each question has five choices, exactly one of which is correct. If C denotes a correct answer and I an incorrect answer, one way in which the student can get six correct answers and two incorrect answers is given by the sequence of letters

$$C\,C\,C\,C\,C\,C\,I\,I$$

Here, the first six questions are answered correctly and the last two are answered incorrectly. Since $P(C) = \frac{1}{5}$ and $P(I) = \frac{4}{5}$, and the trials are all independent, the probability of this particular sequence of correct and incorrect answers is

$$\frac{1}{5}\cdot\frac{1}{5}\cdot\frac{1}{5}\cdot\frac{1}{5}\cdot\frac{1}{5}\cdot\frac{1}{5}\cdot\frac{4}{5}\cdot\frac{4}{5} = \left(\frac{1}{5}\right)^6\left(\frac{4}{5}\right)^2$$

However, there are many other sequences of six C's and two I's, for instance, $CCIICCCC$, $ICICCCCC$, and $CCICCCIC$, and, by the same argument as before, each has the probability

$$\left(\frac{1}{5}\right)^6\left(\frac{4}{5}\right)^2$$

To get the probability of six C's and two I's in any order, we must count the number of ways of arranging six C's and two I's (the number of combinations of eight objects taken six at a time) and multiply by $(\frac{1}{5})^6(\frac{4}{5})^2$. Thus, the result is given by

$$\binom{8}{6}\left(\frac{1}{5}\right)^6\left(\frac{4}{5}\right)^2 = 0.00114688 \approx 0.0011$$

This suggests that, in general, if n is the number of trials, p is the probability of a success for each trial, and the trials are all independent, then the probability of x successes in n trials is

BINOMIAL DISTRIBUTION

$$f(x) = \binom{n}{x}p^x(1-p)^{n-x} \quad \text{for } x = 0, 1, 2, \ldots, \text{ or } n$$

It is customary to say here that the number of successes in n trials is a random variable having the **binomial probability distribution**, or simply the **binomial distribution**. The binomial distribution is called by this name because, for $x = 0, 1, 2, \ldots,$ and n, the values of the probabilities are the $n+1$ successive terms of the binomial expansion of $[(1-p)+p]^n$. Since

$$[(1-p)+p]^n = 1^n = 1$$

the probabilities clearly sum to 1.

EXAMPLE 6.5   Calculate the probability that two of five reviewers will like a new movie, given that the probability that any one of them will like it is 0.70.

Solution   Substituting $x = 2$, $n = 5$, $p = 0.70$ into the formula for the binomial distribution, we get

$$f(2) = \binom{5}{2} (0.70)^2 (1 - 0.70)^{5-2}$$

$$= \frac{5 \cdot 4}{2 \cdot 1} (0.70)^2 (0.30)^3$$

$$= 10(0.70)^2 (0.30)^3 \approx 0.13$$

The following is an example in which we calculate all the probabilities of a binomial distribution.

EXAMPLE 6.6   The probability is 0.60 that a person shopping at a certain market will spend at least $50.

(a) Find the probability that, among five persons shopping at this market, none of them will spend at least $50. Similarly, find the separate probabilities that 1, 2, 3, 4, or 5 will spend at least $50.

(b) Use the results of part (a) to draw a histogram of the binomial distribution with $n = 5$ and $p = 0.60$.

Solution   (a) Substituting $n = 5$, $p = 0.60$, and $x = 0$ into the formula for the binomial distribution, we find the probability that none of them will spend at least $50. This probability is

$$f(0) = \binom{5}{0} (0.60)^0 (1 - 0.60)^{5-0} = 0.01024 \approx 0.010$$

Similarly, using $x = 1, 2, 3, 4$, and 5, we have

$$f(1) = \binom{5}{1} (0.60)^1 (1 - 0.60)^{5-1} \approx 0.077$$

$$f(2) = \binom{5}{2} (0.60)^2 (1 - 0.60)^{5-2} \approx 0.230$$

$$f(3) = \binom{5}{3} (0.60)^3 (1 - 0.60)^{5-3} \approx 0.346$$

$$f(4) = \binom{5}{4} (0.60)^4 (1 - 0.60)^{5-4} \approx 0.259$$

$$f(5) = \binom{5}{5} (0.60)^5 (1 - 0.60)^{5-5} \approx 0.078$$

(b) The histogram is shown in Figure 6.1.

In actual practice, binomial probabilities are seldom obtained by direct substitution into the formula. Sometimes we use the approximations discussed later in this chapter and in Chapter 7. In many cases we can refer to special tables

Figure 6.1
Histogram of binomial
distribution with $n = 5$
and $p = 0.60$.

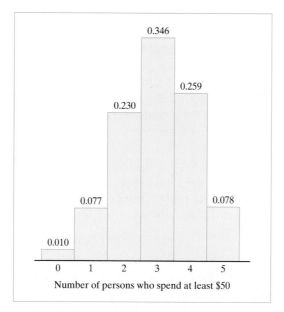

0.346

0.259

0.230

0.077   0.078

0.010

| 0 | 1 | 2 | 3 | 4 | 5 |

Number of persons who spend at least $50

such as Table I at the end of this book. Table I gives the binomial probabilities
for $n = 2$ to $n = 15$ and $p = 0.05, 0.1, 0.2, 0.3, 0.4, 0.5, 0.6, 0.7, 0.8, 0.9$, and 0.95,
all rounded to three decimals. Values omitted in the table are 0.0005 or less and,
therefore, 0.000 rounded to three decimals.

There are also specially programmed statistical calculators and an abundance
of computer programs that give binomial probabilities; because these can deal
with nearly any values of n and p, they turn out to be more useful than tables.

Many of the situations in which binomial random variables are used involve
descriptions like *at most, at least, not more than, less than, more than,* and so on.
Care is needed in dealing with these descriptions. For instance, *less than* 7 refers
to the set of values

$$0, 1, 2, 3, 4, 5, 6$$

while *not more than* 7 refers to the set of values

$$0, 1, 2, 3, 4, 5, 6, 7$$

EXAMPLE 6.7   If the probability is 0.40 that a divorcee will remarry within three years, find the
probabilities that of ten divorcees

   (a) at most three will remarry within three years;
   (b) at least seven will remarry within three years;
   (c) from two to five will remarry within three years;
   (d) at least two will remarry within three years.

Solution   **(a)** From Table I for $n = 10$, the column headed 0.4, and the rows corresponding
to $x = 0, 1, 2$, and 3, the probabilities are 0.006, 0.040, 0.121, and 0.215; thus,
the probability that at most three of ten divorcees will remarry within three
years is

$$0.006 + 0.040 + 0.121 + 0.215 = 0.382$$

(b) From Table I for $n = 10$, the column headed 0.4, and the rows corresponding to $x = 7, 8, 9,$ and 10, the probabilities are 0.042, 0.011, 0.002, and 0.000; thus, the probability that at least seven of ten divorcees will remarry within three years is

$$0.042 + 0.011 + 0.002 + 0.000 = 0.055$$

(c) From Table I for $n = 10$, the column headed 0.4, and the rows corresponding to $x = 2, 3, 4,$ and 5, the probabilities are 0.121, 0.215, 0.251, and 0.201; thus, the probability that from two to five of ten divorcees will remarry within three years is

$$0.121 + 0.215 + 0.251 + 0.201 = 0.788$$

(d) From Table I for $n = 10$, the column headed 0.4, and the rows corresponding to $x = 0$ and 1, the probabilities are 0.006 and 0.040; thus, the probability that at least two of ten divorcees will remarry within three years is

$$1 - (0.006 + 0.040) = 0.954 \qquad \blacksquare$$

If the probability that a divorcee will remarry within three years had been 0.42 instead of 0.40, we could not have worked the preceding example by using Table I. In general, if n is greater than 15 and/or p takes on a value other than 0.05, 0.1, 0.2, ..., 0.9, or 0.95, we will have to use a more detailed table (see the Bibliography), refer to the formula for the binomial distribution, or employ a computer as in the following example.

EXAMPLE 6.8 Use the computer printout of Figure 6.2 to rework the preceding example with $p = 0.42$ instead of $p = 0.40$.

Solution **(a)** Adding the probabilities in the $P(X = k)$ column corresponding to 0, 1, 2, and 3, and rounding to four decimals, we get

$$0.0043 + 0.0312 + 0.1017 + 0.1963 = 0.3335$$

Note, however, that since the printout also gives the cumulative *less than or equal* probabilities, the answer is the value in the $P(X \text{ LESS OR} = k)$ column corresponding to $K = 3$.

(b) Adding the probabilities in the $P(X = k)$ column corresponding to 7, 8, 9, and 10, and rounding to four decimals, we get

$$0.0540 + 0.0147 + 0.0024 + 0.0002 = 0.0713$$

Since the probability of "at least 7" is 1 minus the probability of "6 or less," we get $1 - 0.9288 = 0.0712$, where 0.9288 is the value in the $P(X \text{ LESS OR} = k)$ column corresponding to $K = 6$. The difference between the two results is due to rounding.

(c) Adding the probabilities in the $P(X = k)$ column corresponding to 2, 3, 4, and 5, and rounding to four decimals, we get

$$0.1017 + 0.1963 + 0.2488 + 0.2162 = 0.7630$$

Figure 6.2
MINITAB printout for the binomial distribution with $n = 10$ and $p = 0.42$.

Since the probability of "2 to 5" equals the difference between the probabilities of "5 or less" and "1 or less," the values in the P(X LESS OR = k) column yield $0.7984 - 0.0355 = 0.7629$. Again, the difference between the two results is due to rounding.

(d) Subtracting from 1 the probability of "1 or less," we get

$$1 - (0.0043 + 0.0312) = 0.9645$$

from the column headed P(X = k), and $1 - 0.0355 = 0.9645$ from the column P(X LESS OR = k). In this case, the results are the same.

When we observe a value of a random variable having the binomial distribution—for instance, when we observe the number of heads in 50 flips of a coin, or the number of seeds (in a package of 24 seeds) that germinate, or the number of students (among 200 interviewed) who are opposed to an increase in tuition, or the number of automobile accidents (among 100 investigated) that are due to drunk driving—we say that we are **sampling a binomial population**. This terminology is widely used in statistics. Computers can also be used to simulate sampling binomial populations, that is, to simulate observations of random variables having given binomial distributions. In fact, that is what we did in the illustration on page 130, which illustrated the Law of Large Numbers. In Figure 4.3, instructions "RANDOM 100 C1;" and "INTEGERS 0 1." tell the computer to take 100 random observations of a random variable having the binomial distribution with $n = 1$ and $p = 0.50$.

Exercises 6.1, 6.5, 6.6, 6.7, and 6.18 are practice exercises; their complete solutions are given on pages 233–234.

✓ **6.1**   In each case, determine whether the given values can be looked upon as the values of a probability distribution of a random variable that can take on the values 1, 2, and 3, and explain your answers.

(a) $f(1) = 0.37$, $f(2) = 0.35$, $f(3) = 0.30$;

(b) $f(1) = \dfrac{4}{9}$, $f(2) = \dfrac{4}{9}$, $f(3) = \dfrac{1}{9}$;

(c) $f(1) = 0.57$, $f(2) = -0.59$, $f(3) = 0.16$.

6.2   Check in each case whether the given function can serve as the probability distribution of an appropriate random variable.

(a) $f(x) = \dfrac{1}{5}$ for $x = 0, 1, 2, 3, 4, 5$;

(b) $f(x) = \dfrac{x^2 - 1}{25}$ for $0, 1, 2, 3$;

(c) $f(x) = \dfrac{x^2}{14}$ for $x = 0, 1, 2, 3$.

6.3   In each case determine whether the given values can be looked upon as the values of a probability distribution of a random variable that can take on the values 1, 2, 3, and 4. Explain your answers.

(a) $f(1) = 0.25$, $f(2) = 0.30$, $f(3) = 0.35$, $f(4)\ 0.40$;

(b) $f(1) = 0.05$, $f(2) = 0.10$, $f(3) = 0.15$, $f(4) = 0.20$;

(c) $f(1) = 0.10$, $f(2) = 0.20$, $f(3) = 0.30$, $f(4) = 0.40$;

(d) $f(1) = -0.10$, $f(2) = -0.20$, $f(3) = -0.30$; $f(4) = -0.40$.

6.4   Check in each case whether the given function can serve as the probability distribution of an appropriate random variable.

(a) $f(x) = \dfrac{\binom{2}{x}}{4}$ for $x = 0, 1, 2$;

(b) $f(x) = \dfrac{x - 2}{9}$ for $x = 1, 2, 3, 4, 5, 6$;

(c) $f(x) = \dfrac{x^2 - 6x + 9}{10}$ for $x = 1, 2, 3, 4, 5$;

(d) $f(x) = \dfrac{x^2 - 6x + 8}{5}$ for $x = 1, 2, 3, 4, 5$.

✓ **6.5**   According to the Bureau of Labor Statistics, about 10% of the civilian labor force (both men and women) were 20 to 24 years of age in 1996. Using this data and the formula for the binomial distribution, calculate the probabilities that for a randomly selected group of six members of the civilian labor force in 1996

(a) exactly two were 20 to 24 years of age;

(b) fewer than two were 20 to 24 years of age.

✓ **6.6**   According to the College Board, 70% of high school graduates in the Commonwealth of Massachusetts took the SAT examination in a certain year. Find the probability that two out of four randomly selected high school graduates in Massachusetts took the SAT examination

(a) using the formula for the binomial distribution;

(b) referring to Table I.

✓ **6.7**   A civil service examination is designed so that 80% of all high school graduates can pass. Use Table I to find the probabilities that among 14 high school graduates

(a) at least 12 will pass the test;

(b) at most 10 will pass the test.

6.8   A store advertised a clearance sale, and the probability is 0.20 that any one customer will make a purchase. Use Table I to determine the probability that in a random sample of eight shoppers

(a) three will make a purchase;

(b) two will make a purchase;

(c) one will make a purchase;

(d) none will make a purchase.

6.9   Using the formula for the binomial distribution and the data of Exercise 6.8(a), determine the probability that three customers will make a purchase.

6.10   Find the probability that all of the seven participants in a taste-test of a new beverage will like its flavor, when the probability that any one of them will like it is 0.90. Use Table I to find the answer.

6.11   Using the formula for the binomial distribution and the data of Exercise 6.10, find the probability that all 7 of the participants in the taste-test will like the flavor.

6.12   According to the National Center for Health Statistics, 30% of total deaths in the United States are caused by heart disease. Use Table I to find the probability that, in a random sample of five fatalities, none were caused by heart disease.

6.13   Using the formula for the binomial distribution and the data of Exercise 6.12, find the probability that none of the deaths were caused by heart disease.

6.14   According to a study in a certain city, 50% of workers between the ages of 25 to 34 years were paid hourly rates of $10.00 or more in the year 2003. Find the probabilities that among 10 randomly selected workers in this category

(a) at least five earned wages of $10.00 per hour or more;

(b) at most five workers earned wages of $10.00 per hour or more;

(c) anywhere from four to six workers earned wages of $10.00 per hour or more.

6.15   A study conducted at a certain college shows that 60% of the school's graduates obtain a job in their chosen field within a year after graduation. Use Table I to find the probabilities that, within a year after graduation, among 14 randomly selected graduates of that college

(a) at least 6 will find a job in their chosen field;

(b) at most 3 will find a job in their chosen field;

(c) anywhere from 5 through 8 will find a job in their chosen field.

6.16   Research shows that 30% of all women taking a certain medication do not respond favorably. Use Table I to find the probabilities that among nine randomly selected women taking the medication

(a) at least four will respond favorably;

(b) at least four will not respond favorably.

6.17   A study has shown that 50% of the families in a certain large area have at least two cars. Find the probabilities that among 15 families randomly selected in this area

(a) 7 have at least two cars;

(b) more than 8 have at least two cars;

(c) 6 have fewer than two cars;

(d) more than 9 have fewer than two cars.

✓ **6.18** A quality control engineer wants to check whether (in accordance with specifications) 95% of the products shipped are in perfect condition. To this end, he randomly selects 10 items from each large lot ready to be shipped and passes it only if they are all in perfect condition; otherwise, each item in the lot is checked. Use Table I to find the probabilities that he will commit the error of

(a) holding a lot for further inspection even though 95% of the items are in perfect condition;

(b) letting a lot pass through even though only 90% of the items are in perfect condition;

(c) letting a lot pass through even though only 80% of the items are in perfect condition.

6.19 A food distributor claims that 80% of her 6-ounce cans of mixed nuts contain at least three pecans. To check on this, a consumer testing service decides to examine six of these 6-ounce cans of mixed nuts from a very large production lot and reject the claim if fewer than four of them contain at least three pecans. Use Table I to find the probabilities that the testing service will commit the error of

(a) rejecting the claim even though it is true;

(b) not rejecting the claim when in reality only 60% of the cans of mixed nuts contain at least three pecans;

(c) not rejecting the claim when in reality only 40% of the cans of mixed nuts contain at least three pecans.

6.20 With reference to Exercise 6.15, suppose that the study had shown that 61% of the school's graduates obtain a job in their chosen field within a year after graduation. Use a computer printout of the binomial distribution with $n = 14$ and $p = 0.61$ to rework the three parts of that exercise.

6.21 With reference to Exercise 6.16, suppose that the study had shown that 33% of all women taking the medication do not respond favorably. Use a computer printout of the binomial distribution with $n = 9$ and $p = 0.33$ to rework both parts of that exercise.

6.22 With reference to Exercise 6.17, suppose that the study had shown that 49% of the families in that area have at least two cars. Use a computer printout of the binomial distribution with $n = 15$ and $p = 0.49$ to rework all four parts of that exercise.

6.23 A florist needs 12 exotic potted ferns for a floral exhibition to take place in six months. He knows that the probability is 0.70 that a fern planted now will be suitable for the exhibition in six months. His concern is the minimum number of ferns that he should plant if he wants the probability of having 12 suitable ferns in six months to be at least 0.90.

(a) Will 15 be enough? If he plants 15 ferns, find the probability that he will have at least 12 suitable ones.

(b) Will 20 be enough? If he plants 20 ferns, find the probability that he will have at least 12 suitable ones.

(c) Will 25 be enough? If he plants 25 ferns, find the probability that he will have at least 12 suitable ones.

(d) Find the smallest number of ferns for which the probability of 12 or more suitable ones is 0.90 (or more).

6.3 THE HYPERGEOMETRIC DISTRIBUTION

To illustrate another important kind of probability distribution, let us consider the following problem. A factory ships certain tape recorders in lots of 16, and when they arrive at their destination, three are randomly selected from each lot for inspection. Now suppose that four of the tape recorders in a lot are defective, and we are interested in the probability that among the three tape recorders selected, two will be all right and one will be defective. In other words, we are interested in the probability of "one success (defective tape recorder) in three trials," and we might be tempted to argue that since 4 of the 16 tape recorders are defective, the probability of x objects of type A and $n - x$ objects of type B is

$$f(1) = \binom{3}{1}\left(\frac{1}{4}\right)^1\left(1 - \frac{1}{4}\right)^{3-1} \approx 0.422$$

This result, obtained by means of the formula for the binomial distribution, would be correct if sampling is with replacement, that is, if each tape recorder is replaced in the lot before the next one is selected. In actual practice, we seldom, if ever, sample with replacement. To obtain the correct answer for our problem when sampling is without replacement, we might argue as follows.

There are altogether $\binom{16}{3}$ ways of choosing 3 of the 16 tape recorders, and they are all equiprobable by virtue of the assumption that the selection is random. Since there are $\binom{4}{1}$ ways of choosing 1 of the defective tape recorders and $\binom{12}{2}$ ways of choosing 2 that are not defective, there are $\binom{4}{1} \cdot \binom{12}{2}$ ways of getting "1 success (defective tape recorder) in 3 trials." Thus, it follows by the special formula $\dfrac{s}{n}$ for equiprobable outcomes that the desired probability is

$$\frac{\binom{4}{1}\binom{12}{2}}{\binom{16}{3}} = \frac{4.66}{560} \approx 0.471$$

This suggests that, in general, if n objects are chosen at random from a set consisting of a objects of type A and b objects of type B, and the selection is without replacement, then the probability of "x of type A and $n - x$ of type B" is

HYPERGEOMETRIC
DISTRIBUTION

$$f(x) = \frac{\binom{a}{x} \cdot \binom{b}{n-x}}{\binom{a+b}{n}} \quad \text{for } x = 0, 1, 2, \ldots, \text{ or } n$$

where x cannot exceed a and $n - x$ cannot exceed b, since we cannot get more successes (or failures) than there are in the whole set. Note also that x cannot exceed n. This is the formula for the **hypergeometric distribution**.

EXAMPLE 6.9 A secretary is supposed to send 6 of 15 overseas letters by airmail, but she gets them all mixed up and randomly puts airmail stamps on 6 of the letters. What is the probability that only 3 of the letters that should go by airmail get an airmail stamp?

Solution Since this problem involves sampling without replacement, the situation requires the formula for the hypergeometric distribution. Substituting $a = 6, b = 9, n = 6$, and $x = 3$ into this formula, we get

$$f(3) = \frac{\binom{6}{3}\binom{9}{3}}{\binom{15}{6}} = \frac{20 \cdot 84}{5{,}005} \approx 0.336$$

The following is an example in which we calculate all the probabilities of a hypergeometric distribution.

EXAMPLE 6.10 Among a department store's 16 delivery trucks, 5 have worn brakes. If 3 of the trucks are randomly picked for a general overhaul, find the separate probabilities that these 3 will include 0, 1, 2, or 3 of the trucks with worn brakes.

Solution Again, the problem involves sampling without replacement. Thus, substituting $a = 5, b = 11, n = 3$, and $x = 0, 1, 2$, and 3 into the formula for the hypergeometric distribution, we get

$$f(0) = \frac{\binom{5}{0}\binom{11}{3}}{\binom{16}{3}} = \frac{1 \cdot 165}{560} \approx 0.295$$

$$f(1) = \frac{\binom{5}{1}\binom{11}{2}}{\binom{16}{3}} = \frac{5 \cdot 55}{560} \approx 0.491$$

$$f(2) = \frac{\binom{5}{2}\binom{11}{1}}{\binom{16}{3}} = \frac{10 \cdot 11}{560} \approx 0.196$$

$$f(3) = \frac{\binom{5}{3}\binom{11}{0}}{\binom{16}{3}} = \frac{10 \cdot 1}{560} \approx 0.018$$

A histogram of this probability distribution is shown in Figure 6.3.

This probability distribution answers many card game questions. Consider the following:

Figure 6.3
Histogram of hyperge-
ometric distribution with
$a = 5, b = 11$, and
$n = 3$.

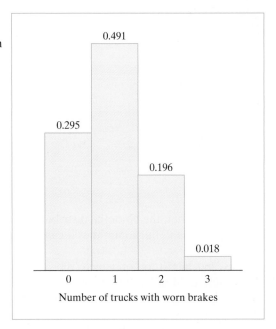

0.491

0.295

0.196

0.018

0   1   2   3

Number of trucks with worn brakes

EXAMPLE 6.11 Suppose that 5 cards are dealt, without replacement, from a standard deck of 52. What is the probability that the 5 cards will include exactly two spades?

Solution Here the spades correspond to type A and the nonspades correspond to type B. Therefore, $a = 13, b = 39, n = 5$, and $x = 2$. We find that

$$f(2) = \frac{\binom{13}{2}\binom{39}{3}}{\binom{52}{5}} = \frac{78 \cdot 9,139}{2,598,960} \approx 0.274$$

EXAMPLE 6.12 A bridge hand consists of 13 cards dealt at random from the deck of 52. Find the probabilities that a bridge hand will have $0, 1, 2, \ldots, 13$ spades.

Solution The probability that a bridge hand will have exactly x spades is

$$f(x) = \frac{\binom{13}{x}\binom{39}{13-x}}{\binom{52}{13}}$$

for $x = 0, 1, 2, \ldots$, or 13.

This expression could be calculated directly for each value of x, but the task is much more easily done with a computer. As can be seen from the computer output in Figure 6.4, bridge hands commonly (about 73% of the time) have two, three, or four spades, while bridge hands with seven or more spades are very unlikely.

Figure 6.4
MINITAB printout of
the hypergeometric dis-
tribution with $a = 13$,
$b = 39$, and $n = 13$.

Worksheet 1 ***						
↓	C1	C2	C3	C4	C5	C6
	x	P(X=x)				
1	0	0.012791				
2	1	0.080062				
3	2	0.205873				
4	3	0.286330				
5	4	0.238608				
6	5	0.124692				
7	6	0.041564				
8	7	0.008817				
9	8	0.001167				
10	9	0.000093				
11	10	0.000004				
12	11	0.000000				
13	12	0.000000				
14	13	0.000000				

As we have pointed out, the binomial distribution does not apply when sampling is without replacement, but sometimes it can be used as an approximation.

EXAMPLE 6.13   A laboratory has 260 hamsters, of which 100 are male and 160 are female. (Hamsters are widely used in laboratory work since they reproduce faster than any other mammal, having very short gestation periods and, frequently, have large litters.)

If six hamsters are randomly selected for an experiment, find the probability that only one is male by using

(a) the formula for the hypergeometric distribution; and

(b) the formula for the binomial distribution, with $n = 6$ and $p = \dfrac{100}{260} = \dfrac{5}{13}$.

Solution   (a) Substituting $a = 100$, $b = 160$, $n = 6$, and $x = 1$ into the formula for the hypergeometric distribution, we get

$$f(1) = \frac{\binom{100}{1}\binom{160}{5}}{\binom{260}{6}} = \frac{(100)\left(\dfrac{160 \cdot 159 \cdot 158 \cdot 157 \cdot 156}{5!}\right)}{\left(\dfrac{260 \cdot 259 \cdot 258 \cdot 257 \cdot 256 \cdot 255}{6!}\right)}$$

$$= \frac{(100)(820{,}384{,}032)}{404{,}830{,}840{,}960} = 0.2026$$

(b) Substituting $n = 6$, $p = \dfrac{100}{260} \approx 0.3846$, and $x = 1$ into the formula for the binomial distribution, we get

$$f(1) = \binom{6}{1}(0.3846)^1(1 - 0.3846)^{6-1}$$

$$= (6)(0.3846)(0.0883) = 0.2038.$$

Since the difference between the two values is only 0.0012, the approximation is very good.

Observe that the hypergeometric situation involves a population of finite size $a + b$. The binomial approximation does not give an exact answer to problems with a finite population. It is generally agreed, though, that the approximation illustrated above may be used as long as the sample does not exceed 5% of the population, namely, when

CONDITION FOR BINOMIAL APPROXIMATION TO HYPERGEOMETRIC DISTRIBUTION

$$n \leq (0.05)(a + b)$$

In our example, $n = 5$ is less than $0.05(100 + 140) = 12$, so that the condition is satisfied. The main advantage of the binomial approximation to the hypergeometric distribution is that the binomial distribution appears much more extensively in published tables and computer programs. Also, the binomial calculations are generally less complicated. Observe that the binomial distribution is described by two parameters (n and p), while the hypergeometric distribution requires three (a, b, and n).

EXERCISES

Exercises 6.24, 6.25, 6.32, and 6.33 are practice exercises; their complete solutions are given on pages 234 and 235.

✓ **6.24** A produce market has 16 tomatoes of which 7 have minor blemishes. If 5 of these are randomly selected for a customer, what is the probability that exactly 3 of them will have minor blemishes?

✓ **6.25** Among the 15 used cars on a used car lot, 5 have faulty starters. If 2 of these cars are randomly selected for a newspaper advertisement, find the probabilities that

   (a) neither car will have a faulty starter;

   (b) only one of the cars will have a faulty starter;

   (c) both cars will have faulty starters.

6.26 Among 18 antique chairs for sale at an antique show, 15 are genuine and 3 are reproductions. If 3 chairs are randomly selected for consideration by a prospective purchaser, find the separate probabilities that these will include 0, 1, 2, or 3 reproductions.

6.27 Among 15 applicants for sales positions in a corporation, 8 have college degrees and 7 do not have college degrees. If 4 are randomly selected for interviews, what is the probability that all have college degrees?

6.28 A certain squad of the U.S. Army has seven male and five female enlisted personnel, of whom six are randomly selected to form an honor guard. Use the formula for the hypergeometric distribution and Table X (Binomial Coefficients) to determine the probabilities that

   (a) three personnel of the honor guard will be male, and three will be female;

   (b) five will be male and one will be female.

6.29 Ten seedless grapefruit and eight seedy grapefruit were unintentionally mixed on a store display. If a clerk randomly selects four grapefruit from the display, what is the probability that he will select two seedless and two seedy grapefruit? Use the formula for the hypergeometric distribution and Table X, Binomial Coefficients, to solve this exercise.

6.30 At a birthday party for a child, a birthday cake was brought to the table with 10 burning candles, 5 of which were pink and 5 of which were blue. With one breath, the child randomly blew out the flames of 9 candles. What is the probability that the candle which was not blown out was blue? What is the probability that the candle was pink? Use the formula for the hypergeometric distribution and Table X, Binomial Coefficients, to solve this exercise.

6.31 Three of the 10 smoke detectors in a building are not operative (do not make noise in the presence of smoke). If the building manager randomly inspects 3 smoke detectors, determine the probabilities that

(a) none of the 3 smoke detectors inspected is not operative;

(b) one of the 3 smoke detectors inspected is not operative;

(c) 2 of the smoke detectors inspected are not operative;

(d) all of the smoke detectors inspected are not operative.

✓ **6.32** Check in each case whether the condition for the binomial approximation to the hypergeometric distribution is satisfied.

(a) $a = 40, b = 160$, and $n = 15$;

(b) $a = 380, b = 120$, and $n = 22$;

(c) $a = 400, b = 240$, and $n = 30$.

✓ **6.33** A shipment of 200 burglar alarms contains 6 defectives. If 3 of these burglar alarms are randomly selected and shipped to a customer, find the probability that she will get 1 bad unit using

(a) the formula for the hypergeometric distribution;

(b) the binomial distribution as an approximation.

6.34 Check in each case whether the condition for the binomial approximation to the hypergeometric distribution is satisfied.

(a) $a = 200, b = 200$, and $n = 32$;

(b) $a = 150, b = 180$, and $n = 12$;

(c) $a = 100, b = 120$, and $n = 14$.

6.35 Among 90 cities that a professional group is considering for its next three annual conventions, 36 are in the western part of the United States. If, to avoid arguments, the selection is left to chance (that is, random sampling without replacement), what is the probability that none of the three conventions will be held in the western part of the United States? How big an error would we make if we approximated this probability with the binomial probability of "zero successes in three trials" when $p = \dfrac{36}{90} = 0.40$?

6.36 Among the 120 employees of a company, 45 are members of a clerical workers union, and the others are members of the typesetters union. If 5 of the employees are chosen by lot to serve on a grievance committee, find the probability that 2 of them will be members of the clerical workers union and the other 3 members of the typesetters union, using

(a) the formula for the hypergeometric distribution;

(b) the binomial distribution as an approximation.

6.37 A commercial hatchery has 240 White Plymouth Rock chicks and 60 Rhode Island Red chicks. Use the binomial distribution (and Table I) to approximate the probability that if a poultryman randomly selects 12 chicks for examination, only one will be a Rhode Island Red.

6.4 THE POISSON DISTRIBUTION

When n is large and p is small, binomial probabilities are often approximated by means of the formula

POISSON
APPROXIMATION TO
BINOMIAL
DISTRIBUTION

$$f(x) = e^{-np} \frac{(np)^x}{x!} \quad \text{for } x = 0, 1, 2, 3, \ldots$$

which is a special form of the **Poisson distribution**; the more general form is given on page 220. It is difficult to state precisely what we mean by "n is large and p is small," and there are other situations where the approximation may be used; but we shall use it here only when n is at least 100 and np is less than 10. Symbolically,

$$n \geq 100 \quad \text{and} \quad np < 10$$

To get some idea of the closeness of the Poisson approximation to the binomial distribution, consider the computer printout of Figure 6.5, which gives the binomial distribution with $n = 100$ and $p = 0.015$ and the corresponding Poisson distribution with $np = 100(0.015) = 1.5.$[†] Comparing the probabilities in the columns headed $P(X = k)$, and rounding to four decimals, we find that the maximum difference, corresponding to k = 0, is

$$0.2231 - 0.2206 = 0.0025$$

In the formula for the Poisson distribution, e is the number $2.71828\ldots$ used in connection with natural logarithms, and the values of $e^{-\lambda}$ may be obtained from Table XI. Most scientific calculators have keys that will calculate exponentials, and these are generally more convenient, and often more precise, than tables.

Figure 6.5
MINITAB printout of the binomial distribution with $n = 100$ and $p = 0.015$ and the Poisson distribution with mean 1.5.

[†] In the computer printout of Figure 6.5, the quantity MU = 1.5 is referred to as the **mean of the Poisson distribution**. This will be explained on pages 219 and 226.

For the Poisson distribution, x can take on the countably infinite set of values $0, 1, 2, 3, \ldots$; but this poses no problems, since the probabilities usually become negligible after relatively few values of x.

EXAMPLE 6.14   One percent of the labels on a large shipment of jars of preserves are defective (torn, loose, crooked, etc.). Use the Poisson distribution to find the probability that among 400 randomly inspected jars, only 3 will have defective labels.

Solution   Substituting $x = 3$ and $np = 400(0.01) = 4$ into the formula for the Poisson distribution, and getting $e^{-4} = 0.018$ from Table XI, we obtain

$$f(3) = 0.018 \cdot \frac{4^3}{3!} = 0.1920$$

If a calculator is used in place of Table XI, the value obtained for e^{-4} is slightly more accurate since it is carried out to many more places beyond the decimal point.   ▪

It would have been possible, though very cumbersome, to use the formula for the binomial distribution in this example.

EXAMPLE 6.15   Records show that the probability is 0.00005 that a car will have a flat tire while driving through a certain tunnel. Use the Poisson approximation to the binomial distribution to find the probabilities that among 10,000 cars passing through this tunnel

(a) at least 2 will have a flat tire;

(b) at most 2 will have a flat tire.

Solution   **(a)** For this probability, we shall subtract from 1 the probability that 0 or 1 of the cars will have a flat tire. Substituting

$$np = 10{,}000(0.00005) = 0.5$$

and, respectively, $x = 0$ and $x = 1$ into the formula for the Poisson distribution and getting $e^{-0.5} = 0.607$ from Table XI, we obtain

$$f(0) = 0.607 \frac{0.5^0}{0!} = 0.607$$

$$f(1) = 0.607 \frac{0.5^1}{1!} = 0.304$$

Thus, the answer is $1 - (0.607 + 0.304) = 0.089$.

(b) This probability is $f(0) + f(1) + f(2)$. If we substitute $np = 0.5$, $e^{-0.5} = 0.607$, and $x = 2$ into the formula for the Poisson distribution, we get

$$f(2) = 0.607 \frac{0.5^2}{2!} \approx 0.076$$

Combining this result with the two probabilities calculated in part (a), we find that the answer is

$$0.607 + 0.304 + 0.076 = 0.987$$

A calculator gives $e^{-0.5} \approx 0.606531$. This more precise figure would give 0.090 for (a) and 0.986 for (b).   ■

Since in some cases the hypergeometric distribution can be approximated by a binomial distribution, and the binomial distribution can in some cases be approximated by a Poisson distribution, there will exist situations in which the hypergeometric distribution can be approximated by a Poisson distribution. Consider the following example:

EXAMPLE 6.16 An auditor has been asked to investigate a collection of 4,000 sales invoices, of which 28 contain errors. A sample of 150 invoices is selected. What is the probability that this set of 150 invoices will contain exactly 2 with errors?

Solution This is a hypergeometric problem with $a = 28$, $b = 3,972$, $n = 150$, and $x = 2$. Since $150 < 0.05(4,000) = 200$, it is reasonable to use the binomial approximation. This binomial distribution has

$$p = \frac{a}{a+b} = \frac{28}{4,000} = 0.007, \quad n = 150, \quad \text{and} \quad x = 2$$

Since $n > 100$ and $np = 150(0.007) = 1.05 < 10$, it would also be reasonable to use the Poisson approximation with $np = 1.05$. This yields

$$f(2) = e^{-1.05}\frac{1.05^2}{2!} \approx 0.3505\frac{1.1025}{2} \approx 0.193$$

where 0.3505 is halfway between 0.368 and 0.333, the values given for $e^{-1.0}$ and $e^{-1.1}$ in Table XI.   ■

Since auditors deal frequently with situations with large values of n and even larger values of $a + b$, they find the Poisson distribution to be quite useful. Observe that, aside from x, the hypergeometric distribution involves three parameters (a, b, and n), while the binomial distribution involves two (n and p), and the Poisson distribution involves only one (np). In actual practice, Poisson probabilities are seldom obtained by direct substitution into the formula. Sometimes we refer to tables of Poisson probabilities, which may be found in handbooks of statistical tables, and more frequently we use a computer. For instance, had we used the computer printout of Figure 6.6 in the flat-tire example, we would have found that the answer to part (a) is $1 - 0.9098 = 0.0902$, where 0.9098, rounded to four decimals, is the value in the right-hand column corresponding to k = 1. Similarly, the answer to part (b) is 0.9856, which is the value in the right-hand column corresponding to k = 2. In both instances, the differences between the results obtained here and before are due to rounding. The Poisson distribution also has many important applications that have no direct connection with the binomial distribution. In that case, np is replaced by the parameter λ (Greek lowercase lambda) and we calculate the probability of getting x "successes" by means of the formula

POISSON
DISTRIBUTION
(PARAMETER λ)

$$f(x) = e^{-\lambda} \frac{\lambda^x}{x!} \quad \text{for } x = 0, 1, 2, 3, \ldots$$

where λ is interpreted as the expected, or average, number of successes (see discussion on page 226). This formula applies to many situations where we can expect a fixed number of successes per unit time (or for some other kind of unit), say, when 1.8 accidents can be expected per day at a busy intersection, or when eight small pieces of meat can be expected in a frozen meat pie, or when 4.5 imperfections can be expected per roll of cloth, or when 0.12 complaint per passenger can be expected by an airline, and so on.

Figure 6.6
MINITAB printout of
the Poisson distribution
with $np = 0.5$.

Worksheet 1 ***					
✦	C1	C2	C3	C4	C5	C6
	k	Poisson P(X=k)	Poisson P(X <= k)			
1	0	0.606531	0.60653			
2	1	0.303265	0.90980			
3	2	0.075816	0.98561			
4	3	0.012636	0.99825			
5	4	0.001580	0.99983			
6	5	0.000158	0.99999			
7	6	0.000013	1.00000			
8	7	0.000001	1.00000			
9	8	0.000000	1.00000			
10	9	0.000000	1.00000			
11	10	0.000000	1.00000			
12	11	0.000000	1.00000			
13	12	0.000000	1.00000			
14	13	0.000000	1.00000			
15	14	0.000000	1.00000			

EXAMPLE 6.17   If a bank receives on the average $\lambda = 6$ bad checks per day, what is the probability that it will receive 4 bad checks on any given day?

Solution   Substituting into the formula, we get

$$f(4) = e^{-6} \cdot \frac{6^4}{4!} \approx 0.0025 \cdot \frac{1,296}{24} \approx 0.135$$

where the value of e^{-6} was read from Table XI. Using $e^{-6} \approx 0.002478$ from a calculator would give the final answer as 0.134.   ■

In this chapter we have given a number of approximations relating the hypergeometric, binomial, and Poisson random variables. This simply means that the numerical values produced by the corresponding formulas will be nearly equal. The probability distributions still apply to very distinct situations: the hypergeometric distribution deals with a sample without replacement from a finite population, the binomial distribution deals with repeated independent trials, and the Poisson distribution deals with counted events with no fixed upper limit. The conditions under which the approximations are reasonable, noted on pages 214 and 217, are conservatively stated. The approximations may perform well even when these conditions are violated, but no promises can be made.

Exercises 6.38, 6.40, and 6.46 are practice exercises; their complete solutions are given on page 235.

✓ **6.38**   Check in each case whether the conditions for the Poisson approximation to the binomial distribution are satisfied.

(a) $n = 300$ and $p = \dfrac{1}{50}$;

(b) $n = 900$ and $p = \dfrac{1}{100}$;

(c) $n = 400$ and $p = \dfrac{1}{25}$.

6.39   Check in each case whether the conditions for the Poisson approximation to the binomial distribution are satisfied.

(a) $n = 90$ and $p = \dfrac{1}{10}$;

(b) $n = 200$ and $p = \dfrac{1}{5}$;

(c) $n = 500$ and $p = \dfrac{1}{60}$.

✓ **6.40**   It is known from experience that 2% of the calls received by a switchboard are wrong numbers. Use the Poisson approximation to the binomial distribution to determine the probability that among 250 calls received by the switchboard, 4 will be wrong numbers.

6.41   According to projections by the Bureau of the Census, 1.9% of the population of the United States will be 85 years or older in the year 2010. Assuming that the projection is correct, use the Poisson approximation to the binomial distribution to determine the probability, in 2010, that in a random sample of 100 inhabitants of the United States, 3 will be 85 years of age or older.

6.42   According to the FBI Uniform Crime Reports for a recent year, 2.1% of arrests were for burglary (arrests on suspicion not included). Use the Poisson approximation to the binomial distribution to determine the probability, in the year of the study, that in a random sample of 100 arrests, 3 were for burglary.

6.43   According to the National Center for Health Statistics, U.S. Department of Health and Human Services, 1.2% of patients visited physicians' offices in a recent year because of nasal congestion. Use the Poisson approximation to the binomial distribution to determine the probability that in a random sample of 500 patients, 2 or 3 visited physicians' offices because of nasal congestion.

6.44   An insurance adjuster has been asked to undertake a survey of automobile collision claims. He plans to take a random sample of 200 claims from a computerized record of 8,200 claims. He operates under the premise that exactly 1% of the computerized record consists of blatantly fraudulent claims. He wants to know the probability of getting at most three of these blatantly fraudulent claims in his sample.

(a) Identify the values of a, b, and n for this hypergeometric situation.

(b) The binomial approximation is appropriate here. Identify the values of n and p for the binomial approximation.

(c) Use the Poisson approximation to the binomial distribution, and determine the probability that the sample will have at most three blatantly fraudulent claims.

(d) Suppose that the adjuster decides to change the 1% premise to 1.5%. For this revised premise, use the same approximation as in part (c) to determine the probability that the sample will have at most three blatantly fraudulent claims.

6.45 The conditions for good approximations, as noted on pages 214 and 217, can be combined.

    (a) Under what conditions on a, b, and n would it be reasonable to approximate a hypergeometric distribution by a Poisson distribution?

    (b) In terms of a, b, and n, what is the value of np to be used in the Poisson calculation?

✓ **6.46** The number of minor injuries a football coach can expect during the course of a game is a random variable having the Poisson distribution with $\lambda = 4.4$. Find the probability that during the course of a game there will be at most three minor injuries.

6.47 If an ambulance service receives, on the average, five emergency calls per day, what is the probability that it will receive six emergency calls on any given day?

6.48 In the inspection of a fabric produced in continuous rolls, the number of imperfections spotted by an inspector during a 5-minute period is a random variable having the Poisson distribution with $\lambda = 3.4$. What are the probabilities that during a 5-minute period an inspector will spot more than one imperfection?

6.49 If a police department receives, on the average, $\lambda = 7$ reports of crimes per day, what is the probability that it will receive only 4 or 5 reports of crimes on a given day?

6.50 A credit card company receives, on the average, $\lambda = 6$ requests per day to replace lost or stolen credit cards. What is the probability that it will receive only 4 requests to replace lost or stolen credit cards on a certain day?

6.51 Using the data of Exercise 6.50, what is the probability that the credit card company will receive 7 requests, on a certain day, to replace lost or stolen credit cards?

6.52 Use appropriate computer software or a graphing calculator to rework Exercise 6.41 by determining the binomial probability of "3 successes in 100 trials" with $p = 0.019$. Compare the result with 0.171, the value obtained in Exercise 6.41.

6.53 With reference to Exercise 6.46, use appropriate software or a graphing calculator to determine the probabilities that during the course of a game there will be

    (a) at most six minor injuries;

    (b) at least four minor injuries.

As we indicated in Exercise 6.46, the number of minor injuries during the course of a game is a random variable having the Poisson distribution with $\lambda = 4.4$.

6.54 With reference to Exercise 6.48, use a computer printout of the Poisson distribution with $\lambda = 3.4$ to determine the probabilities that during a 5-minute interval an inspector will spot

    (a) three or four imperfections;

    (b) at most five imperfections;

    (c) at least six imperfections;

    (d) anywhere from one to three imperfections.

Round your answers to four decimals.

6.5 THE MULTINOMIAL DISTRIBUTION

An important generalization of the binomial distribution arises when there are more than two possible outcomes for each trial, the probabilities of the various outcomes remain the same for each trial, and the trials are all independent. This is the case, for example, when we repeatedly roll a die, where each trial has six possible outcomes; or when students are asked whether they like a certain new recording, dislike it, or don't care; or when a U.S. Department of Agriculture inspector grades beef as prime, choice, good, commercial, or utility.

If there are k possible outcomes for each trial and their probabilities are $p_1, p_2, \ldots,$ and p_k, it can be shown that the probability of x_1 outcomes of the first kind, x_2 outcomes of the second kind, $\ldots$, and x_k outcomes of the kth kind in n trials is given by

MULTINOMIAL DISTRIBUTION

$$\frac{n!}{x_1! x_2! \cdot \ldots \cdot x_k!} p_1^{x_1} \cdot p_2^{x_2} \cdot \ldots \cdot p_k^{x_k}$$

In using this expression, it is required that

$$x_1 + x_2 + \cdots + x_k = n \quad \text{and} \quad p_1 + p_2 + \cdots + p_k = 1$$

EXAMPLE 6.18   In a certain city, channel 3 has 50% of the viewing audience on Saturday nights, channel 12 has 30%, and channel 10 has 20%. What is the probability that among eight television viewers randomly selected in that city on a Saturday night, five will be watching channel 3, two will be watching channel 12, and one will be watching channel 10?

Solution   Substituting $n = 8$, $x_1 = 5$, $x_2 = 2$, $x_3 = 1$, $p_1 = 0.50$, $p_2 = 0.30$, and $p_3 = 0.20$ into the formula for the multinomial distribution, we get

$$\frac{8!}{5! \cdot 2! \cdot 1!} (0.50)^5 (0.30)^2 (0.20)^1 = 0.0945$$

Exercise 6.55 is a practice exercise; its complete solution is given on page 235.

✓ **6.55**   The percentage distribution of money incomes in U.S. households in a recent year was

Dollars	Percentage of households
Under 14,999	21.0
15,000–49,999	30.1
50,000–74,999	34.0
75,000 and over	14.9

What is the probability that among twelve randomly chosen households, one has income under $14,999, three have incomes between $15,000 and $49,999, seven have incomes between $50,000 and $74,999, and one has income of $75,000 and over?

6.56   In the service garage of a large automobile dealer, customers spend the following amounts, during each visit, for maintenance and repair of their motor vehicles.

Dollars	Percentage of customers
Less than $200	40
$200 but less than $400	40
$400 or more	20

What is the probability that on a certain day, among six randomly selected customers, three will spend less than $200, two will spend $200 and under $400, and one will spend $400 or more?

6.57   It can readily be shown that the probabilities of getting two heads, a head, and a tail, and two tails when flipping a pair of balanced coins are, respectively, $\frac{1}{4}, \frac{1}{2}$, and $\frac{1}{4}$. What is the probability of getting two heads once, a head and a tail twice, and two tails twice in five flips of a pair of balanced coins?

6.58   At the terminal of a certain airline, records disclose that 10% of their flights arrive early, 70% arrive on time, 12% arrive somewhat late, and 8% arrive very late. Use the formula for the multinomial distribution to determine the probability that in six randomly selected flights, one will arrive early, three will arrive on time, two will be somewhat late, and none will be very late.

6.59   The probability that a randomly selected person contacted by telephone will respond to a set of questions about cosmetics purchases is 0.40.

(a)   Using the binomial distribution, find the probability that exactly 6 of 15 persons selected at random will respond to the set of questions.

(b)   Use the multinomial distribution to answer the same question. Assume that $k = 2$ and that the probabilities of response and nonresponse are 0.40 and 0.60.

(c)   Can you give a relationship between the binomial distribution and the multinomial distribution with $k = 2$?

6.6   THE MEAN OF A PROBABILITY DISTRIBUTION

Let us again direct our attention to the concept of a mathematical expectation, which we explained in Section 4.5. There we used several examples and illustrations of mathematical expectations and made the following statement: When we said that a child ... can *expect* to visit a dentist 1.92 times a year, we referred to the result that was obtained by multiplying $0, 1, 2, 3, \ldots$, by the corresponding probabilities that such a child will visit a dentist that many times a year and then adding all these products.

If we apply the same argument to the probability distribution of rolling a balanced die, we find that the number of points that we can expect in one roll of a die is

$$1 \cdot \frac{1}{6} + 2 \cdot \frac{1}{6} + 3 \cdot \frac{1}{6} + 4 \cdot \frac{1}{6} + 5 \cdot \frac{1}{6} + 6 \cdot \frac{1}{6} = 3\frac{1}{2}$$

Similarly, the number of heads that we can expect in three flips of a balanced coin is

$$0 \cdot \frac{1}{8} + 1 \cdot \frac{3}{8} + 2 \cdot \frac{3}{8} + 3 \cdot \frac{1}{8} = 1\frac{1}{2}$$

Of course, we cannot roll $3\frac{1}{2}$ with a die or get $1\frac{1}{2}$ heads; like all mathematical expectations, these figures must be looked upon as averages. In general, if a random variable takes on the values $x_1, x_2, x_3, \ldots,$ or x_k, with the probabilities $f(x_1), f(x_2), f(x_3), \ldots,$ or $f(x_k)$, its expected value (or its mathematical expectation) is given by

$$x_1 \cdot f(x_1) + x_2 \cdot f(x_2) + x_3 \cdot f(x_3) + \cdots + x_k \cdot f(x_k)$$

We refer to this quantity as the **mean of a probability distribution** of the random variable, and, in the $\sum$ notation, we write

MEAN OF THE
PROBABILITY
DISTRIBUTION

$$\mu = \sum x \cdot f(x)$$

where the summation extends over all values taken on by the random variable. Like the mean of a population, it is denoted by the Greek letter μ (lowercase mu). The notation is the same, for as we pointed out in connection with the binomial distribution, when we observe any random variable, we use its probability distribution to identify the population that we are sampling. For instance, Figure 6.1 pictures the population that we are sampling when we observe a value of a random variable having the binomial distribution with $n = 5$ and $p = 0.60$, and Figure 6.3 pictures the population we are sampling when we observe a value of a random variable having the hypergeometric distribution with $a = 5, b = 11$, and $n = 3$.

EXAMPLE **6.19**   Use the probabilities obtained in the shopping example, Example 6.6, to determine how many of five persons shopping at the given market can be expected to spend at least $50.

Solution   Substituting $x = 0, 1, 2, 3, 4,$ and 5 and the corresponding probabilities into the formula for μ, we get

$$\mu = 0(0.010) + 1(0.077) + 2(0.230) + 3(0.346) + 4(0.259) + 5(0.078)$$
$$= 3.001$$

or approximately 3. Thus, three out of five shoppers can be expected to spend at least $50 at the given market.

When a random variable can take on many different values, the calculation of μ becomes very laborious. For instance, if we want to know how many of 400 persons attending a movie can be expected to buy popcorn (when the probability is, say, 0.30 that any one of them will buy popcorn), we will first have to calculate the 401 probabilities corresponding to $0, 1, 2, \ldots,$ or 400 of them buying popcorn. We might argue, however, that if 30% of all moviegoers buy popcorn, then we can expect that 30% of our 400, or $400(0.30) = 120$, will buy popcorn. Similarly, if a balanced coin is flipped 500 times, we can argue that in the long run heads will come up 50% of the time and, hence, that we can expect to get $500(0.50) = 250$ heads in 500 flips of a balanced coin. These two values are, indeed, correct; both problems deal with random variables having binomial distributions, and it can be shown that in general

MEAN OF THE BINOMIAL DISTRIBUTION

$$\mu = n \cdot p$$

for the **mean of the binomial distribution**. In words, the mean of a binomial distribution is the product of the number of trials and the probability of success on an individual trial.

EXAMPLE 6.20   Use the formula from Example 6.19 to find the mean of the probability distribution of the number of persons, among five, who will spend at least $50 at the given market.

Solution   Since we are dealing with a binomial distribution with $n = 5$ and $p = 0.60$, we find that $\mu = 5(0.60) = 3$, and it should be apparent that the small difference of 0.001 between this exact value and the one obtained before is due to rounding the probabilities to three decimals. ■

EXAMPLE 6.21   Find the mean of the probability distribution of the number of heads obtained in three flips of a balanced coin.

Solution   Since we are dealing with a binomial distribution with $n = 3$ and $p = \dfrac{1}{2}$, we get $\mu = 3 \cdot \dfrac{1}{2} = 1\dfrac{1}{2}$, and this agrees with the result obtained in the beginning of this section. ■

It is important to remember that the formula $\mu = np$ applies only to binomial distributions. There are other formulas for other distributions; for instance, for the hypergeometric distribution, the formula for the mean is $\mu = n\dfrac{a}{a+b}$, and for the Poisson distribution it is $\mu = \lambda$. Proofs of these special formulas may be found in most textbooks on mathematical statistics.

6.7   THE STANDARD DEVIATION OF A PROBABILITY DISTRIBUTION

In Chapter 3 we saw that there are many situations in which we must not only calculate the mean or some other measure of location, but also describe the variability (spread or dispersion) of a set of data. As we indicated in that chapter, the most widely used measure of variation is the variance and its square root, the standard deviation. For probability distributions, we measure variability in almost the same way, but instead of averaging the squared deviations from the mean, we calculate their expected value. If x is a value of a random variable whose probability distribution has the mean μ, its deviation from the mean is $x - \mu$, and we define the **variance of a probability distribution** as the expected value (mathematical expectation) of the squared deviation from the mean, namely, as,

VARIANCE OF A PROBABILITY DISTRIBUTION

$$\sigma^2 = \sum (x - \mu)^2 \cdot f(x)$$

where the summation extends over all values taken on by the random variable. As in the preceding section and for the same reason, we denote descriptions of probability distributions with the same symbols as the corresponding descriptions of populations. The square root of the variance defines the **standard deviation of a probability distribution**, and we write

STANDARD DEVIATION OF A PROBABILITY DISTRIBUTION

$$\sigma = \sqrt{\sum (x - \mu)^2 \cdot f(x)}$$

Note that this formula is like that of the standard deviation of a population as defined on page 70, with the probability $f(x)$ substituted for $\frac{1}{N}$.

EXAMPLE 6.22 Use the probabilities obtained in the shoppers' expenditures of Example 6.6, which are $f(0) = 0.010$, $f(1) = 0.077$, $f(2) = 0.230$, $f(3) = 0.346$, $f(4) = 0.259$, and $f(5) = 0.078$ to determine the standard deviation of the distribution of the number of persons, among five, who will spend at least $50 at the given market.

Solution Since we have already shown that $\mu = 3$, we can arrange the necessary calculations as in the following table:

Number of persons x	$x - \mu$	$(x - \mu)^2$	Probability $f(x)$	$(x - \mu)^2 \cdot f(x)$
0	-3	9	0.010	0.090
1	-2	4	0.077	0.308
2	-1	1	0.230	0.230
3	0	0	0.346	0.000
4	1	1	0.259	0.259
5	2	4	0.078	0.312
			Total	$\sigma^2 = 1.199$

The values in the column on the right were obtained by multiplying each squared deviation from the mean by its probability, and their sum is the variance of the distribution. To find the standard deviation, we must obtain the square root of 1.199. This is found readily on most calculators as

$$\sqrt{1.199} \approx 1.0949885 \approx 1.095$$

The calculations were easy in this example since the deviations from the mean were all small integers. When the deviations from the mean are large numbers, or when they are given to several decimal places, it is usually worthwhile to simplify the calculations by using a special computing formula for the variance, which will be given in Exercise 6.64.

As in the case of the mean, the calculation of the variance or the standard deviation of a probability distribution can generally be simplified when we deal with special kinds of probability distributions. For instance, for the binomial distribution we have the formula

STANDARD DEVIATION OF THE BINOMIAL DISTRIBUTION

$$\sigma = \sqrt{np(1-p)}$$

EXAMPLE 6.23   Use the preceding formula to verify the result obtained in the previous example.

Solution   Since we are dealing with a binomial distribution with $n = 5$ and $p = 0.60$, we get

$$\sigma = \sqrt{5(0.60)(0.40)} = \sqrt{1.20} \approx 1.095$$

This agrees with the results obtained previously.

EXAMPLE 6.24   Find the variance of the probability distribution of the number of heads obtained in three flips of a balanced coin.

Solution   Since we are dealing with a binomial distribution with $n = 3$ and $p = \dfrac{1}{2}$, we note that

$$1 - p = \frac{1}{2} \quad \text{and} \quad \sigma^2 = 3 \cdot \frac{1}{2} \cdot \frac{1}{2} = \frac{3}{4}$$

For the hypergeometric distribution, the variance is given by

$$\sigma^2 = \frac{nab}{(a+b)^2} \cdot \frac{a+b-n}{a+b-1}$$

and the standard deviation is the square root of this. For the Poisson distribution, the standard deviation is simply $\sigma = \sqrt{\lambda}$. The derivations of these formulas may be found in more advanced texts.

EXERCISES

Exercises 6.60, 6.66, and 6.67 are practice exercises; their complete solutions are given on page 235.

✓ **6.60**   The probabilities that a building inspector will observe 0, 1, 2, 3, 4, or 5 violations of the building code in a home built in a large development are given in the following table:

Number of violations	0	1	2	3	4	5
Probability	0.41	0.22	0.17	0.13	0.05	0.02

Find the mean of this probability distribution.

6.61   The probability that Mary will buy lunch at the school cafeteria on 0, 1, 2, 3, 4, or 5 days during a week are given in the following table:

Number of days lunches purchased	0	1	2	3	4	5
Probability	0.05	0.08	0.12	0.25	0.30	0.20

Find the mean of this probability distribution.

6.62   Suppose that the probabilities are 0.50, 0.25, 0.15 and 0.10 that 0, 1, 2, or 3 students are absent from a certain class on any day.

(a) Find the mean of this probability distribution.

(b) Find the variance of this probability distribution.

6.63   The following table gives the probabilities that a jeweler will sell 0, 1, 2, 3, 4, or 5 wrist watches on any given day.

Number of watches	0	1	2	3	4	5
Probability	0.05	0.20	0.30	0.25	0.15	0.05

(a) Find the mean of this probability distribution.

(b) Find the standard deviation of this probability distribution.

6.64   The calculation of the variance and the standard deviation is easy where μ and the deviations from the mean are all integers. When this is not the case, as in Exercise 6.63, the calculations can be somewhat simplified by using the computing formula

$$\sigma^2 = \sum x^2 \cdot f(x) - \mu^2$$

To calculate μ, we add the products obtained by multiplying each value of x by the corresponding probability. Then we add the products obtained by multiplying the square of each x by the corresponding probability and substitute these two sums into the formula for σ^2. Use this computing formula for σ^2 to rework the standard deviation of Exercise 6.63.

6.65   On page 224 we showed that $\mu = 3.5$ for the number of points we roll with a balanced die, that is, for a random variable that takes on the values 1, 2, 3, 4, 5, or 6 with equal probabilities of $\frac{1}{6}$. Find the standard deviation of this probability distribution using

(a) the formula on page 226 that defines σ^2;

(b) the computing formula given in Exercise 6.64.

✓ **6.66**   Find the mean of the binomial distribution with $n = 4$ and $p = 0.10$

(a) by looking up the probabilities in Table I and then using the formula that defines the mean of a probability distribution;

(b) by using the special formula for the mean of the binomial distribution.

✓ **6.67**   Find the standard deviation of the binomial distribution with $n = 4$ and $p = 0.10$

(a) by using the value of μ obtained in part (b) of Exercise 6.66, the probabilities looked up in part (a) of that exercise, and the formula that defines the standard deviation of a probability distribution;

(b) by using the special formula for the standard deviation of the binomial distribution.

6.68   As can easily be verified by listing all possibilities or by using the formula for the binomial distribution, the probabilities of getting 0, 1, 2, 3, 4, or 5 heads in five flips of a balanced coin are given in this table:

Number of heads	0	1	2	3	4	5
Probability	$\frac{1}{32}$	$\frac{5}{32}$	$\frac{10}{32}$	$\frac{10}{32}$	$\frac{5}{32}$	$\frac{1}{32}$

Find the mean of this probability distribution using

    (a) the formula that defines μ;

    (b) the special formula for the mean of a binomial distribution.

6.69 With reference to Exercise 6.68, find the variance of the probability distribution using

    (a) the computing formula given in Exercise 6.64;

    (b) the special formula for the variance of a binomial distribution.

6.70 A traffic study shows that 90% of motor vehicles that come to a stop for a stop signal at a certain intersection have to wait at least 15 seconds before proceeding. Find the mean of the distribution of the number of motor vehicles among 10 randomly selected motor vehicles that come to this stop signal and have to wait at least 15 seconds before proceeding

    (a) by looking up the probabilities in Table I and then using the formula that defines the mean of a probability distribution;

    (b) by using the special formula for the mean of the binomial distribution.

6.71 With reference to Exercise 6.70, where $\mu = 9$ motor vehicles, find the standard deviation of the number of motor vehicles that have to wait at least 15 seconds before proceeding

    (a) using the formula for the standard deviation of a probability distribution;

    (b) using the special formula for the standard deviation of the binomial distribution.

6.72 If 95% of the bottles collected in a recycling bin are suitable for recycling, find the mean and the standard deviation of the distribution of the number of these bottles among 15 selected at random that are suitable for recycling, using the special formulas for the mean and the standard deviation of a binomial distribution.

6.73 In general, the mean of a binomial distribution can be found by using the formula $\mu = np$. Using this formula, find

    (a) the mean number of heads obtained in 5 flips of a balanced coin;

    (b) the mean number of persons in a sample of 10 persons who will miss the departure of a passenger plane if the probability of missing the departure (*no shows*) is 0.05;

    (c) the mean number of *seconds* (slightly flawed) shirts produced among 6 shirts, where the probability of producing a *second* is 0.10.

6.74 The standard deviation of a binomial distribution can be found using the formula $\sigma = \sqrt{np(1-p)}$. Using this formula, find the standard deviation of the data in parts (a), (b), and (c) of Exercise 6.73.

6.75 The probabilities for Figure 6.3 are

Number of trucks	0	1	2	3
Probability	0.295	0.491	0.196	0.018

Use these probabilities to calculate the mean of the given hypergeometric distribution. Verify the result by using the special formula

$$\mu = n \cdot \frac{a}{a+b}$$

6.76   The probabilities that there will be 0, 1, 2, 3, 4, or 5 fires caused by lightning during a summer storm are as follows:

Number of fires caused by lightning	0	1	2	3	4	5
Probability	0.449	0.360	0.144	0.038	0.008	0.001

Calculate the mean of this Poisson distribution with $\lambda = 0.8$, and use the result to verify the special formula $\mu = \lambda$ mentioned on page 226.

6.8 CHEBYSHEV'S THEOREM

Intuitively speaking, the variance and the standard deviation of a probability distribution measure the expected size of the chance fluctuations of a corresponding random variable. When σ is small, the probability is high that we will get a value close to the mean, and when σ is large, we are more likely to get a value far away from the mean. This important idea is expressed formally by Chebyshev's theorem, which we introduced in Section 3.7, as it pertains to numerical data. For probability distributions, **Chebyshev's theorem** can be stated as follows:

CHEBYSHEV'S THEOREM

> *The probability that a random variable will take on a value within k standard deviations of the mean is at least* $1 - \dfrac{1}{k^2}$.

Thus, the probability of getting a value within two standard deviations of the mean (a value between $\mu - 2\sigma$ and $\mu + 2\sigma$) is at least $1 - \dfrac{1}{2^2} = \dfrac{3}{4}$, the probability of getting a value within five standard deviations of the mean (a value between $\mu - 5\sigma$ and $\mu + 5\sigma$) is at least $1 - \dfrac{1}{5^2} = \dfrac{24}{25}$, and so forth.

EXAMPLE 6.25   The number of telephone calls that an answering service receives between 9 A.M. and 10 A.M. is a random variable whose distribution has the mean $\mu = 27.5$ and the standard deviation $\sigma = 3.2$. What does Chebyshev's theorem with $k = 3$ tell us about the number of telephone calls that the answering service may receive between 9 A.M. and 10 A.M.?

Solution   Since

$$\mu - 3\sigma = 27.5 - 3(3.2) = 17.9 \quad \text{and} \quad \mu + 3\sigma = 27.5 + 3(3.2) = 37.1$$

we can assert with a probability of at least

$$1 - \frac{1}{3^2} = \frac{8}{9}$$

or approximately 0.89, that the answering service will receive between 17.9 and 37.1 calls (that is, anywhere from 18 to 37 calls).   ∎

EXAMPLE 6.26   What does Chebyshev's theorem with $k = 6$ tell us about the number of heads that we may get in 400 flips of a balanced coin?

Solution   Here $n = 400$ and $p = \dfrac{1}{2} = 0.5$, the mean and the standard deviation of the number of heads are

$$\mu = n \cdot p = 400 \cdot \frac{1}{2} = 200$$

and

$$\sigma = \sqrt{np(1-p)} = \sqrt{400(0.5)(0.5)} = \sqrt{100} = 10$$

Then

$$\mu - 6\sigma = 200 - 6(10) = 140 \quad \text{and} \quad \mu + 6\sigma = 200 + 6(10) = 260$$

Thus, we can assert with a probability of at least

$$1 - \frac{1}{6^2} = \frac{35}{36}$$

or approximately 0.97, that we will get between 140 and 260 heads.   ∎

If we convert the numbers of heads into proportions in the preceding example, we can assert with a probability of at least $\dfrac{35}{36}$ that the proportion of heads that we get in 400 flips of a balanced coin will lie between

$$\frac{140}{400} = 0.35 \quad \text{and} \quad \frac{260}{400} = 0.65$$

To continue this argument, the reader will be asked to show in Exercise 6.82 that the probability is at least $\dfrac{35}{36}$ that for 10,000 flips of a balanced coin the proportion of heads will lie between 0.47 and 0.53, and that for 1,000,000 flips of a balanced coin it will lie between 0.497 and 0.503. All this provides support for the Law of Large Numbers, which we mentioned in Section 4.5 in connection with the frequency interpretation of probability. In actual practice, Chebyshev's theorem is very rarely used, since the probability $1 - \dfrac{1}{k^2}$ is usually unnecessarily small. For instance, in the preceding example we showed that the probability of getting a value within six standard deviations of the mean is at least 0.97, whereas the actual probability that this will happen for a random variable having the binomial distribution with $n = 400$ and $p = \dfrac{1}{2}$ is about 0.999999998. "At least 0.97" is not wrong, but for most practical purposes it does not tell us quite enough. In spite of the difference between these probabilities, Chebyshev's theorem cannot be improved, because sometimes it is exact, as will be noted in Exercise 6.82.

Exercise 6.77 is a practice exercise; its complete solution is given on page 235.

✓ **6.77**   The number of customers to whom a restaurant serves breakfast on a weekday morning is a random variable with $\mu = 142$ and $\sigma = 12$. According to Chebyshev's theorem, with what probability can we assert that between 94 and 190 customers will have breakfast there on a weekday morning?

6.78   A student answers the 144 questions on a true–false test by flipping a balanced coin.

(a)   Use the special formulas for the mean and the standard deviation of the binomial distribution to find the values of μ and σ for the distribution of the number of correct answers that he or she will get.

(b)   According to Chebyshev's theorem, what can we assert with a probability of at least 0.96 about the number of correct answers that he or she will get?

6.79   Use Chebyshev's theorem to determine the probability of getting a value

(a)   within 4 standard deviations of the mean;

(b)   within 2.5 standard deviations of the mean;

(c)   within 2.8 standard deviations.

6.80   The number of CDs (compact disks) sold daily in a music store is a random variable with $\mu = 130$ and $\sigma = 10$. What does Chebyshev's theorem with $k = 3$ tell us about the number of CDs sold each day?

6.81   The number of dinners served at a restaurant each evening is a random variable with $\mu = 50$ and $\sigma = 5$. Using Chebyshev's theorem, at least how many dinners can the restaurant expect to serve each evening, given that

(a) $k = 2$

(b) $k = 3$?

6.82   Use Chebyshev's theorem to show that the probability is at least $\dfrac{35}{36}$ that

(a)   in 10,000 flips of a balanced coin there will be between 4,700 and 5,300 heads, and hence the proportion of heads will be between 0.47 and 0.53;

(b)   in 1,000,000 flips of a balanced coin there will be between 497,000 and 503,000 heads, and hence the proportion of heads will be between 0.497 and 0.503.

✓ SOLUTIONS TO PRACTICE EXERCISES

6.1   (a)   The values are all on the interval from 0 to 1, but since $0.37 + 0.35 + 0.30 = 1.02$, and not 1, the values cannot serve as the values of a probability distribution.

(b)   The values are all on the interval from 0 to 1, and since their sum is 1, they can serve as the values of a probability distribution.

(c)   Since one of the values is negative, they cannot serve as the values of a probability distribution.

6.5   (a)   Substituting $n = 6$, $p = 0.10$, and $x = 2$ into the formula for the binomial distribution, we get

$$\binom{6}{2}(0.10)^2(0.90)^4 = 15(0.01)(0.6561) = 0.098415 \approx 0.098$$

(b) Following the procedure of part (a), we find that the probabilities for $x = 0$ and $x = 1$ are 0.531441 and 0.354294, respectively. Adding these two values, we get $0.885735 \approx 0.886$.

6.6 (a) Substituting $n = 4$, $p = 0.70$, and $x = 2$ into the formula for the binomial distribution, we get

$$f(2) = \binom{4}{2}(0.70)^2(1 - 0.70)^{4-2} = 6(0.49)(0.09) = 0.2646 \approx 0.265$$

(b) The value in Table I is 0.265.

6.7 (a) $0.250 + 0.154 + 0.044 = 0.448$;

(b) $0.002 + 0.009 + 0.032 + 0.086 + 0.172 = 0.301$.

6.18 (a) The probability of not getting 10 items in perfect condition is $1 - 0.599 = 0.401$.

(b) The probability of getting 10 items in perfect condition is 0.349.

(c) The probability of getting 10 items in perfect condition is 0.107.

6.24

$$\frac{\binom{7}{3}\binom{9}{2}}{\binom{16}{5}} = \frac{35 \cdot 36}{4,368} = \frac{1,260}{4,368} \approx 0.288.$$

6.25 (a)

$$\frac{\binom{5}{0}\binom{10}{2}}{\binom{15}{2}} = \frac{1 \cdot 45}{105} = \frac{45}{105} \approx 0.429;$$

(b)

$$\frac{\binom{5}{1}\binom{10}{1}}{\binom{15}{2}} = \frac{5 \cdot 10}{105} = \frac{50}{105} \approx 0.476;$$

(c)

$$\frac{\binom{5}{2}\binom{10}{0}}{\binom{15}{2}} = \frac{10 \cdot 1}{105} = \frac{10}{105} \approx 0.095.$$

6.32 (a) Since $n = 15$ exceeds $0.05(40 + 160) = 10$, the condition is not satisfied.

(b) Since $n = 22$ does not exceed $0.05(380 + 120) = 25$, the condition is satisfied.

(c) Since $n = 30$ does not exceed $0.05(400 + 240) = 32$, the condition is satisfied.

6.33 (a) Substituting $a = 6$, $b = 194$, $n = 3$, and $x = 1$ into the formula for the hypergeometric distribution, we get

$$f(1) = \frac{\binom{6}{1}\binom{194}{2}}{\binom{200}{3}} = \frac{6 \cdot 18,721}{1,313,\,400} \approx 0.086$$

(b) Substituting $n = 3$, $p = \dfrac{6}{200} = 0.03$, and $x = 1$ into the formula for the binomial distribution, we get

$$f(1) = \binom{3}{1} (0.03)^1 (1 - 0.03)^{3-1} = 3(0.03)(0.97)^2 \approx 0.085$$

6.38 (a) Since $n = 300$ is not less than 100 and $np = 300 \cdot \dfrac{1}{50} = 6$ is less than 10, the conditions are satisfied.

(b) Since $n = 900$ is not less than 100 and $np = 900 \cdot \dfrac{1}{100} = 9.0$ is less than 10, the conditions are satisfied.

(c) $n = 400$ is not less than 100, but since $np = 400 \cdot \dfrac{1}{25} = 16$ exceeds 10, the conditions are not satisfied.

6.40 The Poisson approximation with $np = 250(0.02) = 5.00$ yields

$$f(4) = 0.0067 \cdot \frac{5.0^4}{4!} = 0.0067 \left(\frac{625}{24} \right) = 0.174$$

6.46 Substituting $\lambda = 4.4$ and, respectively, $x = 0, 1, 2$, and 3 into the formula for the Poisson distribution, we get

$$f(0) = \frac{4.4^0 \cdot e^{-4.4}}{0!} = e^{-4.4} \approx 0.012$$

$$f(1) = \frac{4.4^1 \cdot 0.012}{1!} \approx 0.053$$

$$f(2) = \frac{4.4^2 \cdot 0.012}{2!} \approx 0.116$$

$$f(3) = \frac{4.4^3 \cdot 0.012}{3!} \approx 0.170$$

and the answer is $0.012 + 0.053 + 0.116 + 0.170 \approx 0.351$.

6.55 Substituting $n = 12$, $x_1 = 1$, $x_2 = 3$, $x_3 = 7$, and $x_4 = 1$; and $p_1 = 0.210$, $p_2 = 0.301$, $p_3 = 0.340$, and $p_4 = 0.149$ into the formula for the multinomial distribution, we get

$$\frac{12!}{1!3!7!1!} \cdot (0.210)^1 (0.301)^3 (0.340)^7 (0.149)^1$$

$$= (15{,}840)(0.210)(0.02727)(0.000525)(0.149) \approx 0.0071$$

6.60 $\mu = 0(0.41) + 1(0.22) + 2(0.17) + 3(0.13) + 4(0.05) + 5(0.02) = 1.25$.

6.66 (a) $\mu = 0(0.656) + 1(0.292) + 2(0.049) + 3(0.004) = 0.402$;

(b) $\mu = 4(0.10) = 0.40$.

6.67 (a) $\sigma^2 = (0 - 0.4)^2 (0.656) + (1 - 0.4)^2 (0.292) + (2 - 0.4)^2 (0.049)$
$\qquad + (3 - 0.4)^2 (0.004) = 0.36256$
$\qquad \sigma = \sqrt{0.36256} \approx 0.602$;

(b) $\sigma = \sqrt{4(0.10)(0.90)} = 0.60$.

6.77 Since $142 - 12k = 94$ and $142 + 12k = 190$, it follows that $12k = 48$ and $k = 4$; thus, the probability is at least

$$1 - \frac{1}{4^2} = \frac{15}{16}, \text{ or approximately } 0.94$$

7

THE NORMAL DISTRIBUTION

Continuous sample spaces and continuous random variables arise whenever we deal with quantities that are measured on a continuous scale, for instance, when we measure the amount of alcohol in a person's blood, the net weight of a package of frozen food, the amount of tar in a cigarette, the speed of a car, and so forth.

In the continuous case, the place of histograms is taken by continuous curves. These may be pictured as in Figure 7.1, where we can think of histograms with narrower and narrower classes approaching the continuous curve.

Among the many continuous distribution curves used in statistics, by far the most important is the **normal distribution**, which is sometimes described simply as a bell-shaped distribution. This description leaves a good deal to be desired; it is true that normal distribution curves, often called **normal curves**, are bell shaped, but not all bell-shaped distribution curves are necessarily those of normal distributions. The normal distribution dates back to the work of the French-English mathematician Abraham de Moivre (1667–1745), who studied a curve which closely approximates the binomial distribution when the number of trials is very large.

In this chapter we shall study continuous random variables. The concept of a continuous distribution will be introduced in Section 7.1, followed by that of a

Figure 7.1
Continuous distribution
curve.

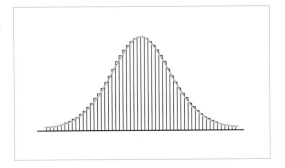

normal distribution in Section 7.2. Various applications of the normal distribution
will be discussed in Sections 7.3 and 7.4.

7.1 CONTINUOUS DISTRIBUTIONS

In all the histograms shown in preceding chapters, the frequencies, percentages,
or probabilities are represented by the heights of the rectangles; but as long as
the class intervals are all equal, we can also say that the frequencies, percentages,
or probabilities are represented by the areas of the rectangles. The latter is
preferable for the work of this chapter, for in the continuous case we also
represent probabilities by areas—not by areas of rectangles, but, as is illustrated
in Figure 7.2, by areas under continuous curves. The diagram on the left shows a
histogram of the probability distribution of a random variable that takes on only
the values 0, 1, 2, ..., and 10, and the probability of getting a 3, for example,
is given by the area of the blue rectangle. The diagram on the right refers to a
continuous random variable, which can take on any value on the interval from
0 to 10, and the probability of getting a value on the interval from 2.5 to 3.5 is
given by the area of the blue region under the curve. Similarly, the area of the
dark blue region under the curve gives the probability of getting a value greater
than 8.

Continuous curves such as the one shown in the right-hand diagram of Figure
7.2 are the graphs of functions that we refer to as **probability densities** or, more
informally, as **continuous distributions**. The first of these terms is borrowed from
the language of physics, where the terms *weight* and *density* are used in very much
the same way in which we use the terms *probability* and *probability density* in
statistics. What characterizes a probability density is the fact that

Figure 7.2
Histogram of a prob-
ability distribution and
graph of a continuous
distribution.

Figure 7.3
Probability density.

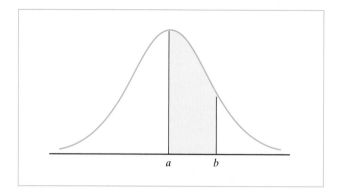

The area under the curve between any two values a and b (see Figure 7.3) gives the probability that a random variable having this continuous distribution will take on a value on the interval from a to b.

Observe that when a and b are very close together the heights of the density at a and at b, and at the midpoint $\dfrac{a+b}{2}$ as well, are nearly equal. Thus the area under the curve between a and b is very nearly equal to

$$\left(\text{height of the density at } \frac{a+b}{2}\right) \cdot (b-a)$$

It follows that the values of a probability density should not be negative and that the total area under the curve, representing the certainty that a random variable must take on one of its values, is always equal to 1. This corresponds to the two rules about probability distributions given near the end of Section 6.1. There is no requirement that the heights of a probability density be less than 1.

EXAMPLE 7.1   If a continuous random variable has the probability density shown in Figure 7.4, find the probabilities that it will take on a value

  (a) between -2 and 3;
  (b) greater than 1;
  (c) greater than or equal to 1.

Solution   **(a)** Since this is the total area under the curve, it must be equal to 1; also, multiplying the base of the rectangle by its height, we get

$$[3 - (-2)] \cdot \frac{1}{5} = 5 \cdot \frac{1}{5} = 1$$

  (b) Multiplying the base of the corresponding rectangle by its height, we get

$$(3 - 1) \cdot \frac{1}{5} = 2 \cdot \frac{1}{5} = \frac{2}{5}$$

  (c) The answer is the same as that to part (b), $\dfrac{2}{5}$.

Figure 7.4
A special probability
density.

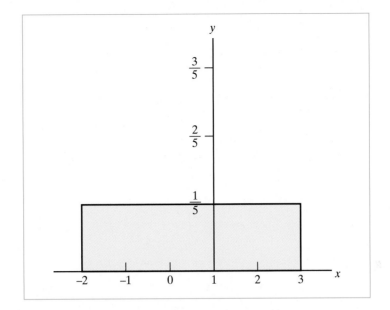

Parts (b) and (c) of the preceding example illustrate the fact that the probability that a random variable having a continuous distribution will take on a particular value is always zero. In this case, the base of the rectangle has shrunk to zero, and so has the area under the curve.

It is a consequence of measuring (rather than counting) that we must assign probability zero to any particular outcome. We claim that the probability must be zero that an individual will have a weight of *exactly* 145.27 pounds or that a horse will run a race in *exactly* 58.442 seconds. However, observe that, even though every particular outcome has probability zero, the process will still produce a value (whether or not we can measure it with extra-fine precision); apparently, then, events of probability zero not only can occur, but *must* occur when dealing with measured random quantities.

Statistical descriptions of continuous distributions, such as the mean and the standard deviation, are as important as descriptions of probability distributions or distributions of observed data. Informally, we can always picture continuous distributions as being approximated by histograms of probability distributions (see Figure 7.1) for which the mean and the standard deviation can be determined by means of the formulas of Sections 6.6 and 6.7. Then, if we choose histograms with narrower and narrower classes, the means and the standard deviations of the corresponding probability distributions will approach the mean and the standard deviation of the continuous distribution.

Since formal definitions of the mean and the standard deviation of a continuous distribution cannot be given without the use of calculus, they will be omitted in this text. Again informally, the mean μ of a continuous distribution is a measure of its *center* or *middle*, and the standard deviation σ of a continuous distribution is a measure of its *dispersion* or *spread*.

7.2 THE NORMAL DISTRIBUTION

The normal distribution is often referred to as the cornerstone of modern statistics. This is due partly to its role in the development of statistical theory and partly to the fact that distributions of observed data often have the same general pattern as normal distributions.

This pattern is shown in Figure 7.5, but it cannot be seen from such a small drawing that the bell-shaped curve extends indefinitely in both directions, coming closer and closer to the horizontal axis without ever reaching it. Fortunately, it is seldom necessary to extend the *tails* of a normal distribution very far, because the area under the curve becomes negligible once we go more than four or five standard deviations away from the mean.

Figure 7.5
Normal distribution.

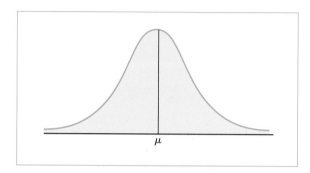

Since the normal distribution has a complicated mathematical equation, let us merely point out that this distribution is completely determined by the values of μ and σ. In other words, there is one and only one normal distribution with a given mean μ and a given standard deviation σ. Part (a) of Figure 7.6 shows two normal distributions with the same mean, but different standard deviations; part (b) shows two normal distributions with the same standard deviation, but different means; and part (c) shows two normal distributions with different means and different standard deviations.

In practice, we find areas under normal curves in special tables, such as Table II at the end of the book. It is not necessary to construct separate tables of normal-curve areas for all conceivable pairs of values of μ and σ. We tabulate these areas only for the normal distribution with $\mu = 0$ and $\sigma = 1$, called the **standard normal distribution**. Then we obtain areas under any normal curve by performing the change of scale (see Figure 7.7) that converts the units of measurement into **standard units** by means of the formula

STANDARD
UNITS

$$z = \frac{x - \mu}{\sigma}$$

Figure 7.6
Three pairs of normal distributions.

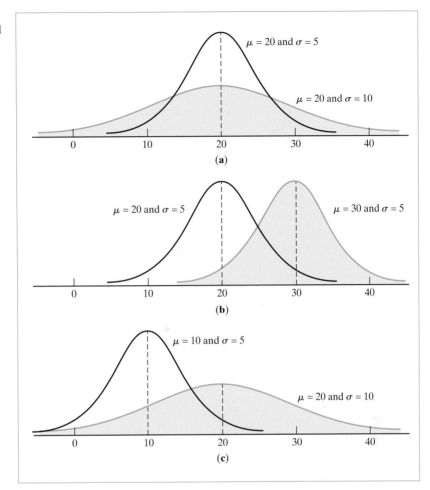

$\mu = 20$ and $\sigma = 5$

$\mu = 20$ and $\sigma = 10$

(a)

$\mu = 20$ and $\sigma = 5$

$\mu = 30$ and $\sigma = 5$

(b)

$\mu = 10$ and $\sigma = 5$

$\mu = 20$ and $\sigma = 10$

(c)

Figure 7.7
Change in scale to standard units.

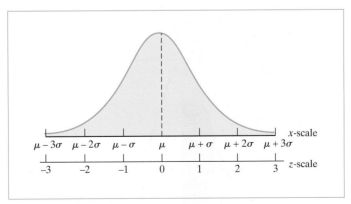

$\mu - 3\sigma$ $\mu - 2\sigma$ $\mu - \sigma$ μ $\mu + \sigma$ $\mu + 2\sigma$ $\mu + 3\sigma$ x-scale

-3 -2 -1 0 1 2 3 z-scale

As pointed out previously, *a* value of *z* simply tells us how many standard deviations the corresponding *x*-value lies above or below the mean.

The entries in Table II are the areas under the standard normal curve (that is, the graph of the standard normal distribution) between the mean, $z = 0$, and $z = 0.00, 0.01, 0.02, \ldots, 3.08$, and 3.09, and also $z = 4.00$, $z = 5.00$, and $z = 6.00$. In other words, the entries in Table II are areas under the standard normal curve like that of the blue region in Figure 7.8.

Table II has no entries corresponding to negative values of *z*, for these are not needed by virtue of the symmetry of any normal curve about its mean.

EXAMPLE 7.2

Find the area under the standard normal curve between $z = -1.20$ and $z = 0$.

*Optional Solution

This solution is provided primarily for users of graphing calculators. It can be omitted by other readers without loss of continuity. Had we wanted to use a graphing calculator in this example, we would have obtained the result shown in Figure 7.9. The "Using Technology: The Graphing Calculator" material that follows Figure 7.9 explains how to obtain this figure with a TI-83 graphing calculator. The numerical solution can be read directly from the figure and the area under the standard normal curve between $z = -1.20$ and $z = 0$ is 0.38493.   ■

Figure 7.8
Tabulated areas under the graph of the standard normal distribution.

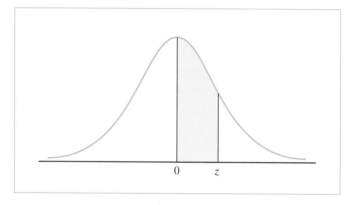

Figure 7.9
Area under normal curve, obtained with a TI-83 graphing calculator.

How to graph the area under the normal curve shown in Figure 7.9

(**Tip**: Before beginning, reset the memory to default setting—see Appendix A: TI-83 Tips for help.)

Step 1. Clear all previous drawings.
Press 2nd PRGM to access the DRAW menu.
Select 1: Clr Draw and press ENTER.

Step 2. Set the viewing window.

```
Press  WINDOW
Set    Xmin = -5
       Xmax = 5
       Xscl = 1
       Ymin = -.1
       Ymax = .4
       Yscl = 1
       Xres = 1
```

Step 3. Select the type of display and enter the interval to be shaded.
Press 2nd, VARS to access the DISTR DRAW menu.
Highlight DRAW.
Select 1: ShadeNorm(and press ENTER.
Enter $-1.20, 0)$
(Be sure to use the $(-)$ key for the negative sign; use the black (,) key; and use the) key to close the expression.)

Step 4. View the graph.
Press ENTER to solve.

Solution (Without a Graphing Calculator) Since the area under the standard normal curve between $z = -1.20$ and $z = 0$ equals the area under the standard normal curve between $z = 0$ and $z = 1.20$, we simply look up the entry corresponding to $z = 1.20$ in Table II, which is 0.3849.

Questions concerning areas under normal curves arise in various ways, and the ability to find any desired area quickly can be a big help. Although the table gives only areas between $z = 0$ and selected positive values of z, we often have to find areas to the left or to the right of given positive or negative values of z. This should not cause any difficulties provided that we remember exactly what areas are represented by the entries in Table II, and also that the standard normal curve is symmetrical about $z = 0$; so that the area under the curve to the left of $z = 0$ and the area under the curve to the right of $z = 0$ are both equal to 0.5000.

EXAMPLE 7.3 Find the area under the standard normal curve that lies

(a) to the right of $z = -0.55$;

(b) to the left of $z = 0.84$;

(c) to the right of $z = 1.69$.

(d) to the left of $z = -0.74$.

(e) between $z = 0.90$ and $z = 1.33$;

(f) between $z = -0.29$ and $z = 0.59$.

Solution

(a) The area to the right of $z = -0.55$ is 0.5000 plus the entry in Table II corresponding to $z = 0.55$:

$$0.5000 + 0.2088 = 0.7088 \text{ (see Figure 7.10)}$$

(b) The area to the left of 0.84 is 0.5000 plus the entry in Table II corresponding to $z = 0.84$, that is,

$$0.5000 + 0.2995 = 0.7995 \text{ (see Figure 7.11)}$$

(c) The area to the right of $z = 1.69$ is 0.5000 minus the entry corresponding to 1.69:

$$0.5000 - 0.4545 = 0.0455 \text{ (see Figure 7.12)}$$

(d) The area to the left of $z = -0.74$ is 0.5000 minus the entry corresponding to $z = -0.74$:

$$0.5000 - 0.2704 = 0.2296 \text{ (see Figure 7.13)}$$

(e) The area between $z = 0.90$ and $z = 1.33$ is the difference between the entries corresponding to $z = 1.33$ and $z = 0.90$:

$$0.4082 - 0.3159 = 0.0923 \text{ (see Figure 7.14)}$$

Figure 7.10
Area to the right of
$z = -0.55$.

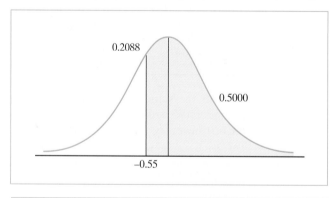

Figure 7.11
Area to the left of
$z = 0.84$.

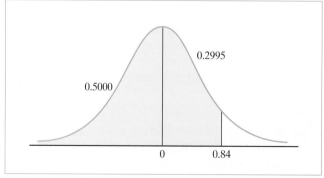

Figure 7.12
Area to the right of
$z = 1.69$.

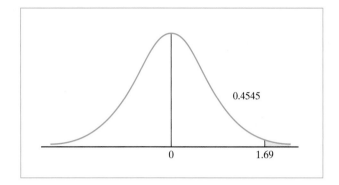

Figure 7.13
Area to the left of
$z = -0.74$.

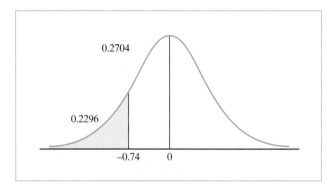

Figure 7.14
Area between $z = 0.90$
and $z = 1.33$.

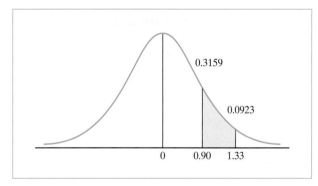

(f) The area between $z = -0.29$ and $z = 0.59$ is the sum of the entries corresponding to $z = -0.29$ and $z = 0.59$:

$$0.1141 + 0.2224 = 0.3365 \text{ (see Figure 7.15)}$$ ■

In both of the preceding examples we dealt directly with the standard normal distribution. Now let us consider an example where μ and σ are not 0 and 1, so that we must first use the formula $z = \dfrac{x - \mu}{\sigma}$ to convert to standard units.

EXAMPLE 7.4 Find the probabilities that a random variable will take on a value between 12 and 15 given that it has a normal distribution with
(a) $\mu = 10$ and $\sigma = 5$; (b) $\mu = 20$ and $\sigma = 10$.

Figure 7.15
Area between $z = -0.29$
and $z = 0.59$.

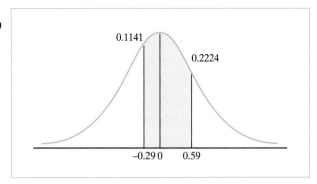

Figure 7.16
Areas under normal
curves.

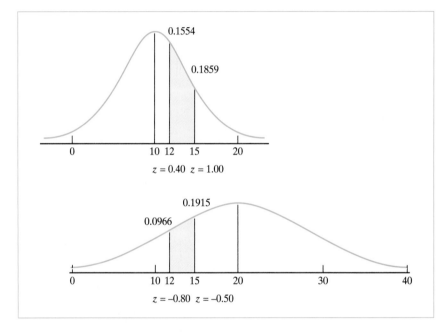

Solution

(a) The probability is given by the area of the blue region of the upper diagram of Figure 7.16. The values of z corresponding to 12 and 15 are

$$z = \frac{12 - 10}{5} = 0.40 \quad \text{and} \quad z = \frac{15 - 10}{5} = 1.00$$

The corresponding entries in Table II are 0.1554 and 0.3413, and the probability that the random variable will take on a value between 12 and 15 is $0.3413 - 0.1554 = 0.1859$.

(b) The probability is given by the area of the blue region of the lower diagram of Figure 7.16. The values of z corresponding to 12 and 15 are

$$z = \frac{12 - 20}{10} = -0.80 \quad \text{and} \quad z = \frac{15 - 20}{10} = -0.50$$

The entries in Table II corresponding to $z = 0.80$ and $z = 0.50$ are 0.2881 and 0.1915, and the probability that the random variable will take on a value between 12 and 15 is

$$0.2881 - 0.1915 = 0.0966$$

■

There are also problems in which we are given areas under normal curves and are asked to find the corresponding values of z. The results of the example that follows will be used extensively in subsequent chapters.

EXAMPLE 7.5 If z_α denotes the value of z for which the area under the standard normal curve to its right is equal to α (Greek lowercase *alpha*), find

(a) $z_{0.01}$;

(b) $z_{0.05}$.

Solution **(a)** As can be seen from Figure 7.17, $z_{0.01}$ corresponds to an entry of $0.5000 - 0.0100 = 0.4900$ in Table II; since the nearest entry is 0.4901, corresponding to $z = 2.33$, we let $z_{0.01} = 2.33$.

Figure 7.17
Diagram for determination of z_α.

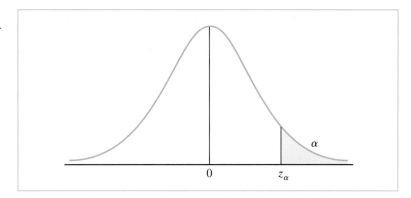

(b) As can be seen from Figure 7.17, $z_{0.05}$ corresponds to an entry of $0.5000 - 0.0500 = 0.4500$ in Table II; since the two nearest entries are 0.4495 and 0.4505, corresponding to $z = 1.64$ and $z = 1.65$, we let $z_{0.05} = 1.645$. Our graphing calculator confirms that $z = 1.644853626$, or $z = 1.645$ rounded to three decimals. ■

Table II also enables us to verify our earlier remark that for frequency distributions having the general shape of the cross section of a bell, about 68% of the values will lie within one standard deviation of the mean, about 95% will lie within two standard deviations of the mean, and about 99.7% will lie within three standard deviations of the mean. These figures apply to frequency distributions having the general shape of normal distributions, and in Exercise 7.16 the reader will be asked to verify these percentages with the use of Table II.

Exercises 7.1, 7.3, 7.5, 7.11, 7.14, and 7.18 are practice exercises; their complete solutions are given on pages 262 and 263. Each of Exercises 7.3 through 7.6 can be answered without written work and without use of the statistical tables.

✓ **7.1** Suppose that a continuous random variable takes on values on the interval from 1 to 5 and that the graph of its probability density is given by the horizontal line of Figure 7.18.

(a) What probability is represented by the blue region of the diagram and what is its value?

(b) What is the probability that the random variable will take on a value greater than 4.5? Would the answer be the same if we asked for the probability that the random variable will take on a value greater than or equal to 4.5?

(c) What is the probability that the random variable will take on a value between 1.8 and 4.2?

7.2 Suppose that a continuous random variable takes on values on the interval from 0 to 4 and that the graph of its probability density is given by the blue line of Figure 7.19.

(a) Verify that the total area under the curve is equal to 1.

(b) What is the probability that this random variable will take on a value less than 3?

(c) What is the probability that this random variable will take on a value between 1 and 2?

Figure 7.18
Diagram for Exercise 7.1.

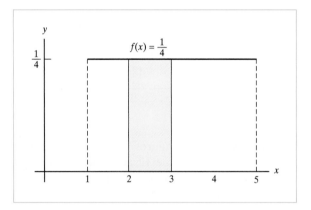

Figure 7.19
Diagram for Exercise 7.2.

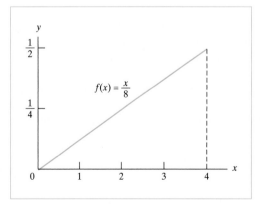

✓ **7.3**   For each of the following cases involving areas under the standard normal curve, decide whether the first area is bigger, the second area is bigger, or the two areas are equal:

(a) the area to the right of $z = +1$ or the area to the right of $z = +1.5$;

(b) the area to the left of $z = -1$ or the area to the left of $z = -1.5$;

(c) the area to the right of $z = +1$ or the area to the left of $z = -1.5$;

(d) the area to the right of $z = +1$ or the area to the left of $z = -1$;

(e) the area to the right of $z = -0.5$ or the area to the right of $z = -1.5$;

(f) the area to the left of $z = 0$ or the area to the right of $z = -0.1$;

(g) the area to the right of $z = 0$ or the area to the left of $z = 0$;

(h) the area to the right of $z = -1.2$ or the area to the left of $z = -1.2$.

7.4   For each of the following cases involving areas under the standard normal curve, decide whether the first area is bigger, the second area is bigger, or the two areas are equal:

(a) the area between $z = 0$ and $z = 0.3$, or the area between $z = 2.0$ and $z = 2.3$;

(b) the area between $z = 1.5$ and $z = 2$, and $z = -1.5$ and $z = -2.0$;

(c) the area to the left of $z = 1.0$ and the area to the left of $z = 1.5$;

(d) the area to the right of $z = 1.0$ or the area to the right of $z = 1.5$;

(e) the area between $z = 1.0$ and $z = -1.0$, or the area between $z = 1.5$ and $z = -1.5$;

(f) the area between 0 and -1, or the area between 1 and 2;

(g) the area to the left of -1 or the area to the right of -1.

✓ **7.5**   For each of the following cases involving random variables with normal distributions, decide whether the first probability is bigger, the second probability is bigger, or the two probabilities are equal:

(a) for a random variable with a normal distribution with $\mu = 100$ and $\sigma = 20$, the probability of a value greater than 120 or the probability of a value less than 90;

(b) for a random variable with a normal distribution with $\mu = 100$ and $\sigma = 20$, the probability of a value greater than 120 or the probability of a value greater than 130;

(c) for a random variable with a normal distribution with $\mu = 200$ and $\sigma = 40$, the probability of a value greater than 250 or the probability of a value less than 140;

(d) for a random variable with a normal distribution with $\mu = 60$ and $\sigma = 12$, the probability of a value between 60 and 84 or the probability of a value between 48 and 72.

7.6   For each of the following cases involving random variables with normal distributions, decide whether the first probability is bigger, the second probability is bigger, or the two probabilities are equal:

(a) the probability that a random variable having the normal distribution with $\mu = 40$ and $\sigma = 5$ takes a value greater than 40, or the probability that a random variable having the normal distribution with $\mu = 50$ and $\sigma = 5$ takes a value greater than 40;

(b) the probability that a random variable having the normal distribution with $\mu = 50$ and $\sigma = 10$ takes a value less than 60, or the probability that a random variable having the normal distribution with $\mu = 500$ and $\sigma = 100$ takes a value less than 600;

(c) the probability that a random variable having the normal distribution with $\mu = 100$ and $\sigma = 5$ takes a value greater than 110, or the probability that a random variable having the normal distribution with $\mu = 108$ and $\sigma = 5$ takes a value greater than 110;

(d) the probability that a random variable having the normal distribution with $\mu = 50$ and $\sigma = 10$ takes a value less than 60, or the probability that a random variable having the normal distribution with $\mu = 50$ and $\sigma = 20$ takes a value less than 60.

7.7   In each given situation, you will receive a dollar amount generated either by process A or process B. For each process, the random variable follows a normal distribution. Decide whether you prefer process A or process B. [Only part (a) has a "right" answer.]

(a) Process A has $\mu = \$50$ and $\sigma = \$10$ and process B has $\mu = \$70$ and $\sigma = \$10$.

(b) Process A has $\mu = \$50$ and $\sigma = \$1$ and process B has $\mu = \$50$ and $\sigma = \$10$.

7.8   Find the area under the standard normal curve that lies

(a) between $z = -0.78$ and $z = 0$;

(b) to the left of $z = 2.50$;

(c) to the left of $z = -1.55$;

(d) between $z = 0.33$ and $z = 0.66$.

7.9   Find the area under the standard normal curve that lies

(a) between $z = 0$ and $z = 0.85$;

(b) to the right of $z = -0.65$;

(c) to the right of $z = 2.01$;

(d) between $z = -1.00$ and $z = -1.10$.

7.10 Find the area under the standard normal curve that lies

(a) between $z = -0.38$ and $z = 0.38$;

(b) between $z = -2.88$ and $z = 3.01$;

(c) between $z = -1.97$ and $z = -1.96$;

(d) between $z = -2.50$ and $z = 2.00$.

✓ **7.11**   Find z if

(a) the normal-curve area between 0 and z is 0.3340;

(b) the normal-curve area to the left of z is 0.6517;

(c) the normal-curve area to the left of z is 0.3085;

(d) the normal-curve area between $-z$ and z is 0.9700.

7.12   Find z if

(a) the normal-curve area between 0 and z is 0.1915;

(b) the normal-curve area to the right of z is 0.8665;

(c) the normal-curve area to the right of z is 0.0228;

(d) the normal-curve area between $-z$ and z is 0.9500.

7.13   Find z if

(a) the normal curve area between 0 and $-z$ is 0.4772;

(b) the normal curve area to the left of $-z$ is 0.1335;

(c) the normal curve area to the left of z is 0.7088;

(d) the normal curve area between z and $-z$ is 0.9980.

✓ **7.14**   A random variable has a normal distribution with $\mu = 69.0$ and $\sigma = 5.1$. What are the probabilities that the random variable will take on a value

    (a) less than 74.1;

    (b) greater than 63.9;

    (c) between 69.0 and 72.3;

    (d) between 66.2 and 71.8?

7.15   A random value has a normal distribution with $\mu = 56.4$ and $\sigma = 4.8$. What are the probabilities that this random variable will take on a value

    (a) less than 70.0;

    (b) less than 50.0;

    (c) between 50.0 and 70.0?

7.16   Find the probabilities that a random variable having a normal distribution will take on a value within

    (a) one standard deviation of the mean;

    (b) two standard deviations of the mean;

    (c) three standard deviations of the mean;

    (d) four standard deviations of the mean.

7.17   According to the definition given on page 247, $z_{\alpha/2}$ denotes the value of z for which the area under the standard normal curve to its right is equal to $\alpha/2$ and, hence, the area between $-z_{\alpha/2}$ and $z_{\alpha/2}$ is equal to $1 - \alpha$ (see Figure 7.20). Verify that

    (a) $z_{0.025} = 1.96$ corresponding to $\alpha = 0.05$;

    (b) $z_{0.005} = 2.575$ corresponding to $\alpha = 0.01$.

Figure 7.20
Diagram for Exercise
7.17.

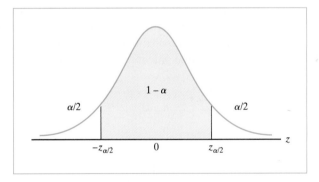

✓ **7.18**   A normal distribution has the mean $\mu = 61.6$. Find its standard deviation if 20% of the total area under the curve lies to the right of 70.0.

    7.19   A normal distribution has the mean $\mu = 74.4$. Find its standard deviation if 10% of the area under the curve lies to the right of 100.0.

    7.20   A random variable has a normal distribution with the standard deviation $\sigma = 10$. Find its mean if the probability is 0.8264 that it will take on a value less than 77.5.

    7.21   For a certain random variable having the normal distribution, the probability is 0.33 that it will take on a value less than 245, and the probability is 0.48 that it will take on a value greater than 260. Find the mean and the standard deviation of this random variable.

7.3 SOME APPLICATIONS

Let us now consider some applied problems in which it will be assumed that the random variables under consideration have normal distributions.

EXAMPLE 7.6   The amount of cosmic radiation to which a person is exposed while flying by jet across the United States is a random variable having a normal distribution with $\mu = 4.35$ mrem and $\sigma = 0.59$ mrem.[†] Find the probabilities that a person on such a flight will be exposed to

(a) more than 5.00 mrem of cosmic radiation;

(b) anywhere from 3.00 to 4.00 mrem of cosmic radiation.

Solution   (a) This probability is given by the area of the blue region of the upper diagram of Figure 7.21, that is, the area under the curve to the right of

$$z = \frac{5.00 - 4.35}{0.59} \approx 1.10$$

Since the entry in Table II corresponding to $z = 1.10$ is 0.3643, we find that the probability is $0.5000 - 0.3643 = 0.1357$, or approximately 0.14, that a person will be exposed to more than 5.00 mrem of cosmic radiation on such a flight.

(b) This probability is given by the area of the blue region of the lower diagram of Figure 7.21, that is, the area under the curve between

$$z = \frac{3.00 - 4.35}{0.59} \approx -2.29 \quad \text{and} \quad z = \frac{4.00 - 4.35}{0.59} \approx -0.59$$

Since the entries in Table II corresponding to $z = -2.29$ and $z = -0.59$ are, respectively, 0.4890 and 0.2224, we find that the probability is $0.4890 - 0.2224 = 0.2666$, or approximately 0.27, that a person will be exposed to anywhere from 3.00 to 4.00 mrem of cosmic radiation on such a flight. ∎

Figure 7.21
Diagrams for cosmic
radiation example.

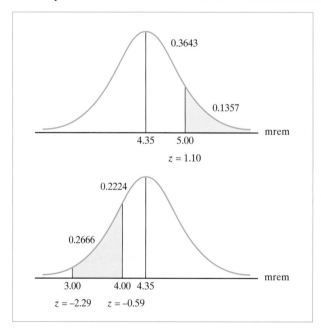

[†] The unit of radiation mrem stands for *milliroentgen equivalent man*.

EXAMPLE 7.7   The actual amount of instant coffee that a filling machine puts into "4-ounce" jars varies from jar to jar, and it may be looked upon as a random variable having a normal distribution with $\sigma = 0.04$ ounce. If only 2% of the jars are to contain less than 4 ounces of coffee, what must be the mean fill of these jars?

Solution   We are given $\sigma = 0.04$, a normal-curve area (that of the blue region of Figure 7.22), and we are asked to find μ. Observe that μ must be safely above the stated weight of 4 ounces. Since the value of z for which the entry in Table II is closest to $0.5000 - 0.0200 = 0.4800$ is 2.05, we have $-2.05 = \dfrac{4.00 - \mu}{0.04}$. Solving for μ, we get $4.00 - \mu = -2.05(0.04) = -0.082$ and then $\mu = 4.00 + 0.082 = 4.082$ ounces or 4.08 ounces to the nearest hundredth of an ounce.   ■

Although the normal distribution is a continuous distribution that applies to continuous random variables, it is often used to approximate distributions of random variables that can take on only a finite number of values or as many values as there are positive integers. In many situations, this will yield satisfactory results, provided that we make the **continuity correction** illustrated in the following example.

EXAMPLE 7.8   In a study of aggressive behavior, male white mice, returned to the group in which they live after four weeks of isolation, averaged 18.6 fights in the first five minutes with a standard deviation of 3.3 fights. If it can be assumed that the distribution of this random variable (the number of fights into which such a mouse gets under the stated conditions) can be approximated closely with a normal distribution, what is the probability that such a mouse will get into at least 15 fights in the first five minutes?

*Optional Solution   This solution is provided primarily for users of graphing calculators. With the exception of the explanation of the continuity correction, it can be omitted by other readers. The answer is read directly from the screen of the graphing calculator shown in Figure 7.23, and the probability is 0.89296. The probability is given by the area of the shaded region of the figure which is to the right of 14.5, not 15. The reason for this is that the number of fights in which such a mouse gets involved is a whole number. Hence, if we want to approximate the distribution of this random variable with a normal curve, we must *spread* its values over a continuous scale, and we do this by representing each whole number k by the interval from $k - \dfrac{1}{2}$ to $k + \dfrac{1}{2}$. For instance, 5 is represented by the interval from

Figure 7.22
Diagram for instant-coffee filling machine example.

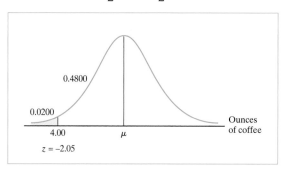

Figure 7.23
Normal-curve area reproduced from the display screen of a TI-83 graphing calculator.

4.5 to 5.5, 10 is represented by the interval from 9.5 to 10.5, 20 is represented by the interval from 19.5 to 20.5, and the probability of 15 or more is given by the area under the curve to the right of 14.5 (the boundary between the white and black areas of the graph). Accordingly, it follows from Figure 7.23 that the area of the black region—the probability that such a mouse will get into at least 15 fights in the first 5 minutes—is 0.89296, or approximately 0.89.

USING TECHNOLOGY: The Graphing Calculator

How to graph the area under the normal curve shown in Figure 7.23
(**Tip**: Before beginning, reset the memory to default setting—see Appendix A: TI-83 Tips.)

Step 1. Clear all previous drawings.
Press 2nd PRGM to access the DRAW menu.
Select 1: Clr Draw and press ENTER.

Step 2. Set the view window.

```
Press WINDOW
Set    Xmin = 2.1
       Xmax = 35.1
       Xscl = 6
       Ymin = -.03
       Ymax = .12
       Yscl = 1
       Xres = 1
```

Step 3. Select the type of display and enter the interval to be shaded.
Press 2nd, VARS to access the DISTR DRAW menu.
Highlight DRAW.
Select 1: ShadeNorm(and press ENTER.
Enter 14.5, 1E99, 18.6, 3.3).
(For instructions on entering "1E99", for infinity, see Appendix A: TI-83 Tips. 14.5 and 1E99 are the upper and lower limits of the interval; 18.6 and 3.3 are the mean and standard deviation of the normal random variable.)

Step 4. View the graph.
Press ENTER to solve.

Solution (Without a Graphing Calculator) To solve without a graphing calculator, we use the continuity correction and substitute $x = 14.5$ in place of 15, $\mu = 18.6$, and $\sigma = 3.3$ into the formula $z = \dfrac{x - \mu}{\sigma}$, obtaining

$$z = \frac{14.5 - 18.6}{3.3} \approx -1.24$$

Entering Table II with $z = 1.24$, we obtain 0.3925. Finally, we combine 0.3925 and 0.5000 and get the probability $0.5000 + 0.3925 = 0.8925$, which is nearly identical to the result obtained with the graphing calculator. ■

All the examples of this section dealt with random variables having normal distributions, or distributions that can be approximated closely with normal curves. Whenever we observe a value of a random variable having a normal distribution, we say that we are **sampling a normal population**.

EXERCISES

Exercises 7.22 and 7.29 are practice exercises; their complete solutions are given on page 263.

✓ **7.22** In an experiment to determine the amount of time required to assemble an "easy to assemble" toy, the assembly time was found to be a random variable having approximately a normal distribution with $\mu = 27.8$ minutes and $\sigma = 4.0$ minutes. What are the probabilities that this kind of toy can be assembled in

(a) less than 25.0 minutes;

(b) anywhere from 26.0 to 29.6 minutes?

7.23 The weights of a large shipment of cast iron bollards (used for securing ropes on docks, etc.) are random variables with $\mu = 50.25$ pounds and $\sigma = 0.63$ pounds. What is the probability that a bollard selected from this shipment will weigh

(a) less than 49.00 pounds;

(b) more than 50.5 pounds;

(c) between 50 and 51 pounds?

7.24 A random value has a normal distribution with the mean 102.9 and the standard deviation $\sigma = 4.7$. What are the probabilities that this random value will also take on a value

(a) less than 110.1;

(b) greater than 95.6;

(c) between 104.5 and 105.9;

(d) between 98.7 and 105.0?

7.25 The daily demand for admission to a swimming pool is a random variable with a distribution that can be approximated closely by a normal distribution with the mean $\mu = 43.3$ and the standard deviation $\sigma = 4.6$. What is the probability that the demand for admission will exceed 50 persons on any given day? Use the continuity correction.

7.26 The lengths of the sardines received by a cannery have a mean of 4.64 inches and a standard deviation of 0.25 inch. If the distribution of these lengths can be approximated closely with a normal distribution, what percentage of all these sardines are

(a) shorter than 4.00 inches;

(b) from 4.40 to 4.80 inches long?

7.27   With reference to Exercise 7.26, above which length lies the longest 15% of the sardines?

7.28   The management of a small ice skating rink knows that the daily numbers of skaters using the rink each morning is a random variable with a distribution that can be approximated closely by a normal distribution with the mean $\mu = 45.7$ and the standard deviation $\sigma = 4.8$. Using the continuity correction, determine the probability that the daily number of skaters using the rink during the morning session will

(a) exceed 50 on any given day;

(b) be less than 40 on any given day.

✓ **7.29**   In a very large class in world history, the final examination grades have a mean of 66.5 and a standard deviation of 12.6. Assuming that it is reasonable to approximate the distribution of these grades with a normal distribution, what percentage of the grades should exceed 74? Assume that the grades are always whole numbers.

7.30   The management of a motel with excess room capacity knows that the daily number of rooms rented is a random variable with a distribution that can be approximated by a normal distribution with the mean $\mu = 53.9$ and the standard deviation 4.5. Using the continuity correction, determine the probability that the number of rentals on a certain day will

(a) exceed 55 rooms;

(b) be less than 45 rooms.

7.31   In a municipal parking lot, the number of parking tickets issued daily for overtime parking is a random variable with $\mu = 15.3$ and $\sigma = 3.4$. What is the probability that on any day exactly 20 tickets will be issued in this parking lot?

7.32   The annual number of tornadoes in a certain state is a random variable with $\mu = 28.2$ and $\sigma = 5.1$. Approximating the distribution of this random variable with a normal distribution, find the probability that there will be at least 24 tornadoes in the state in any given year.

7.33   The number of days that patients are hospitalized at a certain hospital is a random variable having approximately a normal distribution with mean $\mu = 7.1$ and $\sigma = 3.2$. Among 1,000 patients, how many can be expected to be hospitalized anywhere from 5 to 10 days?

7.4 THE NORMAL APPROXIMATION TO THE BINOMIAL DISTRIBUTION

The normal distribution, which is a continuous distribution, provides a close approximation to the binomial distribution when n, the number of trials, is large and p, the probability of success, is not too far from $\frac{1}{2}$. Figure 7.24 shows the histograms of binomial distributions with $p = \frac{1}{2}$ and $n = 2, 5, 10$, and 25, and it can be seen that with increasing n these distributions approach the symmetrical bell-shaped pattern of a normal curve. In fact, a normal distribution with $\mu = np$ and $\sigma = \sqrt{np(1 - p)}$ can often be used to approximate a binomial distribution, even when n is fairly small and p differs from $\frac{1}{2}$.

Figure 7.24
Binomial distributions
with $p = \dfrac{1}{2}$ and $n = 2$,
5, 10, and 25.

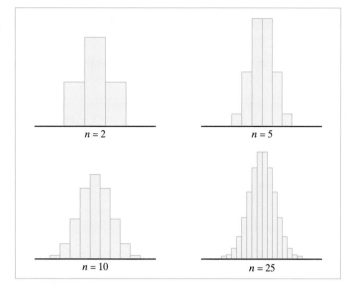

A good rule of thumb is to use this approximation only when np and $n(1 - p)$ are both greater than 5;[†] symbolically, when

CONDITIONS FOR
THE NORMAL
APPROXIMATION TO
THE BINOMIAL
DISTRIBUTION

$$np > 5 \text{ and } n(1 - p) > 5$$

EXAMPLE 7.9   Check in each case whether the conditions for the normal approximation to the binomial distribution are satisfied.

(a) $n = 40$ and $p = \dfrac{1}{4}$;

(b) $n = 100$ and $p = \dfrac{1}{25}$;

(c) $n = 150$ and $p = 0.98$.

Solution   (a) Since $np = 40 \cdot \dfrac{1}{4} = 10$ and $n(1 - p) = 40 \cdot \dfrac{3}{4} = 30$ both exceed 5, the conditions are satisfied.

(b) Since $np = 100 \cdot \dfrac{1}{25} = 4$ does not exceed 5, the conditions are not satisfied.

(c) Since $n(1 - p) = 150(0.02) = 3$ does not exceed 5, the conditions are not satisfied.   ■

EXAMPLE 7.10   Find the exact probability of getting 6 heads and 10 tails in 16 flips of a balanced coin, and also use a normal curve to approximate this binomial probability.

[†] Note that there are values of n and p (for instance, $n = 100$ and $p = 0.08$) for which either the normal approximation to the binomial distribution or the Poisson approximation (see page 216) can be used. This matter is discussed further on page 258.

Solution   The formula for the binomial distribution, repeated here from page 203, is
$f(x) = \binom{n}{x} p^x (1-p)^{n-x}$. Substituting $n = 16$, $p = \dfrac{1}{2}$, and $x = 6$ into the formula
for the binomial distribution, we find that the probability of getting 6 heads and
10 tails in 16 flips of a balanced coin is

$$f(6) = \binom{16}{6} \left(\frac{1}{2}\right)^6 \left(1 - \frac{1}{2}\right)^{16-6} = 8,008 \left(\frac{1}{2}\right)^{16} = \frac{8,008}{65,536} \approx 0.1222$$

To find the normal-curve approximation to this probability, we use the continuity
correction and represent 6 heads by the interval from 5.5 to 6.5 (see Figure 7.25).
Since $\mu = np = 16 \cdot 1/2 = 8$, and since

Figure 7.25
Normal-curve approxi-
mation to the binomial
distribution.

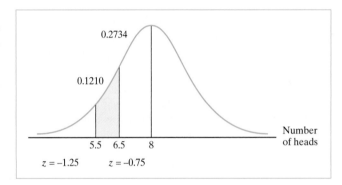

$$\sigma = \sqrt{np(1-p)} = \sqrt{16 \cdot 1/2 \cdot 1/2} = 2$$

converting 5.5 and 6.5 into standard units yields

$$z = \frac{5.5 - 8}{2} = -1.25 \text{ and } z = \frac{6.5 - 8}{2} = -0.75$$

The entries in Table II corresponding to $z = 1.25$ and $z = 0.75$ are 0.3944 and
0.2734, and we get

$$0.3944 - 0.2734 - 0.1210$$

for the normal-curve approximation to the probability of getting 6 heads and 10
tails in 16 flips of a balanced coin. This differs by only 0.0012 from the value
obtained with the formula for the binomial distribution.   ■

The normal-curve approximation to the binomial distribution is particularly
useful in problems where we would otherwise have to use the formula for the
binomial distribution repeatedly to obtain the values of many different terms.

EXAMPLE 7.11   What is the probability that at least 26 of 50 mosquitoes will be killed by a new
insect spray when the probability is 0.60 that any one of them will be killed by the
spray?

Figure 7.26
Normal-curve approximation to the binomial distribution.

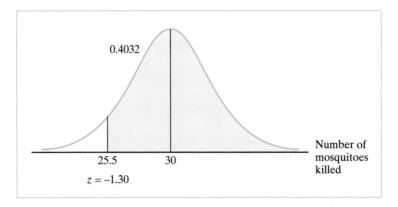

Solution   If we tried to find the answer by using the formula for the binomial distribution, we would have to find the sum of the probabilities corresponding to 26, 27, 28, ..., 49, and 50 successes in 50 trials. Without a computer, this would obviously involve a great deal of work; but by using the normal-curve approximation, we need only find the area of the blue region of Figure 7.26, namely, that to the right of 25.5. Observe that the mean is $0.60 \cdot 50 = 30$, so that we draw 25.5 to the left of the mean. Here, again, we are using the continuity correction since we are dealing with whole numbers. Accordingly, 26 is represented by the interval from 25.5 to 26.5, 27 is represented by the interval from 26.5 to 27.5, and so on.

Since $\mu = 50(0.60) = 30$ and $\sigma = \sqrt{50(0.60)(0.40)} \approx 3.464$, we find that in standard units 25.5 becomes

$$z = \frac{25.5 - 30}{3.464} \approx -1.30$$

The entry in Table II corresponding to $z = 1.30$ is 0.4032, and the probability of at least 26 successes in 50 trials when $p = 0.60$ is

$$0.5000 + 0.4032 = 0.9032$$

To determine the error of the approximation made in this example, let us refer to the computer printout of Figure 7.27, which shows the cumulative "LESS OR =" binomial probabilities for $n = 50$ and $p = 0.60$. Since the probability of "at least 26 successes" is 1 minus the probability of "25 or less," rounding to four decimals, we find that the desired probability is

$$1 - 0.0978 = 0.9022$$

and we find 0.0978 in the row corresponding to k = 25. Thus, the error of the normal approximation is only

$$0.9032 - 0.9022 = 0.0010$$

With the method of this section, we now have four ways of determining binomial probabilities:

Figure 7.27
MINITAB printout of
the binomial distribu-
tion with $n = 50$ and $p = 0.60$.

↓	C1	C2	C3	C4	C5	C6	C7
	k	P(X<=k)					
16	15	0.00002					
17	16	0.00006					
18	17	0.00018					
19	18	0.00052					
20	19	0.00137					
21	20	0.00336					
22	21	0.00762					
23	22	0.01603					
24	23	0.03141					
25	24	0.05734					
26	25	0.09781					
27	26	0.15617					
28	27	0.23398					
29	28	0.32986					
30	29	0.43897					
31	30	0.55352					
32	31	0.66439					
33	32	0.76312					
34	33	0.84391					
35	34	0.90450					
36	35	0.94604					
37	36	0.97201					
38	37	0.98675					
39	38	0.99431					
40	39	0.99780					
41	40	0.99924					
42	41	0.99977					

1. We refer to a table of binomial probabilities or a computer printout of binomial probabilities.
2. We use the formula for the binomial distribution.
3. We use the Poisson approximation to the binomial distribution.
4. We use the normal approximation to the binomial distribution.

The first alternative is by far the most preferable, although it frequently requires that we go beyond the table given at the end of this book. In contrast, the formula for the binomial distribution is used only when absolutely necessary, as when it is the basis of an algebra exercise, rather than a direct computation.

As far as the approximations are concerned, they are not only subject to the rules of thumb stated previously, but their use requires a good deal of professional judgment. For instance, if either approximation can be used, we may be inclined to use the Poisson approximation to determine a probability associated with a single value of a binomial random variable, and we may be inclined to use the normal approximation to determine a probability associated with the tail (and hence many values) of a binomial distribution.

In any case, we have given the Poisson approximation to the binomial distribution mainly to *introduce* the Poisson distribution, and the normal approximation because it will be needed in Chapter 11 for large-sample inferences concerning proportions.

Exercise 7.34 is a practice exercise; its complete solution is given on page 263.

✓ **7.34** If 20% of the loan applications received by a bank are refused, what is the probability that among 225 loan applications at least 50 will be refused?

7.35 Check in each case whether the conditions for the normal approximation to the binomial distribution are satisfied.
(a) $n = 18$ and $p = 0.60$; (b) $n = 90$ and $p = .05$; (c) $n = 10$ and $p = 0.50$.

7.36 Use the normal-curve approximation to find the probability of getting 5 heads and 7 tails in 12 flips of a balanced coin, and compare the result with the corresponding value given in Table I.

7.37 If 62% of all clouds seeded with silver iodide show spectacular growth, what is the probability that among 24 clouds thus seeded exactly 15 will show spectacular growth? If the value given for this probability in the table of binomial probabilities is 0.1661, what is the error of the approximation?

7.38 A dating service finds that 15% of the couples that it matches eventually get married. In the next 50 matches that the service makes, find the probabilities that
(a) at least 6 couples marry; (b) at most 10 couples marry.

7.39 The manager of an air-conditioning company knows that the number of service calls made each day is a random variable having approximately a normal distribution with the mean 34.9 service calls, and the standard deviation 4.8 calls. What are the probabilities that in any given day the company will make

(a) exactly 31 air-conditioning service calls;

(b) at most 31 air-conditioning service calls?

7.40 A student answers each of the 48 questions on a multiple-choice test, each with four possible answers, by randomly drawing a card from an ordinary deck of 52 playing cards and checking the first, second, third, or fourth answer depending on whether the card drawn is a spade, heart, diamond, or club.

(a) Use the normal-curve approximation to find the probability that the student will get exactly 15 correct answers.

(b) Use the normal-curve approximation to find the probability that the student will get at least 15 correct answers.

7.41 What is the probability that fewer than 25 workers out of a work force of 500 workers fail to report to work on time, given that the probability is 0.06 that any one worker will not get to work on time?

7.42 An automobile manufacturer knows that, on the average, 4% of the automobiles produced by the firm will require repairs that are covered by their warranties. What is the probability that among 1,000 automobiles shipped by the manufacturer, at least 30 will require repairs before their warranties expire?

7.43 To illustrate the Law of Large Numbers, which we mentioned in connection with the frequency interpretation of probability, suppose that among all adults living in a large city we know that there are as many men as women. Using the normal-curve approximation, find the probabilities that in a random sample of adults living in this city the proportion of men will be anywhere from 0.49 to 0.51 when the number of persons in the sample is
(a) 100; (b) 1,000; (c) 10,000.

🖥 **7.44** Use a computer printout of the binomial distribution with $n = 50$ and $p = 0.15$ to determine the errors of the approximations of both parts of Exercise 7.38.

7.45   Use a computer printout of the binomial distribution with $n = 48$ and $p = 0.25$ to determine the errors of the approximations of both parts of Exercise 7.40.

7.46   Use a computer printout of the binomial distribution with $n = 40$ and $p = 0.30$ to determine the errors that we would make by using the normal-curve approximation to find the probabilities that a random variable having this binomial distribution will take on

(a) a value greater than or equal to 15;

(b) a value less than or equal to 13;

(c) the value 9.

7.47   The normal-curve approximation can also be used for discrete distributions other than the binomial distribution. For instance, it is reasonable to use this approximation if the Poisson distribution has parameter λ greater than 30. This approximation uses λ as the mean and $\sqrt{\lambda}$ as the standard deviation. Suppose that the number of false alarms received weekly by the fire department of a large city follows a Poisson distribution with $\lambda = 45$. In any given week, find the probability that

(a) there will be 50 or more false alarms;

(b) there will be at most 44 false alarms.

✓ SOLUTIONS TO PRACTICE EXERCISES

7.1   (a) The blue region represents the probability of getting a value on the interval from 2 to 3, and its area is $\frac{1}{4} \cdot 1 = 0.25$.

(b) The probability is $\frac{1}{4} \cdot (5 - 4.5) = 0.125$; the answer would be the same, since the probability is zero that the value of the random variable will actually equal 4.5.

(c) The probability is $\frac{1}{4} \cdot (4.2 - 1.8) = 0.60$.

7.3   (a) First area is bigger;

(b) first area is bigger;

(c) first area is bigger;

(d) areas are equal;

(e) second area is bigger;

(f) second area is bigger;

(g) areas are equal;

(h) first area is bigger.

7.5   (a) Second probability is bigger;

(b) first probability is bigger;

(c) first probability is bigger;

(d) second probability is bigger.

7.11   (a) Since 0.3340 is the entry in Table II corresponding to 0.97, the answer is $z = 0.97$ or $z = -0.97$.

(b) Since $0.6517 - 0.5000 = 0.1517$ is the entry in Table II corresponding to 0.39, the answer is $z = 0.39$.

(c) Since $0.5000 - 0.3085 = 0.1915$ is the entry in Table II corresponding to 0.50, the answer is $z = -0.50$.

(d) Since $\dfrac{0.9700}{2} = 0.4850$ is the entry in Table II corresponding to 2.17, the answer is $z = 2.17$ or $z = -2.17$.

7.14 (a) Since $z = \dfrac{74.1 - 69.0}{5.1} = 1.00$ and the entry corresponding to $z = 1.00$ in Table II is 0.3413, the probability is $0.3413 + 0.5000 = 0.8413$.

(b) Since $z = \dfrac{63.9 - 69.0}{5.1} = -1.00$ and the entry corresponding to $z = 1.00$ in Table II is 0.3413, the probability is $0.3413 + 0.5000 = 0.8413$.

(c) Since $z = \dfrac{72.3 - 69.0}{5.1} \approx 0.65$ and the entry corresponding to $z = 0.65$ in Table II is 0.2422, the probability is 0.2422.

(d) Since $z = \dfrac{66.2 - 69.0}{5.1} \approx -0.55$, and $z = \dfrac{71.8 - 69.0}{5.1} \approx 0.55$ and the entry corresponding to $z = 0.55$ in Table II is 0.2088, the probability is $(0.2088)2 = 0.4176$.

7.18 Since the entry in Table II nearest to $0.5000 - 0.2000 = 0.3000$ is 0.2995, corresponding to 0.84, we get

$$z = \frac{70.0 - 61.6}{\sigma} = 0.84$$

and hence

$$0.84\sigma = 70.0 - 61.6 = 8.4 \text{ and } \sigma = \frac{8.4}{0.84} = 10$$

7.22 (a) $z = \dfrac{25.0 - 27.8}{4.0} = -0.70$, and since the entry in Table II corresponding to 0.70 is 0.2580, the probability is

$$0.5000 - 0.2580 = 0.2420$$

(b) $z = \dfrac{29.6 - 27.8}{4.0} = 0.45$ and $z = \dfrac{26.0 - 27.8}{4.0} = -0.45$; and since the entry in Table II corresponding to 0.45 is 0.1736, the probability is

$$0.1736 + 0.1736 = 0.3472$$

7.29 Using the continuity correction, we must find the area under the curve to the right of 74.5; since $z = \dfrac{74.5 - 66.5}{12.6} \approx 0.63$ and the entry in Table II corresponding to 0.63 is 0.2357, it follows that the answer is

$$0.5000 - 0.2357 = 0.2643, \text{ or } 26.43\%$$

7.34 Using the continuity correction, we must find the area under the curve to the right of 49.5; since $\mu = 225(0.20) = 45$ and $\sigma = \sqrt{225(0.20)(0.80)} = 6$, so that $z = \dfrac{49.5 - 45}{6} = 0.75$, it follows that the entry in Table II corresponding to 0.75 is 0.2734 and that the required probability is

$$0.5000 - 0.2734 = 0.2266$$

8

SAMPLING AND SAMPLING DISTRIBUTIONS

The main objective of most statistical studies is to make sound generalizations, based on samples, about the parameters of populations. Note the word *sound*, because the question of when and under what conditions samples permit such generalizations is not easily answered. For instance, if we want to estimate the average amount of money that persons spend on their vacations, would we take as our sample the amounts spent by deluxe-class passengers on an around-the-world cruise? Would we attempt to predict changes in retail prices of farm products on the basis of changes in the prices of peaches alone? Obviously not, but just which vacationers and which farm products we should include in our samples, and how many of them, is neither intuitively clear nor self-evident.

In most of the methods that we shall study in this book, it will be assumed that we are using **random sampling**, that is, a method of sampling for which every possible sample has the same probability of being selected. We pay this attention to random sampling because it permits valid, or logical, generalizations and hence is widely used in practice. In this chapter we begin with a formal definition of random sampling in Section 8.1. Then, in Section 8.2, we introduce the related concept of a sampling distribution, which tells us how quantities determined from samples may vary from sample to sample; and in Sections 8.3 and 8.4 we learn how such variations can be measured.

8.1 RANDOM SAMPLING

In Section 1.3, we distinguished between populations and samples, saying that a population consists of all conceivably possible observations (instances, or occurrences) of a given phenomenon, while a sample is simply part of a population. For the work that follows, let us also distinguish between **finite populations** and **infinite populations**.

A population is finite if it consists of a finite, or fixed, number of elements (items, objects, measurements, or observations). Examples of finite populations are the net weights of 5,000 cans of baked beans in a production lot, SAT scores of all the freshmen admitted to a certain university in 2004, or the assessed values of all structures in a certain town.

In contrast, a population is infinite if there is no limit to the number of items, or values, that can be observed. This is the case, for example, when we observe repeated flips of a coin, when we sample with replacement from a population, or when we observe values of a continuous random variable, say, when we sample a normal population. The essential distinction between the two kinds of populations is that a finite population has a finite number of elements.

It should be apparent from this distinction between finite and infinite populations that the previous definition of *random sampling* applies only to finite populations—if there are infinitely many possibilities, we cannot very well speak of "every possible sample." Indeed, let us state formally that

A sample of size n from a finite population of size N is random if it is chosen in such a way that each of the $\binom{N}{n}$ possible samples has the same probability, $\dfrac{1}{\binom{N}{n}}$, of being selected.

For instance, if a finite population consists of the $N = 5$ elements a, b, c, d, and e (which might be the incomes of five persons, the weights of five guinea pigs, or the prices of five commodities), there are $\binom{5}{3} = 10$ possible samples of size $n = 3$ consisting, respectively, of the elements *abc, abd, abe, acd, ace, ade, bcd, bce, bde*, and *cde*. If we choose one of these samples in such a way that each sample has the probability $\dfrac{1}{10}$ of being selected, we call this sample a **random sample**.

This leaves the question of how random samples can be drawn in actual practice. In a simple case like the one described in the preceding paragraph, we could write each of the 10 possible samples on a slip of paper, put the slips of paper into a hat, shuffle them thoroughly, and then draw one without looking. Obviously, though, this would be impractical in a more realistically complex situation where N, and hence $\binom{N}{n}$, is quite large. In fact, we have mentioned it here only to stress the point that the selection of a random sample must depend entirely on chance.

Fortunately, we can take random samples from finite populations without having to go through the tedious process of listing all possible samples. One possibility is to write each of the N elements on a slip of paper and then take a random sample by choosing the elements to be included in the sample one at a time, making sure that in each successive drawing the remaining elements all have the same chance of being selected. It is not very difficult to show mathematically that this also leads to the same probability of $\dfrac{1}{\binom{N}{n}}$ for each possible sample.

For instance, to take a random sample of size $n = 12$ from the population that consists of the sales tax collected by a city's 247 drugstores in December 2004, we could write each of the 247 figures on a slip of paper, mix them up thoroughly in a box, and then draw (without looking) 12 of the slips of paper one after the other without replacement.

Even this relatively easy procedure can be simplified further. Usually, the simplest way of taking a random sample from a finite population is to refer to a table of **random numbers**. Such a table contains the digits 0, 1, 2, 3, 4, 5, 6, 7, 8, and 9 set down as they might appear if they had been generated by a gambling device, giving each digit the same probability of $\dfrac{1}{10}$ of appearing at any given place in the table. Table XII at the end of the book is excerpted from such a table.

EXAMPLE 8.1   With reference to the sales tax example, we number the 247 drugstores $001, 002, 003, \ldots, 246$, and 247. Then we arbitrarily pick a starting place in the table in Figure 8.1 and move in any direction, reading off three-digit numbers. For instance, we can read off the digits in the 26th, 27th, and 28th columns of the table, starting with the sixth row and going down the page, to obtain a random sample of 12 of the drugstores and, hence, a random sample of size 12 of the sales tax figures for December 2004.

Solution   Ignoring numbers greater than 247, it can be seen from Figure 8.1 that we get

046   230   079   022   119   150   056   064   193   232   040   146

These are the numbers of the drugstores, and the corresponding sales tax figures constitute the desired random sample. If any number had recurred when reading the values off the table, it would also have been ignored. It is necessary that we can associate the drugstores with the numbers $1, 2, \ldots, 247$. In this case, store 046 was selected, and we have to be able to identify that particular store. ■

Some early tables of random numbers were copied from pages of census data and from tables of 20-place logarithms, but now they are prepared with the use of computers. Indeed, it is not difficult to program a computer so that a person can generate his or her own random numbers.

When lists are available and items are readily numbered, it is easy to draw samples from finite populations with the aid of published or computer-generated random numbers. Unfortunately, though, it is often impossible to proceed as in the preceding example. For instance, if we want to use a sample to estimate the mean diameter of thousands of ball bearings packed in a crate or if we want to estimate the mean height of the trees in a forest, it would be impossible to number the ball bearings or the trees, choose random numbers, and then locate and measure the corresponding ball bearings or trees. In these and in many similar

Figure 8.1
Part of a table of random numbers.

48611	62866	33963	14045	79451	04934	45576
78812	03509	78673	73181	29973	18664	04555
19472	63971	37271	31445	49019	49405	46925
51266	11569	08697	91120	64156	40365	74297
55806	96275	26130	47949	14877	69594	83041
77527	81360	18180	97421	55541	90275	18213
77680	58788	33016	61173	93049	✓04694	43534
15404	96554	88265	34537	38526	67924	40474
14045	22917	60718	66487	46346	30949	03173
68376	43918	77653	04127	69930	43283	35766
93385	13421	67957	20384	58731	53396	59723
09858	52104	32014	53115	03727	98624	84616
93307	34116	49516	42148	57740	31198	70336
04794	01534	92058	03157	91758	80611	45357
86265	49096	97021	92582	61422	75890	86442
65943	79232	45702	67055	39024	57383	44424
90038	94209	04055	27393	61517	✓23002	96560
97283	95943	78363	36498	40662	94188	18202
21913	72958	75637	99936	58715	✓07943	23748
41161	37341	81838	19389	80336	46346	91895
23777	98392	31417	98547	92058	✓02277	50315
59973	08144	61070	73094	27059	69181	55623
82690	74099	77885	23813	10054	✓11900	44653
83854	24715	48866	65745	31131	47636	45137
61980	34997	41825	11623	07320	✓15003	56774
99915	45821	97702	87125	44488	77613	56823
48293	86847	43186	42951	37804	85129	28993
33225	31280	41232	34750	91097	60752	69783
06846	32828	24425	30249	78801	26977	92074
32671	45587	79620	84831	38156	74211	82752
82096	21913	75544	55228	89796	✓05694	91552
51666	10433	10945	55306	78562	89630	41230
54044	67942	24145	42294	27427	84875	37022
66738	60184	75679	38120	17640	36242	99357
55064	17427	89180	74018	44865	53197	74810
69599	60264	84549	78007	88450	✓06488	72274
64756	87759	92354	78694	63638	80939	98644
80817	74533	68407	55862	32476	✓19326	95558
39847	96884	84657	33697	39578	90197	80532
90401	41700	95510	61166	33757	✓23279	85523
78227	90110	81378	96659	37008	✓04050	04228
87240	52716	87697	79433	16336	52862	69149
08486	10951	26832	39763	02485	71688	90936
39338	32169	03713	93510	61244	73774	01245
21188	01850	69689	49426	49128	✓14660	14143
13287	82531	04388	64693	11934	35051	68576
53609	04001	19648	14053	49623	10840	31915
87900	36194	31567	53506	34304	39910	79630
81641	00496	36058	75899	46620	70024	88753
19512	50277	71508	20116	79520	06269	74173

situations, we may have no choice but to proceed according to the alternative dictionary definition of the word *random*, which says that the selection should be "haphazard, without aim or purpose." With some reservations, such samples can often be treated as if they were, in fact, real random samples. We must be careful in such cases to avoid unintentional selection by size. For example, we may favor large trees when making the selection. It therefore helps to have a clear idea as to exactly which trees are to be considered for the sample. For instance, we might decide that all trees with a height of at least 6 feet and diameter of at least 2 inches are eligible to be sampled.

Thus far we have discussed random sampling only in connection with finite populations.

> *For infinite populations, we say that a sample is random if it consists of values of independent random variables having the same distribution.*

By independent, we mean that the distribution of any one of the random variables is in no way affected by the values taken on by the others. For example, this definition applies when we observe honest flips of a balanced coin. If we get, say,

$$\text{H \quad T \quad H \quad T \quad T \quad H \quad H \quad T \quad H \quad H \quad H \quad H \quad T}$$

where H and T denote heads and tails, this result is a random sample as long as the H's and T's are values of independent random variables, each having the binomial distribution with $n = 1$ and $p = \frac{1}{2}$. As we pointed out on page 265, the population that we are sampling here is infinite because there is no limit to the number of times we can flip a coin.

For another example of random sampling from an infinite population, suppose that the weight loss of persons on a certain two-week diet is a random variable having the normal distribution with $\mu = 7.4$ pounds and $\sigma = 1.3$ pounds. If we get, say, 8.3, 5.9, 7.0, 10.5, and 6.8 pounds for the weight loss of five persons who have been on the diet, these figures constitute a random sample as long as they are values of independent random variables, each having the normal distribution with $\mu = 7.4$ pounds and $\sigma = 1.3$ pounds. To judge whether this is actually the case, we would have to see how the persons were chosen, that is, how the data was collected.

EXERCISES

Exercises 8.1 and 8.2 are practice exercises; their complete solutions are given on page 284.

✓ **8.1**   A random sample of size $n = 3$ is drawn from the finite population that consists of the elements a, b, c, d, and e. What is the probability that any specific element, say, element b, will be contained in the sample?

✓ **8.2**   A college president wants to determine the attitude of the college's 7,423 students, who are numbered 0001 through 7,423, toward a small increase in tuition. If the president decides to interview 12 students concerning the proposed change, which ones would be selected if the sample were chosen by using the first four columns of the random digits table (Figure 8.1), beginning with row 5?

8.3   How many different samples of size $n = 3$ can be drawn from a finite population of size

(a) 4;

(b) 6;

(c) 12;

(d) 300?

8.4   How many different samples of size 2 can be drawn from a finite population of size

(a) 4;

(b) 6;

(c) 12;

(d) 18;

(e) 400?

8.5   What is the probability of each possible sample if a random sample of size 3 is to be drawn from a finite population of size

(a) 4;

(b) 6;

(c) 12;

(d) 18?

8.6   What is the probability of each possible sample of size 4 being drawn from a finite population of size 18?

8.7   List the $\binom{4}{3} = 4$ possible samples of size $n = 3$ that can be drawn from the finite population whose elements are denoted by $a, b, c,$ and d.

8.8   List the $\binom{5}{2} = 10$ possible samples of size $n = 2$ that can be drawn from the finite population whose elements are denoted by the letters $a, b, c, d,$ and e.

8.9   The planet Uranus has five moons, which are named Miranda, Ariel, Umbriel, Titania, and Oberon, in order outward from Uranus. For simplicity in this exercise we can label each moon by its initial, M, A, U, T, and O. If an astronomer randomly selects four of these moons for further study:

(a) list all possible combinations of four of these five moons;

(b) find the probability of each possible sample;

(c) find the probability that Miranda will be included in the sample.

8.10   Suppose that the National Basketball Association (NBA) wants to randomly select and inspect 10 of the home courts of their 29 basketball teams. If the home courts are numbered $01, 02, 03, 04, \ldots, 29$, which ones, by number, will be included if the NBA selects them by using the 14th and 15th columns of the table in Figure 8.1, going down the page and beginning with row 6?

8.11   A research scientist wants to randomly select 12 persons from among 24 obese participants who will test the effectiveness of a new weight-reduction pill. The 12 randomly selected persons will not be informed of their selection, but will take harmless, ineffective pills which are placebos, while the remaining 12 will take the experimental new pill. Which ones will be selected for the placebos if the research scientist numbers the participants $1, 2, 3, \ldots, 24$, and selects the 12 persons who will receive the placebo by using the table of Figure 8.1, going down the page using columns 6 and 7, and beginning with the 1st row? If necessary to complete the selection of 12 persons, return to row 1, going down columns 9 and 10.

8.12   Two hundred sales have been recorded by salespersons on sales slips consecutively numbered from 201 to 400, inclusive. If an auditor wants to randomly select six of them for verification, which ones would be chosen by using the 11th, 12th, and 13th columns of the table in Figure 8.1, going down the page and starting with the 10th row?

8.13   A college registrar wants to verify the grade-point averages of a random sample of 15 graduating students in a graduating class of 3,852 students. If the students are numbered 0001, 0002, ... , 3,851, and 3,852, which ones (by number) will the registrar select by choosing columns 21, 22, 23, and 24 of the table in Figure 8.1, going down the page and starting with the second row?

8.14   A laboratory supply warehouse stores its inventory of 6,429 items in bins, compartments and spaces that are consecutively numbered 0001, 0002, ... , 6,428, and 6,429. To check the perpetual inventory of items that is maintained by computer, the manager decides to randomly select 20 storage spaces and manually count the items in them. Which of the bins, compartments, and spaces will be selected if the manager uses the 26th, 27th, 28th, and 29th columns of the table in Figure 8.1, going down the page and starting with the first row?

8.15   Randy and Susan are both members of a population of 60 students. A researcher is going to select 10 students from this population.

(a) What is the probability that Randy will be in the sample?

(b) What is the probability that Susan will be in the sample?

(c) What is the probability that both will be in the sample?

(d) Is your solution to (c) greater than or less than $\left(\dfrac{10}{60}\right)^2 = \dfrac{1}{36}$?

8.2   SAMPLING DISTRIBUTIONS

The sample mean, the sample median, and the sample standard deviation are examples of random variables, whose values will vary from sample to sample. Their distributions, which reflect such chance variations, play an important role in statistics, and they are referred to as **sampling distributions**.

To illustrate the concept of a sampling distribution, let us construct the one for the mean of a random sample of size $n = 2$ from the finite population of size $N = 5$, whose elements are the numbers 1, 3, 5, 7, and 9. The mean and the standard deviation of this population, which we shall need later, are

$$\mu = \frac{\sum x}{N} = \frac{1+3+5+7+9}{5} = 5$$

and

$$\sigma = \sqrt{\frac{\sum (x - \mu)^2}{N}} = \sqrt{\frac{(1-5)^2 + (3-5)^2 + (5-5)^2 + (7-5)^2 + (9-5)^2}{5}}$$

$$= \sqrt{8} \approx 2.828$$

in accordance with formulas on pages 48 and 70.

This illustration is, of course, hypothetical in that we know all N values. In actual applications we never know all the values. Now, if we take a random sample of size $n = 2$ from this population, there are

$$\binom{5}{2} = 10 \text{ possibilities}$$

and they are

1 and 3	3 and 7
1 and 5	3 and 9
1 and 7	5 and 7
1 and 9	5 and 9
3 and 5	7 and 9

The means of these samples are

$$\frac{1+3}{2} = 2, \frac{1+5}{2} = 3, 4, 5, 4, 5, 6, 6, 7, \text{ and } 8$$

If sampling is random, then each sample has the probability $\dfrac{1}{10}$, and we arrive at the following sampling distribution of the mean:

$\bar{x}$	**Probability**
2	$\dfrac{1}{10}$
3	$\dfrac{1}{10}$
4	$\dfrac{2}{10}$
5	$\dfrac{2}{10}$
6	$\dfrac{2}{10}$
7	$\dfrac{1}{10}$
8	$\dfrac{1}{10}$

A histogram of this probability distribution is shown in Figure 8.2.

An examination of this sampling distribution reveals some pertinent information about the chance variations of the mean of a random sample of size $n = 2$ from the given population. For instance, we find that the probability is $\dfrac{2}{10} + \dfrac{2}{10} + \dfrac{2}{10} = \dfrac{6}{10}$ that a sample mean will differ from the population mean $\mu = 5$ by 1 or less, and that the probability is $\dfrac{1}{10} + \dfrac{2}{10} + \dfrac{2}{10} + \dfrac{2}{10} + \dfrac{1}{10} = \dfrac{8}{10}$ that a sample mean will differ from the population mean $\mu = 5$ by 2 or less. (The first case corresponds to $\bar{x} = 4, 5,$ or 6, and the second case corresponds to $\bar{x} = 3, 4, 5, 6,$ or 7.) Thus, if we did not know the mean of the given population and wanted to estimate it with the mean of a random sample of two observations, this would give us some idea about the potential size of our error. Further useful information

Figure 8.2
Sampling distribution
of the mean.

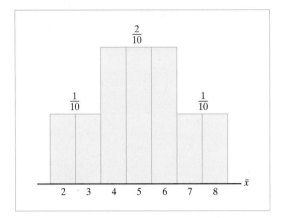

about this sampling distribution of the mean can be obtained by calculating its mean $\mu_{\bar{x}}$ and its standard deviation $\sigma_{\bar{x}}$, where the subscripts serve to distinguish between these parameters and those of the original population. Following the definitions of the mean and the variance of a probability distribution on pages 225 and 226, we get

$$\mu_{\bar{x}} = 2 \cdot \frac{1}{10} + 3 \cdot \frac{1}{10} + 4 \cdot \frac{2}{10} + 5 \cdot \frac{2}{10}$$

$$+ 6 \cdot \frac{2}{10} + 7 \cdot \frac{1}{10} + 8 \cdot \frac{1}{10} = 5$$

and

$$\sigma_{\bar{x}}^2 = (2 - 5)^2 \cdot \frac{1}{10} + (3 - 5)^2 \cdot \frac{1}{10}$$

$$+ (4 - 5)^2 \cdot \frac{2}{10} + (5 - 5)^2 \cdot \frac{2}{10}$$

$$+ (6 - 5)^2 \cdot \frac{2}{10} + (7 - 5)^2 \cdot \frac{1}{10}$$

$$+ (8 - 5)^2 \cdot \frac{1}{10} = 3$$

so that $\sigma_{\bar{x}} = \sqrt{3} \approx 1.732$.

Observe that, at least for this example,

1. $\mu_{\bar{x}}$, the mean of the sampling distribution of $\bar{x}$, equals μ, the mean of the population;
2. $\sigma_{\bar{x}}$, the standard deviation of the sampling distribution of $\bar{x}$, is smaller than σ, the standard deviation of the population.

These relationships are of fundamental importance in statistics, and we shall return to them in Section 8.3.

In the preceding example we took a very small sample from a very small population, but it would be difficult to use the same method to construct the sampling distribution of the mean of a large sample from a large population—we would have to enumerate too many possibilities. To construct the sampling distribution of the mean even for random samples of size $n = 5$ from a finite population of size $N = 100$, we would have to list

$$\binom{100}{5} = 75{,}287{,}520 \text{ possibilities}$$

So, to get an idea about the sampling distribution of the mean of a larger random sample from a larger finite population, we shall use a **computer simulation**. This means that we shall use a computer to take a number of random samples from the given population, calculate their means, and display the results in the form of a histogram. This histogram will, we hope, give us some idea about the key features of the actual sampling distribution of the mean for random samples of the given size from the given population.

8.2   Use a computer to generate 100 random samples of size $n = 15$ from the finite population that consists of the integers from 1 to 1,000. Also, calculate the mean of each of the 100 samples, find their mean and their standard deviation, and group them into a distribution with the class marks $225.5, 275.5, \ldots$, and so on.[†]

Solution   Without a computer, the reader may picture this kind of simulation as follows: First, the numbers from 1 to 1,000 are written on 1,000 slips of paper (poker chips, small balls, or whatever may lend itself to drawing random samples). Then, a random sample of size $n = 15$ is drawn from this population and the values are recorded; each random sample of 15 is replaced before the next sample is drawn. This process is repeated until 100 random samples have been obtained.

Actually using an appropriate computer package, MINITAB in this case, we obtained the printouts shown in Figures 8.3(a)–(d). Our population is the

Figure 8.3(a)
Computer simulation of a sampling distribution of the mean.

[†] These class marks were chosen so that the corresponding class boundaries are "impossible" values; they are values that cannot be taken on by a mean of 15 positive integers.

Figure 8.3(b)

Figure 8.3(c)

Figure 8.3(d)

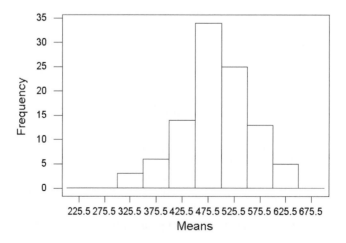

integers from 1 to 1,000. This population has a mean $\mu = 500.5$ and $\sigma = 289$ (to the nearest integer). The previous MINITAB printouts generated 100 samples of $n = 15$ (of which only 9 are actually shown).

We might say that this distribution is fairly symmetrical and bell shaped; in fact, the overall pattern seems to follow quite closely that of a normal curve. All this applies to the distribution that we obtained by means of the computer simulation involving only 100 random samples of size $n = 15$ from the population that consists of the integers from 1 to 1,000.

Note also that the results of this experiment support the two points made on page 272. Although the mean of the one hundred $\bar{x}$'s does not equal $\mu = 500.5$, its value, 493.62, is very close. Also, the standard deviation of the one hundred $\bar{x}$'s, 65.64, is smaller than the population standard deviation, $\sigma = 289$.

8.3 THE STANDARD ERROR OF THE MEAN

In most practical situations we can determine how close a sample mean might be to the mean of the population from which the sample came by referring to two theorems that express essential facts about sampling distributions. The first of these theorems expresses formally what we discovered in connection with the first sampling distribution of Section 8.2: The mean of the sampling distribution of $\bar{x}$ equals the mean of the population, and its standard deviation is smaller than the standard deviation of the population. It may be phrased as follows: For random samples of size n taken from a population having the mean μ and the standard deviation σ, the sampling distribution of $\bar{x}$ has the mean

MEAN OF THE SAMPLING DISTRIBUTION OF $\bar{x}$

$$\mu_{\bar{x}} = \mu$$

and the standard deviation

STANDARD ERROR OF THE MEAN

$$\sigma_{\bar{x}} = \frac{\sigma}{\sqrt{n}} \quad \text{or} \quad \sigma_{\bar{x}} = \frac{\sigma}{\sqrt{n}} \cdot \sqrt{\frac{N-n}{N-1}}$$

depending on whether the population is infinite or finite of size N.

It is customary to refer to $\sigma_{\bar{x}}$, the standard deviation of the sampling distribution of the mean, as the **standard error of the mean**. Its role in statistics is fundamental, since it measures the extent to which sample means can be expected to fluctuate, or vary, due to chance. What determines the size of $\sigma_{\bar{x}}$, and hence the goodness of an estimate, can be seen from the preceding formulas. Both formulas show that the standard error of the mean increases as the variability of the population increases (in fact, it is directly proportional to σ) and that it decreases as the sample size increases.

EXAMPLE 8.3 When we sample from an infinite population, what happens to the standard error of the mean (and hence to the size of the error that we are exposed to when we use $\bar{x}$ as an estimate of μ) if the sample size is increased from $n = 50$ to $n = 200$?

Solution   The ratio of the two standard errors is

$$\frac{\dfrac{\sigma}{\sqrt{200}}}{\dfrac{\sigma}{\sqrt{50}}} = \frac{\sqrt{50}}{\sqrt{200}} = \sqrt{\frac{50}{200}} = \sqrt{\frac{1}{4}} = \frac{1}{2}$$

so that the standard error of the mean is divided by 2.   ■

The factor $\sqrt{\dfrac{N-n}{N-1}}$ in the second formula for $\sigma_{\bar{x}}$ is called the **finite population correction factor**. Without it, the two formulas for $\sigma_{\bar{x}}$ (for infinite and finite populations) are the same. In practice, it is omitted unless the sample constitutes at least 5% of the population, for otherwise it is so close to 1 that it has little effect on the value of $\sigma_{\bar{x}}$.

EXAMPLE 8.4   Find the value of the finite population correction factor for $n = 100$ and

(a) $N = 10,000$;

(b) $N = 200$.

Solution   (a) Substituting $n = 100$ and $N = 10,000$, we get

$$\sqrt{\frac{N-n}{N-1}} = \sqrt{\frac{10,000 - 100}{10,000 - 1}} \approx 0.995$$

and this is so close to 1 that the correction factor can be omitted for most practical purposes.

(b) Substituting $n = 100$ and $N = 200$, we get

$$\sqrt{\frac{N-n}{N-1}} = \sqrt{\frac{200 - 100}{200 - 1}} \approx 0.709$$

so that the finite population correction factor will substantially reduce the standard error of the mean. This reflects the fact that in this case we have a great deal of information about the population; in fact, the sample constitutes half the population.   ■

Since we did not actually prove the two formulas for the standard error of the mean, let us verify the second one with reference to previous illustrations on pages 270 and 275.

EXAMPLE 8.5   With reference to the illustration on page 270, verify that the formula for $\sigma_{\bar{x}}$ for a random sample from a finite population will also yield $\sigma_{\bar{x}} = \sqrt{3}$.

Solution   Substituting $n = 2$, $N = 5$, and $\sigma = \sqrt{8}$ into the second of the two formulas for $\sigma_{\bar{x}}$, we get

$$\sigma_{\bar{x}} = \frac{\sqrt{8}}{\sqrt{2}} \sqrt{\frac{5-2}{5-1}} = \frac{\sqrt{8}}{\sqrt{2}} \cdot \sqrt{\frac{3}{4}} = \sqrt{\frac{8}{2} \cdot \frac{3}{4}} = \sqrt{3}$$ ■

EXAMPLE 8.6   With reference to the computer simulation on page 273, where we had $n = 15$, $N = 1,000$, and $\sigma = 289$, what value might we have expected for the standard deviation of the 100 sample means?

Solution   Substituting $n = 15$, $N = 1,000$, and $\sigma = 289$ into the second of the two formulas for σ, we get

$$\sigma_{\bar{x}} = \frac{289}{\sqrt{15}} \cdot \sqrt{\frac{1,000 - 15}{1,000 - 1}} \approx 74.1$$

and this is fairly close to 65.64, the value that we actually obtained in the printout of Figure 8.3.   ■

EXERCISES

Exercises 8.16 and 8.17 are practice exercises; their complete solutions are given on pages 284 and 285.

✓ **8.16**   For a sample of size $n = 2$, finite population size of $N = 5$, and elements numbered 1, 3, 5, 7, and 9, suppose that sampling is *with replacement*. List the 25 ordered samples 1 and 1, 1 and 3, 1 and 5, 1 and 7, 1 and 9, 3 and 1, 3 and 3, 3 and 5, and so on. Calculate the means of these 25 ordered samples. Then, assuming that each ordered pair has the probability of $\frac{1}{25}$, construct the sampling distribution of the mean for random samples of size $n = 2$ taken with replacement from the given population.

✓ **8.17**   Calculate the standard deviation of the sampling distribution obtained in Exercise 8.16 and verify the result by substituting $n = 2$ and $\sigma = \sqrt{8}$ into the standard error formula $\sigma_{\bar{x}} = \frac{\sigma}{\sqrt{n}}$.

8.18   Random samples of size $n = 2$ are drawn from a finite population that consists of the numbers 2, 4, 6, and 8.

(a) Calculate the mean and the standard deviation of this population.

(b) List the six possible random samples of size $n = 2$ that can be drawn from this population and calculate their means.

(c) Use the results of part (b) to construct the sampling distribution of the mean for random samples of size $n = 2$ from the given population.

(d) Calculate the standard deviation of the sampling distribution obtained in part (c) and verify the result by substituting $n = 2$, $N = 4$, and the value of σ obtained in part (a) into the second of the two standard error formulas on page 275.

8.19   Rework the preceding exercise for sampling with replacement from the given population.

8.20   When we sample from an infinite population, what happens to the standard error of the mean if the sample size is

(a) increased from 5 to 20;

(b) decreased from 20 to 5;

(c) increased from 12 to 48;

(d) decreased from 48 to 12?

8.21   When we sample from an infinite population, what happens to the standard error of the mean if the sample size is

(a) increased from 3 to 27;

(b) decreased from 27 to 3;

(c) increased from 27 to 243;

(d) decreased from 243 to 27?

8.22 What is the value of the finite population correction factor when

(a) $n = 20$ and $N = 200$;

(b) $n = 20$ and $N = 2,000$?

8.23 Find the value of the finite population correction factor for $n = 100$ and $N = 10,000$. Would you use this correction factor in a real problem requiring the standard error of the mean of a finite population? Why or why not?

8.24 For what positive value of n is the finite population correction factor $\sqrt{\dfrac{N-n}{N-1}}$ equal to zero?

8.25 For simplicity, the finite population correction factor $\sqrt{\dfrac{N-n}{N-1}}$ can be replaced by $\sqrt{\dfrac{N-n}{N}} = \sqrt{1-f}$, where $f = \dfrac{n}{N}$ is called the **sampling fraction**. Give the values of $\sqrt{\dfrac{N-n}{N-1}}$ and $\sqrt{1-f}$ in each of the following cases:

(a) $N = 20, n = 5$;

(b) $N = 80, n = 20$;

(c) $N = 1,000, n = 50$;

(d) $N = 350, n = 60$.

8.26 Use a computer package to

(a) generate the population of integers from 1 to 800 and determine its mean and its standard deviation;

(b) take 100 random samples of size $n = 5$ from this population;

(c) determine the means of these 100 samples and calculate their mean and their standard deviation;

(d) construct a histogram of the means obtained in part (c), using the class marks $40.5, 80.5, 120.5, 160.5, \ldots$.
Also,

(e) compare the mean of the 100 sample means with that of the population;

(f) compare the standard deviation of the 100 sample means with the value that we would expect in accordance with the theory on page 275.

(g) discuss the overall shape of the histogram obtained in part (d).

8.27 Repeat Exercise 8.26 using sampling with replacement so that the whole procedure may be looked upon as sampling from an infinite population.

8.28 Use a computer package to

(a) generate the population of integers from 1 to 1,200 and determine its mean and its standard deviation;

(b) take 150 random samples of size $n = 25$ from this population;

(c) determine the means of these 150 samples and calculate their mean and their standard deviation;

(d) construct a histogram of the means obtained in part (c), using the class marks $30.5, 90.5, 150.5, 210.5, \ldots$.
Also,

(e) compare the mean of the 150 sample means with that of the population;

(f) compare the standard deviation of the 150 sample means with the value that we would expect in accordance with the theory on page 275;

(g) discuss the overall shape of the histogram obtained in part (d).

8.29 Repeat Exercise 8.28 using sampling with replacement so that the whole procedure may be looked upon as sampling from an infinite population.

8.4 THE CENTRAL LIMIT THEOREM

When we use a sample mean to estimate the mean of a population, we usually express our confidence in the estimate by attaching a probability statement about the size of our error. For instance, if we apply Chebyshev's theorem (as formulated in Section 6.8) to the sampling distribution of the mean, we can assert with a probability of at least $1 - \dfrac{1}{k^2}$ that the mean of a random sample will differ from the mean of the population from which it came by less than $k \cdot \sigma_{\bar{x}}$. In other words, if we use the mean of a random sample to estimate the mean of the population from which it came, we can assert with a probability of at least $1 - \dfrac{1}{k^2}$ that our error will be less than $k \cdot \sigma_{\bar{x}}$.

EXAMPLE 8.7   Based on Chebyshev's theorem with $k = 2$, what can we assert about the maximum size of our error if we use the mean of a random sample of size $n = 64$ to estimate the mean of an infinite population with $\sigma = 20$?

Solution   Substituting $\sigma = 20$ and $n = 64$ into the first of the two formulas for the standard error of the mean, we get

$$\sigma_{\bar{x}} = \frac{20}{\sqrt{64}} = 2.5$$

and it follows that we can assert with a probability of at least

$$1 - \frac{1}{2^2} = 0.75$$

that the error will be less than $k\sigma_{\bar{x}} = 2 \cdot 2.5 = 5$.   ■

Chebyshev's theorem can be used in problems like this; but if we want to be more specific about the probabilities, we have to refer to another theorem, which is called the **central limit theorem**. It states that for large samples the sampling distribution of the mean can be approximated closely with a normal distribution.

To formulate this more precisely, we must convert $\bar{x}$ to standard units. Since a value of a random variable is converted to standard units by subtracting from it the mean of the distribution of the random variable and then dividing by its standard deviation, we write

$$z = \frac{\bar{x} - \mu}{\sigma_{\bar{x}}}$$

where we used the same notation as in Section 8.3. Recalling that

$$\mu_{\bar{x}} = \mu \quad \text{and} \quad \sigma_{\bar{x}} = \frac{\sigma}{\sqrt{n}}$$

for random samples from infinite populations, we can now say formally that

CENTRAL LIMIT THEOREM

If $\bar{x}$ is the mean of a random sample of size n from an infinite population, with the mean μ and the standard deviation σ, and n is large, then

$$z = \frac{\bar{x} - \mu}{\sigma/\sqrt{n}}$$

has approximately the standard normal distribution.

This theorem is of fundamental importance in statistics, because it justifies the use of normal-curve methods in a wide range of problems; it applies to infinite populations, and also to finite populations when n, though large, constitutes but a small portion of the population. We cannot say precisely how large n must be so that the central limit theorem can be applied, but unless the population distribution has a very unusual shape, $n = 30$ is usually regarded as sufficiently large. When the population that we are sampling has, itself, roughly the shape of a normal curve, the sampling distribution of the mean can be approximated closely with a normal distribution regardless of the size of n.

The central limit theorem can also be used for finite populations, but a precise description of the situations under which this can be done is rather complicated. The most common proper use is the case in which n is large while $\dfrac{n}{N}$ is small. This is the case, for example, in most political polls.

EXAMPLE 8.8   Returning to Example 8.7 on page 279, where we determined on the basis of Chebyshev's theorem that our error was less than 5 with a probability of at least 0.75, with what probability can we assert that our error is less than 5 on the basis of the central limit theorem? In that example we used the mean of a random sample of size $n = 64$ to estimate the mean of an infinite population with $\sigma = 20$.

Solution   The desired probability is given by the area of the blue region under the curve in Figure 8.4, that is, the normal-curve area between

$$z = \frac{-5}{20/\sqrt{64}} = -2 \quad \text{and} \quad z = \frac{5}{20/\sqrt{64}} = 2$$

Figure 8.4
Sampling distribution of the mean.

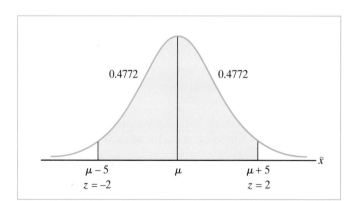

Since the entry in Table II corresponding to $z = 2.00$ is 0.4772, we find that the answer is

$$0.4772 + 0.4772 = 0.9544$$

Compared to "at least 0.75," we can thus make the much stronger statement that the probability is 0.9544 that the mean of a random sample of size $n = 64$ from the given population will differ from the population mean by less than 5. ■

EXAMPLE 8.9   Suppose that $\sigma = 5.5$ tons for the daily sulfur oxides emission of a certain industrial plant. What is the probability that the mean of a random sample of size $n = 40$ will differ from the mean of the population by less than 1.0 ton?

Solution   The desired probability is given by the area of the blue region under the curve in Figure 8.5, namely, the normal-curve area between

$$z = \frac{-1.0}{5.5/\sqrt{40}} \approx -1.15 \quad \text{and} \quad z = \frac{1.0}{5.5/\sqrt{40}} \approx 1.15$$

Figure 8.5
Sampling distribution of the mean.

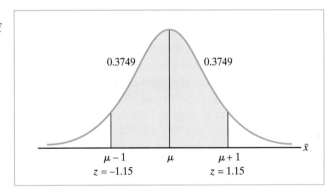

$$0.3749 \qquad 0.3749$$

$$\mu - 1 \qquad \mu \qquad \mu + 1$$
$$z = -1.15 \qquad \qquad z = 1.15$$

Since the entry in Table II corresponding to $z = 1.15$ is 0.3749, we find that the answer is

$$0.3749 + 0.3749 = 0.7498, \text{ or approximately } 0.75$$ ■

In both of the preceding examples, we were concerned with the size of the error; in the example that follows, μ will be given and we will be asked for the probability that the mean of a sample will fall within a certain range.

EXAMPLE 8.10   The time it takes students in a cooking school to learn how to prepare a particular meal is a random variable with the mean $\mu = 3.2$ hours and the standard deviation $\sigma = 1.8$ hours. Find the probability that the average time it will take 36 students to learn how to prepare the meal is less than 3.4 hours.

Solution   Again, we are concerned with the sampling distribution of the mean; and since n is greater than 30, we can approximate it with a normal curve. Thus, the desired probability is given by the area of the blue region under the curve in Figure 8.6, namely, the normal-curve area to the left of

$$z = \frac{3.4 - 3.2}{1.8/\sqrt{36}} = \frac{0.2}{0.3} \approx 0.67$$

Figure 8.6
Sampling distribution
of the mean.

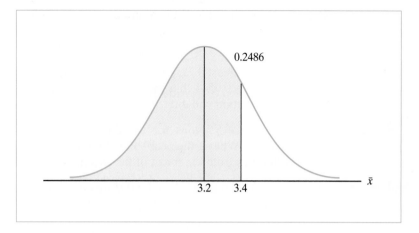

Since the entry in Table II corresponding to $z = 0.67$ is 0.2486, the answer is

$$0.2486 + 0.5000 = 0.7486$$

The primary goal of this chapter has been to introduce the concept of a sampling distribution, and the one that we chose for this purpose was the sampling distribution of the mean. Observe, however, that instead of the mean we could have studied the median or some other statistic and investigated its chance fluctuations. So far as the corresponding theory is concerned, this would, of course, have required different formulas for the standard errors, that is, different formulas for the standard deviations of the respective sampling distributions. For instance, for infinite populations, the **standard error of the median** is approximately

$$\sigma_{\tilde{x}} = 1.25 \, \frac{\sigma}{\sqrt{n}}$$

where n is the size of the sample and σ is the population standard deviation. Note that comparison of the two formulas

$$\sigma_{\bar{x}} = \frac{\sigma}{\sqrt{n}} \quad \text{and} \quad \sigma_{\tilde{x}} = 1.25 \, \frac{\sigma}{\sqrt{n}}$$

reflects the fact that the mean is generally more reliable than the median: it is subject to smaller chance fluctuations (see also Exercise 8.41). There are nonetheless situations, generally with medium-sized samples from nonnormal populations, in which the median is actually more reliable than the mean.

Let us also point out that there have been three occasions so far to work with standard units. In Section 3.8 we used them to illustrate an important application of the standard deviation; in Section 7.2 we used them in connection with the standard normal distribution; and in this section we used them in the formulation of the central limit theorem and in its applications. In general, the conversion of a value of a random variable to standard units is made with the formula

$$z = \frac{\text{(value of random variable)} - \text{(mean of random variable)}}{\text{standard deviation of random variable}}$$

The mean that we subtract in the numerator may be μ or $\mu_{\bar{x}}$, and the standard deviation by which we divide may be σ or $\sigma_{\bar{x}}$, depending on whether we are dealing with one observation or with a mean. They may also be $\mu_{\tilde{x}}$ and $\sigma_{\tilde{x}}$, for example, when we are dealing with the sampling distribution of the median.

Exercises 8.30 and 8.39 are practice exercises; their complete solutions are given on page 285.

✓ **8.30** A biologist wants to estimate the mean weight of a certain kind of animal. He knows that the standard deviation of their weights is 4.8 ounces and decides to use a random sample of size 100. Based on the central limit theorem, with what probability can he assert that his error will be

  (a) less than 0.60 ounce;

  (b) less than 1.20 ounces?

8.31 Using Chebyshev's theorem with $k = 3$, what can we assert about the maximum size of our error if we use the mean of a random sample of size 144 to estimate the mean of a population with $\sigma = 9$?

8.32 Using the data of Exercise 8.31 but changing the sample size from 144 to 72, what can we assert about the maximum size of our error?

8.33 A commuting student wants to use a sample of 36 driving times to estimate the mean time it takes to drive to the college each morning. If the standard deviation of the time required to make the trip is assumed to be 11 minutes, what can the student assert about the probability that his error will be less than 3.3 minutes if he uses

  (a) Chebyshev's theorem;

  (b) the central limit theorem?

8.34 The organizer of a large national political convention wants to use the average monetary expenditures of 100 delegates to estimate the average monetary expenditures of the population of delegates attending the convention. If the standard deviation of the expenditures is assumed to be $150 , what can the organizer assert, using Chebyshev's theorem, about the probability that the error will be less than $18 ?

8.35 The mean of a random sample of size $n = 100$ is used to estimate the mean of an infinite population having the standard deviation $\sigma = 2$. What can be asserted about the probability the error will be less than 0.25 using Chebyshev's theorem?

8.36 The supervisor of a waste disposal truck that collects waste in a residential neighborhood wants to estimate the mean number of cubic feet of waste collected at these residences. Suppose that the supervisor uses a sample of 30 waste pickups and knows from experience that the standard deviation is 3 cubic feet for such data. Based on the central limit theorem, with what probability can the supervisor assert that the error will be

EXERCISES

(a) less than 1 cubic foot;

(b) less than 0.5 cubic foot?

8.37 A college weekly newspaper that accepts advertising wants to estimate the mean number of lines of advertising in its newspapers. Suppose that it uses a sample of 49 weekly editions and that the standard deviation is 250 lines. If the newspaper uses the central limit theorem, with what probability can it assert that its error will be

(a) less than 50 lines;

(b) less than 25 lines?

8.38 During the last week of the semester, students at a certain college spend on the average 4.2 hours using the school's computer terminals with a standard deviation of 1.8 hours. For a random sample of 36 students at that college, find the probabilities that the average time spent using the computer terminals during the last week of the semester is

(a) at least 4.8 hours;

(b) between 4.1 and 4.5 hours.

✓ **8.39** Show that the mean of a random sample of size $n = 64$ is as reliable an estimate of the mean of a symmetrical infinite population as the median of a random sample of size $n = 100$. (For symmetrical populations, the means of the sampling distributions of $\bar{x}$ and $\tilde{x}$ are both equal to the population mean μ.)

8.40 How large a random sample do we have to take so that its mean is as reliable an estimate of the mean of a symmetrical infinite population as the median of a random sample of size $n = 225$?

8.41 How large a random sample do we have to take so that its median is as reliable an estimate of the mean of a symmetrical infinite population as the mean of a random sample of size $n = 1,600$?

8.42 Show that if the mean of a random sample of size n is used to estimate the mean of an infinite population with the standard deviation σ, there is a 50–50 chance that the error will be less than the quantity $0.6745 \cdot \dfrac{\sigma}{\sqrt{n}}$, which is called the **probable error of the mean**. (*Hint*: Interpolate between the entries in Table II corresponding to $z = 0.67$ and $z = 0.68$.)

✓ SOLUTIONS TO PRACTICE EXERCISES

8.1 There are ten possible samples: *abc, abd, abe, acd, ace, ade, bcd, bce, bde*, and *cde*. Six of these include *b*, so that the probability is $\dfrac{6}{10} = 0.6$.

8.2 Reading the values off the table, we get 5580, 1540, 1404, 6837, 0985, 0479, 6594, 2191, 4116, 2377, 5997, and 6198.

8.16 The ordered samples are 1 and 1, 1 and 3, 1 and 5, 1 and 7, 1 and 9, 3 and 1, 3 and 3, 3 and 5, 3 and 7, 3 and 9, 5 and 1, 5 and 3, 5 and 5, 5 and 7, 5 and 9, 7 and 1, 7 and 3, 7 and 5, 7 and 7, 7 and 9, 9 and 1, 9 and 3, 9 and 5, 9 and 7, and 9 and 9. The means are 1, 2, 3, 4, 5, 2, 3, 4, 5, 6, 3, 4, 5, 6, 7, 4, 5, 6, 7, 8, 5, 6, 7, 8, and 9. The sampling distribution of the mean is

$\bar{x}$	Probability
1	$\dfrac{1}{25}$
2	$\dfrac{2}{25}$
3	$\dfrac{3}{25}$
4	$\dfrac{4}{25}$
5	$\dfrac{5}{25}$
6	$\dfrac{4}{25}$
7	$\dfrac{3}{25}$
8	$\dfrac{2}{25}$
9	$\dfrac{1}{25}$

8.17 Since $\mu_{\bar{x}} = \mu = 5$, we get

$$\sigma_{\bar{x}}^2 = (1-5)^2 \cdot \frac{1}{25} + (2-5)^2 \cdot \frac{2}{25} + (3-5)^2 \cdot \frac{3}{25}$$

$$+ (4-5)^2 \cdot \frac{4}{25} + (5-5)^2 \cdot \frac{5}{25} + (6-5)^2 \cdot \frac{4}{25}$$

$$+ (7-5)^2 \cdot \frac{3}{25} + (8-5)^2 \cdot \frac{2}{25} + (9-5)^2 \cdot \frac{1}{25} = 4$$

so $\sigma_{\bar{x}} = 2$. Substitution into the formula yields

$$\sigma_{\bar{x}} = \frac{\sqrt{8}}{\sqrt{2}} = \sqrt{\frac{8}{2}} = \sqrt{4} = 2$$

and the two answers agree.

8.30 (a) $\sigma_{\bar{x}} = \dfrac{4.8}{\sqrt{100}} = 0.48$, $z = \dfrac{0.60}{0.48} = 1.25$, the corresponding entry in Table II is 0.3944, and the answer is

$$0.3944 + 0.3944 = 0.7888$$

(b) $z = \dfrac{1.20}{0.48} = 2.50$, the corresponding entry in Table II is 0.4938, and the answer is

$$0.4938 + 0.4938 = 0.9876$$

8.39 $\sigma_{\bar{x}} = \dfrac{\sigma}{\sqrt{64}} = \dfrac{\sigma}{8}$ and $\sigma_{\tilde{x}} = 1.25 \cdot \dfrac{\sigma}{\sqrt{100}} = 1.25 \cdot \dfrac{\sigma}{10} = \dfrac{\sigma}{8}$; since the two standard errors are equal, the mean of a random sample of size $n = 64$ is as reliable as the median of a random sample of size $n = 100$.

6, 7, & 8

ACHIEVEMENTS

Having read and studied these chapters and having worked a good portion of the exercises, you should be able to

1. Explain what is meant by *random variable* and *probability distribution*.
2. State the two rules that must be satisfied by the values of a probability distribution.
3. List the assumptions that underlie the binomial distribution.
4. Use the formula for the binomial distribution.
5. Use the table of binomial probabilities.
6. Use the formula for the hypergeometric distribution.
7. Approximate hypergeometric probabilities with binomial probabilities and know when this approximation may be used.
8. Approximate binomial probabilities with Poisson probabilities and know when this approximation may be used.
9. Use the formula for the Poisson distribution with the parameter λ.
10. Calculate multinomial probabilities.
11. Determine the mean and the standard deviation of a probability distribution.
12. Find the mean and the standard deviation of a binomial distribution.
13. Find the mean and the standard deviation of a hypergeometric distribution.
14. Find the mean and the standard deviation of a Poisson distribution.
15. Use Chebyshev's theorem as it applies to the distribution of a random variable.
16. Explain what is meant by *probability density* or *continuous distribution*.
17. Discuss the normal distribution and explain what is meant by "standard normal distribution."
18. Use the table of normal-curve areas.
19. Apply the continuity correction where needed.
20. Use the normal distribution to approximate binomial probabilities and know when this approximation may be used.
21. Explain what is meant by *random sample from a finite population*.
22. Use random numbers to select random samples from finite populations.
23. Explain what is meant by *sampling distribution* and *sampling distribution of the mean*.
24. State and apply the theorem about the mean and the standard deviation of the sampling distribution of the mean.
25. Explain what is meant by *standard error*.
26. Determine the finite population correction factor and know when it should be used.
27. State the central limit theorem and use it to calculate probabilities relating to the sampling distribution of the mean.

CHECKLIST OF KEY TERMS (with page references to their definitions)

Binomial distribution, 203
Binomial probability distribution, 203
Central limit theorem, 279
Chebyshev's theorem, 231
Computer simulation, 273
Continuity correction, 253
Continuous distribution, 237
Continuous random variable, 200
Discrete random variable, 200
Finite population, 265
Finite population correction factor, 276
Hypergeometric distribution, 211
Infinite population, 265
Mean of the binomial distribution, 226

Mean of a probability
   distribution, 225
Multinomial distribution, 222
Normal approximation to the
   binomial distribution, 256
Normal curve, 236
Normal distribution, 236
Poisson distribution, 217
Probability density, 237
Probability distribution, 200
Probable error of the mean, 284
Random numbers, 266
Random sample, 265
Random sampling, 264

Random variable, 199
Sampling a binomial population, 207
Sampling distribution, 270
Sampling fraction, 278
Sampling a normal population, 255
Standard deviation of a
   probability distribution, 227
Standard error of the mean, 275
Standard error of the median, 282
Standard normal distribution, 240
Standard units, 240
Variance of a probability
   distribution, 226

R.82 A supervisor at an automobile maintenance and repair garage believes that 4 out of 10 cars come to the garage specifically for an oil change, and for no other purpose. Assuming that this is correct,

(a) use the formula for the binomial distribution repeatedly to find the probabilities that 0, 1, 2, 3, 4, 5, or 6 cars that come to the garage come specifically for an oil change, and for no other purpose; and

(b) verify part (a) of this exercise by using Table I, Binomial Probabilities.

R.83 Using the results obtained in Exercise R.82, what is the probability that among a randomly selected group of 6 cars brought to the garage,

(a) none is brought in for an oil change;

(b) two or more cars are brought to the garage for an oil change;

(c) fewer than two cars are brought to the garage for an oil change?

R.84 With reference to Exercises R.82 and R.83, use a computer printout to work

(a) part (a) of Exercise R.83;

(b) part (b) of Exercise R.83;

(c) part (c) of Exercise R.83.

R.85 A security guard knows, from experience, that the time it takes to make the rounds in a certain warehouse building is a random variable, having an approximately normal distribution with $\mu = 18$ minutes and $\sigma = 3.2$ minutes. (This is a continuous distribution which requires no continuity correction.) What are the probabilities that the security guard will complete a round of the warehouse in

(a) less than 15 minutes.

(b) 15 to 20 minutes.

(c) more than 20 minutes?

R.86 A baker knows that the demand for birthday cakes is a random variable with a distribution which can be approximated closely by a normal distribution with mean $\mu = 43.3$ and standard deviation $\sigma = 4.6$. What is the probability that the demand for these birthday cakes will exceed 50 in any given day? (Note that this is a whole number which requires a continuity correction.)

R.87 Would you use the finite population correction factor in a practical problem where the sample size is $n = 20$ and the population size is $N = 1,000$? Why or why not?

R.88 How many different samples of

(a) $n = 3$ apple pies can be drawn from a population of $N = 6$;

(b) $n = 5$ elective courses can be drawn from a population of $N = 8$;

(c) $n = 4$ sweaters can be drawn from a population of $N = 10$?

R.89 Given that the probability of obtaining a head in a flip of a balanced coin is 0.5, use Table I to find the probabilities that in 15 flips of a balanced coin we will get

(a) up to 2 heads;

(b) 10 or more heads;

(c) exactly 6 heads;

(d) 6 or 7 heads.

R.90 Use the normal curve approximation to find the probability of getting 6 heads in 15 flips of a balanced coin, and compare the result with the value obtained in part (c) of Exercise R.89, where the probability was obtained from Table I.

R.91 Suppose that a continuous random variable takes on values on the interval from 0 to 2 and that the graph of its probability density is as shown in Figure R.1. What are the probabilities that the random variable will take on a value

(a) less than 0.5;

(b) between 1.2 and 1.6?

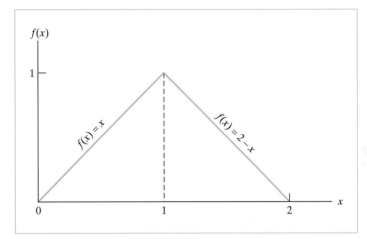

R.92 The manager of an ambulance service knows that the number of emergency calls made by the ambulances each day is a random variable having approximately a normal distribution with the mean $= 36.2$ and the standard deviation $= 5.1$. What are the probabilities that in any given day the ambulance service will make

(a) exactly 30 emergency ambulance service calls;

(b) at most 30 emergency ambulance service calls? (Note that this is a whole number which requires a continuity correction.)

R.93 The owner of a motel wants to randomly select 3 rooms for decorating. The rooms are consecutively numbered from 1 to 24. Which ones, by room number, will be selected if the random sample is chosen by means of random numbers, using columns 6 and 7 of the table in Figure 8.1, going down the table and starting with the 6th row?

R.94 Suppose that in a certain restaurant, the number of minutes which elapse between the time the waiter takes the order and the time the meal is served is a random variable with $\mu = 15.0$ minutes and $\sigma = 2.8$ minutes. Assuming that this random variable can be approximated closely with a normal curve, find the probabilities that the waiter will take

(a) 20 minutes or more to bring a meal to the table after it has been ordered;

(b) between 15 and 20 minutes before bringing the meal to the table after it has been ordered.

No continuity correction is required for this exercise.

R.95 If the number of buds produced by a certain type of lily is a random variable having the Poisson distribution with $\lambda = 3.5$, what are the probabilities that a lily of this type will produce

(a) no buds;

(b) two buds;

(c) four buds;

(d) six buds?

R.96 Find the area under the standard normal curve that lies

(a) between $z = -1.0$ and $z = 1.0$;

(b) to the right of $z = -2.0$;

(c) to the left of $z = 3.0$;

(d) between $z = -4.0$ and $z = 4.0$.

R.97 Find the mean and the variance of the binomial distribution with $n = 4$ and $p = 0.20$

(a) by looking up the probabilities in Table I and then using the formula for the mean and the variance of a probability distribution;

(b) by using the special formulas for the mean and the variance of the binomial distribution.

R.98 Check in each case whether the conditions given in Section 6.4 for the Poisson approximation to the binomial distribution are satisfied:

(a) $n = 150$ and $p = \frac{1}{10}$;

(b) $n = 450$ and $p = \frac{1}{50}$;

(c) $n = 600$ and $p = \frac{1}{100}$.

R.99 A certain type of rope used for nautical purposes lasts, on the average, 36 months, with a standard deviation of 6.0 months. Find the probability that a randomly selected coil of this rope will last, on the average, at least 30 months. The continuity correction is not required in this exercise.

R.100 Suppose that the number of computers sold each day by a computer store is a random variable with $\mu = 30$ computers and $\sigma = 5$ computers. Find the probability that on a certain day

(a) more than 36 computers will be sold;

(b) fewer than 24 computers will be sold;

(c) between 24 and 36 computers, inclusive, will be sold.

Use the continuity correction in this exercise.

R.101 A zoo has a large primate collection including eight baboons and six chimpanzees. If a veterinarian randomly picks four of them for examination, what are the probabilities that these will include

(a) two baboons and two chimpanzees;

(b) one baboon and three chimpanzees?

R.102 Use Table X, Binomial Coefficients, to determine

(a) how many random samples of size 3 can be drawn from a finite population of 12.

(b) Using the result of part (a), what is the probability of each sample?

R.103 In each case, determine whether the given values can be looked upon as the values of a probability distribution of a random variable that can take on the values 1, 2, 3, 4, and 5, and explain your answers:

(a) $f(1) = 0.18$, $f(2) = 0.20$, $f(3) = 0.22$, $f(4) = 0.20$, $f(5) = 0.18$;

(b) $f(1) = 0.05$, $f(2) = 0.05$, $f(3) = 0.05$, $f(4) = -0.05$, $f(5) = 0.90$;

(c) $f(1) = 0.07$, $f(2) = 0.23$, $f(3) = 0.50$, $f(4) = 0.19$, $f(5) = 0.01$.

R.104 A random variable has the normal distribution with $\sigma = 25$. If the probability is 0.6700 that the random variable takes a value less than 284, find the probability that it will take a value greater than 250.

R.105 If 73% of the cars traveling on a certain highway carry only one occupant, what is the probability that among 500 randomly selected cars at most 350 cars will carry only one occupant?

R.106 Determine whether the following can be probability distributions (defined in each case only for the given values of x) and explain your answers:

(a) $f(x) = \dfrac{1}{5}$ for $x = 0, 1, 2, 3, 4, 5$;

(b) $f(x) = \dfrac{x+1}{14}$ for $x = 1, 2, 3, 4$;

(c) $f(x) = \dfrac{x(x-2)}{2}$ for $x = 1, 2, 3$.

R.107 Check in each case whether the conditions for the normal approximation to the binomial distribution are satisfied:

(a) $n = 50$ and $p = \frac{1}{5}$;

(b) $n = 80$ and $p = \frac{1}{40}$;

(c) $n = 100$ and $p = 0.96$;

(d) $n = 200$ and $p = \frac{24}{25}$.

R.108 Find z if

(a) the normal-curve area between zero and z is 0.4783;

(b) the normal-curve area to the right of z is 0.9131;

(c) the normal-curve area to the right of z is 0.1112.

R.109 Find the value of the finite correction factor for $n = 50$ and $N = 100$.

R.110 What is the probability that a telemarketer will make at least 14 sales by telephone calls to 100 persons when the probability is 0.10 that a person receiving a call will make a purchase?

R.111 Check in each case whether the condition for the binomial approximation to the hypergeometric distribution is satisfied:

(a) $a = 150, b = 150$, and $n = 18$;

(b) $a = 100, b = 300$, and $n = 15$;

(c) $a = 400, b = 200$, and $n = 50$.

R.112 Random samples of size $n = 2$ are drawn from the finite population that consists of the numbers 1, 2, 3, 4, 5, and 6. Verify that the mean of this finite population is $\mu = 3\dfrac{1}{2}$ and that its variance is $\sigma^2 = \dfrac{35}{12}$.

R.113 Based on Chebyshev's theorem with $k = 3$, what can we assert about the possible size of our error if we use the mean of a random sample of $n = 81$ to estimate the mean of an infinite population with $\sigma = 25$? Use the formula $\sigma_{\bar{x}} = \dfrac{\sigma}{\sqrt{n}}$.

R.114 During the month of August, the daily number of persons visiting a certain tourist attraction is a random variable with $\mu = 1,200$ and $\sigma = 80$.

(a) What does Chebyshev's theorem with $k = 7$ tell us about the number of persons who will visit the tourist attraction on an August day?

(b) According to Chebyshev's theorem, with what probability can we assert that between 1,000 and 1,400 persons will visit the tourist attraction on an August day?

R.115 Suppose that we draw cards, one at a time and with replacement, from an ordinary deck of 52 playing cards. To determine the probability of getting 5 red kings among 156 cards thus drawn, can we use

(a) the Poisson approximation;

(b) the normal approximation?

R.116 When we sample from an infinite population, what happens to the standard error of the mean if the sample size is

(a) increased from 25 to 100;

(b) decreased from 400 to 40?

R.117 If a newspaper delivery service fails to deliver newspapers to 1.8% of its subscribers, use the Poisson approximation to the binomial distribution to determine the probability that among 250 subscribers, 5 will not receive their newspapers on a given day.

R.118 The average time required to perform job A is 78.5 minutes with a standard deviation of 16.2 minutes, and the average time required to perform job B is 103.2 minutes with a standard deviation of 11.3 minutes. Assuming normal distributions, what proportion of the time will job A take longer than the average job B, and what proportion of the time will job B take less time than the average job A?

PROBLEMS OF ESTIMATION

n statistical inference we make generalizations based on samples and, traditionally, such inferences have been divided into **problems of estimation** and **tests of hypotheses**. In problems of estimation, we assign a numerical value to a population parameter, expecting or hoping that our estimate (based on a sample) will be close; or we assign an interval of values (based on a sample), expecting or hoping that it will contain the population parameter in question. In problems of hypothesis testing, we accept or reject assumptions about the parameters of populations, or perhaps their shape or form, expecting or hoping that what we do will not be in error.

In this chapter we shall concentrate on problems of estimation which are easy to illustrate because they arise everywhere—in science, in business, and in everyday life. In science, a biologist may want to determine the average wingspan of an insect; in business, a retailer may want to determine the average income of all families living within a mile of a proposed new shopping center; and in everyday life, we may want to know how long it takes on the average to iron a shirt or peel a half-pound potato.

Sections 9.1, 9.2, and 9.3 are devoted to the estimation of the means of populations, Section 9.4 to the estimation of their standard deviations, and Section 9.5 to the estimation of proportions, percentages, or probabilities. Tests of hypotheses will be taken up in subsequent chapters.

9.1   THE ESTIMATION OF MEANS

To illustrate some of the problems we face in the estimation of means, let us refer to a study in which a doctor wants to determine the true average increase in the pulse rate of a person performing a certain strenuous task. The following is the data (increases in pulse rate in beats per minute) that the doctor obtained for 32 persons who performed the given task:

$$
\begin{array}{cccccccc}
27 & 25 & 19 & 28 & 35 & 23 & 24 & 22 \\
14 & 30 & 32 & 34 & 23 & 26 & 29 & 27 \\
27 & 24 & 31 & 22 & 23 & 38 & 25 & 16 \\
32 & 29 & 26 & 25 & 28 & 26 & 21 & 28
\end{array}
$$

The mean of this sample is $\bar{x} = 26.2$ beats per minute, and in the absence of any other information, this figure may well have to serve as an estimate of the population average μ, the true average increase in the pulse rate of persons performing the given task. Observe that the estimate is the statistic $\bar{x}$, which in this example has the numerical value 26.2; it estimates the unknown parameter μ.

An estimate like this is called a **point estimate**, since it consists of a single number, or a single point on the real number scale. Although this is the most common way in which estimates are expressed, it leaves room for many questions. For instance, it does not tell us on how much information the estimate is based, and it does not tell us anything about the possible size of the error. And, of course, we must expect an error. This should be clear from our discussion of the sampling distribution of the mean in Chapter 8, where we saw that the chance fluctuations of the mean (and hence its reliability as an estimate of μ) depend on two things—the size of the sample and the size of the population standard deviation σ. Thus, we might supplement 26.2, the foregoing estimate of the true average increase in the pulse rate of persons performing the given task, with the information that the sample size is $n = 32$ and that the sample standard deviation is $s = 5.15$, as can readily be verified. Although this does not tell us the actual value of σ, the sample standard deviation can serve as an estimate of this quantity.

Scientific reports often present sample means in this way, together with the values of n and s, but this does not supply the reader of the report with a coherent picture unless he or she has had some formal training in statistics. To make the supplementary information meaningful also to the lay person, let us refer to the theory of Sections 8.3 and 8.4 and Example 7.5 on page 247 (see also Figure 7.20 on page 251, according to which $z_{\alpha/2}$ is such that the area under the standard normal curve between $-z_{\alpha/2}$ and $z_{\alpha/2}$ is equal to $1 - \alpha$). Thus, making use of the fact that the sampling distribution of the mean of a large random sample from an infinite population is approximately a normal distribution with

$$
\mu_{\bar{x}} = \mu \text{ and } \sigma_{\bar{x}} = \frac{\sigma}{\sqrt{n}}
$$

we find from Figure 9.1 that the probability is $1 - \alpha$ that a sample mean will differ from the population mean μ by at most $z_{\alpha/2} \cdot \dfrac{\sigma}{\sqrt{n}}$. In other words,

Figure 9.1
Sampling distribution of the mean.

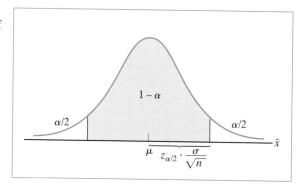

MAXIMUM ERROR
OF ESTIMATE
($n \geq 30$)

> *When we use $\bar{x}$ as an estimate of μ, the probability is $1 - \alpha$ that this estimate will be "off" either way by at most*
>
> $$E = z_{\alpha/2} \cdot \frac{\sigma}{\sqrt{n}}$$

This result applies when n is large ($n \geq 30$) and the population is infinite, or large enough so that the finite population correction factor (see page 276) need not be used. The two values that are most commonly used for $1 - \alpha$ are 0.95 and 0.99. The reader was asked to verify in Exercise 7.17 on page 251 that $z_{0.025} = 1.96$, corresponding to $1 - \alpha = 0.95$, and $z_{0.005} = 2.575$, corresponding to $1 - \alpha = 0.99$.

EXAMPLE 9.1   A supervisor of a garment manufacturing firm intends to use the mean time required to sew a sample of 150 women's coats to estimate the average time required to sew all such women's coats. If, based on experience, she can assume that $\sigma = 6.2$ minutes, what can she assume with probability 0.99 about the maximum error of her estimate?

Solution   Substituting 150 coats, $\sigma = 6.2$ minutes, and $z = 0.005 = 2.575$ into the formula for the maximum error, we get

$$E = 2.575 \cdot \frac{6.2}{\sqrt{150}} = 1.30$$

Thus, the supervisor can assert, with probability 0.99, that her error will be, at most, 1.30 minutes from the value of the mean. ■

The probability assertion of Example 1 applies to the *method* that the supervisor used to determine the maximum error (getting the sample data and using the formula for E) and not directly to the parameter she is trying to estimate.

To make this distinction, it has become the custom to use the word *confidence* here instead of *probability*.

> *In general, we make probability statements about future values of random variables (say, the potential error of an estimate) and confidence statements once the data has been obtained.*

Accordingly, we would say in Example 1 that the supervisor can be 99% confident that the error of her estimate on the mean is 1.30 minutes.

In some situations, we discuss ahead of time statements that will be made after the data is collected, as in Exercises 9.12 through 9.17. In such cases it is difficult to distinguish between confidence and probability.

EXAMPLE 9.2 A quality-control specialist wants to use the mean of a random sample of size $n = 35$ to determine the average fat content of a large shipment of quarter-pound hamburgers. The value of σ is not known, but the specialist maintains that the value of σ could not possibly be larger than 0.25 ounce. What can he assert with probability 0.95 about the maximum size of his error?

Solution Placing $n = 35$, $\sigma = 0.25$, and $z_{0.025} = 1.96$ into the formula for E, we find that

$$E = 1.96 \cdot \frac{0.25}{\sqrt{35}} = 0.083$$

The quality-control specialist will now assert, with probability 0.95, that the error in using $\bar{x}$ to estimate μ will be at most 0.083 ounce. ■

In the preceding example the value of σ was not known, but it was possible to give a *worst-case* value. Accordingly, the maximum error of 0.083 ounce is *conservative*, and the quality-control expert can assert with a probability of at least 0.95 that the error is at most 0.083 ounce.

To be able to judge the size of the error that we might make when we use $\bar{x}$ as an estimate of μ, we must know the value of the population standard deviation σ. Since this is not the case in most practical situations, we have a serious complication. In the previous example it was possible to give a worst-case value for σ. In other situations, it may not be possible to do this, and we have no choice but to replace σ with an estimate, usually the sample standard deviation s. In general, this is considered to be reasonable provided the sample is sufficiently large, and by sufficiently large we mean again $n \geq 30$.

EXAMPLE 9.3 With reference to the pulse rates on page 294, where we had $n = 32$, $\bar{x} = 26.2$, and $s = 5.15$, what can we assert with 95% confidence about the maximum error if we use 26.2 as an estimate of the true average increase in the pulse rate of a person performing the given task?

Solution Substituting $n = 32$, $s = 5.15$ for σ, and $z_{0.025} = 1.96$ into the formula for E, we find that we can assert with 95% confidence that the error is at most

$$E = 1.96 \cdot \frac{5.15}{\sqrt{32}} \approx 1.78 \text{ beats per minute}$$ ■

The formula for E can also be used to determine the sample size that is needed to attain a desired degree of precision. Suppose that we want to use the mean of a large random sample to estimate the mean of a population, and we want to be able to assert with probability $1 - \alpha$ that the error of this estimate will be at most some prescribed quantity E. As before, we write $E = z_{\alpha/2} \cdot \dfrac{\sigma}{\sqrt{n}}$, and upon solving this equation for n we get

SAMPLE SIZE FOR
ESTIMATING
$\mu(n \geq 30)$

$$n = \left(\frac{z_{\alpha/2} \cdot \sigma}{E}\right)^2$$

EXAMPLE 9.4 The superintendent of an irrigation service which uses river water to irrigate local farms wants to use the mean of a random sample to estimate the number of acre feet of water pumped each day and wants the estimate to be in error by, at most, 0.50 acre foot per day (an acre foot covers an acre of land to a depth of 1 foot, which is 43,560 cubic feet). The superintendent knows, from previous studies of a similar kind, that it is reasonable to let $\sigma = 0.75$ acre foot. How large a sample does the superintendent have to take?

Solution Substituting $z_{0.025} = 1.96$, $\sigma = 0.75$, and $E = 0.50$ into the formula for n, we get

$$n = \left(\frac{1.96 \cdot 0.75}{0.50}\right)^2 = 8.64$$

which should be rounded up to 9.

Exercises 9.1, 9.2, and 9.12 are practice exercises; their complete solutions are given on page 313.

EXERCISES

✓ **9.1** A school district official intends to use the mean of a random sample of 100 fifth graders to estimate the mean score that all the fifth graders in the district would get if they took a certain arithmetic achievement test. If, based on experience, the official knows that $\sigma = 9.2$ for such data, what can he assert with probability 0.95 about the maximum error?

✓ **9.2** A random sample of 50 cans of peach halves has a mean weight of 16.1 ounces and a standard deviation of 0.4 ounce. If $\bar{x} = 16.1$ ounces is used as an estimate of the average weight of all the cans of peach halves in the large lot from which the sample came, with what confidence can we assert that the error of this estimate is at most 0.1 ounce?

9.3 A specialist in precision machinery wants to estimate the mean expansion of certain pistons (in inches) on the basis of a sample of 42 of these pistons. The expansion is caused by the heat generated after the engines have been started, and it is known that $\sigma = 0.020$ inch. If the specialist considers the data to constitute a random sample, what can he assert with probability 0.99 about the maximum error if he uses the mean of his sample as an estimate of the actual mean expansion of such pistons.

9.4 A buyer for a large wholesaler of vegetables plans to estimate the mean weight of the Hubbard squash in a freight car load of these squash. To obtain this estimate the buyer will determine the weight of a gross (144) of the squash from this shipment. The buyer knows, from experience and from inspection, that the standard deviation is $\sigma = 3.00$ pounds. What can the buyer assert, with probability 0.95, about the maximum error of the estimate?

9.5 A mailing supervisor wants to use the mean of a random sample of the time it takes to prepare and address 100 catalogs for mailing. If, based on experience, the supervisor can assume that $\sigma = 0.50$ minutes for such data, what can he assert with probability 0.99 about the maximum size of the error?

9.6   Using the data of Exercise 9.5, what can the supervisor assume with probability 0.95 about the maximum size of the error?

9.7   To estimate the average daily production of chairs by a small manufacturer, the shop manager examined a random sample of 56 daily production reports, obtaining a mean of 32.5 chairs and a standard deviation of 2.9 chairs. What can the manager assert with probability of 99% about the maximum error?

9.8   With the information provided in the preceding exercise, what can the manager assume with probability of 95% about the maximum error?

9.9   A bakery wants to determine the average amount of honey used in a certain type of honey cake. An examination of 64 honey cakes revealed that they contained an average of 8.0 ounces of honey, with $\sigma = 0.5$ ounce. If 8 ounces of honey is used as an estimate of the actual mean number of ounces per cake, what can we assert about the maximum error

(a)   with 95% confidence;

(b)   with 99% confidence?

9.10   A study conducted by an airline showed that 120 of its passengers disembarking at Kennedy airport, a random sample, had to wait on the average 9.45 minutes with a standard deviation of 1.84 minutes to get their luggage. What can it say with 99% confidence about the maximum error if the airline uses $\bar{x} = 9.45$ minutes as an estimate of the true average time it takes its passengers to get their luggage when disembarking at Kennedy airport?

9.11   A sample of 200 randomly selected customers shopped in a certain hardware store for an average of 19 minutes, with a standard deviation of $s = 5.5$ minutes. What can the store manager say about the average shopping time of customers with 99% confidence about the maximum error if $x = 19$ minutes is used as an estimate of the shopping time of customers who patronize the store?

✓ **9.12**   In a study of television viewing habits, it is desired to estimate the average number of hours that teenagers spend watching per week. If it is reasonable to assume that $\sigma' = 3.2$ hours for data of this kind, how large a sample is needed so that one will be able to assert with 99% confidence that the sample mean is off by at most half an hour?

9.13   The marketing manager of a toy company needs to know, with reasonable accuracy, how long it takes a child of a certain age to assemble a model airplane. If preliminary studies have shown that the standard deviation is 15 minutes, how large a sample will the manager need to be able to assert with a probability of 0.95 that the sample mean will be off by at most 3 minutes?

9.14   A park ranger wants to know the average size of trout taken from a certain lake. How large a sample of trout must be taken to be able to assert with a probability of 0.98 that a sample mean will not be off by more than 0.5 inch? Assume that it is known from previous studies that $\sigma = 2.5$ inches.

9.15   Suppose that we want to estimate the average speed of cars traveling on a highway, and we want to be able to assert with probability 0.99 that the error of our estimate, the mean speed of a random sample of these cars, will be at most 3 miles per hour. How large a sample will we need if it can be assumed that $\sigma = 7.1$ miles per hour?

9.16   Before purchasing a large shipment of ground meat, a sausage manufacturer wants to be 95% confident that he is in error by no more than 2 grams in estimating the fat content (per 100 grams). If the standard deviation of the fat content (per 100 grams) is assumed to be 8 grams, on how large a sample should he base his estimate?

9.17   The ratio $E^* = \dfrac{E}{\sigma}$ is the error in standard deviation units. In some situations, lack of knowledge about σ makes it difficult to establish realistic values for E, and specifying E^* is a useful alternative. For example, requiring that E^* be 0.30 or less asks for an error of 0.30σ or less, whatever the value of σ. If a school administrator wants to estimate the average reading achievement score for her whole school by using a sample of children, how large a sample should she take if she wishes to assert with 95% confidence that E^* is at most 0.30?

9.2 CONFIDENCE INTERVALS FOR MEANS

Let us now introduce a different way of presenting the information provided by a sample mean together with an assessment of the error we might make when we use it to estimate the mean of the population from which the sample came. In what follows, we shall make use of the fact that for large random samples from infinite populations the sampling distribution of the mean is approximately normal, with the mean μ and the standard deviation $\dfrac{\sigma}{\sqrt{n}}$; that is,

$$z = \frac{\bar{x} - \mu}{\sigma/\sqrt{n}}$$

is a random variable having approximately the standard normal distribution. Since the probability is $1 - \alpha$ that a random variable having the standard normal distribution will take on a value between $-z_{\alpha/2}$ and $z_{\alpha/2}$ (see Figure 7.20), that is,

$$-z_{\alpha/2} < z < z_{\alpha/2}$$

we can substitute into this inequality the foregoing expression for z and get

$$-z_{\alpha/2} < \frac{\bar{x} - \mu}{\alpha/\sqrt{n}} < z_{\alpha/2}$$

If we multiply each term by $\dfrac{\sigma}{\sqrt{n}}$, subtract $\bar{x}$ from each term, and multiply each term by -1, we get

$$\bar{x} + z_{\alpha/2} \cdot \frac{\sigma}{\sqrt{n}} > \mu > \bar{x} - z_{\alpha/2} \cdot \frac{\sigma}{\sqrt{n}}$$

Note that, as a result of multiplying by -1, we had to reverse the inequality signs, which is always the case when we multiply the expressions on both sides of an inequality by a negative number. For instance, where 5 is greater than (to the right of) 2, -5 is less than (to the left of) -2.

The result we obtained can also be written as

LARGE-SAMPLE CONFIDENCE INTERVAL FOR μ

$$\bar{x} - z_{\alpha/2} \cdot \frac{\sigma}{\sqrt{n}} < \mu < \bar{x} + z_{\alpha/2} \cdot \frac{\sigma}{\sqrt{n}}$$

and we can assert with probability $1 - \alpha$ that it will be satisfied for any given sample.[†] In other words, we can assert with $(1 - \alpha)$ 100% confidence that the interval from

$$\bar{x} - z_{\alpha/2} \cdot \frac{\sigma}{\sqrt{n}} \quad \text{to} \quad \bar{x} + z_{\alpha/2} \cdot \frac{\sigma}{\sqrt{n}}$$

[†] Since the probability is also $1 - \alpha$ that $-z_{\alpha/2} \leq z \leq z_{\alpha/2}$, some authors substitute $\leq$ for $<$ in this confidence-interval formula. Practically speaking, this does not make any difference.

determined on the basis of a large random sample, contains the population mean we are trying to estimate. When σ is unknown and n is at least 30, we replace σ by the sample standard deviation s. If n is at least 30, then the use of s is recommended, even if the value of σ is claimed to be known. An interval like this is called a **confidence interval**, its end points are called **confidence limits**, and the probability $1 - \alpha$ is called the **degree of confidence**. As in Section 9.1, the values most commonly used for $1 - \alpha$ are 0.95 and 0.99, the corresponding values of $z_{\alpha/2}$ are 1.96 and 2.575, and the resulting confidence intervals are referred to as 95% and 99% confidence intervals for μ.

EXAMPLE 9.5   With reference to the previously given pulse rate data, where we had $n = 32$, $\bar{x} = 26.2$, and $s = 5.15$, construct a 95% confidence interval for the true average increase in the pulse rate of persons performing the given task.

Solution   Substituting $n = 32$, $\bar{x} = 26.2$, $s = 5.15$ for σ, and $z_{0.025} = 1.96$ into the confidence interval formula, we get

$$26.2 - 1.96 \cdot \frac{5.15}{\sqrt{32}} < \mu < 26.2 + 1.96 \cdot \frac{5.15}{\sqrt{32}}$$

or

$$24.4 < \mu < 28.0$$

for the true average increase in the pulse rate of persons performing the given task. Of course, the interval from 24.4 to 28.0 contains μ or it does not, and we really don't know which, but the 95% confidence implies that the interval was obtained by a method that works 95% of the time. To put it differently, the interval may contain μ or it may not, but if we had to bet, 95 to 5 (or 19 to 1) would be fair odds that it does.

Had we wanted to calculate a 99% confidence interval in this example, we would have substituted 2.575 instead of 1.96 for $z_{\alpha/2}$, and we would have obtained $23.9 < \mu < 28.5$. Comparison of the 95% and 99% confidence intervals shows that

> When we increase the degree of certainty, namely the degree of confidence, the confidence interval becomes wider and thus tells us less about the quantity we are trying to estimate.

In other words, "the surer we want to be, the less we have to be sure of."

When we use a confidence interval to estimate the mean of a population, we call this kind of estimate an **interval estimate**. In contrast to point estimates, interval estimates require no further elaboration about their reliability; this is taken care of indirectly by their width and the degree of confidence.

9.3 CONFIDENCE INTERVALS FOR MEANS (SMALL SAMPLES)

In the preceding section we assumed that the sample is large enough, $n \geq 30$, to treat the sampling distribution of the mean as if it were a normal distribution and, when necessary, to replace σ with s. To develop corresponding theory that applies also to small samples, it will be necessary to assume that the population we are sampling has roughly the shape of a normal distribution. We can then base our method on the *t* **statistic**

$$t = \frac{\bar{x} - \mu}{s/\sqrt{n}}$$

whose sampling distribution is a continuous distribution called the *t* **distribution**. As is shown in Figure 9.2, this distribution is symmetrical and bell shaped with zero mean; in fact, its shape is very similar to that of the standard normal distribution. The exact shape of the *t* distribution depends on a parameter called the **number of degrees of freedom**, or simply the **degrees of freedom**, often abbreviated as df. As the distribution will be used here, the number of degrees of freedom is given by $n - 1$, the sample size less 1. For the *t* distribution, we define $t_{\alpha/2}$ in the same way in which we defined $z_{\alpha/2}$, so that the area under the curve to the right of $t_{\alpha/2}$ is equal to $\frac{\alpha}{2}$. However, $t_{\alpha/2}$ depends on $n - 1$, the number of degrees of freedom, and its value will have to be found in each case from Table III at the end of the book. Now we proceed as on page 299. Making use of the fact that the *t* distribution is symmetrical about $t = 0$, we find that the probability is $1 - \alpha$ that a random variable having the *t* distribution will take on a value between $-t_{\alpha/2}$ and $t_{\alpha/2}$ (see Figure 9.3); that is,

$$-t_{\alpha/2} < t < t_{\alpha/2}$$

Then, substituting into this inequality the foregoing expression for *t*, we get

$$-t_{\alpha/2} < \frac{\bar{x} - \mu}{s/\sqrt{n}} < t_{\alpha/2}$$

Figure 9.2
Standard normal
distribution and *t*
distribution.

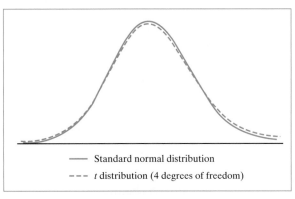

Standard normal distribution

--- *t* distribution (4 degrees of freedom)

Figure 9.3
t distribution.

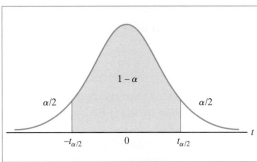

and using the same steps as on page 299, we arrive at the following **small-sample confidence interval for** μ:

S MALL-SAMPLE
CONFIDENCE
INTERVAL FOR μ

$$\bar{x} - t_{\alpha/2} \cdot \frac{s}{\sqrt{n}} < \mu < \bar{x} + t_{\alpha/2} \cdot \frac{s}{\sqrt{n}}$$

The degree of confidence is $1 - \alpha$, and the only differences between this confidence interval formula and the large-sample formula (with s substituted for σ) is that $t_{\alpha/2}$ takes the place of $z_{\alpha/2}$.

EXAMPLE 9.6 To test the durability of a new paint for white center lines, a highway department painted test strips across heavily traveled roads in eight different locations, and electronic counters showed that they deteriorated after having been crossed by 14.26, 16.78, 13.65, 10.83, 12.64, 13.37, 16.20, and 14.94 million cars. Construct a 95% confidence interval for the average amount of traffic (car crossings) this paint can withstand before it deteriorates.

Solution The mean and the standard deviation of these values are $\bar{x} = 14.08$ and $s = 1.92$. For $8 - 1 = 7$ degrees of freedom, we find that $t_{0.025}$ equals 2.365, and substitution into the formula yields

$$14.08 - 2.365 \cdot \frac{1.92}{\sqrt{8}} < \mu < 14.08 + 2.365 \cdot \frac{1.92}{\sqrt{8}}$$

or

$$12.47 < \mu < 15.69$$

This is the desired 95% confidence interval estimate of the average amount of traffic (millions of car crossings) that the paint can withstand before it deteriorates. ∎

A computer printout of the preceding example is shown in Figure 9.4. The differences between the confidence limits given in the printout and those given above are due to rounding.

If we had wanted the degree of confidence to be 0.97, we could not have worked this example by referring to Table III. Of course, we could have used a computer with the same software as for the printout shown in Figure 9.4, but let us illustrate here the construction of a confidence interval for the mean with the use of a graphing calculator. As before, all information concerning the graphing calculator is optional and can be omitted by readers without the loss of continuity.

In Figure 9.5 the 97% confidence interval is given as (12.238, 15.929), which rounded to two decimals becomes

$$12.24 < \mu < 15.93$$

Note that this interval is wider than the corresponding 95% confidence interval.

Instructions for the use of the TI-83 graphing calculator are found in the Using Technology box on page 304.

The method we used in Section 9.1 to determine the maximum error that we risk with probability $1 - \alpha$, when we use a sample mean to estimate the mean of a

Figure 9.4
MINITAB printout for
small-sample confidence
interval for μ.

```
One-Sample T: C1

Variable     N    Mean    StDev  SE Mean      95.0% CI
C1           8  14.084    1.923    0.680  ( 12.475, 15.692)
```

Worksheet 1 ***

	C1	C2	C3	C4	C5	C6	C7	C8	C9
1	14.26								
2	16.78								
3	13.65								
4	10.83								
5	12.64								
6	13.37								
7	16.20								
8	14.94								

Figure 9.5
Small-sample 97% con-
fidence interval for the
mean reproduced from
the display screen of a
TI-83 graphing
calculator.

```
TInterval
(12.238,15.929)
x̄=14.08375
Sx=1.922847126
n=8
```

population, can readily be adapted to small samples, provided that the population
we are sampling has roughly the shape of a normal distribution. All we have to
do is substitute s for σ and $t_{\alpha/2}$ for $z_{\alpha/2}$ in the formula for E on page 295.

EXAMPLE 9.7   The Appalachian Trail is a well-traveled footpath and camping area which extends
from Mount Katahdin, Maine, to Mount Oglethorpe, Georgia, a distance of 2,050
miles. Six randomly selected college students hiked a certain small segment of
the trail in 7.4, 8.3, 9.0, 7.9, 8.1, and 7.9 hours, with standard deviation of 0.53
hours. What can we assert with 95% confidence about the maximum error if we
use $\bar{x} = 8.10$ hours as an estimate of the true average time for randomly selected
college students to hike this segment of the Trail? The values of $\bar{x}$ and s can
readily be verified from the data.

Solution   Substituting $s = 0.53$, $n = 6$, and $t_{0.025} = 2.571$ (the entry in Table III for $6 - 1 = 5$
degrees of freedom) into the modified formula for E, we get

$$E = t_{\alpha/2} \cdot \frac{s}{\sqrt{n}} = 2.5791 \frac{0.53}{\sqrt{6}} = 0.56$$

Thus, if we use the mean $\bar{x} = 8.1$ hours as an estimate of the true time it takes
for randomly selected college students to hike this segment of the Trail, and it is

reasonable to assume that the data constitutes a random sample from a normal population, we can assert with 95% confidence that the error is, at most, 0.56 hour (or about 33.6 minutes). ■

USING TECHNOLOGY: The Graphing Calculator

How to obtain the small-sample 97% confidence interval for Figure 9.5

(TIP: Before beginning, see Appendix A for clearing lists and entering data.)

STEP 1. Clear list L1.
STEP 2. Input the example data into list L1.
STEP 3. Select the type of statistical test you wish to use.
         Press STAT.
         Highlight TESTS.
         Select   8: T Interval, press ENTER.
         Highlight DATA.
         Set   List: L1
                  Freq: 1
                  C-Level: .97
STEP 4. Solve.
         Highlight Calculate, press ENTER.

EXERCISES

Exercises 9.18 and 9.24 are practice exercises; their complete solutions are given on page 313.

✓ **9.18** With reference to Exercise 9.1, where we had $n = 100$ and $\sigma = 9.2$, suppose that the fifth graders in the sample averaged 62.7 on the test. Construct a 99% confidence interval for the average score that all the fifth graders in the district would get if they took the arithmetic achievement test.

9.19 A study of a chemical factory's discharge of pollutants over 49 randomly selected days showed that the mean daily discharge was $\bar{x} = 25.1$ tons with standard deviation $s = 8.5$ tons. Construct a confidence interval for the true mean of the factory's discharge of contaminants
(a) using a 0.95 degree of confidence;
(b) using a 0.99 degree of confidence.

9.20 A dairy farmer's records showed that, on average, $\bar{x} = 6,300$ pounds of milk per cow were produced annually with standard deviation $s = 500$ pounds by a random sample of $n = 50$ cows of the same breed. Construct a 98% confidence interval for the true average number of pounds of milk produced annually per cow in the entire herd.

9.21 Using the data of Exercise 9.20, but changing the sample size from $n = 50$ (large sample) to $n = 10$ (small sample), construct a 98% confidence interval for the true average number of pounds of milk produced annually per cow in the entire herd.

9.22   A beauty parlor finds that in $n = 60$ randomly selected days it had, on average, $\bar{x} = 72.3$ patrons per day with a standard deviation of $s = 8.5$ patrons.

(a) Construct a 90% confidence interval for the actual daily average number of beauty parlor patrons.

(b) Using the data of this exercise, but changing $\bar{x} = 72.3$ to $\bar{x} = 82.3$, recalculate the 90% confidence interval for the actual daily average number of beauty parlor patrons.

9.23   A hospital wants to estimate the mean number of blood tests given to patients each day. Hospital records show that in a randomly selected sample of $n = 90$ days, the average number of blood tests was $\bar{x} = 85$, with a standard deviation of $s = 9.2$. Construct a 95% confidence interval for the mean number of blood tests given to patients each day.

✓ **9.24**   In an air pollution study, an experimental station obtained a mean of 2.36 micrograms of suspended benzene-soluble organic matter per cubic meter with a standard deviation of 0.48 from a random sample of size $n = 10$.

(a) Construct a 99% confidence interval for the mean of the population sampled.

(b) What can be asserted with 95% confidence about the maximum error if $\bar{x} = 2.36$ micrograms is used as an estimate of the mean of the population sampled?

9.25   Determine the value of t from Table III, where

(a) $n = 11$ and the degree of confidence is 0.90;

(b) $n = 12$ and the degree of confidence is 0.99;

(c) $n = 13$ and the degree of confidence is 0.95;

(d) $n = 14$ and the degree of confidence is 0.98.

9.26   In a study of automobile insurance costs, a random sample of $n = 70$ body repair costs on a particular kind of damage had a mean of $\bar{x} = \$529.35$ and a standard deviation of $s = \$66.23$. If $\$529.35$ is used as an estimate of the true average cost of such repairs, what can we say about the possible error with a probability of 0.95?

9.27   Ten randomly selected turkeys in a large flock of domestic turkeys at a poultry farm were weighed by a farmer and were, on the average, 16.1 pounds each with a standard deviation of $s = 1.1$ pounds. What can we assert with 99% confidence about the maximum error if we wish to use $\bar{x} = 16.1$ as an estimate of the true average weight of the turkeys in this flock?

9.28   With reference to Exercise 9.27, construct a 99% confidence interval for the average weight of the turkeys in this flock of domestic turkeys.

9.29   In six attempts it took a locksmith 9, 14, 7, 8, 11, and 5 seconds to open a certain kind of lock. Verify that $\bar{x} = 9$ and $s = 3.16$ for these data, and construct a 95% confidence interval for the average time it takes the locksmith to open this kind of lock.

9.30   In establishing the authenticity of an ancient coin, its weight is often of critical importance. If four experts independently weighed a Phoenician tetra-drachm and obtained 14.28, 14.34, 14.26, and 14.32 grams, verify that $\bar{x} = 14.30$ and $s = 0.0365$ for the given data. What can one assert with 99% confidence about the maximum error if $\bar{x} = 14.30$ grams is used as an estimate of the actual weight of the coin?

9.31   If a random sample of six students took 18, 19, 23, 19, 21, and 20 minutes to complete the registration forms for courses offered in the next semester, construct a 95%

confidence interval for the mean time that it takes to complete such registration forms.

9.32　There is often some uncertainty about the level of confidence that should be used in constructing intervals. It is claimed that, in general, 99% confidence intervals are about 35% longer than the corresponding 95% intervals. Investigate this claim with the data given in Exercise 9.31, where $n = 6$.

9.33　The excess length of the 99% small-sample confidence interval over the corresponding 95% interval, expressed as a percentage of the length of the 95% interval (see Exercise 9.32), is, in fact, determined only by the sample size n. What is the excess length when
　(a)　$n = 10$;
　(b)　$n = 20$;
　(c)　$n = 30$?

9.34　Use a computer package or a graphing calculator to rework Exercise 9.29.

9.35　With reference to Exercise 9.30, use a computer package or a graphing calculator to construct a 99% confidence interval for the true weight of the coin.

9.36　Use a computer package or a graphing calculator to rework Exercise 9.31.

*9.4 CONFIDENCE INTERVALS FOR STANDARD DEVIATIONS

On many occasions we must estimate population standard deviations, as with the substitution of s for σ to calculate a large-sample confidence interval for the mean. Let us now show how confidence intervals for σ may be determined on the basis of large samples. A technique for small samples exists, but it will not be discussed in this book.

For large samples, $n \geq 30$, we base our method on the theory that the sampling distribution of s is approximately normal with the mean σ and the standard deviation $\sigma_s = \dfrac{\sigma}{\sqrt{2n}}$, which is called the **standard error of s**. Thus, the sampling distribution of

$$z = \frac{s - \sigma}{\sigma/\sqrt{2n}}$$

is approximately the standard normal distribution, and if we substitute this expression for z into the inequality $-z_{\alpha/2} < z < z_{\alpha/2}$ and use steps similar to those on page 299, we arrive at the following large-sample confidence interval for σ:

LARGE-SAMPLE CONFIDENCE INTERVAL FOR σ

$$\frac{s}{1 + \dfrac{z_{\alpha/2}}{\sqrt{2n}}} < \sigma < \frac{s}{1 - \dfrac{z_{\alpha/2}}{\sqrt{2n}}}$$

As before, the degree of confidence is $1 - \alpha$.

EXAMPLE 9.8　With reference to the illustration in Section 9.1 on page 294, where we obtained $\bar{x} = 26.2$ and $s = 5.15$ for the increase in the pulse rate of $n = 32$ persons performing a certain strenuous task, construct a 95% confidence interval for the standard deviation of the population sampled.

Solution   Substituting $n = 32$, $s = 5.15$, and $z_{0.025} = 1.96$ into the large-sample confidence interval formula for σ, we get

$$\frac{5.15}{1 + \dfrac{1.96}{\sqrt{64}}} < \sigma < \frac{5.15}{1 - \dfrac{1.96}{\sqrt{64}}}$$

or

$$4.14 < \sigma < 6.82$$

This is an estimate of the true variability of the increase in the pulse rate of persons performing the given task.

Exercise 9.37 is a practice exercise; its complete solution is given on page 314.

✓ ***9.37**   In Exercise 9.2 we stated, "A random sample of 50 cans of peach halves has a mean weight of 16.1 ounces and a standard deviation of $s = 0.4$ ounce." Using $n = 50$ and $s = 0.4$, construct a 99% confidence interval for the standard deviation of the weights of all the cans of peach halves from which the sample came.

*9.38   Find the large sample confidence interval for the population standard deviation where

(a) $s = 10.89$, $n = 60$, and the confidence interval for the population standard deviation is 95%;

(b) $s = 1.29$, $n = 100$, and the confidence interval for the population standard deviation is 98%;

(c) $s = 2.95$, $n = 200$, and the confidence interval for the population standard deviation is 99%.

*9.39   Referring to the pulse rate illustration on page 293, where $s = 5.15$ and $n = 32$, construct a 95% confidence interval for the standard deviation of the population sampled.

*9.40   A random sample of gasoline sales to 81 passenger cars shows average sales of 9.5 gallons of gasoline with standard deviation of 2.8 gallons. Construct a 99% confidence interval for the standard deviation of the population of gasoline sales from which this sample came.

*9.41   With reference to Exercise 9.10, where we had $n = 120$ and $s = 1.84$ minutes, construct a 99% confidence interval for the true standard deviation of the amount of time that passengers disembarking at Kennedy airport have to wait for their luggage.

9.5 THE ESTIMATION OF PROPORTIONS

The information that is usually available for the estimation of a true proportion is a **sample proportion** $\dfrac{x}{n}$, where x is the number of times that an event has occurred in n trials. For instance, if 63 of 150 television viewers (interviewed in a sample survey) liked a certain new situation comedy, then

$$\frac{x}{n} = \frac{63}{150} = 0.42$$

and we can use this figure as an estimate of the true proportion of television viewers who like the new show. Since a percentage is just a proportion multiplied by 100 and a probability may be interpreted as a proportion in the long run, we could also say that we estimate that 42% of all television viewers like the new situation comedy, or that the probability is 0.42 that any one television viewer will like the new show. We have made this point to impress on the reader that the problem of estimating a true percentage or a true probability is essentially the same as that of estimating a true proportion.

Throughout this section it will be assumed that the situations satisfy (at least approximately) the conditions underlying the binomial distribution; that is, our information will consist of the number of successes observed in a given number of independent trials, and it will be assumed that for each trial the probability of a success—the parameter we want to estimate—has the constant value p. Thus, the sampling distribution of the counts on which our methods will be based is the binomial distribution with the mean $\mu = np$ and the standard deviation $\sigma = \sqrt{np(1 - p)}$.

We also know that this distribution can be approximated with a normal curve when np and $n(1 - p)$ are both greater than 5 (see page 257), and it follows that under these conditions

$$z = \frac{x - np}{\sqrt{np(1 - p)}}$$

has approximately the standard normal distribution. If we substitute this expression for z into the inequality $-z_{\alpha/2} < z < z_{\alpha/2}$ (as on page 299) and use some relatively simple algebra, we arrive at the inequality

$$\frac{x}{n} - z_{\alpha/2}\sqrt{\frac{p(1 - p)}{n}} < p < \frac{x}{n} + z_{\alpha/2}\sqrt{\frac{p(1 - p)}{n}}$$

This may look like a confidence interval formula for p and, indeed, the inequality will be satisfied with probability $1 - \alpha$, but it cannot be used in practice because the unknown parameter p appears also in $\sqrt{\frac{p(1 - p)}{n}}$ to the left of the first inequality sign and to the right of the second. This quantity $\sqrt{\frac{p(1 - p)}{n}}$ is called the **standard error of a proportion**, as it is, in fact, the standard deviation of the sampling distribution of a sample proportion (see Exercise 9.58). To get around this difficulty, we substitute for p in $\sqrt{\frac{p(1 - p)}{n}}$ the sample proportion $\frac{x}{n}$, and we have thus arrived at the following large-sample confidence interval for p:

LARGE-SAMPLE CONFIDENCE INTERVAL FOR p

$$\frac{x}{n} - z_{\alpha/2}\sqrt{\frac{\frac{x}{n}\left(1 - \frac{x}{n}\right)}{n}} < p < \frac{x}{n} + z_{\alpha/2}\sqrt{\frac{\frac{x}{n}\left(1 - \frac{x}{n}\right)}{n}}$$

The degree of confidence is $1 - \alpha$. As before, the confidence interval is referred to as a $(1 - \alpha)100\%$ confidence interval. There are other methods for constructing a confidence interval for p, but this is the most common.

EXAMPLE 9.9 An inspector examined a random sample of 400 bottles of shampoo and found that 152 of the bottles had loose bottle-caps that required tightening. Construct a 95% confidence interval for the true proportion of bottle-caps that needed tightening, in the population sampled.

Solution Substituting $n = 400$, $\dfrac{x}{n} = \dfrac{152}{400} = 0.38$, and $z_{0.025} = 1.96$ into the large-sample confidence interval formula, we get

$$0.38 - 1.96\sqrt{\frac{(0.38)(0.62)}{400}} < p < 0.38 + 1.96\sqrt{\frac{(0.38)(0.62)}{400}}$$

$$0.33 < p < 0.43$$

We note that the interval from 0.33 to 0.43 contains the true proportion p or it does not, and we really don't know which. But the 95% confidence implies that the interval was obtained by a method that works 95% of the time. Note also that for n equals 400 and p on the interval from 0.33 to 0.43, np and $n(1 - p)$ are much greater than 5, so that there can be no question about n being large enough to use the normal approximation to the binomial distribution (see page 257). ■

Although the calculations in this example were quite easy, it may be desirable to use a computer or a graphing calculator. Thus, we shall illustrate here the construction of a confidence interval for p with the use of a graphing calculator. As before, all information concerning the graphing calculator is optional and can be omitted by readers without the loss of continuity.

In Figure 9.6 the 95% confidence interval is given as (.33243, .42757), which rounded to two decimals becomes

$$0.33 < p < 0.43$$

Figure 9.6
Large-sample confidence interval for p reproduced from the display screen of a T1-83 graphing calculator.

```
1-PropZInt
(.33243,.42757)
p̂=.38
n=400
```

As should have been expected, this is identical with the result obtained before.

Directions for obtaining the large-sample confidence interval for p are given in the Using Technology box on page 310.

How to construct a confidence interval for p, Figure 9.6

STEP 1. Select the type of statistical test you wish to use.
Press STAT.
Highlight TESTS.
Scroll down to A: 1-PropZint, press ENTER.
Set   x:152
          n: 400
          C-level: .95

STEP 2. Solve. Highlight Calculate, press ENTER.

The large-sample theory that we have presented here can also be used to assess the size of the error we may be making when we use a sample proportion $\frac{x}{n}$ as a point estimate of a population proportion p. In this connection, we can assert with $(1 - \alpha)100\%$ confidence, approximately, that our error is at most

APPROXIMATE
MAXIMUM ERROR OF
ESTIMATE IN USING
$\frac{x}{n}$ TO ESTIMATE p

$$E = z_{\alpha/2}\sqrt{\frac{\frac{x}{n}\left(1 - \frac{x}{n}\right)}{n}}$$

Once we have observed the value of x, we can use this formula to assess the size of our error.

EXAMPLE 9.10 With reference to the illustration on page 307, in which 63 of 150 television viewers liked a new situation comedy, what can we say with 99% confidence about the maximum error if we use $\frac{x}{n} = \frac{63}{150} = 0.42$ as an estimate of the true proportion of television viewers who like the new show?

Solution Substituting $n = 150$, $\frac{x}{n} = 0.42$, and $z_{0.005} = 2.575$ into the formula for E, we get

$$E = 2.575\sqrt{\frac{(0.42)(0.58)}{150}} \approx 0.10$$

The preceding formula cannot be used to select a sample size n; the following formulas will do that. As in the estimation of means, we can use the expression for the maximum error to determine how large a sample is needed to attain a desired degree of precision. If we want to assert with probability $1 - \alpha$ that a sample proportion will differ from the true proportion by not more than E, we solve the equation

$$E = z_{\alpha/2}\sqrt{\frac{p(1 - p)}{n}}$$

for n, and we get

SAMPLE SIZE FOR ESTIMATING p (WITH SOME INFORMATION ABOUT p)

$$n = p(1 - p)\left[\frac{z_{\alpha/2}}{E}\right]^2$$

Since this formula involves p, it cannot be used unless we have some information about the possible values that p might assume. In that case, we substitute for p whichever of its values is closest to $\frac{1}{2}$. Without such information, we make use of the fact that $p(1 - p)$ cannot exceed $\frac{1}{4}$ (which it equals for $p = \frac{1}{2}$) and use the formula

SAMPLE SIZE FOR ESTIMATING p (WITHOUT INFORMATION ABOUT p)

$$n = \frac{1}{4}\left[\frac{z_{\alpha/2}}{E}\right]^2$$

This may make the sample unnecessarily large, but, on the other hand, we can assert with a probability of *at least* $1 - \alpha$ that the error will not exceed E.

EXAMPLE 9.11 Suppose that we want to estimate what proportion of the adult population of the United States has high blood pressure, and we want to be 99% sure that the error of our estimate will not exceed 0.02. How large a sample will we need if

(a) we have no idea what the true proportion might be;
(b) we know that the true proportion lies on the interval from 0.05 to 0.20?

Solution (a) Substituting $E = 0.02$ and $z_{0.005} = 2.575$ into the second formula, we get

$$n = \frac{1}{4}\left[\frac{2.575}{0.02}\right]^2 \approx 4{,}145$$

rounded up to the nearest integer.
(b) Substituting these same values together with $p = 0.20$ (the plausible value that is closest to $\frac{1}{2}$) into the first formula, we get

$$n = (0.20)(0.80)\left[\frac{2.575}{0.02}\right]^2 \approx 2{,}653$$

rounded up to the nearest integer. This shows how some knowledge about the values that p might take on can substantially reduce the required sample size. ∎

The methods discussed in this section are all large-sample techniques. For small samples, confidence intervals for proportions can be based on special tables, which may be found in more advanced texts.

EXERCISES

Exercises 9.42, 9.43, 9.52, and 9.53 are practice exercises; their complete solutions are given on page 314.

✓ **9.42**   In a sample survey, 140 of 500 persons interviewed in a large city said that they shop in the downtown area at least once a week. Construct a 99% confidence interval for the corresponding true proportion.

✓ **9.43**   With reference to Exercise 9.42, what can we say with 95% confidence about the maximum error if we use the sample proportion to estimate the true proportion of persons in the given city who shop in the downtown area at least once a week?

9.44   Among 100 randomly selected voters in a certain town, 30 were opposed to floating a bond issue to build a new school building. If we use the sample proportion $\frac{30}{100} = 0.30$ to estimate the corresponding true proportion, what can we assert with 99% confidence about the maximum error?

9.45   With reference to Exercise 9.44, construct a 95% confidence interval for the true proportion of voters in the town who are opposed to floating a bond issue to build a new school building.

9.46   In a random sample of 500 grade school students, 75 were enrolled in private schools. Construct a 95% confidence interval for the corresponding true proportion.

9.47   With reference to Exercise 9.46, what can we say with 99% confidence about the maximum error if we use the sample proportion $\frac{75}{500} = 0.15$ to estimate the corresponding true proportion?

9.48   In a survey of 1,829 voters prior to a mayoral election, 1,050 voters stated that they would vote for the incumbent mayor seeking reelection. Use $\frac{1,050}{1,829} \cdot 100 \approx 57.41\%$ and construct a 99% confidence interval for the corresponding true proportion of voters for the incumbent mayor.

9.49   In a study of 2,500 workers (who had earnings), 1,030 had pension plan coverage. If $\frac{1,030}{2,500} \cdot 100 \approx 41\%$ is used as an estimate of the true percentage of workers (who had earnings) who have pension plan benefits, what can one assert with 95% confidence about the maximum error?

9.50   In a sample survey of 2,500 commuters, 900 said that they regularly use public transportation. Construct a 95% confidence interval for the corresponding true proportion of commuters who regularly use public transportation.

9.51   Using the data of Exercise 9.50, what can we say with 95% confidence about the maximum error if we estimate that the probability that any such commuter will probably use public transportation is $\frac{900}{2,500} = 0.36$?

✓ **9.52**   Suppose that we want to estimate what proportion of all drivers exceed the 65-mph speed limit on a stretch of road between Reno and Las Vegas. How large a sample will we need to be able to assert with probability 0.95 that the error of our estimate, the sample proportion, will be no more than 0.04?

✓ **9.53**   Rework Exercise 9.52, given that the proportion of all drivers who exceed the 65-mph speed limit on the given stretch of road is at least 0.60.

9.54   A private opinion poll is engaged by a politician to estimate what proportion of her constituents favor the decriminalization of certain narcotics violations. How large a sample will the poll have to take to be able to assert with probability 0.99 that the sample proportion will be off by no more than 0.02?

9.55 Rework Exercise 9.54, given that the poll has reason to believe that the proportion will not exceed 0.30.

9.56 A large personal computer manufacturer wants to determine from a sample what proportion of households intend to purchase personal computers within the next 12 months. How large a sample will the manufacturer need to be able to assert with a probability of at least 0.95 that the sample proportion will not differ from the true proportion by more than 0.05?

9.57 With reference to Exercise 9.56, how large a sample will the manufacturer need to be able to assert with probability 0.95 that the sample proportion will be off by no more than 0.05, if he has good reason to believe that the true proportion is somewhere between 0.10 and 0.20?

9.58 The proportion of successes is simply the number of successes divided by n. Thus, the mean and the standard deviation of the sampling distribution of the sample proportion may be obtained by dividing by n the mean and the standard deviation of the sampling distribution of the number of successes. Use this argument to verify the standard error formula given on page 308.

✓ SOLUTIONS TO PRACTICE EXERCISES

9.1 The probability is 0.95 that the maximum error will be

$$E = 1.96 \cdot \frac{9.2}{\sqrt{100}} \approx 1.80$$

9.2 Substituting $n = 50$, $s = 0.4$, and $E = 0.1$ into the formula for E, we get

$$0.1 = z_{\alpha/2} \cdot \frac{0.4}{\sqrt{50}}$$

so that $z_{\alpha/2} \approx 1.77$, the corresponding entry in Table II is 0.4616, $1 - \alpha = 2(0.4616) = 0.9232$, and we can assert with 92.32% confidence that the error is at most 0.1 ounce.

9.12 Substituting $\sigma = 3.2$ and $E = 0.5$ into the formula for n, we get

$$n = \left[\frac{2.575(3.2)}{0.5}\right]^2 \approx 271.59$$

and we use the next larger integer, 272 hours.

9.18 Substituting $n = 100$, $\bar{x} = 62.7$, and $\sigma = 9.2$ into the confidence interval formula, we get

$$62.7 - 2.575 \cdot \frac{9.2}{\sqrt{100}} < \mu < 62.7 + 2.575 \cdot \frac{9.2}{\sqrt{100}}$$

$$62.7 - 2.37 < \mu < 62.7 + 2.37$$

$$60.3 < \mu < 65.1$$

9.24 (a) Substituting $\bar{x} = 2.36$, $s = 0.48$, $n = 10$, and $t_{0.005} = 3.250$ (for 9 degrees of freedom) into the small-sample confidence interval formula, we get

$$2.36 - 3.250 \cdot \frac{0.48}{\sqrt{10}} < \mu < 2.36 + 3.250 \cdot \frac{0.48}{\sqrt{10}}$$

$$1.87 \text{ micrograms} < \mu < 2.85 \text{ micrograms}$$

(b) Substituting $s = 0.48$, $n = 10$, and $t_{0.025} = 2.262$ (for 9 degrees of freedom) into the formula for E, we get

$$E = 2.262 \cdot \frac{0.48}{\sqrt{10}} \approx 0.34 \text{ microgram}$$

9.37 Substituting $s = 0.4$, $n = 50$, and $z_{0.005} = 2.575$ into the large-sample confidence interval formula for σ, we get

$$\frac{0.4}{1 + \dfrac{2.575}{\sqrt{100}}} < \sigma < \frac{0.4}{1 - \dfrac{2.575}{\sqrt{100}}}$$

$$0.32 \text{ ounce} < \sigma < 0.54 \text{ ounce}$$

9.42 Substituting $n = 500$, $\dfrac{x}{n} = \dfrac{140}{500} = 0.28$, and $z_{0.005} = 2.575$ into the large-sample confidence interval formula, we get

$$0.28 - 2.575\sqrt{\frac{(0.28)(0.72)}{500}} < p < 0.28 + 2.575\sqrt{\frac{(0.28)(0.72)}{500}}$$

$$0.228 < p < 0.332$$

or $0.23 < p < 0.33$ rounded to two decimals.

9.43 Substituting $n = 500$, $\dfrac{x}{n} = \dfrac{140}{500} = 0.28$, and $z_{0.025} = 1.96$ into the formula for E, we get

$$E = 1.96\sqrt{\frac{(0.28)(0.72)}{500}} \approx 0.039$$

We can assert with 95% confidence that the error is at most 0.039.

9.52 Substituting $E = 0.04$ and $z_{0.025} = 1.96$ into the second boxed formula on page 311, we get

$$n = \frac{1}{4}\left(\frac{1.96}{0.04}\right)^2 \approx 601$$

rounded up to the nearest integer.

9.53 Substituting $E = 0.04$, $z_{0.025} = 1.96$, and $p = 0.60$ into the first boxed formula on page 311, we get

$$n = (0.60)(0.40)\left(\frac{1.96}{0.04}\right)^2 \approx 577$$

rounded up to the nearest integer.

10

TESTS CONCERNING MEANS

The problems we studied in Chapter 9 were primarily problems of estimation. Now, however, we shall consider whether a particular population parameter equals a prescribed value. We shall call such decision procedures *tests of hypotheses.*

In the introduction to Chapter 9 we gave three illustrations of problems of estimation—one from science, one from business, and one from everyday life. They would have been tests of **hypothesis**, however, if the biologist had wanted to check another scientist's claim that the average wingspan of the insects is 13.4 mm, if the retailer had wanted to find out whether the average income of all families living within 1 mile of the proposed shopping center is at least $38,000, or if we wanted to check whether it really takes only 2.6 minutes to iron a shirt.

After a general introduction to tests of hypotheses in Sections 10.1 and 10.2, Sections 10.3 and 10.4 are devoted to tests of hypotheses concerning the mean of *one* population; Sections 10.5, 10.6, and 10.7 are devoted to tests of hypotheses concerning the means of *two* populations; and the optional Sections 10.8 and 10.9 are devoted to tests of hypotheses concerning the means of *more than two* populations (analysis of variance).

10.1 TESTS OF HYPOTHESES

In Sections 9.1 through 9.3 we learned how to estimate a population mean μ by giving a confidence interval or by accompanying the point estimate $\bar{x}$ with an assessment of the possible error. Now we shall learn how to test a hypothesis about a population mean μ; that is, we shall present methods for deciding whether to accept or reject an assertion about a particular value, or particular values, of μ. As we already indicated on page 315, the biologist would be testing a hypothesis if he wanted to check (see whether to accept or reject) the claim that the average wingspan of certain insects is $\mu = 13.4$ mm. In general,

A statistical hypothesis is an assertion or conjecture about the parameter, or parameters, of a population. It may also concern the type, or nature, of a population.

Insofar as the second part of this definition is concerned, we shall test in Section 11.6 whether it is reasonable to treat a random variable as having a binomial distribution, or perhaps a Poisson distribution, and whether it is reasonable to treat a set of data as coming from a normal population. In Chapter 10 we shall be concerned only with hypotheses concerning the population mean μ.

To know exactly what to expect when a hypothesis is true, we often have to hypothesize the opposite of what we hope to prove. For instance, if we want to find out whether one method of teaching computer programming is more effective than another, we hypothesize that the two methods are equally effective. Also, if we want to determine whether one method of irrigating the soil is more expensive than another, we hypothesize that the two methods are equally expensive; and if we want to see whether a new steel alloy is stronger than ordinary steel, we hypothesize that their strength is the same. Since we hypothesize in each case that there is no difference—no difference in the effectiveness of the two teaching methods, no difference in the cost of the two methods of irrigation, and no difference in the strength of the two kinds of steel—we refer to hypotheses like these as **null hypotheses** and denote them by H_0. Actually, the term *null hypothesis* is used for any hypothesis set up primarily to see whether it can be rejected, and the idea of setting up a null hypothesis is common even in nonstatistical thinking. It is precisely what we do in criminal proceedings, where an accused is assumed to be innocent unless his or her guilt is established beyond a reasonable doubt. The assumption that the accused is not guilty is a null hypothesis. To know when to reject a null hypothesis, we must also formulate an **alternative hypothesis**. Denoted by H_A, this is the hypothesis that we accept when the null hypothesis can be rejected. For instance, in the wingspan example the alternative hypothesis might be $\mu \neq 13.4$ mm or $\mu > 13.4$ mm, and in the criminal proceedings referred to previously, the alternative hypothesis would be that the accused is guilty.

EXAMPLE 10.1 A research worker wants to compare the effectiveness of advertising a department store sale with newspaper ads or with television ads. Formulate a null hypothesis and an alternative hypothesis appropriate for this situation.

Solution

H_0:   The two methods of advertising are equally effective.

H_A: The two methods of advertising are not equally effective.   ■

There is no question here about H_0, but depending on what the research worker hopes to show, the alternative hypothesis might also be

H_A: Newspaper ads are more effective.

or

H_A: Television ads are more effective.

It might even be that newspaper ads are twice as effective as advertising on television.

To illustrate the problems we face when testing a statistical hypothesis, let us refer again to the wingspan example, and let us suppose that the biologist wants to test the null hypothesis

$$H_0: \quad \mu = 13.4 \text{ mm}$$

against the alternative hypothesis

$$H_A: \quad \mu \neq 13.4 \text{ mm}$$

where μ is the mean wingspan of the given kind of insect. To perform this test, he decides to take a random sample of size $n = 40$ with the intention of accepting the null hypothesis if the mean of the sample falls anywhere between 13.2 and 13.6 mm; otherwise, the null hypothesis is rejected.

This provides a clear-cut criterion for accepting or rejecting the null hypothesis, but unfortunately it is not infallible. Since the decision is based on a sample, there is the possibility that the sample mean will be greater than or equal to 13.6 mm or less than or equal to 13.2 mm, even though the true mean is 13.4 mm. There is also the possibility that the sample mean will fall in the interval from 13.2 mm to 13.6 mm, even though the true mean is, say, 13.7 mm. Thus, before adopting the criterion (and, for that matter, any decision criterion) it would seem wise to investigate the chances that it will lead to a wrong decision.

Assuming that it is known from similar studies that $\sigma = 0.8$ mm for this kind of data, let us first investigate the possibility of falsely rejecting the null hypothesis. That is, assume for the sake of argument that the true mean is really 13.4 mm; then find the probability that the sample mean will be greater than or equal to 13.6 mm or less than or equal to 13.2 mm. The probability that this will happen purely due to chance is given by the combined area of the blue regions of Figure 10.1, and it can readily be determined by approximating the sampling distribution of the mean with a normal distribution. Assuming that the population sampled is large enough to be treated as if it were infinite, which seems very reasonable in this case, we have

$$\sigma_{\bar{x}} = \frac{\sigma}{\sqrt{n}} = \frac{0.8}{\sqrt{40}} \approx 0.126$$

and it follows that the dividing lines of the criterion, in standard units, are

$$z = \frac{13.2 - 13.4}{0.126} \approx -1.59 \quad \text{and} \quad z = \frac{13.6 - 13.4}{0.126} \approx 1.59$$

According to Table II, the area in each tail of the sampling distribution of Figure 10.1 is $0.5000 - 0.4441 = 0.0559$, and hence the probability of getting a value in either tail of the sampling distribution is

$$0.0559 + 0.0559 = 0.1118 \approx 0.11$$

This is the probability of erroneously rejecting the null hypothesis, that is, the probability of rejecting it when it is true. Whether this is an acceptable risk is for the biologist to decide, and it will have to depend on the consequences of making this kind of error.

Let us now look at the other possibility, where the test fails to detect that the null hypothesis is false, that is, $\mu \neq 13.4$ mm. Assuming again that $\mu = 13.7$ mm, purely for the sake of argument, we find that the probability that the sample mean will fall between 13.2 and 13.6 is given by the area of the blue region of Figure 10.2, the area under the curve between 13.2 and 13.6. The mean of the

Figure 10.1
Test criterion where "accept the claim" and "reject the claim" refer to the corresponding intervals of $\bar{x}$.

Figure 10.2
Test criterion.

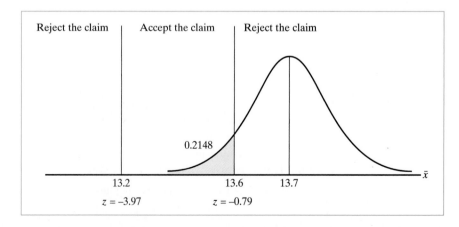

sampling distribution is now 13.7 and its standard deviation is again

$$\sigma_{\bar{x}} = \frac{0.8}{\sqrt{40}} \approx 0.126$$

so that the dividing lines of the criterion, in standard units, are now

$$z = \frac{13.2 - 13.7}{0.126} \approx -3.97 \quad \text{and} \quad z = \frac{13.6 - 13.7}{0.126} \approx -0.79$$

Since the area under the curve to the left of $z = -3.97$ is negligible, it follows from Table II that the area of the blue region of Figure 10.2 is $0.5000 - 0.2852 = 0.2148 \approx 0.21$. This is the probability of erroneously accepting the null hypothesis, that is, the probability of accepting it when it is false with $\mu = 13.7$ mm. Again, it is up to the biologist to decide whether this is an acceptable risk.

The situation described in this example is typical of testing a statistical hypothesis, and it may be summarized as in the following table:

	Accept H_0	*Reject H_0*
H_0 *is true*	Correct decision	Type I error
H_0 *is false*	Type II error	Correct decision

If the null hypothesis H_0 is true and accepted, or false and rejected, the decision is in either case correct; if it is true and rejected, or false and accepted, the decision is in either case in error.

> *The error of rejecting H_0 when it is true is called a **Type I error**, and the probability of committing such an error is denoted by α (Greek lowercase alpha).*

> *The error of accepting H_0 when it is false is called a **Type II error**, and the probability of committing such an error is denoted by β (Greek lowercase beta).*

The terms *alpha error* and *beta error* are sometimes used for Type I and Type II errors.

In our example we showed that for the given test criterion $\alpha = 0.11$. We also showed that $\beta = 0.21$ when the mean wingspan of the given kind of insect is actually 13.7 mm.

EXAMPLE 10.2 Suppose that in our example $\mu = 13.0$ mm, which the biologist does not know, and that the mean of his sample is $\bar{x} = 13.3$ mm. What type of error will he commit if he uses the criterion of Figure 10.1?

Solution Since $\bar{x} = 13.3$ mm falls in the interval from 13.2 to 13.6 mm, he will accept the null hypothesis $\mu = 13.4$ mm even though it is false; this is a Type II error. ∎

In calculating the probability of a Type II error in our example, we chose for illustrative purposes the alternative value $\mu = 13.7$. However, in this problem, as in most others, there are infinitely many other alternatives, and for each of them there is a positive probability β of erroneously accepting H_0. So, in practice we choose some key alternative values and calculate the probabilities of committing a Type II error, as we calculated it for $\mu = 13.7$, or we sidestep the issue by using a method that will be explained in Section 10.2. If we do calculate β for various alternative values and plot these probabilities as in Figure 10.3, the resulting curve is called an **operating characteristic curve**, or simply an **OC-curve**. Such a curve provides a good overall picture of the risks to which one is exposed by a test criterion.

The operating characteristic curve shown in Figure 10.3 pertains to the wingspan example, and in Exercise 10.8 the reader will be asked to verify some of the other values of β. Since the probability of a Type II error is the probability of accepting H_0 when it is false, we "completed the picture" in Figure 10.3 by labeling the vertical scale "Probability of accepting H_0" and plotting at $\mu = 13.4$ the probability $1 - \alpha = 1 - 0.11 = 0.89$.

Figure 10.3
Operating characteristic curve.

The problems we met in this section are not limited to the particular example dealing with the wingspan of the given kind of insect; the same questions would also have arisen if we had wanted to test the hypothesis that the mean age of divorced women at the time of their divorce is 35.6 years, the hypothesis that an antibiotic is 82% effective, the hypothesis that a computer-assisted method of instruction will on the average raise a student's score on a standard achievement test by 8.2 points, and so forth.

10.2 SIGNIFICANCE TESTS

In the example of the preceding section, we had less trouble with Type I errors than with Type II errors, because we formulated H_0 as a **simple hypothesis** about μ. That is, with H_0 giving μ a single value, the probability of a Type I error could readily be calculated. Had we formulated instead the **composite**

hypothesis 13.0 mm $\leq \mu \leq$ 13.8 mm, the composite hypothesis $\mu \leq$ 13.4 mm, or the composite hypothesis $\mu \geq$ 13.4 mm, where in each case μ can take on more than one possible value, we could not have calculated the probability of a Type I error without specifying by how much μ differs from, is less than, or is greater than 13.4 mm.

As we saw in Section 10.1, the probability of a Type I error can easily be calculated when we are given a simple hypothesis, an unambiguous criterion, and enough information to apply the sampling theory of Chapter 8. We also saw how the probabilities of Type II errors may have to be calculated for specific alternative values of μ. The following illustrates how we can sidestep Type II errors altogether. Suppose that a retailer wants to find out whether the average income of families living within 1 mile of a proposed shopping center is greater than $48,000. He has a number of these families interviewed and bases his decision on the following criterion:

> *Reject the null hypothesis $\mu = \$48,000$ and accept the alternative $\mu > \$48,000$ if the sample mean is greater than or equal to, say, $48,800; if not, reserve judgment (perhaps pending further checks).*

If we reserve judgment, there is no possibility of committing a Type II error—no matter what happens, the null hypothesis is never really accepted. This is all right in our example, because the retailer wants to know primarily whether the null hypothesis can be rejected.

The procedure we have outlined here is called a **significance test**. If the difference between what we expect under the null hypothesis and what we observe in a sample is too large to be reasonably attributed to chance, we reject the null hypothesis. If the difference between what we expect and what we observe is so small that it may well be attributed to chance, we say that the results are **not statistically significant**.

With respect to the wingspan example, the biologist could convert the criterion on page 317 into that of a significance test by writing

> *Reject the null hypothesis $\mu = 13.4\,mm$ (and accept the alternative $\mu \neq 13.4\,mm$) if the mean of a random sample of size $n = 40$ is greater than or equal to 13.6 mm or less than or equal to 13.2 mm; otherwise, reserve judgment.*

Insofar as the rejection of the null hypothesis is concerned, the criterion has remained unchanged and the probability of a Type I error is still 0.11. However, so far as its acceptance is concerned, the biologist is now playing it safe by reserving judgment.

Reserving judgment in a significance test is similar to what happens in court proceedings when the prosecution does not have sufficient evidence to get a conviction, but it would be going too far to say that the defendant definitely did not commit the crime. In general, whether one can afford the luxury of reserving judgment in any given situation depends entirely on the nature of the problem. If a decision must be reached one way or the other, there is no way of avoiding the risk of committing a Type II error.

Since the general problem of testing hypotheses and constructing statistical decision criteria may seem confusing, at least to the beginner, it will help to proceed systematically as outlined in the following five steps:

1. We formulate a simple null hypothesis and an appropriate alternative hypothesis that is to be accepted when the null hypothesis must be rejected.

In the wingspan example, the null hypothesis was $\mu = 13.4$ mm and the alternative hypothesis was $\mu \neq 13.4$ mm (since the biologist wanted to reject 13.4 mm if this value is too high or too low). We refer to this kind of alternative as a **two-sided alternative**. In the family income example, the null hypothesis was $\mu = \$48,000$ and the alternative hypothesis was $\mu > \$48,000$ (since the retailer wants to know whether average family income in the given area exceeds $48,000). This is called a **one-sided alternative**. We can also write a one-sided alternative with the inequality going the other way. For instance, if we hope to be able to show that the mean time required to do a certain job is less than 20 minutes, we would test the null hypothesis $\mu = 20$ minutes against the alternative hypothesis $\mu < 20$ minutes.

As in the examples of the preceding paragraph, alternative hypotheses usually specify that the population mean (or whatever other parameter may be of concern) is not equal to, greater than, or less than the value assumed under the null hypothesis. For any given problem, the choice of an appropriate alternative hypothesis depends mostly on what we hope to be able to show, or better, perhaps, where we want to put the burden of proof.

EXAMPLE 10.3 A metal parts producer whose stamping machines average 160 pieces per work shift is considering the purchase of new machines. Against what alternative hypothesis would the producer test the null hypothesis $\mu = 160$ pieces if

(a) the producer does not want to buy new machines unless they will increase the output;

(b) the producer wants to buy new machines (that have some new features) unless they will decrease the output?

Solution   **(a)** The producer would use the alternative hypothesis $\mu > 160$ and buy the new machines only if the null hypothesis can be rejected.
(b) The producer would use the alternative hypothesis $\mu < 160$ and buy the new machines unless the null hypothesis is rejected. ■

EXAMPLE 10.4 A company uses a production process designed to make machine components that have an average thickness of 5 inches. The company suspects that the process is not maintaining its intended average.

(a) If the company wants to modify its process if the average thickness is smaller than 5 inches, what null hypothesis and alternative hypothesis should it use?

(b) If the company wants to modify its process if the average thickness is different from 5 inches, what null hypothesis and alternative hypothesis should it use?

Solution   **(a)** The words *smaller than* suggest that the hypothesis $\mu < 5$ inches is needed together with the hypothesis $\mu = 5$ inches. Only the second of these hypotheses is a simple hypothesis, where μ takes on a single value, so it must be the null hypothesis and we write

$$H_0: \quad \mu = 5 \text{ inches}$$

$$H_A: \quad \mu < 5 \text{ inches}$$

(b) The words *different from* suggest that the hypothesis $\mu \neq 5$ inches is needed together with the hypothesis $\mu = 5$ inches. Again, only the second of the two hypotheses is a simple hypothesis, where μ takes on a single value. Using it as the null hypothesis, we write

$$H_0: \quad \mu = 5 \text{ inches}$$

$$H_A: \quad \mu \neq 5 \text{ inches}$$

1. In each of these examples, the hypotheses H_0 and H_A were formulated without experimental data. It is essential that H_A be expressed according to what we hope to demonstrate with subsequent data. In particular, the choice of a one-sided or two-sided version of H_A must not be suggested by the data.

It commonly happens, however, that we are presented with the data before we have had a chance to contemplate the hypotheses; in such a situation we must try to assess the motives without using the numerical data. If there is any doubt as to whether the situation calls for a one-sided or a two-sided alternative, the scrupulous action calls for the two-sided alternative. Exercise 10.17 deals with this difficulty.

2. We specify the probability of a Type I error; if possible, desired, or necessary, we may also make some specifications about the probabilities of Type II errors for specific alternatives.

The probability of a Type I error is also called the **level of significance**, and it is usually set at $\alpha = 0.05$ or $\alpha = 0.01$. Testing a null hypothesis at the level of significance $\alpha = 0.05$ simply means that we are fixing the probability of rejecting the null hypothesis even though it is true at 0.05. The decision to use $\alpha = 0.05$, $\alpha = 0.01$, or some other value depends mostly on the consequences of committing a Type I error in the given situation. Observe, however, that we cannot make the probability of a Type I error too small, for this will have the tendency to make the probability of serious Type II errors too large. In actual practice, experimenters choose α depending on the risks; in the exercises in this text, the level of significance will usually be specified.

3. Based on the sampling distribution of an appropriate statistic and the choice of the level of significance, we construct a test criterion for testing the null hypothesis against the given alternative.
4. We calculate from the data the value of the statistic on which the decision is to be based.
5. We decide whether to reject the null hypothesis, whether to accept it, or whether to reserve judgment.

Insofar as step 5 is concerned, in the wingspan example we based the criterion on the normal-curve approximation to the sampling distribution of the mean. In general, this step depends on the statistic on which we want to base the decision and on its sampling distribution. We call this statistic the **test statistic**.

Looking back at our two illustrations—the wingspan example and the family income example—we find that the construction of a test criterion depends also on the alternative hypothesis that we happen to choose. In the wingspan example, we used a **two-sided criterion** with the two-sided alternative hypothesis $\mu \neq 13.4$ mm, rejecting the null hypothesis for large or small values of the sample mean; in the family income example, we used a one-sided criterion with the one-sided alternative hypothesis $\mu > \$48,000$, rejecting the null hypothesis only for large values of the sample mean.

In general, a test is called **two-sided** or **two-tailed** if the null hypothesis is rejected for values of the test statistic falling into either tail of its sampling distribution, and it is called **one-sided** or **one-tailed** if the null hypothesis is rejected only for values of the test statistic falling into one specified tail of its sampling distribution. A two-tailed test usually goes with a two-sided criterion, and a one-tailed test usually goes with a one-sided criterion, but there are exceptions (see, for example, the tests of Section 10.9).

In connection with step 5, let us point out that we often accept null hypotheses with the tacit hope that we are not exposed to overly high risks of committing serious Type II errors. Of course, if it is necessary we can calculate enough probabilities of Type II errors to get an overall picture from the operating characteristic curve of the test criterion.

As at the end of Section 10.1, let us also point out that the concepts we have introduced here are not limited to tests concerning population means; they apply equally to tests concerning other parameters or tests concerning the nature, or form, of populations.

With the general availability of computers and statistical software, the five steps outlined previously may be modified to allow for more freedom in the choice of the level of significance. How this is done will be explained in Section 10.3.

Exercises 10.1, 10.6, and 10.10 are practice exercises; their complete solutions are given on pages 360 and 361.

✓ **10.1** A pharmaceutical inspector has to examine shipments of antibiotics to check whether they possess full potency.
  (a) What type of error would he commit if he erroneously rejects the hypothesis that a given bottle of the antibiotic possesses full potency?
  (b) What type of error would he commit if he erroneously accepts the hypothesis that a given bottle of the antibiotic possesses full potency?

10.2 An oceanographer wants to test the null hypothesis that a cubic yard of ocean water contains about 45.6 pounds of salt (which it does). Explain under what conditions the oceanographer would commit a Type I error and under what conditions the oceanographer would commit a type II error.

10.3 The manager of a large gasoline station claims that the average amount of gasoline purchased by automobile drivers at the station is 9.3 gallons, and we want to test whether he is correct.
  (a) What hypothesis would we be testing if we commit a Type I error by erroneously concluding that the average amount of gasoline purchased by automobile drivers at the station is 9.3 gallons?

EXERCISES

(b) What hypothesis would we be testing if we commit a Type II error by erroneously concluding that the average amount of gasoline purchased by automobile drivers at the station is 9.3 gallons?

10.4   If we want to test the hypothesis that telemarketing (telephone sales) increases the number of sales made by automobile salespersons, under what conditions would we commit

(a) a Type I error;

(b) a Type II error?

10.5   Whether an error is a Type I error or a Type II error depends on how the hypothesis to be tested is formulated. To illustrate this, reword the hypothesis of the preceding exercise so that a Type I error becomes a Type II error, and vice versa.

✓ **10.6**   With reference to the wingspan illustration on page 317, suppose that the biologist increases the sample size to $n = 60$ while everything else remains unchanged.

(a) Show that this decreases the probability of a Type I error from 0.11 to 0.05.

(b) Show that this decreases the probability of a Type II error when $\mu = 13.7$ mm from 0.21 to 0.17.

10.7   With reference to the wingspan illustration on page 317, suppose that the biologist changes the criterion so that the hypothesis $\mu = 13.4$ mm is accepted if the sample mean falls anywhere from 13.1 to 13.7 mm, while otherwise the hypothesis will be rejected. If everything else remains the same, show that

(a) this will decrease the probability of a Type I error from 0.11 to 0.02;

(b) this will increase the probability of a Type II error when $\mu = 13.7$ from 0.21 to 0.50.

10.8   With reference to the wingspan illustration on page 317, suppose that the biologist increases the sample size to $n = 100$ while everything else remains unchanged.

(a) What is the probability that the biologist will erroneously reject the null hypothesis $\mu = 13.4$ mm?

(b) What is the probability that the biologist will erroneously accept the null hypothesis $\mu = 13.4$ mm when actually $\mu = 13.7$ mm?

10.9   With reference to the operating characteristic curve of Figure 10.3, verify that the probabilities of Type II errors are

(a) 0.78 when $\mu = 13.5$ mm or $\mu = 13.3$ mm;

(b) 0.50 when $\mu = 13.6$ mm or $\mu = 13.2$ mm;

(c) 0.06 when $\mu = 13.8$ mm or $\mu = 13.0$ mm.

✓ **10.10**   A department store has a salesperson whom it suspects of making more mistakes than the average of all its salespersons.

(a) If the department store decides to let the salesperson go if this suspicion is confirmed, what null hypothesis and what alternative hypothesis should it use?

(b) If the department store decides to let the salesperson go unless he actually makes fewer mistakes than the average of all its salespersons, what null hypothesis and what alternative hypothesis should it use?

10.3 TESTS CONCERNING MEANS

Having used tests concerning means to illustrate the basic principle of hypothesis testing, let us now see how we proceed in practice. Actually, we will depart somewhat from the procedure used in the illustrations given earlier in this chapter. In the wingspan illustration and also in the family income illustration, we used $\bar{x}$ as the test statistic and stated the test criterion in terms of values of $\bar{x}$. Now we will base the criterion on the statistic

STATISTIC FOR A LARGE-SAMPLE TEST CONCERNING THE MEAN

$$z = \frac{\bar{x} - \mu_0}{\sigma/\sqrt{n}}$$

where μ_0 is the value of the mean assumed under the null hypothesis. The reason for working with this statistic, which amounts to using standard units, is that it enables us to formulate criteria that are applicable to a great variety of problems, not just one.

We refer to tests based on this statistic as **large-sample tests**, because we are using the normal-curve approximation to the sampling distribution of the mean. Thus, the test requires that $n \geq 30$, unless the population we are sampling is normal. If we are able to assume that the population is normal, the test may be used for any value of n.

EXAMPLE 10.5   In the introduction to Chapter 9 we mentioned the retailer who "may want to determine the average income of all families living within a mile of a proposed new shopping center." Let us assume that the retailer wants to know whether the true mean income in the area is really $29,400 per year, as claimed by others. Using the following data, determine what the retailer can conclude at the 0.05 level of significance if the average income of 100 families, randomly selected and living in the prescribed area, is $\bar{x} = \$30,000$, and the standard deviation is $s = \$2,500$. Let the alternative be two sided because the retailer wants to reject the hypothesis if $\mu = \$29,400$ is too high or too low.

Solution   1. *Hypotheses*

$$H_0: \quad \mu = \$29,400$$
$$H_A: \quad \mu = \$29,400$$

2. *Level of significance*

$$\alpha = 0.05$$

3. *Criterion*   Putting half of 0.05 into each tail of the sampling distribution and making use of the fact that $z_{0.025} = 1.96$, reject the null hypothesis if $z \leq -1.96$ or $z \geq 1.96$, where

$$z = \frac{\bar{x} - \mu_0}{\sigma/\sqrt{n}}$$

Otherwise, accept the null hypothesis or reserve judgment (see also Figure 10.4). We include the dividing lines (here -1.96 and $+1.96$) among the values for which the null hypothesis is rejected.[†]

[†] Some authors exclude the dividing lines from the values for which the null hypothesis is rejected, but this is of no practical consequence since the standard normal distribution is a continuous distribution.

Figure 10.4
Test criterion for retailer
of Example 10.5.

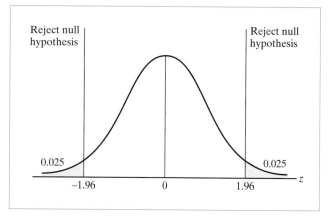

4. *Calculations*   Substituting $n = 100, \bar{x} = \$30{,}000, \mu_0 = 29{,}400,$ and $s = 2{,}500$ (for σ) into the formula for z, we get

$$z = \frac{30{,}000 - 29{,}400}{2{,}500/\sqrt{100}} = 2.40$$

5. *Decision*   Since $z = 2.40$ exceeds 1.96, the null hypothesis must be rejected. In other words, the difference between \$29,400 and \$30,000 is too large to be attributed to chance, and this allows the retailer to conclude that the 29,400 figure must be wrong. It appears from the data that the true average income is actually greater than \$29,400

In this example, the value s was used in place of σ in the calculation of the test statistic in step 4. We prefer the data based value of s over the prespecified assessment of σ. This is especially so when the sample is large (over 30). ■

In general, for tests concerning means, the dividing lines of the criteria are as shown in Figure 10.5.

1. For the one-sided alternative, $\mu < \mu_0$, the test is one tailed, α is placed in the left-hand tail, and the dividing line is $-z_\alpha$.
2. For the one-sided alternative, $\mu > \mu_0$, the test is one tailed, α is placed in the right-hand tail, and the dividing line is z_α.
3. For the two-sided alternative, $\mu \neq \mu_0$, the test is two tailed, α is divided equally between the two tails, and the dividing lines are $-z_{\alpha/2}$ and $z_{\alpha/2}$.

If $\alpha = 0.05$, the dividing lines, or **critical values**, of the criteria are -1.645 for the one-sided alternative in situation 1, 1.645 for the one-sided alternative in situation 2, and -1.96 and 1.96 for the two-sided alternative in situation 3. Similarly, if $\alpha = 0.01$, the dividing lines of the criteria are -2.33 for the one-sided alternative in situation 1, 2.33 for the one-sided alternative in situation 2, and -2.575 and 2.575 for the two-sided alternative in situation 3. All these values come directly from Table II (see Exercise 7.17 and Example 7.5 on page 247).

In most practical situations where σ is unknown, we must substitute the sample standard deviation s. Again, this is permissible when the sample is large, that is, when $n \geq 30$.

Figure 10.5
Test criteria.

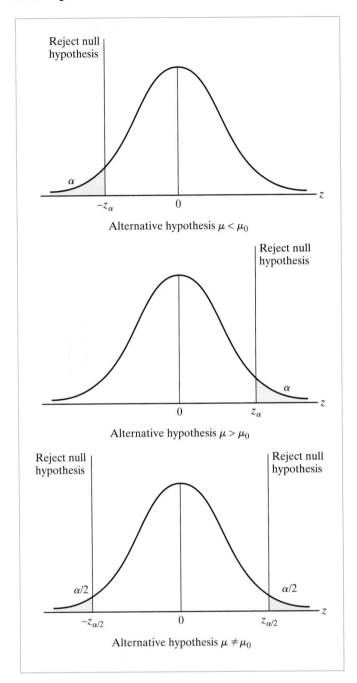

Reject null
hypothesis

α

$-z_\alpha$

0

z

Alternative hypothesis $\mu < \mu_0$

Reject null
hypothesis

0

z_α

α

z

Alternative hypothesis $\mu > \mu_0$

Reject null
hypothesis

Reject null
hypothesis

$\alpha/2$

$\alpha/2$

$-z_{\alpha/2}$

0

$z_{\alpha/2}$

z

Alternative hypothesis $\mu \neq \mu_0$

EXAMPLE 10.6   A taxicab firm suspects that the average lifetime of 25,000 miles claimed for certain tires is too high. To test the claim, the firm puts a random sample of 40 of these tires on its taxicabs and later finds that their mean lifetime is 24,421 miles and the standard deviation is 1,349 miles. What can the firm conclude at the 0.01 level of significance?

Solution Using the five-step procedure shown in the preceding example, we get

1. H_0: $\mu = 25,000$ miles.
   H_A: $\mu < 25,000$ miles.
2. $\alpha = 0.01$.
3. Reject the null hypothesis if $z \le -2.33$, where

$$z = \frac{\bar{x} - \mu_0}{\sigma/\sqrt{n}}$$

with $s = 1,349$ substituted for σ; otherwise accept the null hypothesis or reserve judgment.

4. Substituting $\bar{x} = 24,421$, $\mu_0 = 25,000$, $n = 40$, and $s = 1,349$ for σ into the formula for z, we get

$$z = \frac{24,421 - 25,000}{1,349/\sqrt{40}} = -2.71$$

5. Since $z = -2.71$ is actually less than -2.33, the null hypothesis must be rejected. In other words, the taxicab firm can conclude that the average lifetime of the tires is not as high as claimed.

With reference to the family income of Example 10.5, let us now introduce the alternative procedure for conducting tests of significance, which we mentioned on page 322. Instead of comparing $z = 2.40$ with 1.96 (and tacitly with -1.96), we compare the total areas of the blue regions of Figure 10.6, $0.0082 + 0.0082 = 0.0164$, with the specified level of significance $\alpha = 0.05$. This area is referred to as the **p-value**, prob value, tail probability, or observed level of significance corresponding to the observed value of the test statistic. When the alternative hypothesis is two-sided, the p-value is usually the sum of two areas, as in Figure 10.6. When the alternative hypothesis is one sided, the p-value is simply the area under the curve to the right (or left) of the observed value of the test statistic.

In general, p-values may be defined as follows: Corresponding to an observed value of a test statistic, the p-value is the lowest level of significance at which the null hypothesis could have been rejected. In our example, the p-value was 0.0082 and we could have rejected the null hypothesis at the 0.0082 level of significance.

Figure 10.6
The p-value for the family income of Example 10.5.

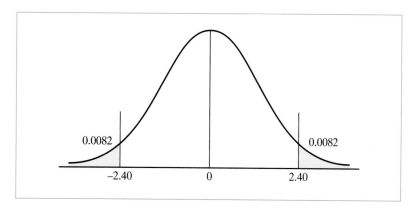

0.0082   0.0082

−2.40   0   2.40

We certainly would have rejected the null hypothesis at the actual specified level of significance $\alpha = 0.05$.

If we wish to base tests of significance on p-values, steps 1 and 2 remain the same, but steps 3, 4, and 5 must be modified as follows:

3′. We specify the test statistic.

4′. We calculate the value of the specified test statistic and the corresponding p-value from the sample data.

5′. We compare the p-value obtained in step 4′ with the level of significance α specified in step 2. If the p-value is *less than or equal to* α, the null hypothesis must be rejected; otherwise, we accept the null hypothesis or reserve judgment.[†] Sometimes the p-value is determined in this step.

Let us illustrate this procedure with reference to the taxicab tires example. In the solution, steps 1 and 2 remain the same, but steps 3, 4, and 5 are replaced by

3′. *Test statistic* The test statistic is

$$z = \frac{\bar{x} - \mu_0}{\sigma/\sqrt{n}} \text{ with } \sigma \text{ replaced by } s$$

4′. *Calculation* Substituting $n = 40, \bar{x} = 24{,}421, \mu_0 = 25{,}000,$ and $s = 1{,}349$ for σ into the formula, we get

$$z = \frac{24{,}421 - 25{,}000}{1{,}349/\sqrt{40}} = -2.71$$

and from Table II we find that the p-value, the area under the curve to the left of z, is $0.5000 - 0.4966 = 0.0034$.

5′. *Decision* Since 0.0034 is less than $\alpha = 0.01$, the null hypothesis must be rejected. As before, the firm's suspicions are confirmed.

In this taxicab firm example, Example 10.6, the firm may decide to change the tires it now uses for another brand. The end result may be a decision about adopting another brand of tires. The decision should be reached, however, by specifying the level of significance, and using the p-value to supplement the results.

On the other hand, the social scientist does not have to make any policy decision. He might not even specify a level of significance, finding the p-value a more convenient way of handling the data, and letting it go at that.

Many experimenters relate the p-value to the common levels of significance $\alpha = 0.05$ and $\alpha = 0.01$, by means of statements such as the following: $p < 0.01$ or $0.01 < p < 0.05$. When the p-value is obtained from the normal distribution, as in Table II, this is readily done. In many areas of research, the result $0.01 < p < 0.05$ is indicated by attaching a single asterisk to the observed value of the statistic. Correspondingly, the result $p < 0.01$ is indicated by attaching a double asterisk

---

[†] If test criteria are modified as in the footnote on page 326, step 5′ will call for rejection of the null hypothesis when the p-value is *less than* α.

(**) to the test statistic. For instance, in the preceding example we could have written $z = -2.71^*$ as part of step 4'.

10.4 TESTS CONCERNING MEANS (Small Samples)

When we do not know the value of the population standard deviation σ and the sample is small, $n < 30$, we assume, as on page 294, that the population we are sampling has roughly the shape of a normal distribution and base our decision on the statistic

STATISTIC FOR A
SMALL-SAMPLE TEST
CONCERNING THE
MEAN

$$t = \frac{\bar{x} - \mu_0}{s/\sqrt{n}}$$

whose sampling distribution is the t distribution (see page 301) with $n - 1$ degrees of freedom. Of course, if it cannot be assumed that the population we are sampling has roughly the shape of a normal distribution, this small-sample procedure cannot be used.

The criteria we use for this test are those of Figure 10.5 with z replaced by t and z_α and $z_{\alpha/2}$ replaced by t_α and $t_{\alpha/2}$. As was explained on page 301, for given degrees of freedom, t_α and $t_{\alpha/2}$ are values for which the areas to their right under the corresponding t distributions are, respectively, α and $\alpha/2$. All the critical values for this **one-sample t test** may be read from Table III, with the number of degrees of freedom equal to $n - 1$.

EXAMPLE 10.7   Suppose that we want to test, on the basis of a random sample of size $n = 5$, whether the fat content of a certain kind of processed meat exceeds 30%. What can we conclude at the 0.01 level of significance if the sample values are 31.9, 30.3, 32.1, 31.7, and 30.9%?

Solution   The sample size $n = 5$ is small, and we must begin by stating that the work rests on the assumption that the values are sampled from a population that has roughly the shape of a normal distribution.

1. *Hypotheses*

$$H_0: \quad \mu = 30\%$$

$$H_A: \quad \mu > 30\%$$

The alternative is one sided because our concern is that the fat content might exceed 30%.

2. *Level of significance* $\alpha = 0.01$

3. *Criterion*   Reject the null hypothesis if $t \geq 3.747$, the value of $t_{0.01}$ for $5 - 1 = 4$ degrees of freedom, where

$$t = \frac{\bar{x} - \mu_0}{s/\sqrt{n}}$$

(see also Figure 10.7); otherwise, accept the null hypothesis or reserve judgment.

Figure 10.7
Test criterion for fat-content example.

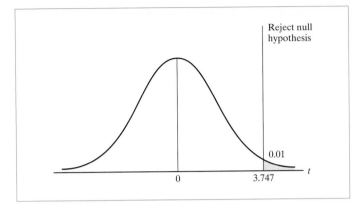

4. *Calculations*   First calculating the mean and the standard deviation of the sample, we get $\bar{x} = 31.38$ and $s = 0.756$, and substituting these values together with $n = 5$ into the formula for t, we find that

$$t = \frac{31.38 - 30}{0.756/\sqrt{5}} \approx 4.08$$

5. *Decision*   Since $t = 4.08$ exceeds 3.747, the null hypothesis must be rejected. We believe that the mean fat content of the given kind of processed meat exceeds 30%.

Table III shows that the value obtained for t in the preceding example, 4.08, falls between $t_{0.010} = 3.747$ and $t_{0.005} = 4.604$. Thus, the corresponding p-value falls between 0.005 and 0.010, but to determine its actual value we would have to use a computer package or a special calculator. For instance, using a TI-83 graphing calculator, we obtain the result shown in Figure 10.8, which is reproduced from the display screen of the graphing calculator. The graphing calculator information shown here and elsewhere in this book is entirely optional and can be omitted by the reader without the loss of continuity. The p-value shown here is $p = .007548869$, or $.0075$ rounded to four decimals. Thus, if the level of significance had been, say, $\alpha = 0.009$, we would have found that the p-value, 0.0075, is less than α and, hence, that the null hypothesis could also have been rejected.

Directions for obtaining the p-value given in Figure 10.8 are in the following Using Technology box.

Figure 10.8
The p-value for a one-sample t test reproduced from the display screen of a TI-83 graphing calculator.

How to obtain the p-value given in Figure 10.8

(**TIP**: Before beginning, see Appendix A for clearing lists and entering data.)

STEP 1. Clear list L1.

STEP 2. Input the fat percentage data from Example 10.7 into list L1.

STEP 3. Select the type of statistical test you wish to use.
    Press STAT.
    Highlight TESTS.
    Select 2: T-TEST, press ENTER.
    Highlight: DATA
    Set $\mu_0 = 30$
          List: L1
          Freq: 1
    Highlight μ: $> \mu_0$

STEP 4. Solve.
    Highlight Calculate, press ENTER.

When the sample size is small, it is important to state the assumption that the population sampled has roughly the shape of a normal distribution. Merely stating this assumption does not make it true, of course. The assumption has minimal impact on statistical practice since the appropriateness of the normal distribution is difficult to verify with a small sample and, moreover, the t statistic generally makes correct decisions even when the population does not have the shape of a normal distribution. Nevertheless, sound statistical practice recommends that the assumption be stated.

EXERCISES

Exercises 10.11 and 10.21 are practice exercises; their complete solutions are given on page 361.

✓ **10.11** A law student, who wants to check a professor's claim that convicted embezzlers spend on the average 12.8 months in jail, takes a random sample of 60 such cases from court files, finding $\bar{x} = 11.2$ months and $s = 3.5$ months.
    (a) Test the null hypothesis $\mu = 12.8$ months at the 0.01 level of significance.
    (b) Find the p-value.
    (c) If you base the hypothesis test on the p-value, do you reach the same conclusion with regard to significance at the 0.01 level?

10.12 The "reasonable and customary" fee for anesthesia services for surgery lasting less than 1 hour in a certain region is \$425, as specified by an insurer. A particular anesthesiology office is suspected of charging, on average, more than this amount, and a random sample of 45 of their billings shows $\bar{x} = \$455$ and $s = \$44$. Can we conclude at the 0.05 level of significance that this particular anesthesiology office charges on average more than $425?

10.13 According to a production engineer, a trained worker should be able to assemble a certain type of chair in 15.0 minutes. Is this figure substantiated by the carefully timed assembly of 150 randomly selected chairs whose average assembly time is 15.4 minutes with standard deviation of 2.4 minutes? Use a two-sided test with the level of significance

(a) $\alpha = 0.05$;

(b) $\alpha = 0.01$.

10.14 Solve Exercise 10.13 using a one-sided test with the alternative hypothesis H_A: $\mu > 15$.

10.15 The manager of the women's dress department of a department store wants to know whether the true average number of women's dresses sold per day is 24. If in a random sample of 36 days the average number of dresses sold is 23 with a standard deviation of seven dresses, is there, at the 0.05 level of significance, sufficient evidence to reject the null hypothesis that $\mu = 24$?

10.16 The nurse in a dentist's office claims that he treats 20 patients per day. To verify this claim, the nurse randomly selects a sample of records for 100 days and concludes that the mean number of patients is 18.1 with a standard deviation of 5.4. Use the 0.05 level of significance to test the null hypothesis $\mu = 20$ patients against the alternative hypothesis $\mu < 20$ patients.

10.17 While investigating a claim that a coin-operated hot coffee machine is dispensing too much hot coffee into hand-held cups (creating a dangerous situation for holders of cups), the owner of the machine takes a random sample of 49 "8-ounce servings" from a large number of servings and finds that the mean is 8.2 fluid ounces with a standard deviation of 0.4 fluid ounces. Is this evidence of overfilling at the 0.01 level of significance?

10.18 According to the norms established for a history test, eighth graders should average 81.7 with a standard deviation of 8.5. If 100 randomly selected eighth graders from a certain school district averaged 79.6 on this test, can we conclude at the 0.05 level of significance that eighth graders from this school district can be expected to average less than the norm of 81.7 on this test? Find the p-value and use it to make the hypothesis test. Do you reach the same conclusion with regard to significance at the 0.05 level?

10.19 The average running time of a certain variety of nickel–cadmium rechargeable flashlight battery is known to be 8.5 hours. A change in the production method for this battery has been proposed, and a sample of 60 batteries produced by the new method has mean running time of 8.62 hours and a standard deviation of 0.55 hour.

(a) Show that for the null hypothesis H_0: $\mu = 8.5$ the test statistic has the value $z = 1.69$.

(b) If the level of significance is 0.05 and if the alternative is H_A: $\mu \neq 8.5$, what conclusion do you reach?

(c) A chemical engineer examines the data and makes the inference that the change in production was made for the purpose of improving the battery running time. He claims that the alternative must be H_A: $\mu > 8.5$. What conclusion does he reach?

(d) A second chemical engineer notes that the production method change might have been made for reasons such as reducing cost or increasing the number of times that the battery can be recharged. She claims that the major concern must be that the running time has not significantly worsened and that the alternative must be H_A: $\mu < 8.5$. What conclusion does she reach?

(e) Suppose that you were given the data with no explanation as to why the change in production method was done. Suppose also that you were given no advice about the level of significance. What would you do?

10.20 A cook at a busy restaurant asserts that he can take a dinner order from a waitress and return it, ready to serve, in 10 minutes. What can we conclude at the level of significance $\alpha = 0.05$ about the cook's assertion concerning the time required to fill these orders if a random sample of 60 orders was filled by the cook in a mean time of $\bar{x} = 10.4$ minutes, with standard deviation of $s = 2.0$ minutes? Use a two-sided alternative and a level of significance of 0.05.

✓ **10.21** In an experiment with a new tranquilizer, the pulse rates (per minute) of 12 patients were determined before they were given the tranquilizer and again 5 minutes later, and their pulse rates were found to be reduced on the average by 7.2 beats with a standard deviation of 1.8. At the level of significance 0.05, do we have significant evidence that the mean pulse reduction with this tranquilizer is less than 9.0 beats?

10.22 In nine test jumps, a newly designed track shoe enables a high-jumper to jump an average of 7.02 feet with a standard deviation of 0.24 foot. What does this tell us about the athlete's claim that with these new shoes he can jump on the average 7.1 feet? Test at the 0.05 level of significance whether his true average jump is less than 7.1 feet.

10.23 A new weight-reducing program that includes diet, exercise, and medication helped 10 overweight people lose 7, 9, 10, 8, 9, 12, 10, 7, 8, and 10 pounds during a three-month test period. Test at the level of significance 0.05 whether this supports the claim that participants in this program lose 10.0 pounds in a three-month period.

10.24 A randomly selected sample of 10 persons who borrowed books at a local public library showed that they borrowed 1, 5, 4, 4, 3, 6, 6, 2, 5, and 4 books. Test at the 0.05 level whether this supports the comment that the library is not lending, on average, 4.2 books per patron.

10.25 A manufacturer guarantees a certain ball bearing to have a mean outside diameter of 0.7500 inch with a standard deviation of 0.0030. If a random sample of 10 such bearings has a mean outside diameter of 0.7510 and standard deviation 0.0030, can we reject the manufacturer's guarantee with regard to the mean outside diameter at the level of significance 0.01?

10.26 The yield of alfalfa from six test plots is 1.2, 2.2, 1.9, 1.1, 1.8, and 1.4 tons per acre. Test at the level of significance $\alpha = 0.05$ whether this supports the contention that the true average yield for this kind of alfalfa is less than 2.0 tons per acre.

10.27 A random sample from a company's very extensive files shows that orders for a certain piece of machinery were filled, respectively, in 12, 10, 17, 14, 13, 18, 11, and 9 days. Use the 0.01 level of significance to test the claim that, on average, such orders are filled in 9.5 days. Choose the alternative hypothesis in such a way that rejection of the null hypothesis $\mu = 9.5$ days implies that it takes longer than that. Find the p-value and indicate whether use of the p-value leads you to the same conclusion regarding significance at the 0.01 level.

10.28 A florist is considering the effects of a new fertilizer on the fruit size of ornamental potted orange trees. The use of standard fertilizers produces, on average, fruits of diameter 4.4 cm. As an experiment, she tries the fertilizer on 20 potted orange trees and then randomly selects one orange from each tree. The diameters of these 20 selected oranges have mean 4.2 cm and standard deviation 0.5 cm.

(a) Using the 0.05 level of significance, can she conclude that the fertilizer has any effect on the diameters of the oranges?

(b) After observing that the oranges that received the new fertilizer are on average smaller, she realizes that it might be possible to advertise these as "dainty petite" oranges. Accordingly, she reformulates the alternative hypothesis as $\mu < 4.4$. At the 0.05 level of significance, does this revised alternative allow her to claim that the fertilized oranges are significantly smaller?

(c) Comment on the appropriateness of selecting H_A after seeing the data.

10.29 The standard formulation used by a certain company for making wall paint will enable a 1-gallon can to cover 250 square feet. The company's research laboratory has produced a new formulation, and the management is anxious to show that it does not cover significantly less area. Accordingly, they test the hypothesis H_0: $\mu = 250$ square feet against the alternative H_A: $\mu < 250$ square feet. The experiment involves 20 cans of the new formulation, and this sample gets mean coverage of 237 square feet and a standard deviation of 25 square feet.

(a) Give the value of the test statistic.

(b) What is the p-value associated with this statistic?

10.30 Suppose that an unscrupulous manufacturer wants "scientific proof" that a totally useless chemical additive will improve the mileage yield of gasoline.

(a) If a research group investigates this additive with one experiment, what is the probability that they will come up with a "significant result" (which they will then use to promote the additive with "scientific claims"), even though the additive is totally ineffective? (Assume that $\alpha = 0.05$ is used.)

(b) If two independent research groups investigate the additive, what is the probability that at least one of them will come up with a "significant result," even though the additive is totally ineffective? (Assume that both use $\alpha = 0.05$.)

(c) If 32 independent research groups investigate the additive, what is the probability that at least one of them will come up with a "significant result," even though the additive is totally ineffective? (Assume that all use $\alpha = 0.05$.)

10.31 Suppose that a manufacturer of pharmaceuticals would like to find a new ointment to reduce swellings. It tries 20 different medications and tests for each whether it reduces swellings at the 0.10 level of significance.

(a) What is the probability that at least one of them will "prove effective," even though all of them are totally useless?

(b) What is the probability that more than one will "prove effective," even though all of them are totally useless?

10.32 Use a computer package or a graphing calculator to rework Exercise 10.24.

10.33 Use a computer package or a graphing calculator to rework Exercise 10.26. If the results include a p-value, note whether it is smaller than the level of significance α.

10.34 Use a computer package or a graphing calculator to rework Exercise 10.27. If the results include a p-value, note whether it is smaller than the level of significance α.

10.5 DIFFERENCES BETWEEN MEANS

There are many problems in which we must decide whether an observed difference between two sample means can be attributed to chance, or whether it is indicative of the fact that the two samples came from populations with unequal means. For instance, we may want to know whether there really is a difference in the mean gasoline consumption of two kinds of cars, when sample data shows that one kind averaged 24.6 miles per gallon while, under the same conditions, the other kind averaged 25.7 miles per gallon. Similarly, we may want to decide on the basis of sample data whether men can perform a certain task faster than women, whether one kind of ceramic insulator is more brittle than another, whether the average diet in one country is more nutritious than that in another country, and so on.

The method we shall use to test whether an observed difference between two sample means can be attributed to chance or whether it is statistically significant is based on the following theory: If $\bar{x}_1$ and $\bar{x}_2$ are the means of two **independent random samples**, then the sampling distribution of the statistic $\bar{x}_1 - \bar{x}_2$ has the mean

$$\mu_1 - \mu_2$$

and the standard deviation

$$\sqrt{\frac{\sigma_1^2}{n_1} + \frac{\sigma_2^2}{n_2}}$$

where μ_1, μ_2, σ_1, and σ_2 are the means and the standard deviations of the two populations sampled. It is customary to refer to the standard deviation of this sampling distribution as the **standard error of the difference between two means**.

By *independent* samples, we mean that the selection of one sample is in no way affected by the selection of the other. Thus, the theory does not apply to "before and after" kinds of comparisons, nor does it apply, say, if we want to compare the caloric value of the food intake of husbands and wives. A special method for comparing the means of dependent samples is explained in Section 10.7.

To base tests of the significance between two sample means on the normal distribution, we shall have to convert to standard units, writing

$$z = \frac{\bar{x}_1 - \bar{x}_2 - (\mu_1 - \mu_2)}{\sqrt{\frac{\sigma_1^2}{n_1} + \frac{\sigma_2^2}{n_2}}}$$

where we subtracted from $\bar{x}_1 - \bar{x}_2$ the mean of its sampling distribution and then divided by the standard deviation of its sampling distribution.

Then, if we limit ourselves to large samples, $n_1 \geq 30$ and $n_2 \geq 30$, we can base the test of the null hypothesis $\mu_1 = \mu_2$ on the foregoing z statistic with $\mu_1 - \mu_2 = 0$, that is, on the statistic

S TATISTIC FOR A LARGE-SAMPLE TEST CONCERNING THE DIFFERENCE BETWEEN TWO MEANS

$$z = \frac{\bar{x}_1 - \bar{x}_2}{\sqrt{\frac{\sigma_1^2}{n_1} + \frac{\sigma_2^2}{n_2}}}$$

which has approximately the standard normal distribution.

Depending on whether the alternative hypothesis is $\mu_1 < \mu_2, \mu_1 > \mu_2$, or $\mu_1 \neq \mu_2$, the criteria we use for these significance tests of the difference between two means are again those of Figure 10.5, with $\mu_1 - \mu_2$ substituted for μ and 0 substituted for μ_0. We can refer to Figure 10.5, even though we are concerned with the sampling distribution of the difference between two means instead of the sampling distribution of the mean, because the criteria are all given in standard units.

The test we have described here is essentially a large-sample test; it is exact only when both of the populations sampled are normal. In most practical situations, where σ_1 and σ_2 are unknown, we must make the further approximation of substituting for them the sample standard deviations s_1 and s_2.

EXAMPLE 10.8   In a study designed to test whether its electric light bulbs are superior to those of a competitor, a manufacturer tested the burning life of 40 of its own randomly selected light bulbs which had an average of $\bar{x} = 628$ hours of continuous use and a standard deviation of $s = 27$ hours. Another sample of 30 randomly selected light bulbs produced by the competitor had an average of $\bar{x} = 619$ hours of continuous use with a standard deviation of $s = 25$ hours. Use the level of significance 0.05 to test the null hypothesis that the corresponding population means are equal against the alternative hypothesis that they are not equal.

Solution   1. *Hypotheses*

$$H_0: \quad \mu_1 = \mu_2$$

$$H_A: \quad \mu_1 \neq \mu_2$$

where μ_1 and μ_2 are the unknown means of the two populations.

2. *Level of significance* $\alpha = 0.05$

3. Reject the null hypothesis if $z \leq -1.96$ or $z \geq 1.96$. where

$$z = \frac{\bar{x}_1 - \bar{x}_2}{\sqrt{\dfrac{\sigma_1^2}{n_1} + \dfrac{\sigma_2^2}{n_2}}}$$

with s_1 and s_2 substituted for σ_1 and σ_2; otherwise, accept the null hypothesis or reserve judgment.

4. *Calculations*   Substituting $n_1 = 40, n_2 = 30, \bar{x}_1 = 628, \bar{x}_2 = 619, s_1 = 27$, and $s_2 = 25$ into the formula for z, we get

$$z = \frac{628 - 619}{\sqrt{\dfrac{(27)^2}{40} + \dfrac{(25)^2}{30}}} = 1.44$$

5. *Decision*   Since $z = 1.44$ lies between -1.96 and 1.96, the null hypothesis cannot be rejected. There is no real evidence that either one of the two brands of light bulbs has a longer burning life than the other.   ◼

EXAMPLE 10.9   Let us now rework Example 10.8, using the *p*-value, which requires only that we modify steps 3, 4, and 5.

Solution

1. *Hypothesis* ...
2. *Level of significance* ...
3. *Eliminate the verbiage.* The formula is unchanged.
4. *Calculations.* Unchanged. The solution is, of course, also unchanged.
5. *Decision.* The area under the normal curve to the right of $z = 1.44$ is .0749. The area to the left of $z = -1.44$ is also .0749. The p-value is $.0749 + .0749 = 0.1498$. This is larger than $\alpha = 0.05$ so that the null hypothesis cannot be rejected.[†] ■

It should be noted that there is a certain awkwardness about comparing means when σ_1 and σ_2 have different values (or when the sample values s_1 and s_2 are not close). Consider, for example, two normal populations with means $\mu_1 = 50$ and $\mu_2 = 52$ and standard deviations $\sigma_1 = 5$ and $\sigma_2 = 15$. Though the second population has a larger mean, it is much more likely to produce a value below 40. An investigator faced with situations like this must decide whether the comparison of μ_1 and μ_2 really addresses the relevant problem.

10.6 DIFFERENCES BETWEEN MEANS (Small Samples)

As in Section 10.4, a small-sample test of the significance of the difference between two means may be based on an appropriate t statistic. For this test, which is used when $n_1 < 30$ or $n_2 < 30$, we must assume that we have independent random samples from populations that can be approximated closely by normal distributions with the same standard deviation. Then we can base our decision on the statistic

STATISTIC FOR A SMALL-SAMPLE TEST CONCERNING THE DIFFERENCE BETWEEN TWO MEANS

$$t = \frac{\bar{x}_1 - \bar{x}_2}{s_p\sqrt{\dfrac{1}{n_1} + \dfrac{1}{n_2}}} \quad \text{where } s_p = \sqrt{\frac{(n_1 - 1)s_1^2 + (n_2 - 1)s_2^2}{n_1 + n_2 - 2}}$$

whose sampling distribution is the t distribution with $(n_1 - 1) + (n_2 - 1) = n_1 + n_2 - 2$ degrees of freedom. Here s_p is the **pooled sample standard deviation**. Observe that t can be rewritten algebraically in various ways. Of course, if we cannot assume that the two populations sampled are approximately normal and that they have the same standard deviation, this small-sample procedure cannot be used.

[†] If the alternative hypothesis is $\mu < \mu_0$, the p-value is given by the area under the standard normal curve to the left of the observed value of z; if the alternative hypothesis is $\mu > \mu_0$, the p-value is given by the area under the standard normal curve to the right of the observed value of z. If the alternative hypothesis is $\mu \neq \mu_0$ and the observed value of z is positive, the p-value is given by *twice* the area of the standard normal curve to the right of the observed value of z; if the alternative is $\mu \neq \mu_0$ and the observed value of z is negative, the p-value is given by *twice* the area under the standard normal curve to the left of the observed value of z. These arguments apply to the large-sample test concerning means, but they can easily be adapted to tests concerning other population parameters based on other test statistics.

The criteria we use for this **two-sample t test** are again those of Figure 10.5, with t, t_α, and $t_{\alpha/2}$ substituted for z, z_α, and $z_{\alpha/2}$, with $\mu_1 - \mu_2$ substituted for μ and 0 substituted for μ_0.

EXAMPLE 10.10   In five games with a 15-pound ball, a professional bowler scored 205, 220, 200, 210, and 201, and in five games with a 16-pound ball he scored 218, 204, 223, 198, and 211. At the 0.05 level of significance, can we conclude that on the average he will score higher with the 16-pound ball?

Solution   Since the sample sizes are small, the work rests on the assumption that the populations sampled have roughly the shape of normal distributions with equal standard deviations.

1. *Hypotheses*

$$H_0: \quad \mu_1 = \mu_2$$

$$H_A: \quad \mu_1 < \mu_2$$

2. *Level of significance* $\alpha = 0.05$
3. *Criterion*   Reject the null hypothesis if $t \leq -1.860$, where 1.860 is the value of $t_{0.05}$ for 8 degrees of freedom and where t is calculated by means of the formula given above; otherwise, accept the null hypothesis or reserve judgment.
4. *Calculations*   First calculating $\bar{x}_1, \bar{x}_2, s_1$, and s_2 for the given data, we substitute these values together with $n_1 = n_2 = 5$ into the formula for s_p, getting

$$S_p = \sqrt{\frac{4(8.17)^2 + 4(10.13)^2}{8}}$$

$$\approx 9.20$$

and

$$t = \frac{207.20 - 210.80}{9.20\sqrt{\dfrac{1}{5} + \dfrac{1}{5}}}$$

$$\approx -0.62$$

5. *Decision*   Since $t = -0.62$ is greater than -1.860, the null hypothesis cannot be rejected. There is no real evidence that the bowler's performance with the two balls is not equally good.   ∎

Had the level of significance been, say, $\alpha = 0.15$ in the preceding example, we could not have obtained the value of $t_{0.15}$ from Table III. To determine the p-value and, at the same time, graph the corresponding area under the curve, we enter the data into our TI-83 graphing calculator and solve, obtaining the results shown in Figure 10.9. The graphing calculator information shown here and elsewhere in this book is entirely optional and can be omitted by the reader without the loss of continuity.

Figure 10.9
The *p*-value for a two-sample *t* test reproduced from the display screen of a TI-83 graphing calculator.

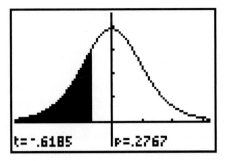

Since the *p*-value of .2767 in Figure 10.9 exceeds $\alpha = 0.15$, we conclude that at this level of significance, too, the null hypothesis cannot be rejected. Directions for obtaining the graph in Figure 10.9 are given in the following Using Technology box.

USING TECHNOLOGY: The Graphing Calculator

How to obtain the *p*-value and graph the area under the curve in Figure 10.9

(**TIP**: Before beginning, see Appendix A for clearing lists and entering data.)

STEP 1. Clear list L1 and L2.

STEP 2. Input the scores from the games with the 15-pound ball into C1 and the scores from the games with the 16-pound ball into C2.

STEP 3. Select the type of statistical test you wish to use.
Press STAT.
Highlight TESTS.
Select 4: 2-SampTTEST, press ENTER.
Highlight: DATA
Set   List1: L1
        List2: L2
        Freq1: 1
        Freq2: 1
        Highlight μ_1: $< \mu_2$
        Highlight Pooled: Yes
        Hit the down arrow.

STEP 4. Solve.
Highlight DRAW, press ENTER.

Finally, we note that the two-sample *t* test is one of the most commonly used statistical procedures and is a fundamental tool of scientific research.

10.7 DIFFERENCES BETWEEN MEANS (Paired Data)

The methods of Sections 10.5 and 10.6 apply only when the samples are independent. For instance, they cannot be used for "before and after" kinds of comparisons or studies of differences between husbands and wives in the caloric intake of their diets. To handle data of this kind, we work with the (signed) differences of the paired data and test whether these differences may be looked on as a random sample from a population that has the mean $\mu = 0$. If the sample is small, we make the assumption that the population sampled has roughly the shape of a normal distribution and use the one-sample t test of Section 10.4; otherwise, we use the large-sample test of Section 10.3.

EXAMPLE 10.11 Use the 0.05 level of significance to test the effectiveness of an industrial safety program on the basis of the following data on the average weekly loss of labor hours due to accidents in 10 plants before and after the program was put into operation:

Plant	Before	After	Difference
1	45	36	9
2	73	60	13
3	46	44	2
4	124	119	5
5	33	35	−2
6	57	51	6
7	83	77	6
8	34	29	5
9	26	24	2
10	17	11	6

The difference column represents before minus after. It can easily be verified that $\bar{x}$ and s for the 10 differences are 5.02 and 4.08, respectively.

Solution Since the sample size is small, we begin by stating that the work rests on the assumption that the differences constitute a sample from a population having roughly the shape of a normal distribution.

1. *Hypotheses*

$$H_0: \quad \mu = 0$$

$$H_A: \quad \mu > 0$$

where μ is the mean of the population of differences.

2. *Level of significance* $\quad \alpha = 0.05$

3. *Criterion* Reject the null hypothesis if $t \geq 1.833$ (obtained from Table III), the value of $t_{0.05}$ for $10 - 1 = 9$ degrees or freedom.

$$t = \frac{\bar{x} - \mu_0}{s/\sqrt{n}}$$

Otherwise, accept the null hypothesis or reserve judgment.

4. *Calculations*   Since $\bar{x} = 5.2$ and $s = 4.08$ for the 10 differences, substitution of these values together with $n = 10$ and $\mu_0 = 0$ into the formula for t yields

$$t = \frac{5.2 - 0}{4.08/\sqrt{10}} \approx 4.03$$

5. *Decision*   Since $t = 4.03$ exceeds 1.833, the null hypothesis must be rejected, and we conclude that the industrial safety program is effective.

We can also use the *p*-value in this example. We would replace steps 3, 4, and 5 with these:

3′. *Test statistic*   The test statistic is

$$t = \frac{\bar{x} - \mu_0}{s/\sqrt{n}} \quad \text{which has 9 degrees of freedom}$$

4′. *Calculations*   Find $t = 4.03$, as previously.

5′. *Decision*   Since $4.03 > t_{0.005} = 3.250$, using 9 degrees of freedom, we conclude that $p < 0.005$. Thus, the *p*-value is less than the specified level of significance, $\alpha = 0.05$, and the null hypothesis must be rejected.

EXERCISES

Exercises 10.35, 10.43, and 10.45 are practice exercises; their complete solutions are given on pages 362 and 363.

✓ **10.35** A sample study was made of the number of business lunches that executives claim as deductible expenses per month. If 60 executives in the insurance industry average 12.8 such deductions with a standard deviation of 5.4 in a given month, while 50 bank executives averaged 9.2 with a standard deviation of 6.3, test at the 0.01 level of significance whether the difference between these two sample means is significant.

10.36 A consumer testing organization wants to know whether two brands of work shoes have equally long-lasting heels. Tests show that 50 randomly selected workers wore out the heels of Brand X shoes, on average, in 96 days with a standard deviation of 6.4 days, while 50 randomly selected workers wore out the heels of Brand Y shoes, on average, in 99 days with a standard deviation of 6.1 days.

(a) Test at the level of significance $\alpha = 0.05$ whether the difference between the two average wearing lives is significant.

(b) Compute the *p*-value and use it to decide whether the difference between the sample means is significant at the 0.05 level of significance.

10.37 Repeat Exercise 10.36 after changing the level of significance from 0.05 to 0.01. Also, compute the *p*-value and use it to decide whether the difference between the sample means is significant at the 0.01 level.

10.38 Determine the *p*-value corresponding to z for the observed level of the statistic in one-tailed tests

(a) $z = 1.64$

(b) $z = -1.64$

(c) $z = 2.76$

(d) $z = -2.76$

(e) $z = 0.00$

(f) $z = -1.645$

(g) $z = 3.09$

(h) $z = -4.0$

(*Hint:* Find the area of each z-value from Table II. Normal-Curve Areas, and subtract from 0.5000.)

10.39 Determine the p-value corresponding to z for the observed level of the statistic in two-tailed tests.

(a) $z = 2.02$

(b) $z = -2.02$

(c) $z = 1.98$

(d) $z = -1.98$

(e) $z = 0.55$

(f) $z = -2.80$

(g) $z = 0.00$

(h) $z = 5.0$

(*Hint:* Find the area of each z-value from Table II. Subtract this area from 0.5000, as in Exercise 10.38. Multiply the result by 2.)

10.40 Suppose we want to base our decisions on the p-value obtained for the test statistic with the level of significance $\alpha = 0.05$. Do we reject; or do we accept (or reserve judgment) the null hypothesis in the following situations?

(a) $p = 0.0083$

(b) $p = 0.0564$

(c) $p = 0.0198$

(d) $p = 0.0100$

(e) $p = 0.0436$

(f) $p = 0.0500$

(*Hint:* Reject the null hypothesis if $p \leq 0.05$.)

10.41 Suppose that the advertisements of a patent medicine manufacturer claim that its product "stops stomach distress quicker" when, in fact, its product does not stop stomach distress quicker than a competing brand. If 20 testing companies each conduct a significance test of the effectiveness of the advertised medication at the 0.05 level of significance, what is the probability that at least one of them will come up with a spuriously significant result which supports the advertiser's claim?

10.42 A study was made of the number of gallons of gasoline purchased by customers for their automobiles at two gasoline stations. If 35 randomly selected purchases made in station A average 9.6 gallons with standard deviation of 1.75 gallons and 51 randomly selected sales at station B average 8.3 gallons with a standard deviation of 2.18, test at the level of significance $\alpha = 0.01$ whether the difference between these two sample means is significant.

✓ **10.43** Measurements of the heat-producing capacity of coal from two mines yielded the following results:

$$n_1 = 5 \quad \bar{x}_1 = 8{,}160 \quad s_1 = 252$$
$$n_2 = 5 \quad \bar{x}_2 = 7{,}730 \quad s_2 = 207$$

where the measurements are in millions of calories per ton. At the level of significance 0.05, can we conclude that the mean heat-producing capacity of coal from the two mines is not the same? Do you reach the same conclusion if you use the p-value?

10.44 Twelve measurements each of the hydrogen content (in percent number of atoms) of gases collected from the eruptions of two volcanos yielded $\bar{x}_1 = 41.5, \bar{x}_2 = 46.1, s_1 = 5.2$, and $s_2 = 6.7$. Use the level of significance $\alpha = 0.05$ to test the null hypothesis that there is no difference (with regard to hydrogen content) in the composition of the gases from the two eruptions.

✓ **10.45** In a study of the effectiveness of physical exercise in weight reduction, a group of 32 persons engaged in a prescribed program of physical exercise for one month showed the following results (in pounds):

Weight before	Weight after	Weight before	Weight after
209	196	170	164
178	171	153	152
169	170	183	179
212	207	165	162
180	177	201	199
192	190	179	173
158	159	243	231
180	180	144	140
211	203	179	180
193	183	202	197
245	229	169	175
188	190	187	190
201	194	213	205
222	219	174	170
190	195	196	197
199	197	201	201

Use a one-sided test and a 0.01 level of significance to determine whether the prescribed program of exercise is effective.

10.46 Suppose that a delivery truck, while on its route on six randomly selected days, made an average of 14 deliveries per day with a standard deviation of 2 deliveries. A second delivery truck, while on another route on seven randomly selected days, made an average of 12 deliveries per day with a standard deviation of 4 deliveries. Can we assert at the level of significance $\alpha = 0.05$ that the first delivery truck on its route makes a larger number of deliveries than the second delivery truck on its route?

10.47 A random sample of 10 high-school students in Phoenix scored $81, 85, 75, 79, 83, 76, 78, 81, 84$, and 78, and a random sample of 10 high-school students in Tucson scored $82, 87, 94, 88, 79, 80, 87, 83, 75$, and 85 on the same social studies examination. Use a computer package or a graphing calculator to test whether the difference between the two sample means is significant at the 0.01 level of significance.

10.48 To compare two kinds of bumper guards, six of each kind were mounted on the same kind of compact car and crashed into a wall at 5 miles per hour. The repair costs for bumper guard A were 345, 290, 360, 380, 310, and 260 dollars, and those for bumper guard B were 310, 290, 315, 385, 280, and 305 dollars. Use a computer package or a graphing calculator to test at the 0.05 level of significance whether the difference between the two sample means is significant.

*10.8 DIFFERENCES AMONG k MEANS

In the three preceding sections we developed procedures for testing the hypothesis that an observed difference between two sample means can be attributed to chance. Now we consider the problem of deciding whether observed differences among more than two sample means can be attributed to chance, or whether they indicate actual differences among the means of the populations sampled. For instance, we may want to decide on the basis of sample data whether there really is a difference in the effectiveness of three methods of teaching computer programming, or we may want to compare the yield of four varieties of wheat, or we may want to see whether there really is a difference in the average mileage obtained with five kinds of gasoline, or we may want to judge whether there really is a difference in the performance of eight different hair driers, and so on. Suppose that we actually want to compare the effectiveness of three methods of teaching the programming of a certain computer: method 1, which is straight teaching-machine instruction; method 2, which involves the personal attention of an instructor and some direct experience working with the computer; and method 3, which involves the personal attention of an instructor, but no work with the computer itself. Suppose, furthermore, that random samples of size 4 are taken from large groups of students taught by the three methods and that these students obtained the following scores in an appropriate achievement test:

Method 1:   71,   75,   65,   69
Method 2:   90,   80,   86,   84
Method 3:   72,   77,   76,   79

The means of these three samples are $\bar{x}_1 = 70$, $\bar{x}_2 = 85$, and $\bar{x}_3 = 76$, and we would like to know whether the differences among them are significant or whether they can be attributed to chance.

If μ_1, μ_2, and μ_3 are the means of the three populations sampled in this example, we shall want to test the null hypothesis $\mu_1 = \mu_2 = \mu_3$ against the alternative hypothesis that these means are not all equal. It is possible to consider specific special alternatives such as $\mu_1 < \mu_2 < \mu_3$, but we shall not do so here. This null hypothesis would be supported if the differences among the sample means, $\bar{x}_1, \bar{x}_2$, and $\bar{x}_3$, are small; the alternative hypothesis would be supported if at least some of the differences among the sample means are large. Thus, we need a precise measure of the discrepancies among the $\bar{x}$'s and, with it, a rule that tells us when the discrepancies are so large that the null hypothesis can be rejected.

Possible choices for such a measure are the standard deviation of the $\bar{x}$'s or their variance. To determine the latter, we first calculate the mean of the three $\bar{x}$'s, getting

$$\frac{70 + 85 + 76}{3} = 77$$

and then we find that

$$s_{\bar{x}}^2 = \frac{(70 - 77)^2 + (85 - 77)^2 + (76 - 77)^2}{3 - 1} = 57$$

where the subscript $\bar{x}$ serves to indicate that $s_{\bar{x}}^2$ measures the variation of the sample means. Let us now make two assumptions that are critical to the method by which we shall continue the analysis of our problem:

1. The populations we are sampling can be approximated closely with normal distributions.
2. These populations all have the same standard deviation σ.

With reference to our example, this means that we are assuming that (1) the test scores, for each method of teaching, are values of a random variable having (at least approximately) a normal distribution, and that (2) these random variables all have the same standard deviation σ.

With these assumptions, and if the null hypothesis $\mu_1 = \mu_2 = \mu_3$ is true, we can look upon the three samples as if they came from the same (normal) population and, hence, upon the variance of their means, $s_{\bar{x}}^2$, as an estimate of $\sigma_{\bar{x}}^2$, the square of the standard error of the mean.

Now, since $\sigma_{\bar{x}} = \dfrac{\sigma}{\sqrt{n}}$ for samples from infinite populations, we can look upon $s_{\bar{x}}^2$ as an estimate of

$$\sigma_{\bar{x}}^2 = \left(\frac{\sigma}{\sqrt{n}}\right)^2 = \frac{\sigma^2}{n}$$

where n is the size of each sample. Then, multiplying by n, we can look upon $n \cdot s_{\bar{x}}^2$ as an estimate of σ^2, and it is important to note that this estimate is based on the variation among the sample means. For our example, we thus have

$$n \cdot s_{\bar{x}}^2 = 4 \cdot 57 = 228$$

as an estimate of σ^2, the common variance of the three populations.

If σ^2 were known, we could compare $n \cdot s_{\bar{x}}^2$ with σ^2 and reject the null hypothesis that the population averages are all equal if $n \cdot s_{\bar{x}}^2$ is much larger than σ^2. However, in most practical situations σ^2 is not known, and we have no choice but to estimate it on the basis of the sample data.

Since we assumed under the null hypothesis that our three samples come from identical populations, we could use any one of the sample variances s_1^2, s_2^2, or s_3^2 as an estimate of σ^2, and we could also use their mean. Averaging, or pooling, the three sample variances in our example, we get

$$
\begin{aligned}
\frac{s_1^2 + s_2^2 + s_3^2}{3} = \frac{1}{3} & \left(\frac{(71-70)^2 + (75-70)^2 + (65-70)^2 + (69-70)^2}{4-1} \right. \\
& + \frac{(90-85)^2 + (80-85)^2 + (86-85)^2 + (84-85)^2}{4-1} \\
& \left. + \frac{(72-76)^2 + (77-76)^2 + (76-76)^2 + (79-76)^2}{4-1} \right) \\
= & \frac{130}{9} \approx 14.44
\end{aligned}
$$

We now have the following two estimates of σ^2:

$$n \cdot s_{\bar{x}}^2 = 228 \quad \text{and} \quad \frac{s_1^2 + s_2^2 + s_3^2}{3} \approx 14.44$$

It should be observed that, whereas the first estimate measures the variation among the sample means, the second estimate measures the variation within the three samples, that is, chance variation. If the $\bar{x}$'s are far apart and the first of these two estimates is much larger than the second, it thus stands to reason that the null hypothesis ought to be rejected. To put this comparison on a rigorous basis, we use the statistic

STATISTIC FOR THE TEST CONCERNING THE DIFFERENCES AMONG MEANS

$$F = \frac{\text{variation among the samples}}{\text{variation within the samples}}$$

where the variation among the samples is measured by $n \cdot s_{\bar{x}}^2$ and the variation within the samples is measured by the mean of the three sample variances.

If the null hypothesis is true and the assumptions we made are valid, the sampling distribution of this statistic is the **F distribution**, a theoretical distribution that depends on two parameters called the **numerator and denominator degrees of freedom**. When the **F statistic** is used to compare the means of k samples of size n, the numerator and denominator degrees of freedom are, respectively, $k - 1$ and $k(n - 1)$.

In general, if F is close to 1 (that is, if the variation among the samples just about equals the variation within the samples and hence reflects only chance variation), the null hypothesis of equal population means cannot be rejected. However, if F is quite large (that is, if the variation among the samples is greater than we would expect it to be if it were due only to chance), the null hypothesis of equal population means will have to be rejected. To determine how large an F we need to reject the null hypothesis, we base our decision on the criterion of Figure 10.10, where F_α is such that the area under the curve to its right is equal to α. For $\alpha = 0.05$ and $\alpha = 0.01$, the values of F_α may be found in Table V at the end of the book.

Returning to our numerical example, we find that

$$F = \frac{228}{14.44} \approx 15.8$$

and since this exceeds 8.02, the value of $F_{0.01}$ for $k - 1 = 3 - 1 = 2$ and $k(n - 1) = 3(4 - 1) = 9$ degrees of freedom, the null hypothesis must be rejected at the 0.01

Figure 10.10
F distribution.

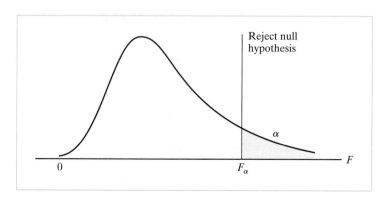

level of significance. In other words, we conclude that the differences among the sample means are too large to be attributed to chance.

Table V can be used to make p-value statements. In our example with $F = 15.8$, we observe that F exceeds $F_{0.01}$ for 2 and 9 degrees of freedom. Thus, $p < 0.01$.

Exercise 10.49 is a practice exercise; its complete solution is given on page 363.

✓ ***10.49** An agronomist planted three test plots each with four varieties of wheat and obtained the following yields (in pounds per plot):

Variety A:	60,	61,	56
Variety B:	59,	52,	51
Variety C:	55,	55,	52
Variety D:	58,	58,	55

(a) Calculate $n \cdot s_{\bar{x}}^2$ for these data, the mean of the variances of the four samples, and the value of F.

(b) Use the 0.05 level of significance to test whether the differences among the four sample means can be attributed to chance.

*10.50 Use Table V to find the value of $F_{0.05}$ for

(a) a set of 5 samples consisting of 4 values each;

(b) a set of 4 samples consisting of 3 values each;

(c) a set of 10 samples consisting of 7 values each.

[*Hint:* $(k - 1)$ provides degrees of freedom for numerator; $k(n - 1)$ provides degrees of freedom for denominator, where k is the number of samples and n is the number of values in each sample.]

*10.51 The following are the mileages that a test driver got with four one-gallon trials for each of five brands of gasoline:

Brand A:	30,	25,	27,	26
Brand B:	29,	26,	29,	28
Brand C:	32,	32,	35,	37
Brand D:	29,	34,	32,	33
Brand E:	32,	26,	31,	27

(a) Calculate $n \cdot s_{\bar{x}}^2$ for these data, the mean of the variances of the five samples, and the value of F.

(b) Use the 0.01 level of significance to test whether the differences among the five sample means can be attributed to chance.

*10.9 ANALYSIS OF VARIANCE

Let us now look at the example of the preceding section from a different point of view, that of an **analysis of variance**. The basic idea of an analysis of variance is to express a measure of the total variation of a set of data as a sum of terms, which can be attributed to specific sources, or causes, of variation. With reference to our example, two such sources of variation would be (1) actual differences in the effectiveness of the three teaching techniques and (2) chance. In problems like this, chance variation is generally referred to as the **experimental error**, which is sometimes described also as statistical *noise*.

One's ability to perform an analysis of variance can be of great importance in scientific work, particularly when many factors produce certain results and we are interested in their individual contributions. For instance, we may want to see whether observed differences in cleansing action are due to the use of different detergents, differences in water temperature, differences in the hardness of the water, differences among the washing machines used in the experiment, and perhaps even differences among the instruments used to obtain the necessary readings. The method by which we analyze a complex situation like this is beyond the scope of this textbook, but it is a direct generalization of the work of this section.

As a measure of the total variation of kn observations consisting of k samples of size n, we shall use the **total sum of squares (SST)**.[†]

$$SST = \sum_{i=1}^{k} \sum_{j=1}^{n} (x_{ij} - \bar{x}_{..})^2$$

where x_{ij} is the jth observation of the ith sample, $i = 1, 2, \ldots, k$, and $j = 1, 2, \ldots, n$. The mean of all the kn measurements or observations is $\bar{x}_{..}$, which is called the **grand mean**. Note that if we divide the total sum of squares by $kn - 1$, we get the variance of the combined data.

Letting $\bar{x}_{i.}$ denote the mean of the ith sample, $i = 1, 2, \ldots, k$, we can write the following identity, which forms the basis of a **one-way analysis of variance**:

$$SST = n \cdot \sum_{i=1}^{k} (\bar{x}_{i.} - \bar{x}_{..})^2 + \sum_{i=1}^{k} \sum_{j=1}^{n} (x_{ij} - \bar{x}_{i.})^2$$

It is customary to refer to the first term on the right-hand side of the equation, which measures the variation among the sample means, as the **treatment sum of squares, SS(Tr)**, and to the second term, which measures the variation within the samples, as the **error sum of squares, SSE**. Use of the word *treatment* is explained by the origin of many analysis-of-variance techniques in agricultural experiments in which different fertilizers, for example, were regarded as different **treatments** applied to the soil. So, we shall refer to the three teaching methods on page 346 as three different treatments, and in other problems we may refer to five nationalities as five different treatments, four different levels of education as four treatments, and so on. The word *error* in *error sum of squares* pertains to the experimental error, or chance.

The identity reads $SST = SS(Tr) + SSE$, and it can be described as a *partition* of SST. The proof of the identity requires a good deal of algebraic manipulation; however, we can easily provide a numerical illustration.

EXAMPLE 10.12 Use the scores of the students taught computer programming by three different methods to verify the identity $SST = SS(Tr) + SSE$.

[†] The use of double subscripts and double summations is explained briefly in Section 3.10.

Solution Substituting the scores, the three sample means 70, 85, and 76, and the grand mean 77 into the expressions for the three sums of squares, we get

$$SST = (71 - 77)^2 + (75 - 77)^2 + (65 - 77)^2 + (69 - 77)^2$$
$$+ (90 - 77)^2 + (80 - 77)^2 + (86 - 77)^2 + (84 - 77)^2$$
$$+ (72 - 77)^2 + (77 - 77)^2 + (76 - 77)^2 + (79 - 77)^2$$
$$= 586$$
$$SS(Tr) = 4[(70 - 77)^2 + (85 - 77)^2 + (76 - 77)^2] = 456$$

and

$$SSE = (71 - 70)^2 + (75 - 70)^2 + (65 - 70)^2 + (69 - 70)^2$$
$$+ (90 - 85)^2 + (80 - 85)^2 + (86 - 85)^2 + (84 - 85)^2$$
$$+ (72 - 76)^2 + (77 - 76)^2 + (76 - 76)^2 + (79 - 76)^2$$
$$= 130$$

Since $586 = 456 + 130$, we have thus shown that $SST = SS(Tr) + SSE$ for the given data. These calculations are difficult in large problems, and a simpler method will be described on page 353. ■

Examining the two terms into which the total sum of squares SST has been partitioned, we note that if we divide $SS(Tr)$ by $k - 1$, we obtain the quantity that we denoted by $n \cdot s_{\bar{x}}^2$ on page 346. Clearly,

$$\frac{SS(Tr)}{k - 1} = \frac{n \cdot \sum_{i=1}^{k} (\bar{x}_{i.} - \bar{x}_{..})^2}{k - 1} = n \cdot \left(\frac{\sum_{i=1}^{k} (\bar{x}_{i.} - \bar{x}_{..})^2}{k - 1} \right) = n \cdot s_{\bar{x}}^2$$

This quantity, which measures the variation among the samples, is called the **treatment mean square**, and it is denoted by **MS(Tr)**. That is,

$$MS(Tr) = \frac{SS(Tr)}{k - 1}$$

Similarly, if we divide SSE by $k(n - 1)$, we obtain the mean of the k sample variances, for we can write

$$\frac{SSE}{k(n - 1)} = \frac{\sum_{i=1}^{k} \sum_{j=1}^{n} (x_{ij} - \bar{x}_{i.})^2}{k(n - 1)} = \frac{1}{k} \cdot \sum_{i=1}^{k} \left(\frac{\sum_{j=1}^{n} (x_{ij} - \bar{x}_{i.})^2}{n - 1} \right)$$

which equals

$$\frac{1}{k} \cdot (s_1^2 + s_2^2 + \cdots + s_k^2)$$

the pooled variance from page 347. This quantity, which measures the variation within the samples, is called the **error mean square**, and it is denoted by **MSE**. Thus,

$$MSE = \frac{SSE}{k(n-1)}$$

Since F was defined previously as the ratio of these two measures of the variation among and within the samples, we can now write

STATISTIC FOR A
TEST CONCERNING
THE DIFFERENCES
AMONG MEANS

$$F = \frac{MS(Tr)}{MSE}$$

In practice, we display the work required for the determination of F in the following kind of table, called an **analysis-of-variance table**:

Source of Variation	Degrees of Freedom	Sum of Squares	Mean Square	F
Treatments	$k-1$	SS(Tr)	$MS(Tr) = \dfrac{SS(Tr)}{k-1}$	$\dfrac{MS(Tr)}{MSE}$
Error	$k(n-1)$	SSE	$MSE = \dfrac{SSE}{k(n-1)}$	
Total	$kn-1$	SST		

The degrees of freedom for treatments and error are the numerator and denominator degrees of freedom referred to on page 348. Note that they are also the quantities we divide into the sums of squares to obtain the corresponding mean squares. Observe that there is only one entry in the F column. By custom, we do not write a number in the Total row under Mean Square.

If we make the same assumptions as in Section 10.8, the significance test is as we described it on page 348: We reject the null hypothesis $\mu_1 = \mu_2 == \mu_k$ against the alternative that these μ's are not all equal, if the value we get for F equals or exceeds F_α for $k-1$ and $k(n-1)$ degrees of freedom.

EXAMPLE 10.13 Use the sums of squares from page 351 to construct an analysis-of-variance table for our numerical example, and test the null hypothesis that the three methods of teaching computer programming are equally effective against the alternative that they are not all equally effective at the 0.01 level of significance.

Solution   Copying the values of the sums of squares from page 351, we get

$$\text{MS(Tr)} = \frac{456}{2} = 228, \quad \text{MSE} = \frac{130}{9} \approx 14.44, \quad F = \frac{228}{14.44} \approx 15.8$$

and hence

Source of Variation	Degrees of Freedom	Sum of Squares	Mean Square	F
Treatments	2	456	228	15.8
Error	9	130	14.44	
Total	11	586		

Since $F = 15.8$ exceeds 8.02, the value of $F_{0.01}$ for $3 - 1 = 2$ and $3(4 - 1) = 9$ degrees of freedom obtained from Table V, the null hypothesis must be rejected; in other words, we conclude that the three methods of teaching computer programming are not all equally effective. ■

The numbers that we used in our illustration were intentionally chosen so that the calculations would be easy. In actual practice, the calculation of the sums of squares can be quite tedious unless we use the following computing formulas, in which $T_{i.}$ denotes the sum of the values in the ith sample, and $T_{..}$ denotes the grand total of all the data:

COMPUTING
FORMULAS FOR
SUMS OF SQUARES
(SAMPLE SIZES
EQUAL)

$$\text{SST} = \sum_{i=1}^{k} \sum_{j=1}^{n} x_{ij}^2 - \frac{1}{kn} \cdot T_{..}^2$$

$$\text{SS(Tr)} = \frac{1}{n} \cdot \sum_{i=1}^{k} T_{i.}^2 - \frac{1}{kn} \cdot T_{..}^2$$

and by subtraction $\text{SSE} = \text{SST} - \text{SS(Tr)}$

EXAMPLE 10.14   Use these computing formulas to verify the sums of squares obtained on page 351.

Solution   Substituting $k = 3$, $n = 4$, $T_{1.} = 280$, $T_{2.} = 340$, $T_{3.} = 304$, $T_{..} = 924$, and $\sum\sum x_{ij}^2 = 71{,}734$ into the computing formulas for the three sums of squares, we get

$$\text{SST} = 71{,}734 - \frac{1}{12}(924)^2 = 71{,}734 - 71{,}148 = 586$$

$$\text{SS(Tr)} = \frac{1}{4}(280^2 + 340^2 + 304^2) - \frac{1}{12}(924)^2$$

$$= 71{,}604 - 71{,}148 = 456$$

Figure 10.11
SAS printout for analysis of variance.

```
data a;
input meth $ score @@;
cards;
1 71 1 75 1 65 1 69
2 90 2 80 2 86 2 84
3 72 3 77 3 76 3 79
;
proc anova;
class meth;
model score=meth;




                SAS
 ANALYSIS OF VARIANCE PROCEDURE

  CLASS LEVEL INFORMATION
 CLASS     LEVELS     VALUES
 METH         3        1 2 3


 NUMBER OF OBSERVATIONS IN DATA SET = 12

                    SAS
         ANALYSIS OF VARIANCE PROCEDURE
 DEPENDENT VARIABLE: SCORE

 SOURCE           DF    SUM OF SQUARES    MEAN SQUARE    F VALUE    PR > F
 MODEL             2       456.00000000   228.00000000     15.78    0.0011
 ERROR             9       130.00000000    14.44444444
 CORRECTED TOTAL  11       586.00000000

    R-SQUARE      C.V.
    0.778157      4.9358

    ROOT MSE        SCORE MEAN
    3.80058475      77.00000000
```

and

$$SSE = 586 - 456 = 130$$

These values are identical to those obtained on page 351.

A computer printout generated by the SAS program for our analysis-of-variance example is shown in Figure 10.11, where the values corresponding to SST = 586, SS(Tr) = 456, and SSE = 130 are given in the column headed SUM OF SQUARES. Besides the degrees of freedom, the sums of squares, the mean squares, the value of F, and the p-value 0.0011, it provides other information that we shall not consider here.

Figure 10.12
Analysis of variance reproduced from the display screen of a TI-83 graphing calculator.

Analysis-of-variance problems like the one we have discussed can also be handled quite readily by means of graphing calculators. Upon entering the numerical data that we used to obtain the analysis-of-variance table of Figure 10.4, we get the results shown in Figure 10.12. As we can observe, the graphing calculator confirms the computer printout and also our manual calculations. Graphing calculator information shown here and elsewhere in this book is entirely optional and can be omitted by the reader without the loss of continuity.

Directions for obtaining the results shown in Figure 10.12 with a TI-83 graphing calculator are given in the Using Technology box on page 356.

The method we have discussed here applies only when the sample sizes are all equal, but minor modifications make it applicable also when the sample sizes are not all equal. If the ith sample is of size n_i, the computing formulas for the sums of squares become

$$\text{SST} = \sum_{i=1}^{k} \sum_{j=1}^{n_i} x_{ij}^2 - \frac{1}{N} \cdot T_{..}^2$$

$$\text{SS(Tr)} = \sum_{i=1}^{k} \frac{T_{i.}^2}{n_i} - \frac{1}{N} \cdot T_{..}^2$$

$$\text{SSE} = \text{SST} - \text{SS(Tr)}$$

where $N = n_1 + n_2 + \cdots + n_k$. The only other change is that the total number of degrees of freedom is $N - 1$, and the degrees of freedom for treatments and error are, respectively, $k - 1$ and $N - k$.

How to obtain the analysis of variance shown in Figure 10.12.

(TIP: Before beginning, see Appendix A for clearing lists and entering data.)

STEP 1. Clear list L1, L2, and L3.

STEP 2. Input the student computer programming scores from the running example into list L1, L2, and L3.

$$L1 = (71, 75, 65, 69)$$
$$L2 = (90, 80, 86, 84)$$
$$L3 = (72, 77, 76, 79)$$

STEP 3. Select the type of statistical test you wish to use.
Press STAT.
Highlight TESTS.
Scroll down to the 2nd page of the menu.
Select F: ANOVA (, press ENTER.
Enter the lists in which the data is stored by pressing 2nd 1 for L1, 2nd 2 for L2, and 2nd 3 for L3 (Use the black, key; and use the) key to Close the expression.)

STEP 4. Solve.
Press ENTER.
Use the ▼ button to scroll down through the output.

EXAMPLE 10.15 A laboratory technician wants to compare the breaking strength of three kinds of thread, and originally he planned to repeat each determination six times. Not having enough time, however, he has to base his analysis on the following results (in ounces):

Thread 1:	18.0,	16.4,	15.7,	19.6,	16.5,	18.2
Thread 2:	21.1,	17.8,	18.6,	20.8,	17.9,	19.0
Thread 3:	16.5,	17.8,	16.1			

Perform an analysis of variance to test at the 0.05 level of significance whether the differences among the sample means are significant.

Solution

1. *Hypotheses*

$$H_0: \quad \mu_1 = \mu_2 = \mu_3$$

$$H_A: \quad \text{the} \mu\text{'s are not all equal}$$

where μ_1, μ_2, and μ_3 are the means of the three populations.

2. *Level of significance* $\alpha = 0.05$
3. *Criterion*   Reject the null hypothesis if $F \geq 3.89$, the value of $F_{0.05}$ for $k - 1 = 3 - 1 = 2$ and $N - k = 15 - 3 = 12$ degrees of freedom, where F is to be determined by an analysis of variance. Otherwise, accept H_0 or reserve judgment.
4. *Calculations*

$$T_{1.} = 104.4, \ T_{2.} = 115.2, \ T_{3.} = 50.4, \ T_{..} = 270.0,$$

and

$$\sum \sum x_{ij}^2 = 4{,}897.46$$

Then, substituting these totals together with $n_1 = 6, n_2 = 6, n_3 = 3$, and $N = 15$ into the computing formulas for the sums of squares, we get

$$\text{SST} = 4{,}897.46 - \frac{1}{15}(270.0)^2 = 4{,}897.46 - 4{,}860 = 37.46$$

$$\text{SS(Tr)} = \frac{104.4^2}{6} + \frac{115.2^2}{6} + \frac{50.4^2}{3} - \frac{1}{15}(270.0)^2$$
$$= 4{,}875.12 - 4{,}860 \approx 15.12$$

and

$$\text{SSE} = 37.46 - 15.12 = 22.34$$

Then

$$\text{MS(Tr)} = \frac{15.12}{2} = 7.56, \quad \text{MSE} = \frac{22.34}{12} \approx 1.86,$$

and

$$F = \frac{7.56}{1.86} \approx 4.06$$

All these results are shown in the following analysis-of-variance table:

Source of Variation	Degrees of Freedom	Sum of Squares	Mean Square	F
Treatments	2	15.12	7.56	4.06
Error	12	22.34	1.86	
Total	14	37.46		

5. *Decision*   Since $F = 4.06$ exceeds 3.89, the null hypothesis must be rejected; in other words, we conclude that there is a difference in the strength of the three kinds of thread.

This example could also utilize the *p*-value. The final three steps would be these:

3'. *Test statistic*   The statistic is

$$F = \frac{MS(Tr)}{MSE}$$

4'. *Calculations*   Find

$$F = \frac{7.56}{1.86} \approx 4.06$$

5'. *Decision*   Since $F = 4.06$ falls between 3.89 and 6.93, the values of $F_{0.05}$ and $F_{0.01}$ for 2 and 12 degrees of freedom, we find that $0.01 < p < 0.05$. Since the level of significance was specified as $\alpha = 0.05$, we conclude, as before, that the null hypothesis must be rejected. Had the level of significance not been specified, we might simply have stated that $0.01 < p < 0.05$ or written the result as $F = 4.06*$. ■

To get the actual size of the *p*-value for the preceding example, we would have to use a computer package or a special calculator. Using a TI-83 graphing calculator and the instructions in the Using Technology box on p. 356 for the ANOVA, notice that the *p*-value is part of the output. You can also use the T1-83 to get the *p*-value and, at the same time, graph the corresponding area under the curve. Entering the requisite information into the calculator, we obtain the result reproduced in Figure 10.13. As can be seen, the *p*-value is .045011, or .045 rounded to three decimals. This agrees with the previous comment that $0.01 < p < 0.05$. Graphical calculator information shown here and elsewhere in this book is optional and can be omitted by the reader without the loss of continuity.

Figure 10.13
F distribution reproduced from the display screen of a TI-83 graphing calculator.

Directions for obtaining the graph shown in Figure 10.13 with a TI-83 graphing calculator are given in the Using Technology box on page 359.

How to obtain the actual size of the *p*-value and graph the area under the curve shown in Figure 10.13

(**TIP**: Before beginning: Reset the memory to default settings and clear all plots—see Appendix A: TI-83 Tips.)

STEP 1. Clear all previous drawings.
Press 2nd PRGM to access the DRAW menu.
Select 1: ClrDraw and press ENTER.

STEP 2. Set the viewing window.

```
Press WINDOW
Set   Xmin = 0
      Xmax = 6
      Xscl = 1
      Ymin = -.25
      Ymax = .9
      Yscl = 1
      Xres = 1
```

STEP 3. Select the type of display and enter the interval to be displayed.
Press 2nd, VARS to access the DISTR DRAW menu.
Highlight DRAW.
Select 4: ShadeF(, press ENTER.
Enter 4.06, 1E99, 2, 12)
(For instructions on entering "1E99," for infinity, see Appendix A: TI-83 Tips; 4.6 and 1E99 (for ∞) are the lower and upper limits of the interval; 2 and 12 are the degrees of freedom.)

STEP 4. View the graph. Press ENTER to solve.

USING TECHNOLOGY: The Graphing Calculator

Exercises 10.52 and 10.54 are practice exercises; their complete solutions are given on page 364.

✓ ***10.52** To study the effectiveness of five different kinds of packaging, a processor of breakfast foods obtained the following data on the numbers of sales on five different days:

Packaging I:	60,	52,	56,	52,	65
Packaging II:	54,	64,	66,	54,	57
Packaging III:	55,	66,	68,	57,	55
Packaging IV:	55,	56,	70,	58,	56
Packaging V:	71,	65,	60,	59,	62

Perform an analysis of variance to test at the 0.05 level of significance whether the differences among the five sample means can be attributed to chance.

EXERCISES

*10.53 The following are eight consecutive weeks' earnings (in dollars) of three door-to-door cosmetics salespersons employed by a firm:

Salesperson A: 309,   293,   284,   300,   306,   288,   312,   276
Salesperson B: 295,   280,   299,   310,   298,   284,   293,   287
Salesperson C: 311,   289,   296,   323,   287,   280,   303,   264

Perform an analysis of variance to test at the 0.01 level of significance whether the differences among the average weekly earnings of the three salespersons are significant. Provide also an inequality that describes the p-value.

✓ ***10.54** The following are the weight losses, in milligrams, of certain machine parts due to friction, when they were used with three different lubricants:

Lubricant X: 12,   11,   7,   13,   9,   11,   12,   9
Lubricant Y: 8,   10,   7,   5,   6,   10,   7,   8,   11,   7,   8
Lubricant Z: 9,   3,   7,   8,   4,   6,   6,   5

Perform an analysis of variance to test at the 0.01 level of significance whether the differences among the three sample means can be attributed to chance.

*10.55 The following are the numbers of words per minute that a secretary typed on several occasions on four different computer keyboards:

Keyboard C: 71,   75,   69,   77,   61,   72,   71,   78
Keyboard D: 68,   71,   74,   66,   69,   67,   70,   62
Keyboard E: 75,   70,   81,   73,   78,   72
Keyboard F: 62,   59,   71,   68,   63,   65,   72,   60,   64

Perform an analysis of variance to test at the 0.01 level of significance whether the differences among the four sample means can be attributed to chance. Give also an inequality that describes the p-value.

*10.56 To study the performance of a newly designed motorboat, it was timed over a marked course under various wind and water conditions, and the following data (in minutes) was obtained:

Calm conditions: 25,   18,   15,   21
Moderate conditions: 24,   27,   24,   19,   17,   22
Choppy conditions: 22,   24,   27,   25,   30

Perform an analysis of variance to test at the 0.05 level of significance whether the differences among the three sample means are significant.

*10.57 Use a computer package or a graphing calculator to rework Exercise 10.52.

*10.58 Use a computer package or a graphing calculator to rework Exercise 10.53.

*10.59 Use a computer package or a graphing calculator to rework Exercise 10.54.

*10.60 Use a computer package or a graphing calculator to rework Exercise 10.55.

✓ SOLUTIONS TO PRACTICE EXERCISES

10.1 We take as the null hypothesis the statement that the bottle of the antibiotic possesses full potency, since this is the usual state of affairs.

(a) Erroneously rejecting a true null hypothesis is a Type I error.

(b) Erroneously accepting a false null hypothesis is a Type II error.

10.6 (a) Since $\sigma_{\bar{x}} = \dfrac{0.8}{\sqrt{60}} \approx 0.103$, the dividing lines of the criterion in standard units are

$$z = \frac{13.2 - 13.4}{0.103} \approx -1.94 \quad \text{and} \quad z = \frac{13.6 - 13.4}{0.103} \approx 1.94$$

and the probability of a Type I error is

$$2(0.5000 - 0.4738) = 0.0524 \approx 0.05$$

(b) The dividing lines of the criterion in standard units are

$$z = \frac{13.2 - 13.7}{0.103} \approx -4.85 \quad \text{and} \quad z = \frac{13.6 - 13.7}{0.103} \approx -0.97$$

and the probability of a Type II error is

$$0.5000 - 0.3340 = 0.1660 \approx 0.17$$

10.10 (a) The null hypothesis $\mu = \mu_0$, where μ is the actual number of mistakes averaged by the salesperson and μ_0 is the number of mistakes averaged by all the salespersons; the alternative hypothesis is $\mu > \mu_0$; the salesperson will be fired if the null hypothesis can be rejected.

(b) The null hypothesis $\mu = \mu_0$, and the alternative hypothesis $\mu < \mu_0$; the salesperson will be fired unless the null hypothesis can be rejected.

10.11 Here μ denotes the true average time that convicted embezzlers spend in jail.

(a) 1. *Hypotheses*

$$H_0: \quad \mu = 12.8$$

$$H_A: \quad \mu \neq 12.8$$

The alternative is two-sided because there is no indication of a one-sided objective.

2. *Level of significance* $\alpha = 0.01$

3. *Criterion*   Reject the null hypothesis if $z \leq -2.575$ or $z \geq 2.575$, where

$$z = \frac{\bar{x} - \mu_0}{\sigma/\sqrt{n}}$$

with s substituted for σ. Otherwise, accept the null hypothesis or reserve judgment.

4. *Calculations*

$$z = \frac{11.2 - 12.8}{3.5/\sqrt{60}} \approx -3.54$$

5. *Decision*   Since $z = -3.54$ is less than -2.575, the null hypothesis must be rejected; in other words, convicted embezzlers do not spend on the average 12.8 months in jail.

(b) Note that the value $z = -3.54$ is off the main part of Table II. We can use the figures at the bottom of the table to determine that the p-value is between $2(0.5000 - 0.4990) = 0.002$ and $2(0.5000 - 0.49997) = 0.00006$.

(c) The p-value is consistent with rejecting the null hypothesis with $\alpha = 0.01$.

10.21 The sample size $n = 12$ is small, and we must begin by stating that the work rests on the assumption that the pulse reduction values are sampled from a population that has roughly the shape of a normal distribution.

1. *Hypotheses*

$$H_0 : \mu = 9.0$$

$$H_A : \mu < 9.0$$

2. *Level of significance* $\alpha = 0.05$

3. *Criterion*   Reject the null hypothesis if $t \leq -1.796$, which is the value of $-t_{0.05}$ for $12 - 1 = 11$ degrees of freedom, where

$$t = \frac{\bar{x} - \mu_0}{s/\sqrt{n}}$$

4. *Calculations*

$$t = \frac{7.2 - 9.0}{1.8/\sqrt{12}} \approx -3.46$$

5. *Decision*   Since $t = -3.46$ is less than -1.796, the null hypothesis must be rejected; in other words, the tranquilizer reduces the pulse rate on the average by less than 9.0 beats.

10.35 1. *Hypotheses*

$$H_0 : \mu_1 = \mu_2$$

$$H_A : \mu_1 \neq \mu_2$$

2. *Level of significance* $\alpha = 0.01$

3. *Criterion*   Reject the null hypothesis if $z \leq -2.575$ or $z \geq 2.575$, where

$$z = \frac{\bar{x}_1 - \bar{x}_2}{\sqrt{\dfrac{\sigma_1^2}{n_1} + \dfrac{\sigma_2^2}{n_2}}}$$

with s_1 and s_2 substituted for σ_1 and σ_2. Otherwise, state that the difference between the two sample means is not significant.

4. *Calculations*

$$z = \frac{12.8 - 9.2}{\sqrt{\dfrac{5.4^2}{60} + \dfrac{6.3^2}{50}}} \approx 3.18$$

5. *Decision*   Since $z = 3.18$ exceeds 2.575, the null hypothesis must be rejected; in other words, the two kinds of executives do not average equally many business lunches.

10.43 1. *Hypotheses*

$$H_0 : \mu_1 = \mu_2$$

$$H_A : \mu_1 \neq \mu_2$$

2. *Level of significance* $\alpha = 0.05$

3. *Criterion*   Reject the null hypothesis if $t \leq -2.306$ or $t \geq 2.306$, where 2.306 is the value of $t_{0.025}$ for $5 + 5 - 2 = 8$ degrees of freedom and t is given by the formula on page 339.

4. *Calculations*

$$s_p = \sqrt{\frac{4(252)^2 + 4(207)^2}{8}} \approx 230.6 \quad \text{and} \quad t = \frac{8,106 - 7,730}{230.6\sqrt{\dfrac{1}{5} + \dfrac{1}{5}}} \approx 2.58$$

5. *Decision*   Since $t = 2.58$ exceeds 2.306, the null hypothesis must be rejected; in other words, the average heat-producing capacity of coal from the two mines is not the same.

The p-value can also be used and the results are consistent with rejecting the null hypothesis with $\alpha = 0.05$.

10.45 The differences are 13, 7, −1, 5, 3, 2, −1, 0, 8, 10, 16, −2, 7, 3, −5, 2, 6, 1, 4, 3, 2, 6, 12, 4, −1, 5, −6, −3, 8, 4, −1, and 0; their mean is $\bar{x} \approx 3.47$, their standard deviation is $s \approx 5.07$, and for these data we perform the following test:

1. *Hypotheses*

$$H_0: \quad \mu = 0$$

$$H_A: \quad \mu > 0$$

where μ is the mean of the population of differences.

2. *Level of significance* $\alpha = 0.01$

3. *Criterion*   Reject the null hypothesis if $z \geq 2.33$, where

$$z = \frac{\bar{x} - \mu_0}{\sigma/\sqrt{n}}$$

using $\mu_0 = 0$ as the comparison value and with s substituted for σ; otherwise, we accept the null hypothesis or reserve judgment.

4. *Calculations*

$$z \approx \frac{3.47 - 0}{5.07/\sqrt{32}} \approx 3.87$$

5. *Decision*   Since $z \approx 3.87$ exceeds 2.33, the null hypothesis must be rejected; in other words, the prescribed program of exercise is effective.

10.49 (a) The four sample means are $\bar{x}_1 = 59, \bar{x}_2 = 54, \bar{x}_3 = 54$, and $\bar{x}_4 = 57$; their mean is

$$\frac{59 + 54 + 54 + 57}{4} = 56$$

so that

$$s_{\bar{x}}^2 = \frac{(59 - 56)^2 + (54 - 56)^2 + (54 - 56)^2 + (57 - 56)^2}{4 - 1} = 6$$

and $n \cdot s_{\bar{x}}^2 = 3 \cdot 6 = 18$; also,

$$s_1^2 = \frac{(60 - 59)^2 + (61 - 59)^2 + (56 - 59)^2}{3 - 1} = 7$$

and, similarly, $s_2^2 = 19, s_3^2 = 3$, and $s_4^2 = 3$; so that

$$\frac{1}{4}(s_1^2 + s_2^2 + s_3^2 + s_4^2) = \frac{7 + 19 + 3 + 3}{4} = 8$$

and $F = \frac{18}{8} = 2.25$.

(b) Since $F = 2.25$ does not exceed 4.07, the value of $F_{0.05}$ for $k - 1 = 4 - 1 = 3$ and $k(n - 1) = 4(3 - 1) = 8$ degrees of freedom, the null hypothesis cannot be rejected; in other words, the differences among the four sample means are not significant.

10.52 1. *Hypotheses*

$$H_0: \quad \mu_1 = \mu_2 = \mu_3 = \mu_4 = \mu_5$$

$$H_A: \quad \text{the } \mu\text{'s are not all equal}$$

2. *Level of significance* $\alpha = 0.05$
3. *Criterion* Reject the null hypothesis if $F \geq 2.87$, the value of $F_{0.05}$ for $5 - 1 = 4$ and $5(5 - 1) = 20$ degrees of freedom; otherwise accept H_0 or reserve judgment.
4. *Calculations* Since $T_1 = 285$, $T_2 = 295$, $T_3 = 301$, $T_4 = 295$, $T_5 = 317$, $T = 1{,}493$, and $\sum\sum x^2 = 89{,}933$, we get

$$\text{SST} = 89{,}933 - \frac{(1{,}493)^2}{25} = 89{,}933 - 89{,}161.96 = 771.04$$

$$\text{SS(Tr)} = \frac{285^2 + 295^2 + 301^2 + 295^2 + 317^2}{5} - 89{,}161.96$$

$$= 89{,}273 - 89{,}161.96 = 111.04$$

$$\text{SSE} = 771.04 - 111.04 = 660$$

The remainder of the arithmetic can be displayed easily in an analysis-of-variance table:

Source of Variation	Degrees of Freedom	Sum of Squares	Mean Square	F
Treatments	4	111.04	27.76	0.84
Error	20	660.00	33	
Total	24	771.04		

5. *Decision* Since $F = 0.84$ does not exceed 2.87, the null hypothesis cannot be rejected.

*10.54 1. *Hypotheses*

$$H_0 : \mu_1 = \mu_2 = \mu_3$$

$$H_A : \text{the } \mu\text{'s are not all equal}$$

2. *Level of significance* $\alpha = 0.01$
3. *Criterion* Reject the null hypothesis if $F \geq 5.61$, the value of $F_{0.01}$ for $k - 1 = 3 - 1 = 2$ and $N - k = 27 - 3 = 24$ degrees of freedom, where F is to be determined by an analysis of variance.
4. *Calculations* $T_{1.} = 84$, $T_{2.} = 87$, $T_{3.} = 48$, $T_{..} = 219$, and $\sum\sum x_{ij}^2 = 1{,}947$. Substituting these values together with $n_1 = 8$, $n_2 = 11$, $n_3 = 8$, and $N = 27$ into the computing formulas for the sums of squares, we get

$$\text{SST} = 1{,}947 - \frac{1}{27}(219)^2 \approx 1{,}947 - 1{,}776.33 = 170.67$$

$$\text{SS(Tr)} = \frac{84^2}{8} + \frac{87^2}{11} + \frac{48^2}{8} - \frac{1}{27}(219)^2$$

$$\approx 1{,}858.09 - 1{,}776.33 = 81.76$$

$$\text{SSE} \approx 170.67 - 81.76 = 88.91$$

The remainder of the work is shown in the following analysis-of-variance table:

Source of Variation	Degrees of Freedom	Sum of Squares	Mean Square	F
Treatments	2	81.76	40.88	11.05
Error	24	88.91	3.70	
Total	26	170.67		

5. *Decision*   Since $F = 11.05$ exceeds 5.61, the null hypothesis must be rejected; in other words, we conclude that there is a difference in the effectiveness of the lubricants in reducing the weight loss of the machine parts. The p-value approach to this exercise requires that we compare the test statistic $F = 11.05$ to the table points $F = 3.40$ and $F = 5.61$. Since $F = 11.05$, we note that $p < 0.01$. We therefore reject the null hypothesis.

11

TESTS BASED ON COUNT DATA

For the most part, Chapters 9 and 10 dealt with inferences based on measurements. We used measurements to estimate the means of populations and their standard deviations, and we also used them to test hypotheses about population means. The only exception was in Section 9.5, where we used sample proportions for the estimation of population proportions, percentages, or probabilities. Such data is referred to as **count data** or **enumeration data**, since it is obtained by performing counts rather than measurements.

In this chapter we shall use count data in tests of hypotheses. For instance, we may want to test the claim that 24% of all recreational skydivers counted at a certain drop zone are women.

In Sections 11.1 and 11.2, we shall study tests about population proportions, which also serve to test hypotheses about percentages and probabilities. As we pointed out on page 27, a percentage is just a proportion multiplied by 100, and a probability may be interpreted as a proportion in the long run. In Sections 11.3 and 11.4, we shall study tests concerning two or more population proportions, and in Section 11.5 we shall learn how to analyze data tallied into a **two-way classification**. This kind of problem might arise, for example, when cars are rated on their fuel economy as being low, average, or high, and also on their ride being very uncomfortable, somewhat uncomfortable, fairly comfortable, or very comfortable. Finally, Section 11.6 deals with the comparison of observed and expected distributions, that is, the comparison of counts we actually got with those which we might have expected according to theory or assumptions.

11.1 TESTS CONCERNING PROPORTIONS

In this section we shall be concerned with tests of hypotheses that enable us to decide, on the basis of sample data, whether the true value of a proportion, percentage, or probability equals a given constant. These tests make it possible, for example, to determine whether the true proportion of tenth graders who can name the two senators of their state is 0.30, whether it is true that 10% of the answers that the IRS gives to taxpayers' telephone inquiries are in error, or whether the true probability is 0.25 that a flight from Seattle to San Francisco will be late.

Questions of this kind are usually decided on the basis of the observed number of successes in n trials, or the observed proportion of successes, and it will be assumed throughout this section that these trials are independent and that the probability of a success is the same for each trial. In other words, we shall assume that we can use the binomial distribution and that we are, in fact, testing hypotheses about the parameter p of binomial populations.

When n is small, tests concerning true proportions can be based directly on tables of binomial probabilities, as is illustrated by the following examples.

EXAMPLE 11.1   It has been claimed that at least 40% of all seniors at a large university prefer to live off campus. If 4 of 14 seniors interviewed at random prefer to live off campus, test the claim at the 0.05 level of significance. Use an appropriate one-sided criterion.

Solution   1. *Hypotheses*

$$H_0: \quad p = 0.40$$

$$H_A: \quad p < 0.40$$

2. *Level of significance* $\alpha = 0.05$
3'. *Test statistic*   The test statistic is simply the number of successes, here four.
4'. *Calculations*   Table I shows that for $n = 14$ and $p = 0.40$ the probability of four or fewer successes is

$$0.001 + 0.007 + 0.032 + 0.085 + 0.155 = 0.280$$

and this is the sum of the areas of the blue rectangles of Figure 11.1. The p-value is thus 0.280.

5'. *Decision*   Since 0.280 is greater than $\alpha = 0.05$, the null hypothesis cannot be rejected. We accept the claim or reserve judgment.   ∎

Here we used the alternative procedure based on p-values, introduced first on page 328. We did this because, for tests based on binomial probabilities, this can greatly simplify the work, especially when a test is two tailed, as in the example that follows.

Figure 11.1
Binomial distribution
with $p = 0.40$ and
$n = 14$.

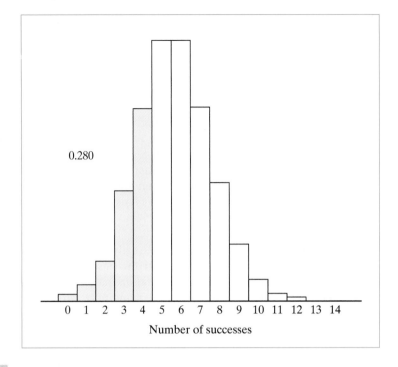

0.280

Number of successes

EXAMPLE 11.2   It has been claimed that 60% of all shoppers can identify a highly advertised trademark. At the 0.05 level of significance, can we reject this claim if, in a random sample of 12 shoppers, only 3 were able to identify the trademark?

Solution   1. *Hypotheses*

$$H_0: \quad p = 0.60$$

$$H_A: \quad p \neq 0.60$$

2. *Level of significance* $\alpha = 0.05$
3′. *Test statistic*   The test statistic is simply the number of successes, here three.
4′. *Calculations*   Table I shows that for $n = 12$ and $p = 0.60$ the probability of three or fewer successes is

$$0.002 + 0.012 = 0.014$$

so that the *p*-value is $2(0.014) = 0.028$.
5′. *Decision*   Since 0.028 is less than 0.05, the null hypothesis must be rejected. We reject the claim that the true percentage of shoppers who can identify the trademark is 60%.

11.2   TESTS CONCERNING PROPORTIONS (LARGE SAMPLES)

For large n, tests concerning proportions are usually based on the normal-curve approximation to the binomial distribution. We base tests of the null hypothesis $p = p_0$ on the value we obtain for

STATISTIC FOR A
LARGE-SAMPLE TEST
CONCERNING
PROPORTION

$$z = \frac{x - np_0}{\sqrt{np_0(1 - p_0)}}$$

which has approximately the standard normal distribution.

To make the same continuity correction as on page 253, some statisticians substitute $x - \frac{1}{2}$ or $x + \frac{1}{2}$ for x into the formula for z, whichever makes z closer to 0. Since the effect of this correction is generally negligible when n is large, we shall not use it here. However, we shall use it later, in Chapter 13, in connection with some small samples. Note that the continuity correction does not even have to be considered, when without it the null hypothesis cannot be rejected.

The test criteria are again those of Figure 10.5 with p and p_0 substituted for μ and μ_0. For the one-sided alternative hypothesis $p < p_0$, we reject the null hypothesis if $z \leq -z_\alpha$, for the one-sided alternative hypothesis $p > p_0$, we reject the null hypothesis if $z \geq z_a$, and for the two-sided alternative hypothesis $p \neq p_0$, we reject the null hypothesis if $z \leq -z_{\alpha/2}$ or $z \geq z_{\alpha/2}$.

EXAMPLE 11.3   Suppose that a nutritionist claims that at least 75% of the preschool children in a certain country have protein-deficient diets, and that a sample survey reveals that this is true for 206 preschool children in a sample of 300. Test the claim at the 0.05 level of significance.

Solution   1. *Hypotheses*

$$H_0: \quad p = 0.75$$

$$H_A: \quad p < 0.75$$

2. *Level of significance* $\alpha = 0.05$
3. *Criterion*   Reject the null hypothesis if $z \leq -1.645$, where

$$z = \frac{x - np_0}{\sqrt{np_0(1 - p_0)}}$$

(see also Figure 11.2); otherwise, accept it or reserve judgment.

Figure 11.2
Test criterion for protein-deficiency example.

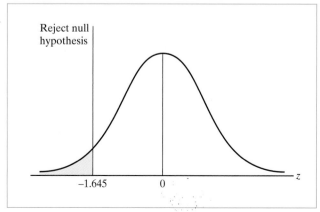

Reject null
hypothesis

−1.645   0

z

4. *Calculations*   Substituting $x = 206$, $n = 300$, and $p_0 = 0.75$ into the formula for z, we get

$$z = \frac{206 - 300(0.75)}{\sqrt{300(0.75)(0.25)}} \approx -2.53$$

5. *Decision*   Since $z = -2.53$ is less than -1.645, the null hypothesis must be rejected; in other words, we conclude that fewer than 75% of the preschool children in the given country have protein-deficient diets.

Had we wanted to use the *p*-value approach in this example, we would have had to determine the area under the curve to the left of $z = -2.53$. Since this area is $0.5000 - 0.4943 = 0.0057$, which is less than $\alpha = 0.05$, we would have concluded (as before) that the null hypothesis must be rejected.   ■

EXERCISES

Exercises 11.1 and 11.8 are practice exercises; their complete solutions are given on pages 390 and 391.

✓ **11.1**   A physicist claims that at most 10% of all persons exposed to a certain amount of radiation will feel any effects. If, in a random sample, 4 of 13 persons feel an effect, test the claim at the 0.05 level of significance.

11.2   Using Table I, Binomial Probabilities, where $n = 14$ and $p = 0.40$, what is the *p*-value of
   (a)   four or fewer successes;
   (b)   four or more successes;
   (c)   more than four successes;
   (d)   fewer than four successes;
   (e)   exactly four successes;
   (f)   no successes?
   [*Hint:* The solution of part (a) is $0.001 + 0.007 + 0.032 + 0.085 + 0.155 = 0.2800$.]

11.3   A manufacturer of athletic shoes wants to test the claim that the probability is 0.60 that a walker will prefer a new type of walking shoe to the old. If, in a random sample, 5 out of 15 walkers prefer the new type of walking shoe to the old, test the null hypothesis $p = 0.60$ against the alternative hypothesis $p < 0.60$ at the 0.05 level of significance.

11.4   A football enthusiast claims that at least 50% of all televised football viewers think that there are too many commercials shown during televised football games. If, in a random sample of 13 viewers of televised football games, 7 complain about the number of commercials, what can we conclude about the enthusiast's claim at the 0.05 level of significance?

11.5   A certain automobile manufacturer wants to test the claim that 40% of new car buyers dislike the floor mats provided in the manufacturer's cars. If 3 of 10 randomly selected new car buyers dislike the floor mats, does this support the claim that 40% of the new car buyers dislike the mats? Use the alternative hypothesis $p \neq 0.40$ and $\alpha = 0.01$.

11.6   Suppose that the football enthusiast of Exercise 11.4 *incorrectly* asserted the sample results. The *correct* sample result was 3 complaints out of a sample of 13 viewers. Rework the example with the data of the corrected sample.

11.7   The fire department of a city claims that 90% of all persons residing in the city are pleased with the quality of service that the fire department provides. To verify this claim, 15 randomly selected residents were interviewed, and 12 were pleased with the quality of services rendered. What can we conclude about the claim of the fire department at the 0.05 level of significance?

✓ **11.8** In a random sample of 400 automobile accidents, it was found that 128 were due at least in part to driver fatigue. Use the 0.01 level of significance to test whether this supports the claim that 35% of all automobile accidents are due at least in part to driver fatigue.

11.9 A self-service gasoline station operator claims that at least 50% of the drivers who patronize the station are women. To check this claim, a random sample was taken, and it was found that 85 of 200 patrons were women. Test the null hypothesis $p = 0.50$ against a suitable alternative hypothesis at the 0.01 level of significance.

11.10 Repeat Exercise 11.9 with the new information that 90 (not 85) of the 200 randomly selected drivers are women. Otherwise, this exercise is the same as Exercise 11.9.

11.11 In a random sample of 500 cars making a left turn at a certain intersection, 169 pulled into the wrong lane. Test the null hypothesis that the actual proportion of drivers who make this mistake (at the given intersection) is 0.30 against the alternative hypothesis that this figure is too low. Use the 0.01 level of significance.

11.12 The alumni association of a state university asserted that at least 23% of their former students left the state within a year after graduation. A random sample was taken which revealed that 72 in a sample of 400 graduates had moved out of state within a year of graduation. Test the alumni association's claim $p = 0.23$ against the alternative $p < 0.23$ at the 0.05 level of significance.

11.13 According to a salesperson in a large department store, 15% of persons who shop in the store make no purchases. In a random sample of 300 shoppers, who were carefully observed, 32 made no purchases. Test the null hypothesis $p = 0.15$ against the alternative $p < 0.15$ at the 0.05 level of significance.

11.3 DIFFERENCES BETWEEN PROPORTIONS

There are many problems in which we must decide whether an observed difference between two sample proportions can be attributed to chance, or whether it is indicative of the fact that the corresponding true proportions are unequal. For instance, we may want to decide on the basis of sample data whether there really is a difference between the proportions of persons with and without flu shots who actually catch the disease, or we may want to check on the basis of samples whether two manufacturers of electronic equipment ship equal proportions of defectives.

The method we shall use to test whether an observed difference between two sample proportions can be attributed to chance or whether it is statistically significant is based on the following theory: If x_1 and x_2 are the numbers of successes obtained in n_1 trials of one kind and n_2 of another, the trials are all independent, and the corresponding probabilities of a success are, respectively, p_1 and p_2, then the sampling distribution of $\frac{x_1}{n_1} - \frac{x_2}{n_2}$ has the mean $p_1 - p_2$ and the standard deviation respectively, p_1 and p_2, then the sampling distribution of $\frac{x_1}{n_1} - \frac{x_2}{n_2}$ has the mean $p_1 - p_2$ and the standard deviation

$$\sqrt{\frac{p_1(1 - p_1)}{n_1} + \frac{p_2(1 - p_2)}{n_2}}$$

It is customary to refer to this standard deviation as the **standard error of the difference between two proportions**.

When we test the null hypothesis $p_1 = p_2 \, (= p)$ against an appropriate alternative hypothesis, the mean of the sampling distribution of the difference between the two sample proportions is $p_1 - p_2 = 0$, and its standard deviation can be written

$$\sqrt{p(1-p)\left(\frac{1}{n_1} + \frac{1}{n_2}\right)}$$

Since the value of p is not known, we must estimate it; this is done by **pooling** the data and substituting for p the combined sample proportion $\hat{p} = \dfrac{x_1 + x_2}{n_1 + n_2}$, which reads "$p$-hat." Then, since for large samples the sampling distribution of the difference between two proportions can be approximated closely with a normal distribution, we base the test on the statistic

STATISTIC FOR A LARGE-SAMPLE TEST CONCERNING THE DIFFERENCE BETWEEN TWO PROPORTIONS

$$z = \frac{\dfrac{x_1}{n_1} - \dfrac{x_2}{n_2}}{\sqrt{\hat{p}(1-\hat{p})\left(\dfrac{1}{n_1} + \dfrac{1}{n_2}\right)}} \quad \text{with} \quad \hat{p} = \frac{x_1 + x_2}{n_1 + n_2}$$

which has approximately the standard normal distribution. The test criteria are again those of Figure 10.5, with $p_1 - p_2$ substituted for μ and 0 substituted for μ_0. For the one-sided alternative hypothesis $p_1 < p_2$, we reject the null hypothesis if $z \leq -z_\alpha$, for the one-sided alternative hypothesis $p_1 > p_2$, we reject the null hypothesis if $z \geq -z_\alpha$, and for the two-sided alternative hypothesis $p_1 \neq p_2$, we reject the null hypothesis if $z \leq -z_{\alpha/2}$ or $z \geq z_{\alpha/2}$.

EXAMPLE 11.4 In a study of unemployment rates in various economies for a recent year, a random sample of 400 persons in the civilian labor force of Germany disclosed that 33 were unemployed; while in a similar study in France, a sample of 100 randomly selected persons in the civilian labor force disclosed that 10 were unemployed. Test the null hypothesis $p_1 = p_2$ (that the two economies had equal unemployment rates) against the alternative hypothesis $p_1 \neq p_2$ at the 0.05 level of significance.

Solution

1. *Hypothesis*

$$H_0: \quad p_1 = p_2$$

$$H_A: \quad p_1 \neq p_2$$

2. *Level of significance* $\alpha = 0.05$
3. *Criterion* Reject the null hypothesis if $z < 1.96$ or $z > 1.96$, where z is given by the appropriate formula; otherwise accept it or reserve judgment.
4. *Calculations* Substituting $x_1 = 33$, $x_2 = 10$, $n_1 = 400$, and $n_2 = 100$, and

$$p = \frac{33 + 10}{400 + 100} = 0.086, \text{ we get}$$

$$z = \frac{\dfrac{33}{400} - \dfrac{10}{100}}{\sqrt{0.086\,(1 - 0.086)\left(\dfrac{1}{400} + \dfrac{1}{100}\right)}} \approx -0.559$$

5. *Decision:*   Since -0.559 falls between $z = -1.96$ and $z = 1.96$, we cannot reject the null hypothesis. Accept the null hypothesis or reserve judgment. (*Note:* Unemployment rates vary substantially over time and, at another time, the results could be quite different.)   ■

EXERCISES

Exercise 11.14 is a practice exercise; its complete solution is given on page 391.

✓ **11.14** In random samples of 200 tractors from one assembly line and 400 tractors from another, there were, respectively, 16 tractors and 20 tractors that required extensive adjustments before they could be shipped. At the 0.05 level of significance, can we conclude that there is a difference in the quality of the work of the two assembly lines?

11.15 To test the effectiveness of a new pain-relieving drug, 80 patients at a clinic were given a pill containing the drug and 80 others were given a placebo. At the 0.01 level of significance, what can we conclude about the effectiveness of the drug if in the first group 56 of the patients felt a beneficial effect, while 38 of those who received the placebo felt a beneficial effect?

11.16 A study of the commuting patterns of the residents of Northboro showed that, in a random sample, 141 of 300 persons commuted by privately owned automobile. At the same time, a random sample of the residents of Southboro showed that 123 of 300 persons commuted by privately owned automobile. Use the 0.05 level of significance to test the null hypothesis that there is no difference between the corresponding proportions of persons in these two large cities who commute by privately owned automobile.

11.17 One production process yielded 28 defective pieces in a random sample of size 400 while another yielded 15 defective pieces in a random sample of size 300. Test the null hypothesis $p_1 = p_2$ (that the two processes yield equal proportions of defectives) against the alternative hypothesis $p_1 \neq p_2$ at the 0.05 level of significance.

11.18 In a recent study of labor force participation rates in the United States, a random sample of 1,000 men (16 years old and older) revealed that 749 participated in the labor force; and a random sample of 1,000 women (16 years old and older) revealed that 593 participated in the labor force. Test at the 0.01 level of significance whether the difference between the two sample proportions, $\dfrac{749}{1,000}$ and $\dfrac{593}{1,000}$, may be attributed to chance.

11.4 DIFFERENCES AMONG PROPORTIONS

Following the pattern of Chapter 10, we shall now consider a test that enables us to decide whether observed differences among more than two sample proportions can be attributed to chance. For instance, if 25 of 200 brand A tires, 21 of 200 brand B tires, 32 of 200 brand C tires, and 18 of 200 brand D tires failed to last

30,000 miles, we may want to know whether the differences among the sample proportions

$$\frac{25}{200} = 0.125, \quad \frac{21}{200} = 0.105, \quad \frac{32}{200} = 0.160, \quad \text{and} \quad \frac{18}{200} = 0.090$$

can be attributed to chance, or whether they are indicative of actual differences in the quality of the tires.

To illustrate the method we use to analyze this kind of data, suppose that a survey in which independent random samples of workers in three parts of the country were asked whether they feel that unemployment or inflation is a more serious problem yielded the results shown in the following table:

	Northeast	Midwest	Southwest
Unemployment	87	73	66
Inflation	113	77	84
	200	150	150

If we let p_1, p_2, and p_3 denote the true proportions of workers in the three parts of the country who feel that unemployment is the more serious economic problem, the hypotheses we shall want to test are

$$H_0: \quad p_1 = p_2 = p_3$$

$$H_A: \quad \text{the } p\text{'s are not all equal}$$

If the null hypothesis is true, we can combine the data and estimate the common proportion of workers in the three parts of the country who feel that unemployment is the more serious economic problem as

$$\hat{p} = \frac{87 + 73 + 66}{200 + 150 + 150} = \frac{226}{500} = 0.452$$

where, again, $\hat{p}$ reads "p-hat."

With this estimate, we would expect $200(0.452) = 90.4$ of the 200 workers in the Northeast, $150(0.452) = 67.8$ of the 150 workers in the Midwest, and $150(0.452) = 67.8$ of the 150 workers in the Southwest to choose unemployment. Also, if we subtract these figures from the sizes of the respective samples, we find that $200 - 90.4 = 109.6$ of the 200 workers in the Northeast, $150 - 67.8 = 82.2$ of the 150 workers in the Midwest, and $150 - 67.8 = 82.2$ of the 150 workers in the Southwest would be expected to choose inflation. These results are summarized in the following table, where the **expected frequencies** are shown in parentheses below the **observed frequencies**:

	Northeast	Midwest	Southwest
Unemployment	87 (90.4)	73 (67.8)	66 (67.8)
Inflation	113 (109.6)	77 (82.2)	84 (82.2)

To test the null hypothesis that the p's are all equal, we then compare the frequencies that were actually observed with the frequencies we can expect if the null hypothesis is true. Clearly, the null hypothesis should be accepted if the discrepancies between the observed and expected frequencies are small, and it should be rejected if the discrepancies between the two sets of frequencies are large.

Denoting the observed frequencies by the letter o and the expected frequencies by the letter e, we base this comparison on the following χ^2 **(chi-square) statistic:**

STATISTIC FOR A
TEST CONCERNING
THE DIFFERENCES
AMONG
PROPORTIONS

$$\chi^2 \sum \frac{(o - e)^2}{e}$$

In words, χ^2 is the sum of the quantities obtained by dividing $(o - e)^2$ by e separately for each **cell** of the table, and for our example we get

$$\chi^2 = \frac{(87 - 90.4)^2}{90.4} + \frac{(73 - 67.8)^2}{67.8} + \frac{(66 - 67.8)^2}{67.8}$$
$$+ \frac{(113 - 109.6)^2}{109.6} + \frac{(77 - 82.2)^2}{82.2} + \frac{(84 - 82.2)^2}{82.2} \approx 1.048$$

It remains to be seen whether this value is large enough to reject the null hypothesis $p_1 = p_2 = p_3$.

If the null hypothesis is true, the sampling distribution of the χ^2 statistic is approximately a theoretical distribution called the **chi-square distribution**. The parameter of this distribution, like that of the t distribution, is called the **number of degrees of freedom**, or simply the **degrees of freedom**, and it equals $k - 1$ when we compare k sample proportions.

Since the null hypothesis is to be rejected only when the value obtained for χ^2 is too large, we base our decision on the criterion of Figure 11.3, where χ^2_α is such that the area under the curve to its right is equal to α. Values of $\chi^2_{0.05}$ and $\chi^2_{0.01}$ for 1, 2, 3, ..., and 30 degrees of freedom are given in Table IV at the end of the book.

Returning to our numerical example and supposing that the level of significance is to be 0.05, we find that $\chi^2_{0.05} = 5.991$ for $k - 1 = 2$ degrees of freedom, and that the null hypothesis cannot be rejected since $\chi^2 = 1.048$ does not exceed 5.991. In other words, the data tends to support, rather than refute, the

Figure 11.3
Chi-square distribution.

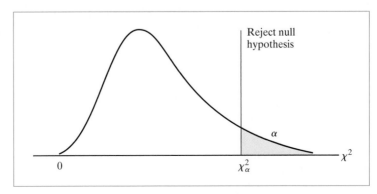

Reject null
hypothesis

α

0

χ^2_α

χ^2

hypothesis that the proportion of workers who feel that unemployment is a more serious economic problem than inflation is the same in all three parts of the country.

In general, if we want to test the null hypothesis $p_1 = p_2 = \cdots = p_k$ on the basis of random samples from k populations, we proceed as follows:

1. *Hypotheses*

$$H_0: \quad p_1 = p_2 = \cdots = p_k$$

$$H_A: \quad \text{the } p\text{'s are not all equal}$$

2. *Level of significance* α
3. *Criterion* Compute

$$\chi^2 = \sum \frac{(o - e)^2}{e}$$

taking the sum over all $2 \cdot k$ cells. Reject the null hypothesis if $\chi^2 \geq \chi_\alpha^2$ for $k - 1$ degrees of freedom. Otherwise, accept the null hypothesis or reserve judgment.

4. *Calculations* We estimate $p_1 = p_2 = \cdots = p_k$, the common population proportion, as

$$\hat{p} = \frac{x_1 + x_2 + \cdots + x_k}{n_1 + n_2 + \cdots + n_k}$$

and then we multiply $\hat{p}$ by the respective sample sizes to obtain the expected frequencies for the first row of the table. Then we determine the expected frequencies for the second row of the table by subtracting those of the first row from the respective sample sizes and substitute all the observed and expected frequencies into the formula for χ^2.

5. *Decision* Depending on the value we get for χ^2, we reject the null hypothesis, or we accept it or reserve judgment.

When we calculate the expected frequencies, there is seldom any need to carry more than two decimals, even though the entries in Table IV are given to three decimals. Also, the test we have been discussing is only an approximate test, and it should not be used when one (or more) of the expected frequencies is less than 5. If this is the case, we can sometimes combine two or more of the samples in such a way that none of the e's is less than 5.

The method we have discussed here can be used only to test the null hypothesis $p_1 = p_2 = \cdots = p_k$ against the alternative hypothesis that the p's are not all equal. However, in the special case where $k = 2$, we can use instead the method of Section 11.3 and test also against either of the alternative hypotheses $p_1 < p_2$ or $p_1 > p_2$. Indeed, for $k = 2$, the two methods are equivalent, as it can be shown that the χ^2 statistic equals the square of the z statistic obtained in accordance with the formula on page 372 (see Exercises 11.24 through 11.26).

Exercises 11.19 and 11.24 are practice exercises; their complete solutions are given on pages 391 and 392.

✓ **11.19** On page 373 we referred to a study in which 25 of 200 brand A tires, 21 of 200 brand B tires, 32 of 200 brand C tires, and 18 of 200 brand D tires fail to last 30,000 miles. Use the 0.05 level of significance to test the null hypothesis that there is no difference in the durability of the four kinds of tires.

11.20 In a survey of investor attitudes, 140 were interviewed in Mexico City, 160 in Toronto, and 250 in New York City, and 91, 104, and 170 investors, respectively, think that the prices of stocks traded on the New York Stock Exchange will rise over the next month. The remaining investors are uncertain or think that stock prices will fall over the next month. Use the level of significance $\alpha = 0.05$ to test the null hypotheses that there is no difference in the true proportion of investors in these three cities who think that stock prices will rise over the next month.

11.21 In three random samples taken from the records for motor vehicle license examinations, 32 of 200 applicants in license examination station A, 69 of 300 applicants in license examination station B, and 19 of 100 applicants in license examination station C failed to pass the license examination. Using the 0.05 level of significance, test the null hypothesis that there is no difference in the true proportion of applicants who fail to pass the license examination in the three stations.

11.22 In studying problems relating to its handling of reservations, an airline takes random samples of 60 of the complaints about reservations filed in each of four cities. If 49 of the complaints from city A, 54 of the complaints from city B, 41 of the complaints from city C, and 48 of the complaints from city D are about overbooking, test at the 0.05 level of significance whether the difference among the corresponding sample proportions can be attributed to chance.

11.23 The following table shows the results of a study in which samples of the members of five labor unions were asked whether they are for or against a certain piece of legislation:

	Union 1	Union 2	Union 3	Union 4	Union 5
For	74	81	69	75	91
Against	26	19	31	25	9

Use the 0.01 level of significance to test the null hypothesis that the corresponding true proportions are all equal.

✓ **11.24** Use the method of this section to rework Exercise 11.14 and verify that the value obtained for χ^2 equals the square of the value obtained originally for z.

11.25 Use the method of this section to rework Exercise 11.16 and verify that the value obtained for χ^2 equals the square of the value obtained originally for z.

11.26 Use the method of this section to rework Exercise 11.18 and verify that the value obtained for χ^2 equals the square of the value obtained originally for z.

Sections 11.5 and 11.6, which follow immediately, are optional. They can be omitted without affecting your understanding of subsequent material in this book. These sections are complete and are included here because they can enhance your knowledge of chi-square, a most useful concept. Exercises are also included here but, for the convenience of readers, their complete solutions are included with the Solutions of Practice Exercises at the end of the chapter.

Students who are using a graphing calculator or computer may wish to examine Figures 11.4, 11.5, 11.6 and the segment on page 384 with the title Using Technology: The Graphing Calculator.

*11.5 CONTINGENCY TABLES

The χ^2 statistic plays an important role in many other problems where information is obtained by counting rather than measuring. The method we shall describe here applies to two kinds of problems, which differ conceptually but are analyzed the same way.

In the first kind of problem we deal with trials permitting more than two possible outcomes. For instance, the weather can get better, remain the same, or get worse; an undergraduate can be a freshman, a sophomore, a junior, or a senior; a movie may be rated G, PG, PG-13, R, or NC-17. In the language of Section 6.5, we could say that we are dealing with multinomial (rather than binomial) trials.

Also, in the illustration of the preceding section, each worker might have been asked whether unemployment is a more serious economic problem than inflation, whether inflation is a more serious economic problem than unemployment, or whether he or she is undecided, and this might have resulted in the following table:

	Northeast	Midwest	Southwest
Unemployment	57	53	44
Undecided	72	40	48
Inflation	71	57	58
	200	150	150

We refer to this kind of table as a 3×3 table (where the notation 3×3 is read "3 by 3"), because it has 3 horizontal rows and 3 vertical columns; more generally, when there are r horizontal rows and c vertical columns, we refer to the table as an **$r \times c$ table**. Here, as in the table analyzed in the preceding section, the column totals, representing the sample sizes, are fixed. On the other hand, the row totals depend on the responses of the persons interviewed and hence on chance.

In the second kind of problem where the method of this section applies, the column totals as well as the row totals depend on chance. To give an example, suppose that a sociologist wants to determine whether there is a relationship between the intelligence of boys who have gone through a special job-training program and their subsequent performance in their jobs, and that a sample of 400 cases taken from very extensive files yielded the following results:

		Performance			
		Poor	Fair	Good	
	Below average	67	64	25	156
Intelligence	Average	42	76	56	174
	Above average	10	23	37	70
		119	163	118	400

This is also a 3×3 table, and it is mainly in connection with problems like this that $r \times c$ tables are referred to as **contingency tables**. In either kind of table, the observed frequencies are referred to as the **observed cell frequencies**.

Before we demonstrate how $r \times c$ tables are analyzed, let us examine what hypotheses we want to test. In the first example we want to test whether the probabilities that a worker will choose "unemployment," "undecided," or "inflation" are the same for the three parts of the country. Formally,

> H_0: For each alternative (unemployment, undecided, and inflation),
>   the probabilities are the same for the three parts of the country.
>
> H_A: For at least one alternative, the probabilities are not the same for
>   the three parts of the country.

In the other example, we want to test whether there is a dependence (relationship) between on the job performance and intelligence, and we write

> H_0: The two variables under consideration are independent.
>
> H_A: The two variables are not independent.

To show how an $r \times c$ table is analyzed, let us refer to the second of our two examples, and let us begin by illustrating the calculation of an **expected cell frequency**.

EXAMPLE 11.5   Assuming that intelligence and on-the-job performance are independent and using the data in the preceding table, how many of the boys in a random sample of 400 can be expected to have below-average intelligence and perform poorly on their jobs?

Solution   Under the assumption of independence, the probability of randomly choosing a boy whose intelligence is below average and whose on-the-job performance is poor is given by the product of the probability of choosing a boy whose intelligence is below average and the probability of choosing a boy whose on-the-job performance is poor. Using the totals of the first row and the first column to estimate these two probabilities, we get

$$\frac{67 + 64 + 25}{400} = \frac{156}{400}$$

for the probability of choosing a boy whose intelligence is below average and

$$\frac{67 + 42 + 10}{400} = \frac{119}{400}$$

for the probability of choosing a boy whose on-the-job performance is poor. Hence, we estimate the probability of choosing a boy whose intelligence is below average and whose on-the-job performance is poor as $\frac{156}{400} \cdot \frac{119}{400}$, and in a sample of size 400 we would expect to find

$$400 \cdot \frac{156}{400} \cdot \frac{119}{400} = \frac{156 \cdot 119}{400} \approx 46.4$$

boys who fit this description.

In the final step of this calculation, $\dfrac{156 \cdot 119}{400}$ is just the product of the of total the first row and the total of the first column divided by the grand total for the entire table. Indeed, the argument that led to this result can be used to show that, in general, under the assumption of independence:

> *The expected frequency for any cell of a contingency table may be obtained by multiplying the total of the row to which it belongs by the total of the column to which it belongs and then dividing by the grand total for the entire table.*

With this rule we get an expected frequency of

$$\frac{156 \cdot 163}{400} \approx 63.6$$

for the second cell of the first row, and

$$\frac{174 \cdot 119}{400} \approx 51.8 \quad \text{and} \quad \frac{174 \cdot 163}{400} \approx 70.9$$

for the first two cells of the second row.

It is not necessary to calculate all the expected cell frequencies in this way, as it can be shown that the sum of the expected frequencies for any row or column must equal the sum of the corresponding observed frequencies. Therefore, we can get some of the expected cell frequencies by subtraction from row or column totals. For instance, for our example we get

$$156 - 46.4 - 63.6 = 46.0$$

for the expected frequency of the third cell of the first row,

$$174 - 51.8 - 70.9 = 51.3$$

for the expected frequency of the third cell of the second row, and

$$119 - 46.4 - 51.8 = 20.8$$
$$163 - 63.6 - 70.9 = 28.5$$

and

$$118 - 46.0 - 51.3 = 20.7$$

for the expected frequencies of the three cells of the third row. These results are summarized in the following table, where the expected frequencies are shown in parentheses below the corresponding observed frequencies:

Performance

		Poor	Fair	Good
	Below average	67 (46.4)	64 (63.6)	25 (46.0)
Intelligence	Average	42 (51.8)	76 (70.9)	56 (51.3)
	Above average	10 (20.8)	23 (28.5)	37 (20.7)

From here on, the work is like that of the preceding section; we calculate the χ^2 statistic according to the formula

STATISTIC FOR AN ANALYSIS OF AN r × c TABLE

$$\chi^2 = \sum \frac{(o - e)^2}{e}$$

with $\frac{(o - e)^2}{e}$ calculated separately for each cell of the table. Then we reject the null hypothesis at the level of significance α if the value obtained for χ^2 exceeds χ^2_α for $(r - 1)(c - 1)$ degrees of freedom. In connection with this formula for the number of degrees of freedom, observe that after $c - 1$ of the expected frequencies have been calculated for each of $r - 1$ rows by means of the rule on page 380, all the other expected frequencies may be obtained by subtraction from row or column totals. For our example, the number of degrees of freedom is $(3 - 1)(3 - 1) = 4$, and it should be observed that after we had calculated two of the expected frequencies for each of the first two rows by means of the rule on page 380, all the others were obtained by subtraction from row or column totals.

EXAMPLE 11.6 With reference to our example, use the 0.01 level of significance to test the null hypothesis that intelligence and on-the-job performance are independent for boys who have gone through the special job-training program.

Solution

1. *Hypotheses*

   H_0: Intelligence and on-the-job performance are independent.
   H_A: Intelligence and on-the-job performance are not independent.

2. *Level of significance* $\alpha = 0.01$
3. *Criterion* Reject the null hypothesis if $\chi^2 \geq 13.277$, the value of $\chi^2_{0.01}$ for $(3 - 1)(3 - 1) = 4$ degrees of freedom, where

$$\chi^2 = \sum \frac{(o - e)^2}{e}$$

(see also Figure 11.4); otherwise, accept it or reserve judgment.

Figure 11.4
Test criterion for intelligence and on-the-job performance example.

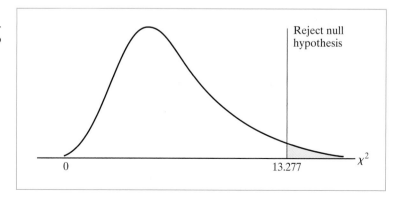

Reject null hypothesis

0   13.277 χ^2

4. *Calculations*   Copying the observed and expected cell frequencies from the table on page 380, we find that

$$\chi^2 = \frac{(67 - 46.4)^2}{46.4} + \frac{(64 - 63.6)^2}{63.6} + \frac{(25 - 46.0)^2}{46.0}$$
$$+ \frac{(42 - 51.8)^2}{51.8} + \frac{(76 - 70.9)^2}{70.9} + \frac{(56 - 51.3)^2}{51.3}$$
$$+ \frac{(10 - 20.8)^2}{20.8} + \frac{(23 - 28.5)^2}{28.5} + \frac{(37 - 20.7)^2}{20.7} \approx 40.89$$

5. *Decision*   Since $\chi^2 = 40.89$ exceeds 13.277, the null hypothesis must be rejected; we conclude that there is a relationship between intelligence and on-the-job performance. ■

A computer printout of the preceding chi-square analysis is shown in Figure 11.5. The difference between the values of χ^2 obtained previously and in Figure 11.5 is due to rounding. Some computer programs also give the probability of getting a value greater than or equal to the observed value of χ^2 when the null hypothesis is true; for our example, it is about 0.00000003. Because the test is based on an approximation, this value should not be taken too literally, although the actual probability of a greater value is certainly quite small.

The method that we have used here to analyze the contingency table applies also when the column totals are fixed sample sizes (as in the illustration on page 378) and do not depend on chance. The rule according to which we multiply the row total by the column total and then divide by the grand total has to be justified in a different way, but this is of no consequence—the expected cell frequencies are determined in exactly the same way.

Figure 11.5
MINITAB printout for analysis of contingency table.

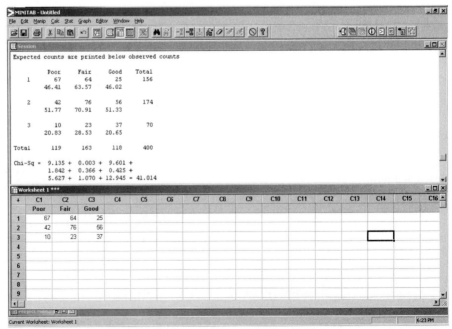

EXAMPLE 11.7   Analyze the 3×3 table on page 378, which pertains to a study in which workers in the Northeast, Midwest, and Southwest are asked whether unemployment or inflation is the more serious economic problem or whether they are undecided. Use the 0.01 level of significance.

Solution   1. *Hypotheses*

H_0: For each alternative (unemployment, undecided, and inflation), the probabilities are the same for the three parts of the country.

H_A: For at least one alternative, the probabilities are not the same for the three parts of the country.

2. *Level of significance* $\alpha = 0.01$
3. *Criterion*   Reject the null hypothesis if $\chi^2 \geq 13.277$, the value of $\chi^2_{0.01}$ for $(3-1)(3-1) = 4$ degrees of freedom, where

$$\chi^2 = \sum \frac{(o-e)^2}{e}$$

Otherwise, accept the null hypothesis or reserve judgment.
4. *Calculations*   The expected frequencies for the first two cells of the first two rows are

$$\frac{154 \cdot 200}{500} = 61.6, \quad \frac{154 \cdot 150}{500} = 46.2, \quad \frac{160 \cdot 200}{500} = 64,$$

and

$$\frac{160 \cdot 150}{500} = 48$$

by subtraction, those for the third cells of the first two rows are 46.2 and 48, and those for the third row are 74.4, 55.8, and 55.8. Then, substituting into the formula for χ^2 we get

$$\chi^2 = \frac{(57-61.6)^2}{61.6} + \frac{(53-46.2)^2}{46.2} + \frac{(44-46.2)^2}{46.2}$$
$$+ \frac{(72-64)^2}{64} + \frac{(40-48)^2}{48} + \frac{(48-48)^2}{48}$$
$$+ \frac{(71-74.4)^2}{74.4} + \frac{(57-55.8)^2}{55.8} + \frac{(58-55.8)^2}{55.8} \approx 4.05$$

5. *Decision*   Since $\chi^2 = 4.05$ does not exceed 13.277, the null hypothesis cannot be rejected; the differences between the observed and expected frequencies may well be due to chance.   ▪

Alternatively, using a graphing calculator we could have obtained the display shown in Figure 11.6. As before, graphing calculator information shown here and elsewhere in this book is entirely optional and can be omitted by the reader without the loss of continuity. The figure shows that the p-value corresponding to $\chi^2 = 4.05$ and 4 degrees of freedom is .399281. Since this p-value exceeds $\alpha = 0.01$, we conclude (as shown earlier) that the null hypothesis cannot be rejected.

Figure 11.6
Chi-square distribution reproduced from the display screen of a TI-83 graphing calculator.

The procedure for obtaining Figure 11.6 is explained in the Using Technology: The Graphing Calculator box, displayed below.

How to determine the *p*-value and graph the area under the curve as shown in Figure 11.6

(**TIP**: Before beginning: Reset the memory to default settings and clear all plots—see Appendix A: TI-83 Tips.)

STEP 1. Clear all previous drawings.
Press 2nd PRGM to access the DRAW menu.
Select 1: ClrDraw and press ENTER.

STEP 2. Set the viewing window.

```
Press WINDOW
Set    Xmin = 0
       Xmax = 18
       Xscl = 1
       Ymin = -.05
       Ymax = .2
       Yscl = 1
       Xres = 1
```

STEP 3. Select the type of display and enter the interval to be displayed.
Press 2nd, VARS to access the DISTR DRAW menu.
Highlight DRAW.
Select 3: Shade χ^2(and press ENTER.
Enter 4.05, 1E99, 4)
(For instructions on entering "1E99," for infinity, see Appendix A: TI-83 Tips; the 4.05 and 1E99 (for ∞) are the lower and upper limits of the interval, and the 4 is the number of degrees of freedom.)

STEP 4. View the graph.
Press ENTER to solve.

USING TECHNOLOGY: The Graphing Calculator

EXERCISES

The complete solutions to Exercises 11.27, 11.28, 11.29, 11.30, 11.31, and 11.32 are found with the Solutions to Practice Exercises on pages 392, 393, 394, and 395.

✓ **11.27** In a survey, 80 single persons, 120 married persons, and 100 widowed or divorced persons were asked whether they feel that friends and social life, job or primary activity, or health and physical condition contributes most to their general happiness. Use the results shown in the following table and the 0.05 level of significance to test whether the probabilities of the three alternatives are the same for persons who are single, married, or widowed or divorced.

	Single	Married	Widowed or divorced
Friends and social life	41	49	42
Job or primary activity	27	50	33
Health and physical condition	12	21	25
Total	80	120	100

✓ **11.28** Suppose that for 120 mental patients who did not receive psychotherapy and 120 mental patients who received psychotherapy a panel of psychiatrists determined after six months whether their condition had deteriorated, remained unchanged, or improved. Based on the results shown in the following table, test at the level of significance $\alpha = 0.05$ whether the therapy is effective.

	No therapy	Therapy
Deteriorated	6	11
Unchanged	65	31
Improved	49	78

✓ **11.29** The following sample data pertains to the number of men's sweaters of various sizes sold in the men's departments of a chain of department stores.

	Department Store 1	Department Store 2	Department Store 3
Small	53	49	72
Medium	108	73	170
Large	48	22	55

Use the 0.01 level of significance to test whether the three department stores sell the same proportion of small, medium, and large sweaters.

✓ **11.30** To determine the income level of borrowers who make personal loans, a large bank examines a random sample of applications for personal loans in four branches, with the following results:

	Branch 1	Branch 2	Branch 3	Branch 4
High income	28	27	22	92
Upper middle income	78	54	67	87
Lower middle income	71	99	82	11
Low income	23	20	29	10
Total	200	200	200	200

Use the 0.05 level of significance to test the null hypothesis that the four bank branches have the same proportion of high, upper middle, lower middle, and low income borrowers.

11.31 A sample survey was designed to show the weight losses, during a certain period of time, of the patrons at three weight-reducing clinics.

	Clinic 1	Clinic 2	Clinic 3
1 to 4 pounds	56	57	72
5 to 9 pounds	32	39	55
10 pounds or more	12	14	13

Use the level of significance $\alpha = 0.05$ to test the null hypothesis that there is no relationship between the number of pounds lost by patrons and the clinics that they patronized.

11.32 Tests of the fidelity and selectivity of 190 radios produced the results shown in the following table:

		Fidelity		
		Low	Average	High
Selectivity	Low	7	12	31
	Average	35	59	18
	High	15	13	0

Use the 0.01 level of significance to test the null hypothesis that fidelity is independent of selectivity.

11.33 Use a computer package or a graphing calculator to rework Exercise 11.27.

11.34 Use a computer package or a graphing calculator to rework Exercise 11.28.

11.35 Use a computer package or a graphing calculator to rework Exercise 11.30.

11.36 A computer program for the analysis of a contingency table or a graphing calculator can be used to test for differences among proportions as in Section 11.4. In that case we are, in fact, analyzing a $2 \times c$ table. Use a computer package or a graphing calculator to rework the illustration of Section 11.4, which dealt with a survey in three parts of the country where workers were asked whether they feel that unemployment or inflation is a more serious problem.

11.37 Use a computer package or a graphing calculator to rework Exercise 11.22.

11.38 Use a computer package or a graphing calculator to rework Exercise 11.23.

11.6 GOODNESS OF FIT

The χ^2 statistic can also be used to compare observed frequency distributions with distributions that we might expect according to theory or assumptions. We refer to such a comparison as a test of **goodness of fit**.

This kind of problem would arise, for example, if a quality-control engineer takes a daily sample of 10 tires coming off an assembly line, and he wants to check, on the basis of the following data, for the numbers of tires with imperfections observed on 200 days, whether it is true that 5% of all the tires have imperfections, that is, whether he is sampling a binomial population with $n = 10$ and $p = 0.05$.

Number with imperfections	Number of samples
0	138
1	53
2 or more	9

If the hypothesis is true, the probabilities of getting 0, 1, or 2 or more tires with imperfections are, respectively, 0.599, 0.315, and $0.075 + 0.010 + 0.001 = 0.086$ according to Table I, and hence the corresponding expected frequencies are $200(0.599) = 119.8$, $200(0.315) = 63.0$, and $200(0.086) = 17.2$.

To test whether the discrepancies between the observed frequencies and the expected frequencies can be attributed to chance, we use the same χ^2 statistic as in the preceding sections:

STATISTIC FOR A
TEST OF GOODNESS
OF FIT

$$\chi^2 = \sum \frac{(o - e)^2}{e}$$

Based on the value of this statistic, we reject the null hypothesis if $\chi^2 > \chi^2_\alpha$ for $k - m - 1$ degrees of freedom, where k is the number of terms in the summation for χ^2 and m is the number of parameters that have to be estimated on the basis of the sample data. The role of m is explained in the following discussion.

EXAMPLE 11.8   Based on the data given previously, test at the 0.05 level of significance whether the quality-control engineer is sampling a binomial population with $n = 10$ and $p = 0.05$.

Solution   1. *Hypotheses*

H_0:   Population is binomial with $n = 10$ and $p = 0.05$.
H_A: Population is not binomial with $n = 10$ and $p = 0.05$.

2. *Level of significance* $\alpha = 0.05$
3. *Criterion*   Reject the null hypothesis if $\chi^2 \geq 5.991$, the value of $\chi^2_{0.05}$ for $k - m - 1 = 3 - 0 - 1 = 2$ degrees of freedom, where

$$\chi^2 = \sum \frac{(o - e)^2}{e}$$

Otherwise, accept the null hypothesis or reserve judgment. Observe that $m = 0$ because no parameters were estimated.

4. *Calculations*   From page 387 the observed frequencies are 138, 53, and 9, and the expected frequencies are 119.8, 63.0, and 17.2. Substituting these values into the formula for x^2, we get

$$x^2 = \frac{(138 - 119.8)^2}{119.8} + \frac{(53 - 63.0)^2}{63.0} + \frac{(9 - 17.2)^2}{17.2} \approx 8.26$$

5. *Decision*   Since $x^2 = 8.26$ exceeds 5.991, the null hypothesis must be rejected; we conclude that the true percentage of tires with imperfections is not 5%. (In Exercise 11.43 the reader will find that the binomial distribution with $n = 10$ and $p = 0.04$ provides a much better fit to the data.)

In the preceding example it was assumed under the null hypothesis that $n = 10$ and $p = 0.05$, so that none of the parameters had to be estimated and the number of degrees of freedom was

$$k - m - 1 = 3 - 0 - 1 = 2$$

However, if we want to test whether the distribution is binomial (without knowing p), we will have to estimate p from the data, and the calculation for the number of degrees of freedom is
$$k - m - 1 = 3 - 1 - 1 = 1$$

We can also test whether or not grouped data has a distribution that can be approximated closely by a normal curve (see Exercises 11.44 and 11.45); in such cases it is required that the parameters μ and σ be estimated directly from the grouped data so that $m = 2$.

Observe also that, as in the tests of the two preceding sections, the sampling distribution of the x^2 statistic of a goodness-of-fit test is only approximately a chi-square distribution. So, if any of the expected frequencies is less than 5, we shall again have to combine some of the data, in this case adjacent classes of the distribution.

Exercises 11.39 through 11.45 are practice exercises; their complete solutions are given on pages 396, 397, 398, and 399.

✓ **11.39**   To see whether a die is balanced, it is rolled 360 times and the following results are obtained: 1 showed 57 times, 2 showed 46 times, 3 showed 68 times, 4 showed 52 times, 5 showed 72 times, and 6 showed 65 times. At the 0.05 level of significance, do these results support the hypothesis that the die is balanced?

✓ **11.40**   Four coins are tossed together 480 times, and 0, 1, 2, 3, and 4 heads showed 26, 104, 171, 142, and 37 times. Using the probabilities $\frac{1}{16}, \frac{4}{16}, \frac{6}{16}, \frac{4}{16}$, and $\frac{1}{16}$, test at the level of significance $\alpha = 0.05$ whether its reasonable to suppose that the coins are balanced and randomly tossed.

✓ **11.41**   To determine whether the customer service department of a large store serves an equal number of customers during each hour of a six-hour afternoon, a record was kept of the number of customers who were provided with service, and the results are

Time	Customers served
1:00–1:59	30
2:00–2:59	51
3:00–3:59	47
4:00–4:59	40
5:00–5:59	51
6:00–6:59	21
Total	240

Letting the expected frequencies be $\dfrac{240}{6} = 40$ customers served for each of the six hours, test at the level of significance $\alpha = 0.05$ whether an equal number of customers are served each hour during the six-hour time period.

✓ **11.42** The following is the distribution of the number of calls received at the switchboard of a government building during 400 five-minute intervals:

Number of calls	Frequency
0	95
1	116
2	112
3	47
4 or more	30

Use the 0.05 level of significance to test whether the number of calls received by the switchboard in a five-minute interval is a random variable having the Poisson distribution with $\lambda = 1.5$; that is, the probabilities for 0, 1, 2, 3, or 4 or more calls are 0.22, 0.33, 0.25, 0.13, and 0.07.

✓ **11.43** If a random variable has the binomial distribution with $n = 10$ and $p = 0.04$, the probabilities of 0, 1, or 2 or more successes are 0.665, 0.277, and 0.058. Based on this information and the data on page 387, test at the 0.05 level of significance whether the quality-control engineer is sampling a binomial population with $n = 10$ and $p = 0.04$.

✓ **11.44** The following is the distribution of the grades that 100 students received on a history test:

Grade	Frequency
65–69	1
70–74	10
75–79	37
80–84	36
85–89	13
90–94	2
95–99	1

As can easily be verified, the mean and the standard deviation of this distribution are $\bar{x} = 80$ and $s = 5$. To test the null hypothesis that this data constitutes a random sample from a normal population with $\mu = 80$ and $\sigma = 5$, proceed with the following steps:

(a) Find the probabilities that a random variable having the normal distribution with $\mu = 80$ and $\sigma = 5$ will take on a value less than 69.5, between 69.5 and 74.5, between 74.5 and 79.5, between 79.5 and 84.5, between 84.5 and 89.5, between 89.5 and 94.5, and greater than 94.5.

(b) Multiply the probabilities obtained in part (a) by 100, the total frequency of the observed data, to find the frequencies that we can expect under the null hypothesis for a random sample of size $n = 100$.

(c) Calculate χ^2 for the observed frequencies shown in the table and the expected frequencies obtained in part (b), and test at the 0.05 level of significance whether the null hypothesis can be rejected.

✓ **11.45** The following is the distribution obtained on page 26 for the amount of time that 80 college students engaged in leisure activities during a typical school week:

Hours	Frequency
10–14	8
15–19	28
20–24	27
25–29	12
30–34	4
35–39	1

As we showed on page 82, the mean and the standard deviation of this distribution are $\bar{x} = 20.7$ and $s = 5.4$. To test the null hypothesis that this data constitutes a random sample from a normal population with $\mu = 20.7$ and $\sigma = 5.4$, proceed with the following steps:

(a) Find the probabilities that a random variable having the normal distribution with $\mu = 20.7$ and $\sigma = 5.4$ will take on a value less than 14.5, between 14.5 and 19.5, between 19.5 and 24.5, between 24.5 and 29.5, between 29.5 and 34.5, and greater than 34.5.

(b) Multiply the probabilities obtained in part (a) by 80, the total frequency of the observed data, to find the frequencies we can expect under the null hypothesis for a random sample of size $n = 80$.

(c) Calculate χ^2 for the observed frequencies shown in the table and the expected frequencies obtained in part (b), and test at the 0.05 level of significance whether the null hypothesis can be rejected.

✓ SOLUTIONS TO PRACTICE EXERCISES

11.1 This solution uses the p-value.

1. *Hypotheses*

$$H_0: \quad p = 0.10$$

$$H_A: \quad p > 0.10$$

2. *Level of significance* $\alpha = 0.05$

3′. *Test statistic*   The test statistic is simply the number feeling an effect, here four.

4′. *Calculations*   Table I shows that for $n = 13$ and $p = 0.10$ the probability of four or more successes is $0.028 + 0.006 + 0.001 = 0.035$.

5′. *Decision*   Since 0.035 is less than or equal to $\alpha = 0.05$, the null hypothesis must be rejected.

11.8 1. *Hypotheses*

$$H_0: \quad p = 0.35$$

$$H_A: \quad p \neq 0.35$$

2. *Level of significance* $\alpha = 0.01$
3. *Criterion*   Reject the null hypothesis if $z \leq -2.575$ or $z \geq 2.575$, where

$$z = \frac{x - np_0}{\sqrt{np_0(1 - p_0)}}$$

Otherwise, accept the null hypothesis or reserve judgment.

4. *Calculations*   Substituting $x = 128$, $n = 400$, and $p_0 = 0.35$ into the formula for z, we get

$$z = \frac{128 - 400(0.35)}{\sqrt{400(0.35)(0.65)}} \approx -1.26$$

5. *Decision*   Since $z = -1.26$ falls between -2.575 and 2.575, the null hypothesis cannot be rejected; the data does not refute the claim.

11.14 1. *Hypotheses*

$$H_0: \quad p_1 = p_2$$

$$H_A: \quad p_1 \neq p_2$$

2. *Level of significance* $\alpha = 0.05$
3. *Criterion*   Reject the null hypothesis if $z \leq -1.96$ or $z \geq 1.96$, where z is given by the formula on page 372. Otherwise, accept the null hypothesis or reserve judgment.

4. *Calculations*   Substituting $x_1 = 16$, $x_2 = 20$, $n_1 = 200$, $n_2 = 400$, and $\hat{p} = \dfrac{16 + 20}{200 + 400} = 0.06$ into the formula for z, we get

$$z = \frac{\dfrac{16}{200} - \dfrac{20}{400}}{\sqrt{(0.06)(0.94)\left(\dfrac{1}{200} + \dfrac{1}{400}\right)}} \approx 1.46$$

5. *Decision*   Since $z = 1.46$ falls between -1.96 and 1.96, the null hypothesis cannot be rejected; in other words, the difference between the two sample proportions is not significant.

11.19 1. *Hypotheses*

$$H_0: \quad p_1 = p_2 = p_3 = p_4$$

$$H_A: \quad \text{the } p\text{'s are not all equal}$$

2. *Level of significance* $\alpha = 0.05$
3. *Criterion*   Reject the null hypothesis if $\chi^2 \geq 7.815$, the value of $\chi^2_{0.05}$ for $4 - 1 = 3$ degrees of freedom, where

$$\chi^2 = \sum \frac{(o - e)^2}{e}$$

Otherwise, accept it or reserve judgment.

4. *Calculations*   Since

$$\hat{p} = \frac{25 + 21 + 32 + 18}{200 + 200 + 200 + 200} = \frac{96}{800} = 0.12$$

we get $200(0.12) = 24$ for each expected frequency in the first row and $200 - 24 = 176$ for each expected frequency in the second row. Then, substituting into the formula for χ^2, we get

$$\chi^2 = \frac{(25 - 24)^2}{24} + \frac{(21 - 24)^2}{24} + \frac{(32 - 24)^2}{24} + \frac{(18 - 24)^2}{24}$$

$$+ \frac{(175 - 176)^2}{176} + \frac{(179 - 176)^2}{176} + \frac{(168 - 176)^2}{176} + \frac{(182 - 176)^2}{176} \approx 5.21$$

5. *Decision*   Since $\chi^2 = 5.21$ does not exceed 7.815, the null hypothesis cannot be rejected; we cannot conclude that there is a real difference in the durability of the tires.

11.24 1. *Hypotheses*

$$H_0: \quad p_1 = p_2$$

$$H_A: \quad p_1 \neq p_2$$

2. *Level of significance* $\alpha = 0.05$

3. *Criterion*   Reject the null hypothesis if $\chi^2 \geq 3.841$, the value of $\chi^2_{0.05}$ for $2 - 1 = 1$ degrees of freedom, where

$$\chi^2 = \sum \frac{(o - e)^2}{e}$$

Otherwise, accept it or reserve judgment.

4. *Calculations*   Since $\hat{p} = \frac{16 + 20}{200 + 400} = 0.06$, we get $200(0.06) = 12$ and $400(0.06) = 24$ for the expected frequencies for the first row, and $200 - 12 = 188$ and $400 - 24 = 376$ for the expected frequencies for the second row. Then, substituting into the formula for χ^2, we get

$$\chi^2 = \frac{(16 - 12)^2}{12} + \frac{(20 - 24)^2}{24} + \frac{(184 - 188)^2}{188} + \frac{(380 - 376)^2}{376} \approx 2.13$$

5. *Decision*   Since $\chi^2 = 2.13$ does not exceed 3.841, the null hypothesis cannot be rejected; in other words, the difference between the two sample proportions is not significant.

If we square 1.46, the value obtained for z in Exercise 11.14, we get $(1.46)^2 = 2.1316$, and, except for rounding, this equals the value that we obtained for χ^2.

11.27 1. *Hypotheses*

H_0:   For each alternative, the probabilities are the same for persons who are single, married, and widowed or divorced.

H_A: For at least one alternative, the probabilities are not the same for the three kinds of persons.

2. *Level of significance* $\alpha = 0.05$

3. *Criterion*   Reject the null hypothesis if $\chi^2 \geq 9.488$, the value of $\chi^2_{0.05}$ for $(3-1)(3-1) = 4$ degrees of freedom, where

$$\chi^2 = \sum \frac{(o-e)^2}{e}$$

Otherwise, accept it or reserve judgment.

4. *Calculations*   The expected frequencies for the first two cells of the first row are

$$\frac{132 \cdot 80}{300} = 35.2 \quad \text{and} \quad \frac{132 \cdot 120}{300} = 52.8$$

those for the first two cells of the second row are

$$\frac{110 \cdot 80}{300} \approx 29.3 \quad \text{and} \quad \frac{110 \cdot 120}{300} = 44.0$$

and, by subtraction, those for the third cells of the first two rows are 44.0 and 36.7; and those for the third row are 15.5, 23.2, and 19.3. Thus, substitution into the formula for χ^2 yields

$$\chi^2 = \frac{(41-35.2)^2}{35.2} + \frac{(49-52.8)^2}{52.8} + \frac{(42-44.0)^2}{44.0}$$
$$+ \frac{(27-29.3)^2}{29.3} + \frac{(50-44.0)^2}{44.0} + \frac{(33-36.7)^2}{36.7}$$
$$+ \frac{(12-15.5)^2}{15.5} + \frac{(21-23.2)^2}{23.2} + \frac{(25-19.3)^2}{19.3} \approx 5.37$$

5. *Decision*   Since $\chi^2 = 5.37$ does not exceed 9.488, the null hypothesis cannot be rejected; there is no real evidence that the probabilities of the three alternatives are not the same for the three kinds of persons.

11.28 1. H_0: The probabilities that a patient's condition will deteriorate, remain unchanged or improved are that same for patients who receive therapy and those who do not.

H_A: The probabilities are not all the same for the two groups.

2. $\alpha = 0.05$

3. Reject the null hypothesis if $\chi^2 \geq 5.991$, which is the value of $\chi^2_{0.05}$ for $(3-1)(2-1) = 2$ degrees of freedom; otherwise, accept H_0 or reserve judgment.

4. The row totals are 17, 96, and 127; the column totals are 120 and 120; and the grand total is 240. Therefore, the expected frequencies for the first row are both $\frac{17 \cdot 120}{240} = 8.5$; those for the second row are both $\frac{96 \cdot 120}{240} = 48$; those for the third row are both $120 - (8.5 + 48) = 63.5$; and

$$\chi^2 = \frac{(6-8.5)^2}{8.5} + \frac{(11-8.5)^2}{8.5} + \frac{(65-48)^2}{48} + \frac{(31-48)^2}{48} + \frac{(49-63.5)^2}{63.5}$$
$$+ \frac{(78-63.5)^2}{63.5} \approx 20.13$$

5. *Decision*   Since $\chi^2 = 20.13$ exceeds 5.991, the null hypothesis must be rejected; we conclude that the probabilities of the three outcomes are not the same for the two groups.

11.29 1. H_0: The probabilities that the department stores will sell the same proportion of small, medium, and large sweaters are the same.

   H_A: The probability that the proportions are not the same for the three department stores.

2. $\alpha = 0.01$

3. Reject the null hypothesis if $\chi^2 \geq 13.277$, which is the value of $\chi^2_{0.01}$ for $(3-1)(3-1) = 4$ degrees of freedom; otherwise, accept H_0 or reserve judgment.

4. The row totals are 174, 351, and 125; the column totals are 209, 144, and 297; and the grand total is 650. Therefore, the expected frequencies for the first row are $\dfrac{174 \cdot 209}{650} \approx 55.9$, $\dfrac{174 \cdot 144}{650} \approx 38.5$, and $174 - (55.9 + 38.5) = 79.6$; those for the second row are $\dfrac{351 \cdot 209}{650} \approx 112.9$, $\dfrac{351 \cdot 144}{650} \approx 77.8$, and $351 - (112.9 + 77.8) = 160.3$; and those for the third row are $\dfrac{125 \cdot 209}{650} \approx 40.2$, $\dfrac{125 \cdot 144}{650} \approx 27.7$, and $125 - (40.2 + 27.7) = 57.1$. It follows that

$$\chi^2 = \frac{(53 - 55.9)^2}{55.9} + \frac{(49 - 38.5)^2}{38.5} + \frac{(72 - 79.6)^2}{79.5} + \frac{(108 - 112.9)^2}{112.9}$$
$$+ \frac{(73 - 77.8)^2}{77.8} + \frac{(170 - 160.3)^2}{160.3} + \frac{(48 - 40.2)^2}{40.2} + \frac{(22 - 27.7)^2}{27.7}$$
$$+ \frac{(55 - 57.1)^2}{57.1} = 7.60$$

5. Since $\chi^2 = 7.60$ does not exceed 13.277, the null hypothesis cannot be rejected; there is no real evidence that the probabilities are not the same for all department stores.

11.30 1. H_0: The probabilities that the banks will have the same proportions of high, upper middle, lower middle, and low income borrowers are all the same.

   H_A: The probabilities that the banks will have the same proportions of high, upper middle, lower middle, and low income borrowers are not all the same.

2. $\alpha = 0.05$

3. Reject the null hypothesis if $\chi^2 \geq 16.919$, which is the value of $\chi^2_{0.05}$ for $(4-1)(4-1) = 9$ degrees of freedom; otherwise, accept H_0 or reserve judgment.

4. The row totals are 169, 286, 263, and 82; the column totals are all 200; and the grand total is 800. Therefore, the expected frequencies for the first row are all $\dfrac{169 \cdot 200}{800} \approx 42.2$; those for the second row are all $\dfrac{286 \cdot 200}{800} \approx 71.5$; those for the third row are all $\dfrac{263 \cdot 200}{800} \approx 65.8$; and those for the fourth row are all $200 - 42.2 - 71.5 - 65.8 = 20.5$. It follows that

$$\chi^2 = \frac{(28 - 42.2)^2}{42.2} + \frac{(27 - 42.2)^2}{42.2} + \frac{(22 - 42.2)^2}{42.2} + \frac{(92 - 42.2)^2}{42.2}$$
$$+ \frac{(78 - 71.5)^2}{71.5} + \frac{(54 - 71.5)^2}{71.5} + \frac{(67 - 71.5)^2}{71.5} + \frac{(87 - 71.5)^2}{71.5}$$

$$+ \frac{(71 - 65.8)^2}{65.8} + \frac{(99 - 65.8)^2}{65.8} + \frac{(82 - 65.8)^2}{65.8} + \frac{(11 - 65.8)^2}{65.8}$$

$$+ \frac{(23 - 20.5)^2}{20.5} + \frac{(20 - 20.5)^2}{20.5} + \frac{(29 - 20.5)^2}{20.5}$$

$$+ \frac{(10 - 20.5)^2}{20.5} \approx 163.22$$

5. Since $\chi^2 = 163.22$ is greatly in excess of 16.919, the null hypothesis must be rejected; we conclude that the income levels of the borrowers at the four banks are different.

11.31 1. H_0: The number of pounds lost is independent of the clinic.
   H_A: The number of pounds lost is not independent of the clinic.

2. $\alpha = 0.05$

3. Reject the null hypothesis if $\chi^2 \geq 9.488$, which is the value of $\chi^2_{0.05}$ for $(3 - 1)(3 - 1) = 4$ degrees of freedom; otherwise, accept H_0 or reserve judgment.

4. The row totals are 185, 126, and 39; the column totals are 100, 110, and 140; and the grand total is 350. Therefore, the expected frequencies for the first row are $\frac{185 \cdot 100}{350} \approx 52.9$, $\frac{185 \cdot 110}{350} \approx 58.1$, and $185 - (52.9 - 58.1) = 74.0$; those for the second row are $\frac{126 \cdot 100}{350} \approx 36.0$, $\frac{126 \cdot 110}{350} \approx 39.6$, and $126 - (36.0 + 39.6) = 50.4$; and those for the third row are $100 - (36.0 + 52.9) = 11.1$, $110 - (39.6 + 58.1) = 12.3$, and $140 - (50.4 + 74.0) = 15.6$. It follows that

$$\chi^2 = \frac{(56 - 52.9)^2}{52.9} + \frac{(57 - 58.1)^2}{58.1} + \frac{(72 - 74)^2}{74} + \frac{(32 - 36)^2}{36} + \frac{(39 - 39.6)^2}{39.6}$$

$$+ \frac{(55 - 50.4)^2}{50.4} + \frac{(12 - 11.1)^2}{11.1} + \frac{(14 - 12.3)^2}{12.3} + \frac{(13 - 15.6)^2}{15.6} \approx 1.87$$

5. Since $\chi^2 = 1.87$ does not exceed 9.488, the null hypothesis cannot be rejected. Accept the null hypothesis that the number of pounds is independent of the clinics, or reserve judgment.

11.32 1. H_0: Fidelity is independent of selectivity.
   H_A: Fidelity is not independent of selectivity.

2. $\alpha = 0.01$

3. Reject the null hypothesis if $\chi^2 \geq 13.277$, which is the value of $\chi^2_{0.01}$ for $(3 - 1)(3 - 1) = 4$ degrees of freedom; otherwise, accept H_0 or reserve judgment.

4. The row totals are 50, 112, and 28; the column totals are 57, 84, and 49; and the grand total is 190. Therefore, the expected frequencies for the first row are $\frac{50 \cdot 57}{190} = 15$, $\frac{50 \cdot 84}{190} = 22.1$, and $50 - (15 + 22.1) = 12.9$; those for the second row are $\frac{112 \cdot 57}{190} = 33.6$, $\frac{112 \cdot 84}{190} = 49.5$, and $112 - (33.6 + 49.5) = 28.9$; and those for the third row are $57 - (15 + 33.6) = 8.4$, $84 - (22.1 + 49.5) = 12.4$, and $49 - (12.9 + 28.9) = 7.2$. It follows that

$$\chi^2 = \frac{(7 - 15)^2}{15} + \frac{(12 - 22.1)^2}{22.1} + \cdots + \frac{(13 - 12.4)^2}{12.4} + \frac{(0 - 7.2)^2}{7.2} = 52.7$$

5. Since $\chi^2 = 52.7$ exceeds 13.277, the null hypothesis must be rejected; we conclude that there is a dependence between fidelity and selectivity.

11.33 Computer exercise.

11.34 Computer exercise.

11.35 Computer exercise.

11.36 Computer exercise.

11.37 Computer exercise.

11.38 Computer exercise.

11.39 1. *Hypotheses*

$$H_0: \quad \text{The die is balanced.}$$

$$H_A: \quad \text{The die is not balanced.}$$

2. *Level of significance* $\alpha = 0.05$

3. *Criterion*   Reject the null hypothesis if $\chi^2 \geq 11.070$, the value of $\chi^2_{0.05}$ for $6 - 1 = 5$ degrees of freedom, where

$$\chi^2 = \sum \frac{(o - e)^2}{e}$$

Otherwise, accept it or reserve judgment.

4. *Calculations*   The expected frequencies are all $\dfrac{360}{6} = 60$, so that

$$\chi^2 = \frac{(57 - 60)^2}{60} + \frac{(46 - 60)^2}{60} + \frac{(68 - 60)^2}{60} + \frac{(52 - 60)^2}{60} + \frac{(72 - 60)^2}{60}$$

$$+ \frac{(65 - 60)^2}{60} \approx 8.37$$

5. *Decision*   Since $\chi^2 = 8.37$ does not exceed 11.070, the null hypothesis cannot be rejected; we conclude that there is no real evidence that the die is not balanced.

11.40 1. H_0:   The coins are all balanced and randomly tossed.
   H_A: The coins are not all balanced and randomly tossed.

2. $\alpha = 0.05$

3. Reject the null hypothesis if $\chi^2 \geq 9.488$, which is the value of $\chi^2_{0.05}$ for $5 - 1$ degrees of freedom; otherwise, accept H_0 or reserve judgment.

4. Since the probabilities for 0, 1, 2, 3, or 4 heads are $\dfrac{1}{16}, \dfrac{4}{16}, \dfrac{6}{16}, \dfrac{4}{16}$, and $\dfrac{1}{16}$, the corresponding expected frequencies are

$$480\left(\frac{1}{16}\right) = 30, \, 480\left(\frac{4}{16}\right) = 120, \, 480\left(\frac{6}{16}\right) = 180, \, 480\left(\frac{4}{16}\right) = 120, \text{ and}$$

$$480\left(\frac{1}{16}\right) = 30$$

It follows that

$$\chi^2 = \frac{(26 - 30)^2}{30} + \frac{(104 - 120)^2}{120} + \frac{(171 - 180)^2}{180} + \frac{(142 - 120)^2}{120}$$

$$+ \frac{(37 - 30)^2}{30} \approx 8.78$$

5. Since $\chi^2 = 8.78$ does not exceed 9.488, the null hypothesis cannot be rejected. Accept the null hypothesis or reserve judgment.

11.41 1. H_0: Same number of customers are served each hour.
   H_A: Number of customers served is not the same for all hours.

2. $\alpha = 0.05$

3. Reject the null hypothesis if $\chi^2 \geq 11.070$, which is the value of $\chi^2_{0.05}$ for $6 - 1 = 5$ degrees of freedom; otherwise, accept H_0 or reserve judgment.

4. Since

o	e
30	40
51	40
47	40
40	40
51	40
21	40
240	240

we get

$$\chi^2 = \frac{(30-40)^2}{40} + \frac{(51-40)^2}{40} + \frac{(47-40)^2}{40} + \frac{(40-40)^2}{40}$$
$$+ \frac{(51-40)^2}{40} + \frac{(21-40)^2}{40} = 18.80$$

5. Since $\chi^2 = 18.80$ exceeds 11.070, we reject the null hypothesis. The number of customers served each hour is not the same for all six hours.

11.42 1. H_0: Number of calls has Poisson distribution with $\lambda = 1.5$.
   H_A: Number of calls does not have Poisson distribution with $\lambda = 1.5$.

2. $\alpha = 0.05$

3. Reject the null hypothesis if $\chi^2 \geq 9.488$, which is the value of $\chi^2_{0.05}$ for $5 - 1 = 4$ degrees of freedom; otherwise, accept H_0 or reserve judgment.

4. Since the expected frequencies are $400(0.22) = 88, 400(0.33) = 132$, $400(0.25) = 100, 400(0.13) = 52$, and $400(0.07) = 28$, we get

$$\chi^2 = \frac{(95-88)^2}{88} + \frac{(116-132)^2}{132} + \frac{(112-100)^2}{100} + \frac{(47-52)^2}{52}$$
$$+ \frac{(30-28)^2}{28} \approx 4.56$$

5. Since $\chi^2 = 4.56$ does not exceed 9.488, the null hypothesis cannot be rejected; there is no real evidence that the number of calls does not have the Poisson distribution with $\lambda = 1.5$.

11.43 1. H_0: The engineer is sampling a binomial population with $n = 10$ and $p = 0.04$.
   H_A: The engineer is not sampling a binomial population with $n = 10$ and $p = 0.04$.

2. $\alpha = 0.05$

3. Reject the null hypothesis if $\chi^2 \geq 5.991$, which is the value of $\chi^2_{0.05}$ for $3 - 1 = 2$ degrees of freedom; otherwise, accept H_0 or reserve judgment.

4. Since the expected frequencies are $200(0.665) = 133, 200(0.277) = 55.4$, and $200(0.058) = 11.6$, we get

$$\chi^2 = \frac{(138 - 133)^2}{133} + \frac{(53 - 55.4)^2}{55.4} + \frac{(9 - 11.6)^2}{11.6} = 0.875$$

5. Since $\chi^2 = 0.875$ does not exceed 5.991, the null hypothesis cannot be rejected; there is no real evidence that the population is not binomial with $n = 10$ and $p = 0.04$.

11.44 (a) In standard units, the class boundaries are $z = \dfrac{69.5 - 80}{5} = -2.10$,

$z = \dfrac{74.5 - 80}{5} = -1.10$, $\quad z = \dfrac{79.5 - 80}{5} = -0.10$, $\quad z = \dfrac{84.5 - 80}{5} = 0.90$,

$z = \dfrac{89.5 - 80}{5} = 1.90$, and $\quad z = \dfrac{94.5 - 80}{5} = 2.90$; the corresponding entries in Table II are 0.4821, 0.3643, 0.0398, 0.3159, 0.4713, and 0.4981 so that the probabilities are $0.5000 - 0.4821 = 0.0179, 0.4821 - 0.3643 = 0.1178, 0.3643 - 0.0398 = 0.3245, 0.0398 + 0.3159 = 0.3557, 0.4713 - 0.3159 = 0.1554, 0.4981 - 0.4713 = 0.0268$, and $0.5000 - 0.4981 = 0.0019$.

(b) The expected normal curve frequencies are $100(0.0179) \approx 1.8, 100(0.1178) \approx 11.8, 100(0.3245) \approx 32.4, 100(0.3557) \approx 35.6, 100(0.1554) \approx 15.5, 100(0.0268) \approx 2.7$, and $100(0.0019) \approx 0.2$. Thus, we have

	o			e	
1			1.8		
10	} 11		11.8	} 13.6	
37			32.4		
36			35.6		
13			15.5		
2	} 16		2.7	} 18.4	
1			0.2		

where we combined some of the classes so that none of the expected frequencies is less than 5.

(c) 1. *Hypotheses*

H_0:   Data comes from a normal population.

H_A: Data does not come from a normal population.

2. *Level of significance* $\alpha = 0.05$

3. *Criterion*   Reject the null hypothesis if $\chi^2 \geq 3.841$, the value of $\chi^2_{0.05}$ for $4 - 2 - 1 = 1$ degree of freedom, where

$$\chi^2 = \sum \frac{(o - e)^2}{e}$$

Otherwise, accept it or reserve judgment.

4. *Calculations*   Substituting into the formula for χ^2, we get

$$\chi^2 = \frac{(11 - 13.6)^2}{13.6} + \frac{(37 - 32.4)^2}{32.4} + \frac{(36 - 35.6)^2}{35.6} + \frac{(16 - 18.4)^2}{18.4}$$

$$\approx 1.47$$

5. *Decision*   Since $\chi^2 = 1.47$ does not exceed 3.841, the null hypothesis cannot be rejected; there is no real evidence to doubt that the data comes from a normal population.

11.45 (a) In standard units the class boundaries are $\dfrac{14.5 - 20.7}{5.4} \approx -1.15$, $\dfrac{19.5 - 20.7}{5.4} \approx$ -0.22, $\dfrac{24.5 - 20.7}{5.4} \approx 0.70$, $\dfrac{29.5 - 20.7}{5.4} \approx 1.63$, and $\dfrac{34.5 - 20.7}{5.4} \approx 2.56$. Since the corresponding entries in Table II are 0.3749, 0.0871, 0.2580, 0.4484, and 0.4948, the probabilities are $0.5000 - 0.3749 = 0.1251, 0.3749 - 0.0871 = 0.2878, 0.0871 + 0.2580 = 0.3451, 0.4484 - 0.2580 = 0.1904, 0.4948 - 0.4484 = 0.0464$, and $0.5000 - 0.4948 = 0.0052$.

(b) The expected frequencies are $80(0.1251) \approx 10.0, 80(0.2878) \approx 23.0, 80(0.3451) \approx 27.6, 80(0.1904) \approx 15.2, 80(0.0464) \approx 3.7$, and $80(0.0052) \approx 0.4$.

(c) 1. H_0:   The population sampled is normal.

   H_A: The population sampled is not normal.

2. $\alpha = 0.05$

3. Since we have to combine the fourth, fifth, and sixth classes, we reject the null hypothesis if $\chi^2 \geq 3.841$, which is the value of $\chi^2_{0.05}$ for $4 - 2 - 1 = 1$ degree of freedom; otherwise accept H_0 or reserve judgment.

4. The combined classes have an observed frequency of 17 and an expected frequency of 19.3, so we get

$$\chi^2 = \frac{(8 - 10.0)^2}{10.0} + \frac{(28 - 23.0)^2}{23.0} + \frac{(27 - 27.6)^2}{27.6} + \frac{(17 - 19.3)^2}{19.3} \approx 1.77$$

5. Since $\chi^2 = 1.77$ does not exceed 3.841, the null hypothesis cannot be rejected; we conclude that there is no real evidence that the population is not normal.

9, 10, & 11

Review •

ACHIEVEMENTS

Having read and studied these chapters and having worked a good portion of the exercises, you should be able to

1. Distinguish between point estimates and interval estimates.
2. Explain the difference between *probability* and *confidence*.
3. Make confidence statements about the maximum error when $\bar{x}$ is used as an estimate of μ.
4. Determine the sample size needed so that a sample mean will have a desired precision.
5. Explain what is meant by *confidence interval* and *degree of confidence*.
6. Construct large-sample confidence intervals for μ.
7. Construct small-sample confidence intervals for μ.
8. Construct large-sample confidence intervals for σ.
9. Explain what is meant by *statistical hypothesis, null hypothesis*, and *alternative hypothesis*.
10. Distinguish between Type I and Type II errors.
11. Explain what is meant by the *operating characteristic curve*.
12. Distinguish between simple and composite hypotheses about parameters.
13. Explain what is meant by *significance test, level of significance*, and *statistically significant*.
14. List the five steps of the outline for testing a statistical hypothesis.
15. Distinguish between one-sided and two-sided alternatives and between one-tailed and two-tailed tests.
16. Explain what is meant by *p*-value.
17. Relate the value of a test statistic to its *p*-value.
18. Perform large-sample tests of the null hypothesis $\mu = \mu_0$.
19. Perform small-sample tests of the null hypothesis $\mu = \mu_0$.
20. Perform large-sample tests of the null hypothesis that two populations have equal means.
21. Perform small-sample tests of the null hypothesis that two populations have equal means.
22. Test the significance of the difference between the means of paired data.
23. Perform a one-way analysis of variance when the sample sizes are all equal.
24. Perform a one-way analysis of variance when the sample sizes are not all equal.
25. Construct large-sample confidence intervals for population proportions.
26. Make confidence statements about the maximum error when $\dfrac{x}{n}$ is used as an estimate of p.
27. Determine the sample size needed so that a sample proportion will have a desired precision (with and without some information about p).

28. Perform small-sample tests of the null hypothesis $p = p_0$.
29. Perform large-sample tests of the null hypothesis $p = p_0$.
30. Test for the equality of two population proportions.
31. Test for the equality of k population proportions.
32. Analyze an $r \times c$ (contingency) table. (Optional)
33. Test for goodness of fit. (Optional)

CHECKLIST OF KEY TERMS (with page references to their definitions)

Alternative hypothesis, 316
Analysis of variance, 349
Analysis of variance table, 352
Cell, 375
Chi-square distribution, 375
Chi-square statistic, 375
Composite hypothesis, 320
Confidence interval, 300
Confidence limits, 300
Contingency table, 379
Count data, 366
Critical values, 327
Degree of confidence, 300
Degrees of freedom, 301, 375
Enumeration data, 366
Error mean square, 352
Error sum of squares, 350
Expected cell frequencies, 379
Expected frequencies, 374
Experimental error, 349
F distribution, 348
F statistic, 348
Goodness of fit, 387
Grand mean, 350
Hypothesis, 315
Independent random samples, 337

Interval estimate, 300
Level of significance, 323
Mean square for error, 352
Mean square for treatments, 351
Null hypothesis, 316
Number of degrees of freedom values, 375
Numerator and denominator degrees of freedom, 348
Observed cell frequencies, 379
Observed frequencies, 374
One-sample t-test, 331
One-sided alternative, 322
One-sided test, 324
One-tailed test, 324
One-way analysis of variance, 350
Operating characteristic curve, 320
Point estimate, 294
Pooled sample standard deviation, 339
Pooling, 372
Problems of estimation, 293
p-value, 329
$r \times c$ table, 378

Sample proportion, 307
Significance test, 321
Simple hypothesis, 320
Small-sample confidence interval, 302
Standard error of a proportion, 308
Standard error of s, 306
Standard error of the difference between two means, 337
Standard error of the difference between two proportions, 372
Statistically significant, 321
t statistic distribution, 301
Test statistic, 323
Tests of hypotheses, 293
Total sum of squares, 350
Treatment mean square, 351
Treatment sum of squares, 350
Treatments, 350
Two-sample t-test, 340
Two-sided alternative, 322
Two-sided criterion, 324
Two-tailed test, 324
Two-way classification, 366
Type I error, 319
Type II error, 319

R.119 To compare two kinds of baseball bats, 18 baseball players were asked to swing 20 times with each kind of bat at balls pitched by a machine, and the following are the respective numbers of balls that they hit more than 300 feet: 6 and 8, 9 and 5, 4 and 4, 7 and 6, 10 and 8, 5 and 6, 9 and 7, 3 and 4, 5 and 4, 6 and 6, 12 and 9, 8 and 9, 5 and 5, 4 and 6, 9 and 6, 10 and 8, 7 and 7, and 11 and 7. Test at the 0.05 level of significance whether the difference between the means of the numbers of balls hit more than 300 feet with the two kinds of bats is significant.

R.120 An arboriculturist for a tree farm program wants to estimate the height of mature maple trees at a certain tree farm from a random sample of 100 trees. He knows, from experience, that the standard deviation is 4.00 inches for such data. What can be said with probability 0.95 about the maximum error of estimate?

R.121 A pollster randomly sampled 300 voters in a certain community, of whom 258 said that they would vote in favor of a new bond issue. Construct a 95% confidence interval for the proportion of all voters in the community who would vote in favor of the bond issue.

R.122 Referring to Exercise R.121, if a pollster wants to determine the proportion of the voters in the community who favor the bond issue, how large a sample is needed if we want to determine with probability 0.90 that the sample proportion will differ from the population proportion by at most 0.04?

R.123 Determine the value of t from Table III, where

    (a) $n = 11$ and the degree of confidence is 0.90;

    (b) $n = 12$ and the degree of confidence is 0.95;

    (c) $n = 13$ and the degree of confidence is 0.98;

    (d) $n = 14$ and the degree of confidence is 0.99.

R.124 The manager of a gasoline station has kept a record of the number of sales of gasoline made daily at the station. If a sample of these sales for 10 randomly selected days shows that an average of 192 sales of gasoline were made daily, with a standard deviation of 23 sales, construct a 99% confidence interval for the true mean number of sales.

R.125 The manager of a laundry takes a random sample of size $n = 64$ of the times it takes employees to iron a shirt and obtains a mean of 2.6 minutes with $s = 0.4$ minute. Construct a 95% confidence interval for the time it takes an employee of this laundry to iron a shirt.

R.126 Referring to Exercise R.125, let us reduce the sample size from 64 to 25, but let everything else remain the same. Construct a 95% confidence interval for the true mean time it takes an employee of this laundry to iron a shirt.

R.127 Again referring to Exercise R.125, where we were given $s = 0.40$ and used this value as a substitute for σ, construct a 95% confidence interval for the standard deviation of the population sampled.

R.128 Based on the results of $n = 12$ trials, we want to test the null hypothesis $p = 0.20$ against the alternative hypothesis $p > 0.20$. If we reject the null hypothesis when the number of successes is five or more and otherwise we accept it, find

    (a) the probability of a Type I error;

    (b) the probability of a Type II error when $p = 0.30$;

    (c) the probability of a Type II error when $p = 0.50$.

R.129 A national charitable organization would like to know how many hours, on average, it will take a volunteer worker to complete a short, self-study course in telephone solicitation procedures. One hundred randomly selected volunteer workers took an average of 7.9 hours to complete the course, with a standard deviation of 2.1 hours. What can we say with 99% confidence about the amount of time by which the charitable organization would be off if it estimated the true average time it takes a volunteer worker to complete the course as 7.9 hours?

R.130 A sample survey conducted in a metropolitan area showed that 200 families spent, on average, $233.58 per week on food eaten at home, with a standard deviation of $18.82. What can we say with 95% confidence about the maximum error if $233.58 in weekly food expenditures is used as an estimate of the average for all families in the metropolitan area?

R.131 The publishers of a daily newspaper would like to know how much time, on average, readers devote to the Sunday edition. A study discloses that 100 randomly selected readers devoted on average 126.5 minutes to the Sunday edition, with a standard deviation of 26.4. Construct a confidence interval for the true average time readers devote to the Sunday edition

  (a) using a 0.95 degree of confidence;

  (b) using a 0.99 degree of confidence.

R.132 A large hospital finds that in 50 randomly selected days it had, on average, $\bar{x} = 96.4$ patient admissions per day, with a standard deviation of $s = 12.2$. Construct a 90% confidence interval for the actual daily average number of hospital admissions.

R.133 Over a period of 40 weeks, the computers and related equipment operated by a company had average weekly maintenance costs of $126.45 with a standard deviation of $37.15. Construct a 95% confidence interval for the true average weekly maintenance cost of the company's computer system.

R.134 If the standard deviation of the IQ scores of 60 elementary school students is $s = 10.89$, construct a 95% confidence interval for the standard deviation of the population of IQ scores from which the sample came.

R.135 The manager of a supermarket observed that a sample of 100 randomly selected customers waited in line at the checkout counter for varying amounts of time before reaching the cashier, with a standard deviation of $s = 1.29$ minutes. Construct a 98% confidence interval for the standard deviation of the population of waiting times from which this sample came.

R.136 In a random sample of 60 days, a newspaper dealer sold varying numbers of newspapers with a standard deviation of $s = 27$. Construct a 95% confidence interval for the actual standard deviation of the number of newspapers sold daily by this dealer.

R.137 Among 80 fish caught in a certain lake, 28 were inedible as a result of the chemical pollution of their environment. If we use the sample proportion $\frac{28}{80} = 0.35$ to estimate the corresponding true proportion, what can we assert with 99% confidence about the maximum error?

R.138 With reference to Exercise R.137, construct a 95% confidence interval for the true proportion of fish in this lake that are inedible as a result of chemical pollution.

R.139 In a recent survey of 1,419 randomly selected households, 979 gave money in varying amounts to religious, educational, and other types of charitable organizations. Construct a 99% confidence interval for the corresponding true percentage of donors.

R.140 The manager of a sports fishing products company wants to test the null hypothesis that certain lead fishing weights have a mean weight of 4.0 ounces. Explain under what conditions the manager would commit a Type I error and under what conditions the manager would commit a Type II error.

R.141 A real estate agent claims that the average price of a single-family residence in a large city is $195,000.

    (a) What hypothesis is she testing if she would commit a Type I error by erroneously concluding that the average price is $195,000?

    (b) What hypothesis is she testing if she would commit a Type II error by erroneously concluding that the average price is $195,000?

R.142 While investigating a complaint about short weighting at a cannery, an investigator finds that a random sample of 100 cans of pineapple that are labeled "net weight 907 grams" (2 pounds) actually weigh an average of only 899 grams, and it is determined that the standard deviation is $s = 29$ grams. Using an appropriate one-sided alternative and the level of significance $\alpha = 0.05$, test whether this constitutes evidence on which the investigator can base a finding that the cannery is producing cans of pineapple that contain on the average less than 907 grams.

R.143 According to the norms established for a history test, eighth graders should average 81.7 with a standard deviation of 8.5. If 100 randomly selected eighth graders from a certain school district averaged 79.6 on this test, can we conclude at the 0.05 level of significance that eighth graders from this school district can be expected to average less than the norm of 81.7 on this test? Find the p-value and use it to make the hypothesis test. Do you reach the same conclusion with regard to significance at the 0.05 level?

R.144 A city manager wants to know whether the true mean number of temporary secretaries employed per day is 24. If, in a random sample of 60 days, the mean number of temporary secretaries employed is 21.6 with a standard deviation $s = 7.1$, is there, at the 0.05 level of significance, sufficient evidence to reject the null hypothesis $\mu = 24$? Use a two-sided test.

R.145 An office supply store claims that its rubber bands are better than those of a competitor. A study shows that a sample of 70 of its rubber bands of a certain type could stretch, on average, 8.6 inches before breaking, with a standard deviation of 0.75 inch; while a sample of 50 rubber bands of the same type sold by a competitor could stretch, on average, only 8.3 inches with a standard deviation of 0.80 inch before breaking. Use a suitable one-sided alternative and the 0.01 level of significance to test the null hypothesis $\mu_1 = \mu_2$ against the suitable alternative hypothesis $\mu_1 > \mu_2$.

R.146 An investigation conducted in two canneries showed that in one cannery a sample of 48 cans averaged 15.0 grams of protein per quarter-cup serving with a standard deviation of 1.6 grams, while in the other cannery a sample of 48 cans averaged 14.5 grams of protein per quarter-cup serving with a standard deviation of 1.4 grams. Test the null hypothesis $\mu_1 = \mu_2$ against the alternative hypothesis $\mu_1 > \mu_2$ at the level of significance $\alpha = 0.01$.

R.147 Twelve persons with Type II diabetes were given a new oral medication to reduce their blood sugar, and their average fasting (before breakfast) blood sugar readings before they started taking the medication and six weeks later were 128 and 117, 145 and 133, 115 and 108, 156 and 133, 129 and 111, 147 and 129, 130 and 122, 127 and 124, 120 and 103, 145 and 129, 124 and 116, and 135 and 117. Test at the 0.05 level of significance whether this medication will reduce a Type II diabetic's blood sugar by more than 10 points in six weeks. Use the 0.05 level of significance.

R.148 The program director of a television series claims that 60% of the television viewing audience recognizes its theme song and associates it with the program. To verify this claim, a random sample of 14 television viewers was interviewed, of whom 5 associated the theme song with the program. What can be concluded at the level of significance $\alpha = 0.05$ if the sample data is used to test the null hypothesis $p = 0.60$ against the alternative hypothesis $p < 0.60$?

R.149 A television critic claims that at least 80% of all viewers find the noise level of a certain commercial objectionable. If 9 out of 15 persons shown this commercial object to the noise level, what can we conclude about this claim at the 0.05 level of significance?

R.150 The dean of student affairs at a college wants to test the claim that 30% of all undergraduate students reside in the college dormitories. If 5 of 10 randomly selected undergraduate students reside in the dormitories, does this support the claim that 30% of all undergraduate students reside in the dormitories? Use the alternative hypothesis $p \neq 0.30$ and $\alpha = 0.01$.

R.151 A food processor wants to know whether the probability is really 0.60 that a customer will prefer a new kind of packaging to the old kind. If, in a random sample, 6 of 15 customers prefer the new kind of packaging to the old kind, test the null hypothesis $p = 0.60$ against the alternative hypothesis $p \neq 0.60$ at the 0.05 level of significance.

R.152 A representative of a health insurance company asserts that its claims payment procedures satisfy at least 80% of its policyholders. If in a random sample of 14 policyholders, it is determined that 8 are satisfied and 6 are dissatisfied, what can we conclude if we test this claim at the 0.05 level of significance?

R.153 A food chemist notes that 90% of the bricks of a certain cheddar cheese will remain mold free for three months under standard refrigeration. A variation is made in the salt content of this cheese, and a sample of 15 bricks of the modified cheese is kept under standard refrigeration for three months.

  (a) Suppose that only 11 of the bricks remain mold free. At the 5% level of significance, test the null hypothesis $p = 0.90$ against the alternative hypothesis $p \neq 0.90$.

  (b) Suppose that all 15 of the bricks remain mold free. At the 5% level of significance, test the null hypothesis $p = 0.90$ against the alternative hypothesis $p \neq 0.90$. What can you say about this experiment's ability to detect a value p larger than 0.90?

R.154 In a random sample of 250 persons who skipped breakfast, 102 reported that they experienced midmorning fatigue, and in a random sample of 250 persons who ate breakfast, 73 reported that they experienced midmorning fatigue. Use the 0.01 level of significance to test the null hypothesis that there is no difference between the corresponding population proportions against the alternative hypothesis that midmorning fatigue is more prevalent among persons who skip breakfast.

R.155 In a true–false test, a test item is considered to be good if it discriminates between well-prepared students and poorly prepared students. What can we conclude about the merit of a test item that was answered correctly by 246 of 300 well-prepared students and by 165 of 300 poorly prepared students? Use the 0.05 level of significance.

12

REGRESSION AND CORRELATION

Whenever possible scientists strive to express, or approximate, relationships between known quantities and quantities that are to be predicted in terms of mathematical equations. In physics it is known, for example, that at a constant temperature the relationship between the volume, y, of a gas and its pressure, x, is given by the formula

$$y = \frac{k}{x}$$

where k is a numerical constant; also, in biology, it has been discovered that the size of a culture of bacteria, y, can be expressed in terms of the time, x, it has been exposed to certain favorable conditions by means of the formula

$$y = a \cdot b^x$$

where a and b are numerical constants.

Business and economic statisticians have borrowed and continue to borrow liberally from the tools of the natural sciences. For instance, the first equation, $y = \dfrac{k}{x}$, is often used to express the relationship between the demand, y, for a commodity and its price, x, and the second equation, $y = a \cdot b^x$, is often used to express, among other things, how a company's production or sales, y, grow with time, x.

The goal of many statistical investigations is to establish relationships that make it possible to predict one variable in terms of another. For example, we might want to predict the sales of a new product in terms of the amount of money spent advertising it on television, predict family expenditures on entertainment in terms of family income, or predict a college student's grade-point average based on the number of hours he or she spent studying.

It would be ideal if we could predict one quantity exactly in terms of another, but this is seldom, if ever, possible. In most instances we can predict only averages or expected values. For instance, we cannot predict exactly how much money a specific college graduate will earn 10 years after graduation, but given suitable data we can predict the average earnings of all college graduates 10 years after graduation. Similarly, we can predict the average yield of a variety of wheat in terms of the total rainfall in July, and we can predict the expected grade-point average of a student starting law school in terms of his or her IQ. This problem of predicting the average value of one variable in terms of the known value of another variable is called the problem of **regression**.

Since the relationship between the average values of one variable and the known values of another variable can be strong, weak, or even nonexistent, we must concern ourselves also with measuring the strength of such a relationship, that is, the strength of the correlation. Whereas problems of regression are taken up in Sections 12.1 through 12.3, the subject of **correlation** is treated in Sections 12.4 through 12.6.

12.1 CURVE FITTING

Whenever we use observed data to arrive at a mathematical equation that describes the relationship between two variables—a procedure known as **curve fitting**—we must face three kinds of problems: (1) We must decide what kind of an equation we want to use (for instance, that of a straight line or that of some other curve); (2) we must find the particular equation that is "best" in some sense; and (3) we must investigate certain questions regarding the merits of the particular equation and of the predictions made from it. The second of these problems will be discussed in some detail in Section 12.2 and the third in Section 12.3.

The first kind of problem is sometimes decided by theoretical considerations, but more often by direct inspection of the data. We plot the data on graph paper, sometimes on special graph paper with special scales, and we judge visually what kind of curve best describes their overall pattern. So far as the work in this book is concerned, we shall consider only **linear equations in two unknowns**, which are of the form

$$y = a + bx$$

where a is the **y-intercept** (the value of y for $x = 0$) and b is the **slope of the line** (the change in y that accompanies an increase of one unit in x). Ordinarily, the values of a and b are estimated from the data (by the method we shall discuss in Section 12.2), and once they have been determined we can substitute a value of x into the equation and calculate the corresponding predicted value of y.

The term *linear equation* arises from the fact that the graph of $y = a + bx$ is a straight line; that is, all pairs of values of x and y that satisfy an equation of the form $y = a + bx$ constitute points which fall on a straight line. Suppose, for instance, that

$$y = 0.23 + 4.42x$$

expresses the relationship between a certain county's September-through-August rainfall in inches, x, and its yield of wheat in bushels per acre, y. For example, for an annual rainfall of 6 inches, $x = 6$ and $y = 0.23 + 4.42(6) = 26.75$ bushels of wheat per acre, and for an annual rainfall of 12 inches, $x = 12$ and $y = 0.23 + 4.42(12) = 53.27$ bushels of wheat per acre. Observe that the points (6, 26.75) and (12, 53.27) lie on the straight line shown in Figure 12.1, and this is true for all other points obtained in the same way.

Figure 12.1
Graph of a linear equation.

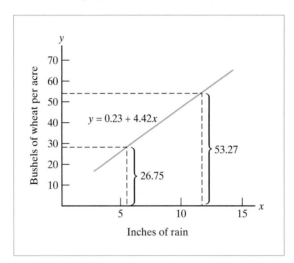

Since we are limiting our discussion here to linear equations, let us point out that **linear equations are useful and important not only because many relationships are actually of this form, but also because they often provide close approximations to relationships that would otherwise be difficult to describe in mathematical terms.** The problem of describing data through other kinds of curves, that is, the problem of *nonlinear curve fitting*, is discussed in the more theoretical texts listed in the bibliography at the end of the book.

12.2 THE METHOD OF LEAST SQUARES

There are many ways in which we can fit a straight line to a set of data points, that is, to a set of paired data. Once we have decided to fit a straight line to a given set of paired data, we then face the problem of finding the equation of the particular line that in some sense provides the best possible fit. To show what is involved, let us consider the following sample data obtained in a study of the relationship between the number of years that applicants for certain foreign service jobs have studied German in high school or college and the grades that they received on a proficiency test in that language.

Number of years x	Grade in test y
3	57
4	78
4	72
2	58
5	89
3	63
4	73
5	84
3	75
2	48

If we plot the points that correspond to these 10 pairs of values, as in Figure 12.2, we observe that, even though the points do not all fall on a straight line, the overall pattern of the relationship is reasonably well described by the blue line. Thus, we feel justified in deciding that a straight line is a plausible description of the underlying relationship.

This takes care of the first kind of problem, and we must now consider the second kind, that of finding the equation of the line that in some sense provides the best fit to the data and that, it is hoped, will later yield the best possible predictions of y for given values of x. Logically, there is no limit to the number of straight lines that we can draw on a piece of paper, say, on the diagram of Figure 12.2. Some of them would fit the data so poorly that we cannot consider them seriously, but many others would come fairly close; and so the problem is to find the one line that fits the data best in some well-defined sense. If all the points actually fell on a straight line, there would be no problem,

Figure 12.2
Data on number of years applicants studied German and their grades on a proficiency test.

but this is an extreme case, which we rarely encounter in practice. In general, we have to be satisfied with a line having certain desirable properties, short of perfection.

The criterion that, today, is used almost exclusively for defining a best fit dates back to the early part of the nineteenth century and the work of the French mathematician Adrien Legendre. It is called the **method of least squares**, and it requires that the sum of the squares of the vertical deviations (distances) from the points to the line be as small as possible.

To explain why we do this, let us refer to Figure 12.3, where we plotted the **data points** corresponding to the data

x	y
1	6
5	5
9	7

and more or less arbitrarily *fit* a straight line; actually, we drew the line through the points $(1, 8)$ and $(9, 4)$.

Had we used this line to predict the values of y for the given values of x, that is, if we had marked $x = 1, x = 5$, and $x = 9$ on the horizontal scale and used the line to read off the corresponding values of y, we would have obtained 8, 6, and 4. The errors of these predictions are $6 - 8 = -2, 5 - 6 = -1$, and $7 - 4 = 3$, and in Figure 12.3 they are the vertical distances from the points to the line.

The sum of these errors is $(-2) + (-1) + 3 = 0$, but this is not indicative of their size, and we find ourselves in a position similar to that on page 70, which led to the definition of the standard deviation. Squaring the errors as we squared the deviation from the mean on page 71, we find that the sum of the squares of the errors is $(-2)^2 + (-1)^2 + 3^2 = 14$.

Figure 12.3
Errors made by using a given line to "predict" the y-values for $x = 1, x = 5$, and $x = 9$.

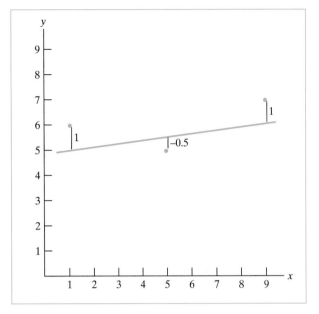

Figure 12.4
Errors made by using
a second line to predict
the y-values for $x = 1$,
$x = 5$, and $x = 9$.

Now let us fit another line to the same data points as before. The one shown in Figure 12.4 actually passes through the points (1, 5) and (9, 6), and judging by eye, it seems to provide a much better fit than the line of Figure 12.3. As can easily be verified, the errors of the predictions, the differences between the observed values of y and the corresponding values read off the line, are now, $6 - 5 = 1, 5 - 5.5 = -0.5$, and $7 - 6 = 1$.

The sum of these errors is $1 + (-0.5) + 1 = 1.5$, which is greater than the sum that we obtained for the errors made with the line of Figure 12.3, but this is not indicative of their size. On the other hand, the sum of the squares of the errors is now $1^2 + (-0.5)^2 + 1^2 = 2.25$, and this is much less than the 14 we obtained for the line of Figure 12.3. Indeed, this is indicative of the fact that the line of Figure 12.4 provides a better fit than the line of Figure 12.3, and we define the *best-fitting* line as the one for which the sum of the squares of the errors is as small as possible. Such a line is called a **least-squares line**.

To show how a least-squares line is actually fit to a set of paired data, let us consider n pairs of numbers $(x_1, y_1), (x_2, y_2), \ldots$, and (x_n, y_n), which might represent the heights and weights of n persons, the heights of n mothers and sons, the monthly food expenditures and mortgage payments of n families, and so on. If we write the equation of the line as $\hat{y} = a + bx$, where the symbol $\hat{y}$ is used to distinguish between the observed values of y and the corresponding values $\hat{y}$, on the line, the method of least squares requires that we minimize the sum of the squares of the differences between the y's and the $\hat{y}$'s (see Figure 12.5). This means that we must find the numerical values of the constants a and b appearing in the equation $\hat{y} = a + bx$ for which

$$\sum(y - \hat{y})^2 = \sum[y - (a + bx)]^2$$

is as small as possible. As it takes calculus or fairly tedious algebra to find the expressions for a and b that minimize this sum, let us merely state the result

Figure 12.5
Least-squares line.

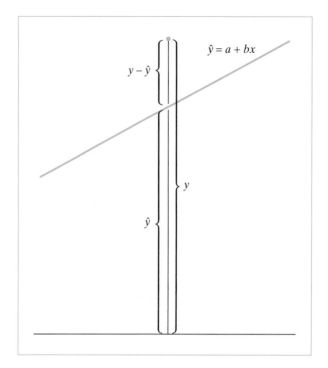

that they are given by the solutions for a and b of the following system of linear equations:

NORMAL
EQUATIONS

$$\sum y = na + b\left(\sum x\right)$$
$$\sum xy = a\left(\sum x\right) + b\left(\sum x^2\right)$$

In these equations, called the **normal equations**, n is the number of pairs of observations, $\sum x$ and $\sum y$ are the sums of the observed x's and y's, $\sum x^2$ is the sum of the squares of the x's, and $\sum xy$ is the sum of the products obtained by multiplying each x by the corresponding y. Since there are easier formulas for computational purposes, we generally do not fit a line numerically using these formulas; the following example illustrates nonetheless how it might be done.

EXAMPLE 12.1   Fit a least-squares line to the data pertaining to the numbers of years that certain applicants for foreign service jobs have studied German in high school or college and the grades that they received in a proficiency test in that language.

Solution   The sums needed for substitution into the normal equations are obtained by performing the calculations shown in the following table:

Number of Years x	Test Grade y	x^2	xy
3	57	9	171
4	78	16	312
4	72	16	288
2	58	4	116
5	89	25	445
3	63	9	189
4	73	16	292
5	84	25	420
3	75	9	225
2	48	4	96
$\sum x = 35$	$\sum y = 697$	$\sum x^2 = 133$	$\sum xy = 2,554$

(There are many handheld calculators on which the various sums can be accumulated directly, so that there is no need to fill in all the details. Indeed, on some calculators the values of a and b can be obtained directly by recording the data and then pressing the appropriate buttons.) Substituting $\sum x = 35$, $\sum y = 697$, $\sum x^2 = 133$, $\sum xy = 2,554$, and $n = 10$ into the two normal equations, we get

$$697 = 10a + 35b$$

$$2,554 = 35a + 133b$$

and we must now solve these two simultaneous linear equations for a and b. There are several ways in which this can be done; the simplest, perhaps, is the **method of elimination**, which the reader may recall from elementary algebra.

To use this method we multiply each term on both sides of the first normal equation by 7. We also multiply on both sides of the second normal equation by 2. We thus get

$$4,879 = 70a + 245b$$

$$5,108 = 70a + 266b$$

We then subtract the *first* equation from the *second* equation. Since a now has the coefficient 70 in both equations, we have thus eliminated a altogether.

$$229 = 21b$$

$$\frac{229}{21} = b \approx 10.90$$

When we substitute 10.90 for b into the first of the normal equations, we get

$$697 = 10a + 35(10.90)$$

$$697 = 10a + 381.5$$

$$\frac{697 - 381.5}{10} = a = 31.35$$

Thus the equation of the least-squares line, $y = a + bx$, is

$$y = 31.55 + 10.90b$$ ∎

As an alternative to this procedure, we will present a number of formulas that will be extremely useful in later calculations, which we will make in connection with least-squares problems.

USEFUL FORMULAS FOR LEAST-SQUARES CALCULATIONS

$$\bar{x} = \frac{\sum x}{n} \qquad \bar{y} = \frac{\sum y}{n}$$

$$S_{xx} = \sum x^2 - \frac{\left(\sum x\right)^2}{n} \qquad S_{yy} = \sum y^2 - \frac{\left(\sum y\right)^2}{n}$$

$$S_{xy} = \sum xy - \frac{\left(\sum x\right) \cdot \left(\sum y\right)}{n}$$

The formulas for $\bar{x}$ and $\bar{y}$ are familiar. The new formulas S_{xx}, S_{xy}, and S_{yy} will appear in many of the formulas that follow. They give us an easy numerical technique for solving the normal equations.

SOLUTIONS OF NORMAL EQUATIONS

$$b = \frac{S_{xy}}{S_{xx}} \qquad a = \bar{y} - b\bar{x}$$

The computational process for finding the least-squares estimates (the solutions of the normal equations) involves several steps. First, find the values of the five sums

$$\sum x, \sum y, \sum x^2, \sum y^2, \quad \text{and} \quad \sum xy$$

Then find the five quantities $\bar{x}$, $\bar{y}$, S_{xx}, S_{xy}, and S_{yy}. Finally, use the formulas above to find b and a. Note that you must find b first, since a is calculated in terms of b. The quantity S_{yy} has not been used yet, but it will appear in Section 12.3.

EXAMPLE 12.2   Rework the previous example, using the formulas given previously.

Solution   In the table on page 413, we found that

$$\sum x = 35, \sum y = 697, \sum x^2 = 133, \quad \text{and} \quad \sum xy = 2{,}554$$

(We do not yet need S_{yy} in this problem, so we can complete the calculation of a and b without either S_{yy} or $\sum y^2$. For future reference, $\sum y^2 = 50{,}085$.)
We note that $n = 10$, so that

$$\bar{x} = \frac{35}{10} = 3.5 \quad \text{and} \quad \bar{y} = \frac{697}{10} = 69.7$$

Also, since

$$S_{xx} = 133 - \frac{(35)^2}{10} = 133 - \frac{1,225}{10} = 10.5$$

and

$$S_{xy} = 2,554 - \frac{35 \cdot 697}{10} = 2,554 - \frac{24,395}{10} = 114.5$$

it follows that

$$b = \frac{S_{xy}}{S_{xx}} = \frac{114.5}{10.5} \approx 10.90$$

and

$$a = \bar{y} - b\bar{x} = 69.7 - 10.90 \cdot 3.5 = 31.55$$

A computer printout of the preceding linear regression problem is shown in Figure 12.6. The values obtained for a and b are 31.533 and 10.905, rounded to three decimals, given in the column headed PARAMETER ESTIMATE. The

Figure 12.6
Computer printout for linear regression.

```
data A;
input x y @ @;
cards;
3 57 4 78 4 72 2 58 5 89 3 63 4 73 5 84 3 75 2 48
;
proc reg;
model y=x;
                            SAS
DEP VARIABLE: Y
                    ANALYSIS OF VARIANCE

                    SUM OF              MEAN
   SOURCE   DF      SQUARES           SQUARE    F VALUE    PROB>F

   MODEL     1     1248.59524       1248.59524   39.094    0.0002
   ERROR     8     255.50476190    31.93809524
   C TOTAL   9     1504.10000

      ROOT MSE          5.65138       R-SQUARE     0.8301
      DEP MEAN          69.7          ADJ R-SQ     0.8089
      C.V.              8.108149


                    PARAMETER ESTIMATES

                    PARAMETER      STANDARD      T FOR HO:
   VARIABLE   DF    ESTIMATE         ERROR   PARAMETER=0   PROB >|T|

   INTERCEP    1   31.53333333    6.36041828    4.958      0.0011
   X           1   10.90476190    1.74405371    6.253      0.0002
```

differences between the results obtained previously and in the printout are due to rounding.

There are relatively inexpensive handheld calculators that are programmed to solve linear regression problems. They yield the values of a and b after we enter the data and press the appropriate buttons. Since it is difficult to check for errors in entering the data and also difficult to correct errors, these calculators should be used only for very small problems.

Once we have determined the equation of a least-squares line, we can use it to make predictions.

EXAMPLE 12.3   Use the least-squares line $\hat{y} = 31.55 + 10.90x$ to predict the proficiency grade of an applicant who has studied German in high school or college for two years.

Solution   Substituting $x = 2$ into the equation, we get

$$\hat{y} = 31.55 + 10.90(2) = 53.35$$

and this is the best prediction we can make in the least-squares sense.   ■

When we make a prediction like this, we cannot really expect that we will always hit the answer right on the nose; in fact, we cannot possibly be right when the answer has to be a whole number, as in our example, and the prediction is 53.35. With reference to this example, it would be very unreasonable to expect that every applicant who had studied German for a given number of years would get the same grade in the proficiency test; indeed, the data on page 409 shows that this is not the case. Thus, to make meaningful predictions based on least-squares lines, we must look at such predictions as averages, or as mathematical expectations. Interpreted in this way, we refer to least-squares lines that enable us to read off, or calculate, expected values of y for given values of x as **regression lines**. Better yet, we refer to such lines as **estimated regression lines**, since the values of a and b are determined on the basis of sample data. Questions relating to the quality of these estimates will be discussed in Section 12.3.

EXERCISES

Exercise 12.1 is a practice exercise; its complete solution is given on page 434.

✓ **12.1**   The following table shows the length of time that six persons have been working at an automobile inspection station and the number of cars each of them checked between noon and 1 o'clock on a given day:

Number of weeks employed x	Cars checked y
5	16
1	15
7	19
9	23
2	14
12	21

(a) Set up the normal equations and solve them by the method of elimination to find the equation of the least-squares line that will enable us to predict y in terms of x.

(b) Use the formulas on page 414 to check the values obtained for a and b in part (a).

(c) If a person has worked at the inspection station for 10 weeks, how many cars can we expect him or her to inspect during the given time period?

12.2 The following data pertains to the chlorine residual in a swimming pool at various times after it has been treated with chemicals:

Number of hours x	Chlorine residual (parts per million) y
2	1.8
4	1.5
6	1.4
8	1.1
10	0.9

(a) Use the formulas for a and b on page 414 to find the equation of the least-squares line from which we can predict the chlorine residual in terms of the time that has elapsed since the pool was treated with chemicals.

(b) Use the equation of the least-squares line to estimate the chlorine residual in the pool five hours after it has been treated with chemicals.

12.3 A firm has six offices with various numbers of employees. Free coffee is provided for the employees and the manager wants to know if there is a relationship between the number of cups of coffee provided and the number of employees in the office. The data is given in the following table. (Use formulas found on page 414.)

Number of employees	Number of cups of coffee
11	18
13	36
15	40
18	50
21	58
24	74

Find

(a) $n, \sum x, \sum y, \sum xy, \bar{x}$, and $\bar{y}$;

(b) S_{xx} and S_{xy};

(c) b and then find a;

(d) the equation of the least-squares line;

(e) a prediction made in the least-squares sense for the number of cups of coffee provided to an office with 20 employees.

12.4 Verify that the linear equation of the illustration on pages 407 and 408 can be obtained by fitting a least-squares line to the following data:

Rain fall (inches)	Yield of wheat (bushels per acre)
12.9	62.5
7.2	28.7
11.3	52.2
18.6	80.6
8.8	41.6
10.3	44.5
15.9	71.3
13.1	54.4

12.5 The records of patients at a children's hospital are maintained by a physician who wants to know the relationship between the ages and heights of boys who are patients. The data for seven randomly selected boys is shown in the following table.

Age to the nearest year	Height in inches
13	62
11	56
11	58
6	46
9	51
7	50
10	54

Using the least-squares equations shown on page 414, estimate the height of a boy patient who is 12 years old.

12.6 The following are the numbers of hours a runner has run during each of eight weeks and the corresponding times in which she ran a mile at the end of the week:

Number of hours run	Time of mile (minutes)
13	5.2
15	5.1
18	4.9
20	4.6
19	4.7
17	4.8
21	4.6
16	4.9

(a) Find the equation of the least-squares line that will allow us to predict the runner's time for the mile from the number of hours that she ran that week.

(b) Use the equation obtained in part (a) to predict how fast she will run a mile at the end of a week in which she ran for 18 hours.

12.7 During its first five years of operation, a company's gross income from sales was 1.2, 1.9, 2.4, 3.3, and 3.5 million dollars. Fit a least-squares line and, assuming that the trend continues, predict this company's gross income from sales during its sixth year of operation.

12.8 The following is data on the IQs of 25 students, the numbers of hours they studied for a certain achievement test, and their scores on the test:

IQ	Number of hours studied	Score on test
105	8	80
98	6	62
112	10	91
102	9	77
107	14	89
95	6	65
100	18	96
110	10	85
102	15	94
96	24	91
115	8	88
105	11	85

IQ	Number of hours studied	Score on test
135	9	99
92	12	76
90	18	83
109	3	70
94	5	61
114	10	86
106	15	93
121	4	92
99	16	82
103	21	95
114	15	90
107	8	79
108	12	88

Use a computer package or a graphing calculator to find the least-squares line that will enable us to predict a student's score on the test in terms of his or her IQ. These data will be used also in Exercises 12.9, 12.10, 12.19 through 12.22, 12.39, and 12.40.

12.9 With reference to the data of Exercise 12.8, use a computer package or a graphing calculator to find the least-squares line that will enable us to predict a student's score on the test in terms of the number of hours that he or she studied for the test.

12.10 With reference to the data of Exercise 12.8, use a computer package or a graphing calculator to find the least-squares line that will enable us to predict how many hours a student will study for the test, given his or her IQ.

*12.3 REGRESSION ANALYSIS

In the preceding section we used a least-squares line to predict that an applicant who has studied German in high school or college for two years will score 53.35 on the proficiency test; but even if we interpret the line correctly as a regression line (that is, treat predictions made from it as averages or expected values), several questions remain to be answered.

How good are the values we found for the constants a and b in the equation. $\hat{y} = a + bx$? After all, $a = 31.55$ and $b = 10.90$ are only estimates based on sample data, and if we base our work on a sample of ten other applicants for the foreign service jobs, the method of least squares would almost certainly lead to different values of a and b.

How good an estimate is 53.35 of the average score of applicants who have had two years of German in high school or college?

In the first of these questions, we said that $a = 31.55$ and $b = 10.90$ are "only estimates based on sample data." This implies the existence of corresponding true values, usually denoted by α and β and referred to as **regression coefficients**. The true regression line is $y = \alpha + \beta x$. To distinguish between a and α and between

Figure 12.7
Distributions of y for given values of x.

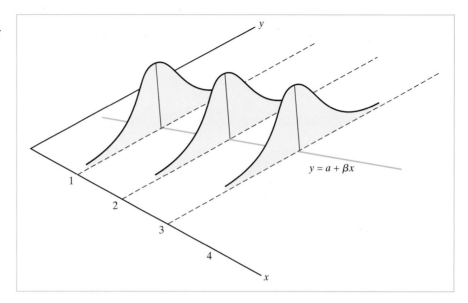

$y = a + \beta x$

x

b and β, we refer to a and b as the **estimated regression coefficients**, and we use the phrase **population regression coefficients** in discussing α and β.

To clarify the idea of a true regression line $y = \alpha + \beta x$, consider Figure 12.7, which shows the distributions of y for several values of x. In connection with our example, these curves represent the distributions of the proficiency scores of persons who have had, respectively, one, two, or three years of German in high school or college. To complete the picture, we can visualize similar curves corresponding to all other values of x within the range of values under consideration.

In **linear regression analysis**, we assume that the x's are known constants, and that for each x the random variable y has a certain distribution (as pictured in Figure 12.7) with the mean $\alpha + \beta x$. In **normal regression analysis**, we assume, furthermore, that these distributions are all normal distributions with the same standard deviation σ. In other words, the distributions pictured in Figure 12.7, as well as those that we add mentally, are normal curves with means on the line $y = \alpha + \beta x$ and the same standard deviation σ.

Based on these assumptions, we can make all sorts of inferences about the regression coefficients α and β. They all require that we estimate σ, the common standard deviation of the normal distributions pictured in Figure 12.7. The estimate we shall use for this purpose is called the **standard error of estimate** and it is denoted by s_e. Its formula is

STANDARD ERROR
OF ESTIMATE
(DEFINING FORMULA)

$$s_e = \sqrt{\frac{\sum (y - \hat{y})^2}{n - 2}}$$

where, again, the y's are the observed values of y, and the $\hat{y}$'s are the corresponding values on the least-squares line. Observe that s_e^2 is the sum of the squares of the vertical deviations from the points to the line in Figure 12.5 (the quantity that we minimized by the method of least squares) divided by $n - 2$.

The preceding formula defines s_e, but in practice we calculate its value by means of the computing formula

STANDARD ERROR
OF ESTIMATE
(COMPUTING
FORMULA)

$$s_e = \sqrt{\frac{S_{yy} - \frac{(S_{xy})^2}{S_{xx}}}{n - 2}}$$

The symbols were defined on page 414.

EXAMPLE 12.4   Calculate s_e for the example on page 409, where we had

	Number of years x	Test grade y
	3	57
	4	78
	4	72
	2	58
	5	89
	3	63
	4	73
	5	84
	3	75
	2	48
Total	35	697

Solution   In working with these numbers, we noted that $n = 10, \bar{x} = 3.5, \bar{y} = 69.7, S_{xx} = 10.5$, and $S_{xy} = 114.5$, but we still need the value of S_{yy}. Since we found that $\sum y^2 = 50{,}085$, it follows that

$$S_{yy} = 50{,}085 - \frac{(697)^2}{10} = 1{,}504.10$$

Substituting these values into the computing formula for s_e, we find that

$$s_e = \sqrt{\frac{1{,}504.10 - \frac{(114.5)^2}{10.5}}{10 - 2}} \approx 5.651$$

Since α is just the y-intercept of the regression line (that is, the value of y that corresponds to $x = 0$), it is not of any special interest in most problems. We shall consider here only inferences about β. They will be based on the statistic

STATISTIC FOR
INFERENCES
CONCERNING β

$$t = \frac{b - \beta}{s_e} \sqrt{S_{xx}}$$

whose sampling distribution is the t distribution with $n - 2$ degrees of freedom.

EXAMPLE 12.5 Based on the data on page 421, test the hypothesis that each additional year of German in high school or college adds another 12.5 points to the expected proficiency score of an applicant. Use the alternative hypothesis $\beta \neq 12.5$ and the level of significance 0.05.

Solution
1. *Hypotheses*
$$H_0: \quad \beta = 12.5$$
$$H_A: \quad \beta \neq 12.5$$

2. *Level of significance* $\alpha = 0.05$
3. *Criterion* Reject the null hypothesis if $t \leq -2.306$ or $t \geq 2.306$, where 2.306 is the value of $t_{0.025}$ for $10 - 2 = 8$ degrees of freedom, and t is given by

$$t = \frac{b - \beta}{s_e} \sqrt{S_{xx}}$$

Otherwise, accept the null hypothesis or reserve judgment.

4. *Calculations* Substituting $b = 10.90$, $\beta = 12.5$, $S_{xx} = 10.5$, and $s_e = 5.651$ into the formula for t, we get

$$t = \frac{10.90 - 12.5}{5.651} \sqrt{10.5} \approx -0.92$$

5. *Decision* Since $t = -0.92$ falls on the interval from -2.306 to 2.306, the null hypothesis cannot be rejected; that is, the difference between $b = 10.90$ and $\beta = 12.5$ may be reasonably attributed to chance. ∎

In using the t statistic given previously, the most common value tested, by far, is $\beta = 0$.

To construct confidence intervals for the regression coefficient β, we substitute for the middle term of $-t_{\alpha/2} < t < t_{\alpha/2}$ the preceding t statistic, and relatively simple algebra yields

CONFIDENCE LIMITS FOR THE REGRESSION COEFFICIENT β

$$b \pm t_{\alpha/2} \cdot \frac{s_e}{\sqrt{S_{xx}}}$$

The degree of confidence is $1 - \alpha$. The value of $t_{\alpha/2}$ for $n - 2$ degrees of freedom may be read from Table III. Note that the quantity

$$\frac{s_e}{\sqrt{S_{xx}}}$$

by which $t_{\alpha/2}$ is multiplied and by which $b - \beta$ is divided in the formula for the t statistic on page 421, is an estimate of the standard deviation of the sampling distribution of the least-squares estimate of β.

EXAMPLE 12.6 Based on the data on page 421, construct a 99% confidence interval for β, the expected increase in the proficiency score of an applicant for each additional year of German in high school or college.

Solution Substituting $t_{0.005} = 3.355$ for $10 - 2 = 8$ degrees of freedom into the confidence-limits formula, as well as the previously determined values of b, s_e, and S_{xx}, we get

$$10.90 \pm 3.355 \cdot \frac{5.651}{\sqrt{10.5}}$$

or 10.90 ± 5.85. Thus, the 99% confidence interval for β is

$$5.05 < \beta < 16.75$$

This interval is rather wide, and this is due to two things—the magnitude of the variation as measured by s_e and the very small sample size n. ∎

Many computer packages have programs for fitting least-squares lines, and most of these programs also provide the answers for the two preceding examples, or at least they greatly facilitate the calculations. The computer printout shown in Figure 12.6 does not actually give the value of the t statistic for testing $\beta = 12.5$ or the confidence limits; but it tells us that s_e, referred to there as ROOT MSE, is 5.651, and in the column headed STANDARD ERROR in the row labeled X that the quantity

$$\frac{s_e}{\sqrt{S_{xx}}}$$

referred to previously is 1.744. The computer printout agrees, within rounding, with the hypothesis test results on page 422 and the preceding confidence interval.

To answer the question asked on page 419 concerning the goodness of an estimate, or prediction, based on a least-squares equation, we use a method that is very similar to the one discussed on page 422. Basing our argument on another t statistic, we arrive at the following confidence limits for $\alpha + \beta x_0$, which is the mean of y when $x = x_0$:

CONFIDENCE LIMITS FOR THE MEAN OF y WHEN $x = x_0$

$$(a + bx_0) \pm t_{\alpha/2} \cdot s_e \sqrt{\frac{1}{n} + \frac{(x_0 - \bar{x})^2}{S_{xx}}}$$

The degree of confidence is $1 - \alpha$ and, as before, $t_{\alpha/2}$ for $n - 2$ degrees of freedom may be read from Table III.

EXAMPLE 12.7 Referring again to the data on page 421, suppose that the original purpose of the study was to estimate the average proficiency score of applicants who have had two years of German in high school or college. Construct a 99% confidence interval for this mean.

Solution The value of $a + bx_0$ is $31.55 + 10.90(2) = 53.35$. Also, $s_e = 5.651$, $\bar{x} = 3.5$, $S_{xx} = 10.5$, $n = 10$, $x_0 = 2$, and 3.355 is the value of $t_{0.005}$ for $10 - 2 = 8$ degrees of freedom. We substitute these values into the confidence interval formula and get

$$53.35 \pm (3.355)(5.651) \sqrt{\frac{1}{10} + \frac{(2 - 3.5)^2}{10.5}}$$

or

$$53.35 \pm 10.63$$

Hence, the 99% confidence interval for the mean proficiency score of applicants who have had two years of German in high school or college is the interval from $53.35 - 10.63 = 42.72$ to $53.35 + 10.63 = 63.98$. ■

Exercises 12.11 and 12.16 are practice exercises; their complete solutions are given on pages 434 and 435.

✓ ***12.11** With reference to Exercise 12.1, test the null hypothesis $\beta = 1.5$ (the hypothesis that each additional week on the job adds 1.5 to the number of cars a person can be expected to inspect in the given period of time) against the alternative hypothesis $\beta < 1.5$. Use the level of significance 0.05.

*12.12 The following table shows the assessed values and the selling prices of eight houses, constituting a random sample of all the houses sold recently in a given metropolitan area:

Assessed value (in $1,000)	Selling price (in $1,000)
40.3	63.4
72.0	118.3
32.5	55.2
44.8	74.0
27.9	48.8
51.6	81.1
80.4	123.2
58.0	92.5

Fit a least-squares line that will enable us to predict the selling prices of a house in terms of its assessed value, and test the null hypothesis $\beta = 1.30$ against the alternative hypothesis $\beta > 1.30$ at the level of significance 0.05.

*12.13 The following data shows the average numbers of hours that six students studied per week and their grade-point indexes:

Hours studied x	Grade-point index y
15	2.0
28	2.7
13	1.3
20	1.9
4	0.9
10	1.7

Test the null hypothesis $\beta = 0.10$ (the hypothesis that each additional hour of study per week will raise the expected grade-point index of the students by 0.10) against the alternative hypothesis $\beta < 0.10$ at the 0.01 level of significance.

*12.14 In a campus bar and grill, hamburger platters account for most of the food sales. The owner would like to evaluate how the price of these platters affects her weekly profit, so she experiments by varying the price during nine different weeks, with the following results:

Price	Profit
$2.00	$2,325
2.50	2,460
3.00	1,600
3.00	1,700
2.00	2,000
2.50	1,800
2.50	2,500
2.00	2,400
3.50	1,700

Fit a least-squares line to these data and construct a 95% confidence interval for the regression coefficient β.

*12.15 With reference to Exercise 12.3, construct a 99% confidence interval for the regression coefficient β.

✓ *12.16 With reference to Exercise 12.13, construct a 99% confidence interval for the mean grade-point index of students who study on the average 12 hours per week.

*12.17 With reference to Exercise 12.12, construct a 95% confidence interval for the average selling price of a house in the given metropolitan area that has an assessed value of $60,000.

*12.18 The following data shows the advertising expenses (expressed as a percentage of total expenses) and the net operating profit (expressed as a percentage of total sales) in a random sample of five furniture stores:

Advertising expenses x	Net operating profit y
1.1	2.5
2.8	3.7
3.1	5.2
1.6	2.9
0.7	1.4

Fit a least-squares line to these data and construct a 99% confidence interval for the mean net operating profit (expressed as a percentage of total sales) when the advertising expenses are 2.0% of total expenses.

*12.19 With reference to Exercise 12.8, use a computer package to test the null hypothesis $\beta = 0.50$ (that each unit increase in IQ adds on the average 0.50 to the score) against the alternative $\beta > 0.50$. Use the 0.05 level of significance.

*12.20 With reference to Exercise 12.8, use a computer package to construct a 95% confidence interval for β.

*12.21 With reference to Exercise 12.9, use a computer package to construct a 99% confidence interval for β (the average increase in the score for each additional hour studied).

*12.22 With reference to Exercise 12.10, use a computer package or a graphing calculator to test the null hypothesis $\beta = 0$ (that a student's IQ does not affect how many hours he or she will study for the test) against the alternative $\beta \neq 0$. Use the 0.05 level of significance.

12.4 THE COEFFICIENT OF CORRELATION

Having learned how to fit a least-squares line to a set of paired data, let us now turn to the problem of determining how well such a line actually fits the data.

Of course, we can get some idea by inspecting a diagram like that of Figure 12.2; but to show how we can be more objective, let us refer to the original data of the example dealing with the foreign service job applicants' proficiency grades in German and the number of years that they have studied the language in high school or college. Copying the data from page 409, we get

Number of years x	Grade on test y
3	57
4	78
4	72
2	58
5	89
3	63
4	73
5	84
3	75
2	48

and it can be seen that there are considerable differences among the y's; the smallest value is 48 and the largest value is 89. However, it can also be seen that the grade of 48 was obtained by an applicant who has studied German for two years, while the grade of 89 was obtained by an applicant who has studied German for five years. This suggests that the differences among the grades may well be due, at least in part, to the fact that the applicants have not all studied German for the same number of years. This raises the following question:

Of the total variation among the y's, how much can be attributed to chance and how much can be attributed to the relationship between the two variables x and y, that is, to the fact that the observed y's correspond to different values of x?

With reference to our example, we would thus want to know what part of the variation among the grades can be attributed to the differences in the number of years that the applicants have studied German, and what part can be attributed to all other factors (the applicants' intelligence, their health or frame of mind on the day they took the test, . . . , and their luck in guessing at some of the answers), which we combine under the general heading of chance.

A convenient measure of the total variation of the observed y's is the quantity $S_{yy} = \sum (y - \bar{y})^2$, which is the **total sum of squares** of the y's and which is simply the variance of the y's multiplied by $n - 1$. We note that for the preceding test grades we had $S_{yy} = 1,504.10$. The calculation was done on page 421.

This takes care of the total variation of the y's, but it remains to be seen how we might measure its two parts, which are attributed, respectively, to the relationship between x and y and to chance. To this end, let us point out that, if the number of years the applicants have studied German is the only thing that affects their grades in the test and the relationship is linear, then all the data points would fall on a straight line. As is apparent from Figure 12.2, this is not the case in our example, and the extent to which the points fluctuate above and below the line provides us with an indication of the size of the chance variations. Thus, chance variation is measured by the sum of the squares of the vertical deviations from the points to the line, that is, by the quantity $\sum(y - \hat{y})^2$, which is called the **residual sum of squares**. This is the quantity that we minimized on page 411 in accordance with the least-squares criterion.

EXAMPLE 12.8   Calculate the residual sum of squares for the 10 test grades.

Solution   To calculate this sum of squares, we must first determine the value of $y = a + bx = 31.55 + 10.90x$ for each of the given values of x, and we get

$$31.55 + 10.90(3) = 64.25$$

$$31.55 + 10.90(4) = 75.15$$

$$\cdots\cdots\cdots\cdots\cdots\cdots$$

$$31.55 + 10.90(2) = 53.35$$

Then

$$\sum(y - \hat{y})^2 = (57 - 64.25)^2 + (78 - 75.15)^2 + \cdots + (48 - 53.35)^2$$

$$\approx 255.50$$

The technique is painful when the sample size is large, and generally we make use of the identity

RESIDUAL SUM OF SQUARES

$$\sum(y - \hat{y})^2 = S_{yy} - \frac{(S_{xy})^2}{S_{xx}}$$

For the preceding problem, this yields

$$\sum(y - \hat{y})^2 = 1,504.10 - \frac{(114.5)^2}{10.5} \approx 255.50$$

From $S_{yy} = \sum(y - \bar{y})^2 = 1,504.10$ and $\sum(y - \hat{y})^2 = 255.50$, it follows that

$$\frac{\sum(y - \hat{y})^2}{\sum(y - \bar{y})^2} = \frac{255.50}{1,504.10} \approx 0.17$$

which is the proportion of the total variation of the grades that can be attributed to chance, while

$$1 - \frac{\sum(y - \hat{y})^2}{\sum(y - \bar{y})^2} = 1 - 0.17 = 0.83$$

is the proportion of the total variation of the grades that can be attributed to the relationship with x, that is, to the differences in the number of years that the applicants have studied German.

If we take the square root of the last proportion (the proportion of the total variation of the y's that can be attributed to the relationship with x), we obtain the statistical measure which is called the **coefficient of correlation**. It is denoted by the letter r, and its sign is chosen so that it is the same as that of the estimated regression coefficient b. Thus, for our example we get

$$r = \sqrt{0.83} \approx 0.91$$

It is of interest to note that all the quantities we have calculated so far in this section can be obtained from the computer printout of Figure 12.6. Under ANALYSIS OF VARIANCE, in the column headed SUM OF SQUARES, we find that the ERROR sum of squares is 255.50 and that the TOTAL sum of squares is 1,504.10. Also, R-SQUARE = 0.83 tells us that the square of the correlation coefficient is 0.83, and this also agrees with the result we obtained before.

It follows from the rule for the sign of r that the correlation coefficient is positive when the least-squares line has an upward slope, that is, when the relationship between x and y is such that small values of y tend to go with small values of x and large values of y tend to go with large values of x. Also, the correlation coefficient is negative when the least-squares line has a downward slope, that is, when large values of y tend to go with small values of x and small values of y tend to go with large values of x. Geometrically, examples of **positive** and **negative correlations** are shown in the first two diagrams of Figure 12.8.

Observe also that since r is defined as $\pm$ the square root of a proportion, its values must lie on the interval from -1 to $+1$. When r equals -1 or $+1$, this means that 100% of the variation among the y's can be attributed to the relationship with x; in other words, $\sum(y - \hat{y})^2 = 0$ and the points must all fall on the least-squares line. When $r = 0$, this means that none of the variation among the y's can be attributed to the relationship with x, and we say that there is **no correlation**. This is pictured in the third diagram of Figure 12.8.

Our definition of r shows clearly the nature, or essence, of the coefficient of correlation, but in actual practice it is seldom used to determine its value. Instead, we use the computing formula

COMPUTING FORMULA FOR THE COEFFICIENT OF CORRELATION

$$r = \frac{S_{xy}}{\sqrt{S_{xx} \cdot S_{yy}}}$$

which has the added advantage that it automatically gives r the correct sign. The quantities used in calculating r by this formula have been used previously and were defined on page 414.

Figure 12.8
Types of correlation.

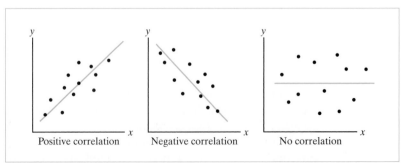

Positive correlation   Negative correlation   No correlation

EXAMPLE 12.9   Use the computing formula to verify the value $r = 0.91$, which we obtained for the data pertaining to the proficiency grades of 10 applicants for certain foreign service jobs and the number of years that they have studied German in high school or college.

Solution   We previously found $S_{xy} = 114.50$, $S_{xx} = 10.50$, and $S_{yy} = 1,504.10$. Substitution into the computing formula for r yields

$$r = \frac{114.50}{\sqrt{10.50 \cdot 1,504.10}} \approx 0.91$$

This agrees, as it should, with the result obtained on page 428.

12.5 THE INTERPRETATION OF r

When r equals $+1$, -1, or 0, there is no question about the interpretation of the coefficient of correlation. As we have already indicated, it is $+1$ or -1 when all the points actually fall on a straight line, and it is 0 when none of the variation among the y's can be attributed to the relationship with x or, in other words, when knowledge of x does not help in the prediction of y. In general, the definition of r tells us that the proportion of the variation of the y's that is due to the relationship with x equals r^2, or that the percentage equals $(100\%) \cdot r^2$, and this is how we interpret the strength of the relationship implied by any value of r.

For instance, if $r = 0.80$ for one set of data and $r = 0.40$ for another, it would be very misleading to say that the correlation of 0.80 is "twice as good" or "twice as strong" as the correlation of 0.40. When $r = 0.80$, then $100\%(0.80)^2 = 64\%$ of the variation of the y's is accounted for by the relationship with x, and when $r = 0.40$, then $100\%(0.40)^2 = 16\%$ of the variation of the y's is accounted for by the relationship with x. Thus, in the sense of "percentage of variation accounted for," we can say that a correlation of 0.80 is *four times as strong* as a correlation of 0.40. By the same token, a correlation of 0.60 is *nine times as strong* as a correlation of 0.20.

There are several pitfalls in the interpretation of r. First, it is often overlooked that r measures only the strength of linear relationships; second, it should be remembered that a strong correlation (a value of r close to $+1$ or -1) does not necessarily imply a cause–effect relationship.

If r is calculated indiscriminantly, for instance, for the three sets of data in Figure 12.9, we get $r = 0.75$ in each case, but it is a meaningful measure of the strength of the relationship only in the first case. In the second case, there is a very strong relationship between the two variables, but it is not linear, and in the third case, six of the seven points actually fall on a straight line; but the seventh point is so far off that it suggests the possibility of a gross error in recording the data. Thus, before we calculate r, we should always plot the data to see whether there is reason to believe that the relationship is, in fact, linear.

The fallacy of interpreting a strong correlation as an indication of a cause–effect relationship is best explained with a few examples. A popular illustration is the strong positive correlation that has been obtained for the annual sales of chewing gum and the incidence of crime in the United States. Obviously, one cannot conclude that the crime rate might be reduced by prohibiting the sale

Figure 12.9
Three sets of paired data
for which $r = 0.75$.

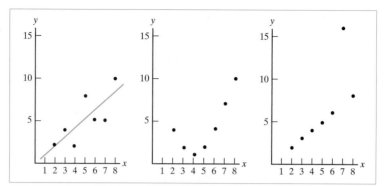

of chewing gum—both variables depend on the size of the population, and it is this mutual relationship with a third variable (population size) that produced the positive correlation. Another example is the strong positive correlation obtained for the number of storks seen nesting in English villages and the number of children born in the same villages. We leave it to the reader's ingenuity to explain why there might be a strong positive correlation in this case in the absence of any cause–effect relationship.

12.6 A SIGNIFICANCE TEST FOR r

It is sometimes overlooked that when r is calculated on the basis of sample data, we may get a fairly strong positive or negative correlation purely by chance, even though there is actually no relationship whatsoever between the two variables under consideration. Suppose, for instance, that we take a pair of dice, one red and one green, roll them a few times, and get the following results:

Red die x	Green die y
3	5
2	2
5	6
3	1
4	3

Calculating r for these data, we get the surprisingly high value $r = 0.66$, and this raises the question of whether anything is wrong with the assumption that there is no relationship here between x and y; after all, one die does not know what the other die is doing. To answer this question, we shall have to see whether this high value of r may be attributed to chance.

When a correlation coefficient is calculated from sample data, as in the example above, the value we obtain for r is only an estimate of a corresponding parameter, the **population correlation coefficient**, which we denote by the Greek letter ρ (lowercase *rho*). What r measures for a sample, ρ measures for a population.

To test the null hypothesis of no correlation, we shall make the same assumptions as in Section 12.3. When these assumptions are met, at least within a reasonable degree of approximation, we reject the null hypothesis of no correlation if $r \leq -r_{\alpha/2}$ or $r \geq r_{\alpha/2}$, where the value of $r_{\alpha/2}$ may be obtained from Table VI for $\alpha = 0.05$ and $\alpha = 0.01$. If the null hypothesis can be rejected, we say that there is a **significant correlation**; otherwise, we say that the value of r is not statistically significant.

EXAMPLE 12.10 Use the 0.05 level of significance to test the null hypothesis of no correlation for the example in which we rolled a pair of dice five times and obtained $r = 0.66$.

Solution   1. *Hypotheses*

$$H_0: \quad \rho = 0$$

$$H_A: \quad \rho \neq 0$$

2. *Level of significance* $\alpha = 0.05$
3. *Criterion*   Reject the null hypothesis if $r \leq -0.878$ or $r \geq 0.878$, where 0.878 is the value of $r_{0.025}$ for $n = 5$; otherwise, state that the value of r is not significant.
4. *Calculations* $r = 0.66$
5. *Decision*   Since $r = 0.66$ falls on the interval from -0.878 to 0.878, the null hypothesis cannot be rejected; in other words, the correlation coefficient of 0.66 is not significant.   ■

EXAMPLE 12.11 Use the 0.01 level of significance to test the null hypothesis of no correlation for the example in which we obtained $r = 0.91$ for the proficiency scores of 10 applicants for certain foreign service jobs and the number of years that they have studied German in high school or college.

Solution   1. *Hypotheses*

$$H_0: \quad \rho = 0$$

$$H_A: \quad \rho \neq 0$$

2. *Level of significance* $\alpha = 0.01$
3. *Criterion*   Reject the null hypothesis if $r \leq -0.765$ or $r \geq 0.765$, where 0.765 is the value of $r_{0.005}$ for $n = 10$; otherwise, state that the value of r is not significant.
4. *Calculations* $r = 0.91$
5. *Decision*   Since $r = 0.91$ exceeds 0.765, the null hypothesis must be rejected; in other words, there is a relationship between the two variables under consideration.   ■

Exercises 12.23, 12.24, and 12.31 are practice exercises; their complete solutions are given on pages 435 and 436.

✓ **12.23** The following data pertains to a study of the effects of environmental pollution on wildlife, in particular, the effect of DDT on the thickness of the eggshells of certain birds:

DDT residues in yolk lipids (parts per million)	Thickness of eggshell (millimeters)
117	0.49
65	0.52
393	0.37
98	0.53
122	0.49

Calculate r for these data.

EXERCISES

✓ **12.24** With reference to Exercise 12.23, test the significance of the value obtained for r at the level of significance 0.01.

12.25 The following are the typing speeds and the reading speeds of nine secretaries:

Typing speed (words per minute)	Reading speed (words per minute)
60	370
56	551
52	528
63	348
70	645
58	454
44	503
79	618
62	500

Measure the extent of the relationship between typing speed and reading speed by calculating the coefficient of correlation.

12.26 With reference to Exercise 12.25, test the null hypothesis of no correlation at the 0.05 level of significance.

12.27 The following table shows the percentages of the vote predicted by a poll for seven candidates for the U.S. Senate in different states, x, and the corresponding percentages of the vote that they actually received, y:

Poll, x	Election, y
42	51
34	31
59	56
41	42
53	58
40	35
55	54

Calculate r for these data.

12.28 With reference to Exercise 12.27, test the null hypothesis $\rho = 0$ at the level of significance $\alpha = 0.05$.

12.29 Calculate r for the grades that 16 students received on final examinations in economics and anthropology.

Economics	Anthropology	Economics	Anthropology
51	74	45	68
68	70	73	87
72	88	93	89
97	93	66	73
55	67	20	33
73	73	91	91
95	99	74	80
74	73	80	86

12.30 With reference to Exercise 12.29, test the null hypothesis $\rho = 0$ at the level of significance $\alpha = 0.05$.

✓ **12.31** With reference to Exercise 12.1, what percentage of the variation of the numbers of cars checked can be attributed to differences in the length of time that the persons have been working at the inspection station?

12.32 With reference to Exercise 12.4, what percentage of the variation in the yield of wheat can be attributed to differences in the amount of rain?

12.33 With reference to Exercise 12.14, what percentage of the variation of the weekly profit can be attributed to the relationship between the weekly profit and the price of the hamburger platters?

12.34 With reference to Exercise 12.25, what percentage of the variation in typing speed can be attributed to the relationship between typing speed and reading speed?

12.35 If we calculate r for each of the following sets of data, should we be surprised if we get $r = 1$ and $r = -1$, respectively? Explain.

(a)	x	y		(b)	x	y
	9	2			4	8
	5	1			7	2

12.36 Check in each case whether the value of r is significant at the 0.05 level of significance:

(a) $n = 12$ and $r = -0.53$;

(b) $n = 20$ and $r = 0.58$;

(c) $n = 15$ and $r = -0.55$;

(d) $n = 9$ and $r = -0.61$.

12.37 Check in each case whether the value of r is significant at the 0.01 level of significance:

(a) $n = 18$ and $r = 0.62$;

(b) $n = 32$ and $r = -0.47$;

(c) $n = 16$ and $r = -0.58$;

(d) $n = 14$ and $r = 0.63$.

12.38 State in each case whether you would expect a positive correlation, a negative correlation, or no correlation:

(a) the ages of husbands and wives;

(b) the amount of rubber on tires and the number of miles that they have been driven;

(c) income and education;

(d) shirt size and sense of humor;

(e) the number of hours that bowlers practice and their scores;

(f) hair color and one's knowledge of foreign affairs.

12.39 With reference to Exercise 12.8, use a computer package to determine the total and residual sums of squares and the value of r measuring the strength of the relationship between students' IQs and their scores on the achievement test.

12.40 With reference to Exercise 12.9, use a computer package to determine the total and residual sums of squares and the value of r measuring the strength of the relationship between the number of hours a student studies for the test and his or her score.

12.41 With reference to Exercise 12.29, use a computer package or a graphing calculator to determine r and also the percentage of the variation of the anthropology scores that can be attributed to the relationship between students' grades in the two subjects.

✓ SOLUTIONS TO PRACTICE EXERCISES

12.1 (a) From the data we get $n = 6$, $\sum x = 36$, $\sum x^2 = 304$, $\sum y = 108$, and $\sum xy = 715$, so the normal equations are

$$108 = 6a + 36b$$
$$715 = 36a + 304b$$

Multiplying the expressions on both sides of the first equation by 6 and leaving the second equation as is, we get

$$648 = 36a + 216b$$
$$715 = 36a + 304b$$

and then, by subtraction, $67 = 88b$ and $b = \dfrac{67}{88} \approx 0.7614$. Substituting this value into the first normal equation, we get $108 = 6a + 36(0.7614)$, which yields $a = 13.4316$. Thus, the equation of the least-squares line is $\hat{y} = 13.4316 + 0.7614x$.

(b) Note first that

$$S_{xx} = \sum x^2 - \frac{\left(\sum x\right)^2}{n} = 304 - \frac{36^2}{6} = 88$$

and that

$$S_{xy} = \sum xy - \frac{\left(\sum x\right) \cdot \left(\sum y\right)}{n} = 715 - \frac{36 \cdot 108}{6} = 67$$

Therefore,

$$b = \frac{67}{88} \approx 0.7614$$

and since $\bar{x} = 6$ and $\bar{y} = 18$, it follows that

$$a = 18 - 0.7614 \cdot 6 = 13.4316$$

(c) Using the values of a and b obtained in part (b), we get

$$\hat{y} = 13.4316 + 0.7614(10) = 21.0456$$

or 21.05 rounded to two decimals.

12.11 1. *Hypotheses*

$$H_0: \quad \beta = 1.5$$
$$H_A: \quad \beta < 1.5$$

2. *Level of significance* $\alpha = 0.05$

3. *Criterion*   Reject the null hypothesis if $t \leq -2.132$, where 2.132 is the value of $t_{0.05}$ for $6 - 2 = 4$ degrees of freedom and

$$t = \frac{b - \beta}{s_e}\sqrt{S_{xx}}$$

Otherwise, accept it or reserve judgment.

4. *Calculations*   To find the value of s_e, we use the formula on page 421, which requires S_{yy}. Since $\sum y^2 = 2{,}008$, we get

$$S_{yy} = 2{,}008 - \frac{108^2}{6} = 64$$

We have already found that $S_{xx} = 88$ and $S_{xy} = 67$, so that

$$s_e = \sqrt{\frac{64 - \dfrac{67^2}{88}}{6 - 2}} \approx 1.8020$$

Then, substituting this value together with $n = 6$, $b = 0.7614$, $\beta = 1.5$, and $S_{xx} = 88$ into the formula for t, we obtain

$$t = \frac{0.7614 - 1.5}{1.8020}\sqrt{88} \approx -3.84$$

5. *Decision*   Since $t = -3.84$ is less than -2.132, the null hypothesis must be rejected; in other words, the slope of the least-squares line is less than 1.5.

12.16   The solution to Exercise 12.13 involved the values $a = 0.7210$, $b = 0.0686$, $S_{xx} = 344$, and $s_e = 0.2720$. We substitute these together with $x_0 = 12$, $\bar{x} = 15$, and 4.604, the value of $t_{0.005}$ for $6 - 2 = 4$ degrees of freedom, into the formula for the confidence limits. This gives

$$[0.7210 + (0.0686)(12)] \pm (4.604)(0.2720)\sqrt{\frac{1}{6} + \frac{(12 - 15)^2}{344}}$$

or

$$1.544 \pm 0.550$$

rounded to three decimals. Hence, the 99% confidence interval for the mean grade-point index of students who study on the average 12 hours per week is the interval from $1.544 - 0.550 = 0.994$ to $1.544 + 0.550 = 2.094$.

12.23   The data gives $n = 5$, $\sum x = 795$, $\sum x^2 = 196{,}851$, $\sum y = 2.40$, $\sum y^2 = 1.1684$, and $\sum xy = 348.26$. From these we obtain

$$S_{xx} = 70{,}446, \quad S_{yy} = 0.0164, \quad \text{and} \quad S_{xy} = -33.34$$

We substitute these into the formula for r to get

$$r = \frac{-33.34}{\sqrt{(70{,}446)(0.0164)}} \approx -0.98$$

12.24   1. *Hypotheses*

$$H_0: \quad \rho = 0$$

$$H_A: \quad \rho \neq 0$$

2. *Level of significance* $\alpha = 0.01$
3. *Criterion*   Reject the null hypothesis if $r \leq -0.959$ or $r \geq 0.959$, where 0.959 is the value of $r_{0.005}$ for $n = 5$; otherwise, state that the value of r is not significant.
4. *Calculations* $r = -0.98$ (from the solution of Exercise 12.23).

5. *Decision*   Since $r = -0.98$ is less than -0.959, the null hypothesis must be rejected; we conclude that there is a relationship between DDT residues and eggshell thickness.

12.31   The data gives $n = 6$, $\sum x = 36$, $\sum x^2 = 304$, $\sum y = 108$, $\sum y^2 = 2{,}008$, and $\sum xy = 715$. From these we obtain

$$S_{xx} = 88, \ S_{yy} = 64, \quad \text{and} \quad S_{xy} = 67$$

We substitute these into the formula for r to get

$$r = \frac{67}{\sqrt{(88)(64)}} \approx 0.893$$

Thus, $100\%(0.893)^2 \approx 79.7\%$ of the variation of the number of cars checked can be attributed to differences in the length of time that the persons have been working at the inspection station.

13

NONPARA-
METRIC TESTS

The tests of hypotheses discussed in the preceding chapters require specific assumptions about the population, or populations, sampled. In many cases we must assume that the population closely resembles the shape of a normal distribution, or that the variances are known to be equal, or that the samples are independent. Since there are many situations in which these assumptions cannot be met, statisticians have developed alternative techniques based on less stringent assumptions, which are known as **nonparametric tests**.

Such tests cannot only be used under more general conditions, but they are often easier to explain and easier to understand than the standard tests that they replace. Moreover, in many nonparametric tests the computational burden is so light that they come under the heading of "quick and easy" or "shortcut" techniques. For these reasons, nonparametric tests have become quite popular, and extensive literature is devoted to their theory and application. In fact, there are entire books dedicated to the subject.

In Sections 13.1 through 13.3, we shall present the sign test as a nonparametric alternative to tests concerning means and differences between means of paired data. In Sections 13.4 through 13.6, we shall study methods based on rank sums, which serve as nonparametric alternatives to the two-sample *t* test and the one-way analysis of variance. Indeed, ranks are often used in nonparametric tests, which makes them applicable for use with ordinal data.

In Sections 13.7 through 13.9, we shall learn how to test the randomness of a sample after the data has been selected; and in Section 13.10, we shall present a nonparametric test concerning the relationship between paired data.

13.1   THE ONE-SAMPLE SIGN TEST

The small-sample tests concerning means and differences between means, which we studied in Sections 10.4 and 10.6, are based on the assumption that the populations sampled have roughly the shape of normal distributions. When this assumption is untenable in practice, these standard tests can be replaced by any one of several nonparametric alternatives, among them the **sign test**, which we shall study in this section and in Sections 13.2 and 13.3.

Small samples usually cannot tell us whether the population sampled follows a normal distribution. A common indicator of a departure from normality is the occurrence of one or more values that straggle far away from the other sample values. If this is the case, most statisticians would be unwilling to assume that they are sampling a normal population. Without a lot of data, decisions about the appropriateness of normal distributions are quite subjective. The nonparametric procedures used in this chapter are valid under fairly general conditions. They should certainly be used when there is any doubt about the appropriateness of assuming normal distributions, and they may even be used when the assumption of normal distributions would be reasonable.

The **one-sample sign test** concerns the median $\tilde{\mu}$ of a continuous population. The probability of getting a sample value less than the median and the probability of getting a sample value greater than the median are both $\frac{1}{2}$. If the population values are symmetrically distributed about the median, then the median $\tilde{\mu}$ and the mean μ are equal. Only occasionally are we in a situation that allows us to assume the symmetry of the population. The procedures that follow will usually be described in terms of the median $\tilde{\mu}$, but we should remember that in certain situations the median and the mean are equal.

Then, to test the null hypothesis $\tilde{\mu} = \tilde{\mu}_0$ against an appropriate alternative on the basis of a random sample of size n, we replace each sample value greater than $\tilde{\mu}_0$ with a plus sign and each sample value less than $\tilde{\mu}_0$ with a minus. Then we test the null hypothesis that the total number of plus signs is the value of a random variable having the binomial distribution with $p = \frac{1}{2}$. If a sample value equals $\tilde{\mu}_0$ exactly, which is certainly possible with rounded data, it is discarded.

EXAMPLE 13.1   In a given month, the average number of gallons of paint used by a (small) random sample of 15 painting contractors was 35.5, 37.3, 35.0, 37.7, 35.2, 33.9, 30.0, 41.3, 43.2, 34.6, 41.1, 33.9, 40.9, 27.8, and 32.0. Use the one-sample sign test to test the null hypothesis $\tilde{\mu} = 35.0$ gallons of paint against the alternative hypothesis $\tilde{\mu} < 35.0$ gallons, at the 0.05 level of significance.

Solution   Since one of the values equals 35 (identical to the value of $\tilde{\mu}_0$) and must therefore be discarded, the sample size for the one-sample sign test is reduced to $n = 14$.

1. *Hypothesis*

$$H_0: \quad \tilde{\mu} = 35.0$$

$$H_A: \quad \tilde{\mu} < 35.0$$

2. *Level of significance* $\alpha = 0.05$
3. *Criterion*   The test statistic is x, the number of plus signs. (In other problems the test may be based on the number of minus signs.)
4. *Calculations*   Replacing each value greater than 35.0 with a plus sign, each value less than 35.0 with a minus sign, and discarding the one value which actually equals 35.0, we get $+ + + + - - + + - + - + - -$. Thus, $x = 8$.
5. *Decision*   Table I shows that for $n = 14$ and $p = 0.50$ the probability that $x \leq 8$ is $0.001 + 0.006 + 0.022 + 0.061 + 0.122 + 0.183 + 0.209 + 0.183 = 0.787$. The p-value is therefore 0.787. Since 0.787 is more than $\alpha = 0.05$ the null hypothesis cannot be rejected. We therefore accept the null hypothesis that $\mu = 35.00$ or reserve judgment. ■

Here we used the procedure based on the p-value. As in Section 11.1, we did this because we were able to refer directly to Table I. For the same reason, we shall do so also in the example of Section 13.2.

13.2 THE PAIRED-SAMPLE SIGN TEST

The sign test can also be used when we deal with paired data as in Section 10.7. In such **paired-sample sign tests**, each pair of sample values is replaced with a plus sign if the first value is greater than the second, with a minus sign if the first value is smaller than the second, and is discarded when the two values are equal. Then we proceed as in Section 13.1.

The sign test for the paired-sample situation asks whether the population of differences between pairs has a median that is equal to zero. In most practical situations, the median is zero only when the two populations have the same distribution. The procedure that follows will therefore be described in terms of comparing the means of the two populations.

EXAMPLE 13.2   A trucking company sold its fleet of eight old trucks, replacing them with eight new trucks in order to reduce the number of mechanical breakdowns which were occurring. Following are the numbers of breakdowns which occurred during the three months before and three months after the new trucks were acquired.

In the following pairs, the first value is the number of breakdowns of an old truck; and the second value is the number of breakdowns of a randomly paired new truck.

4 and 1	2 and 2
3 and 0	4 and 3
3 and 2	0 and 1
1 and 2	2 and 1

Use the paired-sample test to test the null hypothesis that the new trucks are only as effective as the old trucks in minimizing the numbers of mechanical breakdowns. Since one of the pairs, 2 and 2, has to be discarded, the sample size for the paired-sample sign test is only $n = 8 - 1 = 7$.

1. **Hypothesis**

   H_0: $\mu_1 = \mu_2$, where $\mu_1 = \mu_2$ are the mean numbers of mechanical break-downs in 3 months with the old trucks and with the new trucks.

   H_A: $\mu_1 > \mu_2$

2. **Level of significance** $\alpha = 0.05$

3'. **Test statistic**   The statistic is x, the number of plus signs.

4'. **Calculations**   Replacing each pair of values with a plus sign if the first value is greater than the second or with a minus sign if the first value is smaller than the second, the seven unequal sample pairs yield $+ + + - + - +$. Here $x = 5$.

5'. **Decision**   Using Table I for $n = 7$ and $p = 0.50$. the probability of $x \geq 5$ is $0.164 + 0.555 + 0.008 = 0.227$. The p-value is thus 0.227. Since 0.227 exceeds 0.05, the null hypothesis cannot be rejected. In other words, we cannot conclude on the basis of this test that the new trucks have fewer breakdowns.

The test we have described here is only one of several nonparametric ways of analyzing such paired sample data. Other popular tests may be found in the books on nonparametric statistics listed in the bibliography at the end of the book.

13.3   THE SIGN TEST (LARGE SAMPLES)

When np and $n(1 - p)$ are both greater than 5, so that we can use the normal approximation to the binomial distribution, the sign test may be based on the large-sample test of Section 11.2, namely, on the statistic

$$z = \frac{x - np_0}{\sqrt{np_0(1 - p_0)}}$$

with $p_0 = 0.50$, which has approximately the standard normal distribution. When n is small, it may be wise to use the continuity correction suggested on page 369. This is true especially if, without the continuity correction, we could *barely* reject the null hypothesis. Note that the continuity correction does not even have to be considered when *without it* the null hypothesis cannot be rejected.

EXAMPLE 13.3   The following are measurements of the ocean depths in a certain area (in fathoms): 46.4, 48.3, 51.9, 38.8, 46.5, 45.6, 52.1, 41.0, 54.2, 44.9, 52.3, 43.6, 48.7, 42.2, and 44.9. Use the sign test at the level of significance $\alpha = 0.05$ to test the null hypothesis $\tilde{\mu} = 43.0$ (the previously recorded ocean depth in that area) against the alternative hypothesis $\tilde{\mu} \neq 43.0$.

Solution   1. **Hypotheses**

$$H_0:   \tilde{\mu} = 43.0$$

$$H_A: \tilde{\mu} \neq 43.0$$

2. **Level of significance** $\alpha = 0.05$

3. *Criterion*   Reject the null hypothesis if $z \leq -1.96$ or $z \geq 1.96$, where

$$z = \frac{x - np_0}{\sqrt{np_0(1 - p_0)}}$$

with $p_0 = 0.50$; otherwise, accept it or reserve judgment.

4. *Calculations*   Replacing each value greater than 43.0 with a plus sign and each value less than 43.0 with a minus sign, we get

$$+ + + - + + + - + + + + + - +$$

Thus, $x = 12$ and substitution into the formula yields

$$z = \frac{12 - 15(0.50)}{\sqrt{15(0.50)(0.50)}} = 2.32$$

5. *Decision*   Since $z = 2.32$ exceeds 1.96, the null hypothesis must be rejected; in other words, the median ocean depth in the given area is not 43.0 fathoms, as had been previously reported. Had we used the continuity correction, we would have obtained

$$z = \frac{11.5 - 7.5}{\sqrt{15(0.50)(0.50)}} = 2.07$$

and the decision would have been the same.

Exercises 13.1 and 13.7 are practice exercises; their complete solutions are given on page 461.

✓ **13.1**   On 12 occasions, Ms. Brown had to wait 3, 6, 7, 6, 4, 8, 6, 2, 8, 6, 1, and 9 minutes for the bus that takes her to work. Use the sign test based on Table I and the 0.05 level of significance to test the null hypothesis $\tilde{\mu} = 5$ against the alternative hypothesis $\tilde{\mu} \neq 5$.

13.2   Nine women buying new eyeglasses tried on, respectively, 12, 11, 14, 15, 10, 14, 11, 8, and 12 different frames. Use the sign test at the 0.05 level of significance to test the null hypothesis $\tilde{\mu} = 10$ against the alternative hypothesis $\tilde{\mu} > 10$.

13.3   In a sample of 12 randomly selected days an appliance store sells 7, 9, 4, 6, 8, 5, 3, 4, 10, 9, 8, and 7 dishwashing machines. Use the sign test based on Table I and the 0.05 level of significance to test the null hypothesis $\tilde{\mu} = 5$ sales of dishwashing machines daily, against the alternative hypothesis $\tilde{\mu} > 5$.

13.4   A random sample of nine physical therapists employed at a hospital orthopedic center treat 12, 13, 16, 13, 15, 9, 12, 15, and 11 patients during a day. Use the sign test at the 0.05 level of significance to test the null hypothesis $\tilde{\mu} = 11$ (that a physical therapist treats 11 patients per day) against the alternative hypothesis $\tilde{\mu} > 11$.

13.5   Over a 15-day period of time, the number of alarms to which 2 fire departments responded was 5 and 4, 14 and 8, 8 and 7, 13 and 6, 10 and 11, 12 and 14, 12 and 12, 16 and 8, 5 and 4, 8 and 10, 11 and 6, 9 and 5, 11 and 3, 9 and 15, and 7 and 14, respectively. Use the sign test together with Table I and the $\alpha = 0.01$ level of significance to test the null hypothesis that, on the average, the fire departments respond to the same number of alarms against the alternative $\mu_1 > \mu_2$.

13.6 The following are the grades that 15 students received on the midterm and final examinations in a course in European history:

$$
\begin{array}{lllll}
66 \text{ and } 73 & 88 \text{ and } 91 & 75 \text{ and } 78 & 90 \text{ and } 86 & 63 \text{ and } 69 \\
58 \text{ and } 67 & 75 \text{ and } 75 & 82 \text{ and } 80 & 73 \text{ and } 76 & 84 \text{ and } 89 \\
85 \text{ and } 81 & 93 \text{ and } 96 & 70 \text{ and } 76 & 85 \text{ and } 82 & 90 \text{ and } 97
\end{array}
$$

Use the sign test based on Table I and the 0.05 level of significance to test the null hypothesis $\mu_1 = \mu_2$ against the alternative hypothesis $\mu_1 < \mu_2$.

✓ **13.7** The following are the numbers of speeding tickets issued by two police officers on 20 days:

$$
\begin{array}{lllll}
6 \text{ and } 9 & 11 \text{ and } 13 & 12 \text{ and } 12 & 10 \text{ and } 17 & 15 \text{ and } 13 \\
7 \text{ and } 11 & 9 \text{ and } 13 & 7 \text{ and } 12 & 14 \text{ and } 15 & 11 \text{ and } 13 \\
14 \text{ and } 10 & 6 \text{ and } 12 & 9 \text{ and } 9 & 12 \text{ and } 14 & 8 \text{ and } 13 \\
16 \text{ and } 11 & 10 \text{ and } 15 & 12 \text{ and } 14 & 15 \text{ and } 15 & 12 \text{ and } 18
\end{array}
$$

Use the sign test based on the normal approximation to the binomial distribution and the level of significance $\alpha = 0.01$ to test the null hypothesis that on the average the two police officers issue equally many speeding tickets. Work this exercise with and without the continuity correction.

13.8 The following is data for the daily sulfur oxides emission (in tons) of an industrial plant:

$$
\begin{array}{cccccccccc}
17 & 15 & 20 & 29 & 19 & 18 & 22 & 25 & 27 & 9 \\
24 & 20 & 17 & 6 & 24 & 14 & 15 & 23 & 24 & 26 \\
19 & 23 & 28 & 19 & 16 & 22 & 24 & 17 & 20 & 13 \\
19 & 10 & 23 & 18 & 31 & 13 & 20 & 17 & 24 & 14
\end{array}
$$

Use the sign test and the 0.01 level of significance to test the null hypothesis $\tilde{\mu} = 23$ against the alternative hypothesis $\tilde{\mu} < 23$. Work this exercise with and without the continuity correction.

13.9 The following is a list of the weights (in grams) of 14 packages of a certain kind of candy: 101.0, 99.8, 100.9, 103.6, 97.1, 100.0, 102.5, 100.5, 101.0, 98.2, 100.3, 102.6, 100.0, and 100.8. Use the sign test based on Table I and the level of significance $\alpha = 0.05$ to test the null hypothesis $\tilde{\mu} = 100.0$ against the alternative hypothesis $\tilde{\mu} \neq 100.0$.

13.10 Rework Exercise 13.9 using the sign test based on the normal approximation to the binomial distribution.

13.11 The following are the numbers of artifacts dug up by two archeologists at an ancient cliff dwelling on 30 days:

$$
\begin{array}{llllll}
1 \text{ and } 0 & 0 \text{ and } 0 & 2 \text{ and } 1 & 3 \text{ and } 0 & 1 \text{ and } 2 & 0 \text{ and } 0 \\
2 \text{ and } 0 & 2 \text{ and } 1 & 3 \text{ and } 1 & 0 \text{ and } 2 & 1 \text{ and } 0 & 1 \text{ and } 1 \\
4 \text{ and } 2 & 1 \text{ and } 1 & 2 \text{ and } 1 & 1 \text{ and } 0 & 3 \text{ and } 2 & 5 \text{ and } 2 \\
2 \text{ and } 6 & 1 \text{ and } 0 & 3 \text{ and } 2 & 2 \text{ and } 3 & 4 \text{ and } 0 & 1 \text{ and } 2 \\
3 \text{ and } 1 & 2 \text{ and } 0 & 0 \text{ and } 1 & 2 \text{ and } 0 & 4 \text{ and } 1 & 2 \text{ and } 0
\end{array}
$$

Use the sign test at the level of significance $\alpha = 0.01$ to test the null hypothesis that the two archeologists are equally good at finding artifacts against the alternative hypothesis that the first one is better.

13.12 Use the large-sample sign test to rework Exercise 10.45.

13.13 The following are the numbers of employees absent from two departments of a large firm on 25 days:

4 and 3	2 and 5	6 and 6	3 and 6	1 and 4	2 and 4	5 and 2
1 and 4	3 and 4	6 and 5	2 and 5	7 and 1	4 and 6	1 and 3
2 and 5	0 and 3	6 and 5	4 and 6	1 and 2	4 and 1	2 and 4
0 and 1	5 and 3	2 and 3	2 and 4			

Use the sign test at the 0.05 level of significance to test the null hypothesis $\mu_1 = \mu_2$ against the alternative hypothesis $\mu_1 < \mu_2$.

13.4 RANK SUMS: THE *U* TEST

In this section we shall present a nonparametric alternative to the small-sample *t* test concerning the difference between two means. It is called the **U test**, the **Wilcoxon test**, or the **Mann–Whitney test**, named after the statisticians who contributed to its development. The different names refer to different methods of organizing the calculations, but the procedures are logically equivalent. We will be able to test the null hypothesis that the two samples come from identical populations without having to assume that the populations sampled have normal distributions; in fact, the test requires only that the populations sampled be continuous to avoid ties, and in practice it does not even matter whether this assumption is satisfied.

Suppose that you randomly select samples from each of the two populations being compared. The *U* test, as originally defined, is based on the number of times values of the sample from one population are exceeded by values from the other population. This is a somewhat complicated notion, and our work here will be described in terms of the ranks of the jointly ranked data from the two populations.

To illustrate how the *U* test is thus performed, suppose that we want to compare the grain size of sand obtained from two different locations on the moon on the basis of the following diameters (in millimeters):

Location 1: 0.37, 0.70, 0.75, 0.30, 0.45, 0.16, 0.62, 0.73, 0.33
Location 2: 0.86, 0.55, 0.80, 0.42, 0.97, 0.84, 0.24, 0.51, 0.92, 0.69

The means of these two samples are 0.49 and 0.68, and their difference is large, but it remains to be seen whether it is significant.

To perform the *U* test, we first arrange the data jointly, as if it comprises one sample, in an increasing order of magnitude. For our data, we get

0.16	0.24	0.30	0.33	0.37	0.42	0.45	0.51	0.55	0.62
1	2	1	1	1	2	1	2	2	1
0.69	0.70	0.73	0.75	0.80	0.84	0.86	0.92	0.97	
2	1	1	1	2	2	2	2	2	

where we indicated for each value whether it came from location 1 or location 2. Assigning the data, in this order, the ranks 1, 2, 3, ..., and 19, we find that the values of the first sample (location 1) occupy ranks 1, 3, 4, 5, 7, 10, 12, 13, and 14, while those of the second sample (location 2) occupy ranks 2, 6, 8, 9, 11, 15,

16, 17, 18, and 19. There are no ties here, but if there were, we would assign each of the tied observations the mean of the ranks that they jointly occupy. For instance, if the third and fourth values were the same, we would assign each the rank

$$\frac{3+4}{2} = 3.5$$

and if the ninth, tenth, and eleventh values were the same, we would assign each the rank

$$\frac{9+10+11}{3} = 10$$

Now, if there is an appreciable difference between the means of the two populations, most of the lower ranks are likely to go to the values of one sample, while most of the higher ranks are likely to go to the values of the other sample. The test of the null hypothesis that the two samples come from identical populations may thus be based on W_1, the sum of the ranks of the values of the first sample, or on W_2, the sum of the ranks of the values of the second sample. In practice, it does not matter which sample we refer to as sample 1 and which we refer to as sample 2, or whether we base the test on W_1 or W_2.[†] If the sample sizes are n_1 and n_2, the sum of W_1 and W_2 is simply the sum of the first $n_1 + n_2$ positive integers, which is known to equal

$$\frac{(n_1 + n_2)(n_1 + n_2 + 1)}{2}$$

This formula enables us to find W_2 if we know W_1, and vice versa. For our illustration, we get

$$W_1 = 1 + 3 + 4 + 5 + 7 + 10 + 12 + 13 + 14 = 69$$

and since the sum of the first 19 positive integers is

$$\frac{19 \cdot 20}{2} = 190$$

it follows that $W_2 = 190 - 69 = 121$. (This value may be checked by actually adding 2, 6, 8, 9, 11, 15, 16, 17, 18, and 19.)

When the use of **rank sums** was first proposed as a nonparametric alternative to the two-sample t test, the decision was based on W_1 or W_2. Presently, the decision is based on either of the related statistics

U_1 AND U_2
STATISTICS

$$U_1 = W_1 - \frac{n_1(n_1 + 1)}{2} \quad \text{or} \quad U_2 = W_2 - \frac{n_2(n_2 + 1)}{2}$$

[†] When the sample sizes are unequal, it is common practice to let n_1 be the smaller of the two; this is not required, however, for the work in this book.

or on the statistic U, which always equals the smaller of the two. The resulting tests are equivalent to those based on W_1 or W_2, but they have the advantage that they lend themselves more readily to the construction of tables of critical values. Not only do U_1 and U_2 take on values on the interval from 0 to n_1n_2—indeed, their sum is always equal to n_1n_2—but their sampling distributions are symmetrical about $\frac{n_1n_2}{2}$. The use of U, which always equals the smaller of the values of U_1 and U_2, has the added advantage that the resulting test is one-tailed and hence easier to tabulate.

Accordingly, we test the null hypothesis that the two samples come from identical populations against the alternative hypothesis that the two populations have unequal means with the following criterion:

> Reject the null hypothesis if $U \le U'_\alpha$, where U'_α is given in Table VII for $n_1 \le 15, n_2 \le 15,$ and $\alpha = 0.05$ or $\alpha = 0.01$.

In the construction of Table VII, U'_α is the largest value of U for which the probability of $U \le U'_\alpha$ is less than or equal to α, and the blank spaces indicate that the null hypothesis cannot be rejected at the given level of significance, regardless of the value that we obtain for U. More extensive tables may be found in handbooks of statistical tables; but when n_1 and n_2 are both greater than 8, it is generally considered reasonable to use the large-sample test described in Section 13.5.

Observe that the null hypothesis is rejected for small values of U. In every case, either $U = U_1$ or $U = U_2$. When U actually equals U_1, then it is sample 1 that is producing the small values. Similarly, when U is equal to U_2, then sample 2 is producing the small values.

EXAMPLE 13.4 With reference to the previous grain-size data on page 443, use the U test at the 0.05 level of significance to test the null hypothesis that the two samples come from identical populations against the alternative hypothesis that the two populations have unequal means.

Solution   1. *Hypotheses*

$$H_0: \quad \text{populations are identical}$$

$$H_A: \quad \mu_1 \neq \mu_2$$

2. *Level of significance* $\alpha = 0.05$
3. *Criterion*   Reject the null hypothesis if $U \le 20$, which is the value of U'_α for $n_1 = 9, n_1 = 10,$ and $\alpha = 0.05$; otherwise, accept it or reserve judgment.
4. *Calculations*   Having shown that $W_1 = 69$ and $W_2 = 121$, we get

$$U_1 = 69 - \frac{9 \cdot 10}{2} = 24 \quad U_2 = 121 - \frac{10 \cdot 11}{2} = 66$$

and hence $U = 24$. Note that $U_1 + U_2 = 24 + 66 = 90$, which equals $n_1n_2 = 9 \cdot 10$.
5. *Decision*   Since $U = 24$ exceeds 20, the null hypothesis cannot be rejected; in other words, we cannot conclude that there is a real difference in the mean grain size of sand from the two locations on the moon.

The test that we have described here can also be used when the alternative is $\mu_1 < \mu_2$ or $\mu_1 > \mu_2$. However, since the procedure is more complicated in that case, we shall discuss it here only for large samples in the section that follows.

*13.5   RANK SUMS: THE U TEST (LARGE SAMPLES)

The large-sample U test may be based on either U_1 or U_2, as given on page 444. The resulting tests will be equivalent, so that it does not matter which sample we identify as sample 1. In the description that follows, we shall use the statistic U_1.

Under the null hypothesis that the two samples come from identical populations, it can be shown that the mean and the standard deviation of the sampling distribution of U_1 are[†]

MEAN AND STANDARD DEVIATION OF U_1 STATISTIC

$$\mu_{U_1} = \frac{n_1 n_2}{2} \quad \text{and} \quad \sigma_{U_1} = \sqrt{\frac{n_1 n_2 (n_1 + n_2 + 1)}{12}}$$

Furthermore, if n_1 and n_2 are both greater than 8, the sampling distribution of U_1 can be approximated closely with a normal curve. Thus, we base the test of the null hypothesis that the two samples come from identical populations on the statistic

STATISTIC FOR THE LARGE-SAMPLE U TEST

$$z = \frac{U_1 - \mu_{U_1}}{\sigma_{U_1}}$$

which has approximately the standard normal distribution. If the alternative hypothesis is $\mu_1 \neq \mu_2$, we reject the null hypothesis for $z \leq -z_{\alpha/2}$ or $z \geq z_{\alpha/2}$; if the alternative hypothesis is $\mu_1 < \mu_2$, we reject the null hypothesis for $z \leq -z_{\alpha}$, since small values of U_1 correspond to small values of the rank sum W_1. If the alternative hypothesis is $\mu_1 > \mu_2$, we reject the null hypothesis for $z \geq z_{\alpha}$, since large values of U_1 correspond to large values of the rank sum W_1.

Finding ranks when n_1 and n_2 are large can be a surprisingly messy task. However, you can do this easily by first constructing a stem-and-leaf display for the values of the first sample. Then, using a different-colored pencil, continue the stem-and-leaf display for the values of the second sample.

EXAMPLE 13.5   The following are the weight gains (in pounds) of young turkeys, which are fed two different diets but are otherwise kept under identical conditions:

Diet 1:	16.3,	10.1,	10.7,	13.5,	14.9,	11.8,	14.3,	10.2,
	12.0,	14.7,	23.6,	15.1,	14.5,	18.4,	13.2,	14.0
Diet 2:	21.3,	23.8,	15.4,	19.6,	12.0,	13.9,	18.8,	19.2,
	15.3,	20.1,	14.8,	18.9,	20.7,	21.1,	15.8,	16.2

[†] If there are ties in rank, these formulas provide only approximations, but if the number of ties is small, there is no need to make a correction.

Use the large-sample U test at the 0.01 level of significance to test the null hypothesis that the two populations sampled are identical against the alternative hypothesis that, on average, the second diet produces a greater gain in weight.

Solution   1. *Hypotheses*

$$H_0: \text{ populations are identical}$$

$$H_A: \quad \mu_1 < \mu_2$$

2. *Level of significance* $\alpha = 0.01$
3. *Criterion*   Reject the null hypothesis if $z \le -2.33$, where

$$z = \frac{U_1 - \mu_{U_1}}{\sigma_{U_1}}$$

Otherwise, accept it or reserve judgment.
4. *Calculations*   Arranging the data according to size, we get

10.1	10.2	10.7	11.8	12.0	12.0	13.2	13.5	13.9	14.0
1	1	1	1	1	2	1	1	2	1
14.3	14.5	14.7	14.8	14.9	15.1	15.3	15.4	15.8	16.2
1	1	1	2	1	1	2	2	2	2
16.3	18.4	18.8	18.9	19.2	19.6	20.1	20.7	21.1	21.3
1	1	2	2	2	2	2	2	2	2
23.6	23.8								
1	2								

Below the weight gains are listed the samples from which they came. Observe that there is a tie at 12.0 pounds, involving one turkey from each diet. These are the fifth and sixth positions in the list, so each is assigned the rank

$$\frac{5 + 6}{2} = 5.5$$

The sum of the rank positions occupied by the turkeys of the first sample (diet 1) is

$$W_1 = 1 + 2 + 3 + 4 + 5.5 + 7 + 8 + 10$$

$$+ 11 + 12 + 13 + 15 + 16 + 21 + 22 + 31$$

$$= 181.5$$

and

$$U_1 = 181.5 - \frac{16 \cdot 17}{2} = 45.5$$

Since

$$\mu_{U_1} = \frac{16 \cdot 16}{2} = 128 \quad \text{and} \quad \sigma_{U_1} = \sqrt{\frac{16 \cdot 16 \cdot 33}{12}} \approx 26.53$$

it follows that

$$z = \frac{45.5 - 128}{26.53} \approx -3.11$$

5. *Decision* Since $z = -3.11$ is less than -2.33, the null hypothesis must be rejected; in other words, we conclude that, on average, the second diet produces a greater gain in weight. ∎

Exercises 13.14 and 13.19 are practice exercises; their complete solutions are given on page 462.

EXERCISES

✓ **13.14** The following are figures on the number of assaults committed in a city in six weeks in the spring and in six weeks in the fall:

Spring:	46,	37,	42,	48,	38,	45
Fall:	35,	30,	25,	39,	28,	32

Use the U test at the 0.01 level of significance to check the claim that, on average, there are equally many assaults per week in the given city in the spring and in the fall.

13.15 The following are the Rockwell hardness numbers obtained for six aluminum die castings randomly selected from production lot A and for eight aluminum die castings randomly selected from production lot B:

Production lot A:	75,	56,	63,	70,	58,	74	
Production lot B:	63,	85,	77,	80,	86,	76,	72,	82

Use the U test at the 0.05 level of significance to check the claim that the average hardness of die castings from the two production lots is the same.

13.16 Following are the numbers of defective plastic chairs observed by inspectors in randomly selected production runs, using two different molding machines:

Machine 1:	24	18	23	23	27	19
Machine 2:	22	15	18	20	19	17

Use the U test at the 0.05 level of significance to test the null hypothesis that the two machines average equally many defectives per production run against the alternative hypothesis that they do not average equally many defectives per production run.

13.17 The following are the numbers of misprints counted on pages selected at random from two Sunday editions of a newspaper:

May 10:	12,	6,	11,	11,	15,	7
May 24:	10,	3,	6,	8,	7,	5

Use the U test at the level of significance $\alpha = 0.05$ to test the null hypothesis that the two samples come from identical populations against the alternative hypothesis that the two populations have unequal means.

13.18 The following are the numbers of minutes it took a sample of 15 men and 12 women to complete a short screening test given to job applicants at a large bank:

Men: 8.8, 7.8, 6.6, 10.7, 8.9, 8.4, 6.9, 6.4, 6.3, 8.0, 8.6, 8.1, 9.1, 9.7, 9.9
Women: 7.5, 8.7, 8.3, 6.2, 6.5, 7.7, 9.8, 9.6, 9.2, 10.4, 8.2, 8.5

Use the U test based on Table VII and the 0.01 level of significance to test the null hypothesis that the two samples come from identical populations against the alternative hypothesis that the two populations have unequal means.

✓ ***13.19** The following are the scores that samples of students from two minority groups obtained on a current events test:

> Minority group 1: 70, 62, 91, 55, 72, 94, 80, 96, 73, 44, 87, 78
> Minority group 2: 81, 23, 71, 30, 71, 54, 64, 93, 58, 41, 47, 56

Use the large-sample U test at the 0.05 level of significance to test whether students from the two minority groups can be expected to score equally well on this test.

*13.20 Use the normal approximation to the sampling distribution of U to rework the illustration on page 443 which dealt with the grain size of sand from two locations on the moon.

*13.21 Comparing two kinds of emergency flares, a consumer testing service obtained the following burning times (rounded to the nearest tenth of a minute):

> Brand X: 17.2, 18.1, 21.2, 19.3, 14.4, 21.1, 14.6, 19.1, 18.8, 15.2, 20.3, 17.5
> Brand Y: 13.6, 13.7, 11.8, 14.6, 15.2, 14.3, 22.5, 12.3, 13.5, 10.9, 14.4,   8.0

Use the large-sample U test at the 0.01 level of significance to see whether it is reasonable to say that, in general, brand X flares are better (last longer) than brand Y flares.

*13.22 Use the large-sample U test to rework Exercise 13.18.

*13.23 The following are the weekly per-capita food costs (in dollars) of families with two children chosen at random from two suburbs of a large city:

> Suburb A: 78.60, 70.50, 75.38, 86.45, 67.95, 70.78, 67.89, 72.00, 64.19, 71.15
> Suburb B: 62.63, 75.16, 55.35, 78.19, 71.72, 63.12, 75.91, 66.51, 63.76, 60.78

Use the large-sample U test at the 0.05 level of significance to check the claim that, on average, such weekly food expenditures are higher in suburb A than in suburb B.

***13.6** RANK SUMS: THE *H* TEST

The **H test**, or **Kruskal–Wallis test**, is a rank-sum test that serves to test the null hypothesis that k independent random samples come from identical populations against the alternative hypothesis that the means of these populations are not all equal. Unlike the standard test that it replaces, the one-way analysis of variance of Section 10.9, it does not require the assumption that the populations sampled have, at least approximately, normal distributions.

As in the U test, the data is ranked jointly from low to high as though they constitute a single sample. Then, if R_i is the sum of the ranks assigned to the n_i values of the ith sample and $n = n_1 + n_2 + \cdots + n_k$, the H test is based on the statistic

STATISTIC FOR *H* TEST

$$H = \frac{12}{n(n+1)} \sum_{i=1}^{k} \frac{R_i^2}{n_i} - 3(n+1)$$

If the null hypothesis is true and each sample has at least five observations, the sampling distribution of H can be approximated closely with a chi-square distribution with $k - 1$ degrees of freedom. Consequently, we reject the null hypothesis that the populations sampled are identical and accept the alternative hypothesis that the means of these populations are not all equal, if the value we get for H is greater than or equal to χ_α^2 for $k - 1$ degrees of freedom.

EXAMPLE 13.6 The following are the final examination scores of samples of students who are taught German by three different methods (classroom instruction and language laboratory, only classroom instruction, and only self-study in language laboratory):

Method 1:	94,	87,	91,	74,	87,	97
Method 2:	85,	82,	79,	84,	61,	72, 80
Method 3:	89,	67,	72,	76,	69	

Use the H test at the 0.05 level of significance to test the null hypothesis that the three populations sampled are identical against the alternative hypothesis that their means are not all equal.

Solution

1. *Hypotheses*

$$H_0: \quad \text{populations are identical}$$

$$H_A: \quad \text{population means are not all equal}$$

2. *Level of significance* $\alpha = 0.05$

3. *Criterion* Reject the null hypothesis if $H \geq 5.991$, the value of $\chi_{0.05}^2$ for $3 - 1 = 2$ degrees of freedom, where H is calculated in accordance with the formula above. Otherwise, accept it or reserve judgment.

4. *Calculations* Arranging the data jointly according to size, we get 61, 67, 69, 72, 72, 74, 76, 79, 80, 82, 84, 85, 87, 87, 89, 91, 94, and 97. Assigning the data, in this order, the ranks 1, 2, 3, ..., and 18, we find that the values of the first sample occupy ranks 6, 13, 14, 16, 17, and 18, while those of the second sample occupy ranks 1, 4.5, 8, 9, 10, 11, and 12, and those of the third sample occupy ranks 2, 3, 4.5, 7, and 15. (Since the two 87's belong to the same sample, we simply assign them ranks 13 and 14.) Thus,

$$R_1 = 6 + 13 + 14 + 16 + 17 + 18 = 84$$

$$R_2 = 1 + 4.5 + 8 + 9 + 10 + 11 + 12 = 55.5$$

$$R_3 = 2 + 3 + 4.5 + 7 + 15 = 31.5$$

and it follows that

$$H = \frac{12}{18 \cdot 19}\left(\frac{84^2}{6} + \frac{55.5^2}{7} + \frac{31.5^2}{5}\right) - 3 \cdot 19 \approx 6.67$$

5. *Decision* Since $H = 6.67$ exceeds 5.991, the null hypothesis must be rejected; in other words, we conclude that the three methods of teaching German are not all equally effective. ■

Exercise 13.24 is a practice exercise; its complete solution is given on page 463.

✓ ***13.24** The following are the miles per gallon that a test driver got for six tankfuls each of three kinds of gasoline:

Gasoline 1:	28,	23,	26,	31,	14,	29
Gasoline 2:	21,	31,	32,	19,	27,	16
Gasoline 3:	24,	17,	21,	31,	22,	18

Use the *H* test at the level of significance $\alpha = 0.05$ to check the claim that there is no difference in the true average mileage yield of the three kinds of gasoline.

*13.25 A small motion picture theater shows the same film three times a day and the following is the size of the audience on randomly selected days.

First screening:	179	164	181	176	173	178	164
Second screening:	169	177	166	171	181	189	
Third screening:	174	167	168	183	169		

Use the *H* test at the 0.05 level of significance to test the claim that there is no difference in the true average size of the audience in the three daily screenings.

*13.26 Three groups of guinea pigs were injected, respectively, with 0.5 milligram (mg), 1.0 mg, and 1.5 mg of a tranquilizer, and the following are the numbers of seconds it took them to fall asleep:

0.5-mg dose: 8.2, 10.0, 10.2, 13.7, 14.0, 7.8, 12.7, 10.9
1.0-mg dose: 9.7, 13.1, 11.0, 7.5, 13.3, 12.5, 8.8, 12.9, 7.9, 10.5
1.5-mg dose: 12.0, 7.2, 8.0, 9.4, 11.3, 9.0, 11.5, 8.5

Use the *H* test at the level of significance $\alpha = 0.01$ to test the null hypothesis that the differences in dosage have no effect on the length of time it takes guinea pigs to fall asleep.

*13.27 Use the *H* test to rework Exercise 10.53.

13.7 TESTS OF RANDOMNESS: RUNS

All the methods of inference that we have discussed are based on the assumption that the samples are random; yet there are many applications where it is difficult to decide whether this assumption is justifiable. This is true, particularly, when we have little or no control over the selection of the data, as is the case, for example, when we rely on whatever records are available to make long-range predictions of the weather, when we use whatever data is available to estimate the mortality rate of a disease, or when we use sales records for past months to make predictions of a department store's sales. None of this information constitutes a random sample in the strict sense.

There are several methods of judging the randomness of a sample on the basis of the order in which the observations are obtained; they enable us to decide, after the data has been collected, whether patterns that look suspiciously nonrandom may be attributed to chance. The technique we shall describe here and in the next two sections is based on the **theory of runs**.

A **run** is a succession of identical letters (or other kinds of symbols) that is followed and preceded by different letters or no letters at all. To illustrate, consider the following arrangement of healthy, H, and diseased, D, elm trees that were planted many years ago along a country road:

$$\underline{HHHH} \quad \underline{DDD} \quad \underline{HHHHHHH} \quad \underline{DD} \quad \underline{HH} \quad \underline{DDDD}$$

Using underlines to combine the letters that constitute the runs, we find that there is first a run of four H's, then a run of three D's, then a run of seven H's, then a run of two D's, then a run of two H's, and finally a run of four D's.

The **total number of runs** appearing in an arrangement of this kind is often a good indication of a possible lack of randomness. If there are too few runs, we might suspect a definite grouping or clustering, or perhaps a trend; if there are too many runs, we might suspect some sort of repeated alternating, or cyclical, pattern. In the preceding example there seems to be a definite clustering—the diseased trees seem to come in groups—but it remains to be seen whether this is significant or whether it can be attributed to chance.

If there are n_1 letters of one kind, n_2 letters of another kind, and u runs, we base this kind of decision on the following criterion:

Reject the null hypothesis of randomness if

$$u \leq u'_{\alpha/2} \quad or \quad u \geq u_{\alpha/2}$$

where $u'_{\alpha/2}$ and $u_{\alpha/2}$ are given in Table VIII for $n_1 \leq 15, n_2 \leq 15$, and $\alpha = 0.05$ or $\alpha = 0.01$.

In the construction of Table VIII, $u'_{\alpha/2}$ is the largest value of u for which the probability of $u \leq u'_{\alpha/2}$ is less than or equal to $\alpha/2$, $u_{\alpha/2}$ is the smallest value of u for which the probability of $u \geq u_{\alpha/2}$ is less than or equal to $\alpha/2$, and the blank spaces indicate that the null hypothesis of randomness cannot be rejected for values in that tail of the sampling distribution of u, regardless of the value that we obtain for u. More extensive tables for the **u test** may be found in handbooks of statistical tables, but when n_1 and n_2 are both at least 10, it is generally considered reasonable to use the large-sample test described in Section 13.8.

EXAMPLE 13.7   With reference to the arrangement of healthy and diseased elm trees given previously, use the u test at the 0.05 level of significance to test the null hypothesis of randomness against the alternative hypothesis that the arrangement is not random.

Solution   1. *Hypotheses*

$$H_0: \quad \text{arrangement is random}$$

$$H_A: \quad \text{arrangement is not random}$$

2. *Level of significance* $\alpha = 0.05$
3. *Criterion*   Since $n_1 = 13$, $n_2 = 9$, and $\alpha = 0.05$, we get $u'_{0.025} = 6$ and $u_{0.025} = 17$ from Table VIII; thus, the null hypothesis must be rejected if $u \leq 6$ or $u \geq 17$. Otherwise, accept it or reserve judgment.
4. *Calculations* $u = 6$, as can be seen from the data.
5. *Decision*   Since $u = 6$ is less than or equal to 6, the null hypothesis must be rejected; in other words, we conclude that the arrangement of healthy and

diseased elm trees is not random. Indeed, it seems that the diseased trees come in clusters. ▪

*13.8 TESTS OF RANDOMNESS: RUNS (LARGE SAMPLES)

Under the null hypothesis that n_1 letters of one kind and n_2 letters of another kind are arranged at random, it can be shown that the mean and the standard deviation of u, the total number of runs, are

MEAN AND STANDARD DEVIATION OF u

$$\mu_u = \frac{2n_1n_2}{n_1+n_2}+1 \quad \text{and} \quad \sigma_u = \sqrt{\frac{2n_1n_2(2n_1n_2-n_1-n_2)}{(n_1+n_2)^2(n_1+n_2-1)}}$$

Furthermore, if neither n_1 nor n_2 is less than 10, the sampling distribution of u can be approximated closely with a normal curve. Thus, we base the test of the null hypothesis of randomness on the statistic

STATISTIC FOR THE LARGE-SAMPLE u TEST

$$z = \frac{u-\mu_u}{\sigma_u}$$

which has approximately the standard normal distribution. If the alternative hypothesis is that the arrangement is not random, we reject the null hypothesis for $z \le -z_{\alpha/2}$ or $z \ge z_{\alpha/2}$; if the alternative hypothesis is that there is a clustering or a trend, we reject the null hypothesis for $z \le -z_\alpha$; and if the alternative hypothesis is that there is an alternating, or cyclical, pattern, we reject the null hypothesis for $z \ge z_\alpha$.

EXAMPLE 13.8   The following is an arrangement of men, M, and women, W, lined up to purchase tickets for a rock concert:

	M	W	M	W	M	M	M	W	M	W	M	M	M	W	W	M
(cont.)	M	M	M	W	W	M	W	M	M	M	W	M	M	M	W	W
(cont.)	W	M	W	M	M	M	W	M	W	M	M	M	M	W	W	M

Test for randomness at the 0.05 level of significance.

Solution   1. *Hypotheses*

H_0:   arrangement is random

H_A: arrangement is not random

2. *Level of significance* $\alpha = 0.05$
3. *Criterion*   Reject the null hypothesis if $z \le -1.96$ or $z \ge 1.96$, where

$$z = \frac{u-\mu_u}{\sigma_u}$$

Otherwise, accept it or reserve judgment.

4. *Calculations*　Since $n_1 = 30$, $n_2 = 18$, and $u = 27$, we get

$$\mu_u = \frac{2 \cdot 30 \cdot 18}{30 + 18} + 1 = 23.5$$

$$\sigma_u = \sqrt{\frac{2 \cdot 30 \cdot 18(2 \cdot 30 \cdot 18 - 30 - 18)}{(30 + 18)^2(30 + 18 - 1)}} \approx 3.21$$

and hence,

$$z = \frac{27 - 23.5}{3.21} \approx 1.09$$

5. *Decision*　Since $z = 1.09$ falls between -1.96 and 1.96, the null hypothesis of randomness cannot be rejected; there is no real evidence of any lack of randomness. ◾

13.9　TESTS OF RANDOMNESS: RUNS ABOVE AND BELOW THE MEDIAN

The u test is not limited to tests of the randomness of sequences of attributes, such as the H's and D's or M's and W's of our examples. Any sample consisting of numerical measurements or observations can be treated similarly by using the letters a and b to denote, respectively, values falling above and below the median of the sample. Numbers equal to the median are omitted. The resulting sequence of a's and b's (representing the data in its original order) can then be tested for randomness on the basis of the total number of runs of a's and b's, that is, the total number of **runs above and below the median**. Depending on the size of n_1 and n_2, we use Table VIII or the large-sample test of Section 13.8.

EXAMPLE 13.9　On 24 successive trips between two cities, a bus carried

24	19	32	28	21	23	26	17	20	28	30	24
13	35	26	21	19	29	27	18	26	14	21	23

passengers. Use the total number of runs above and below the median and the level of significance $\alpha = 0.01$ to decide whether it is reasonable to treat the data as if it constitutes a random sample.

Solution　Since the median is 23.5, as can easily be verified, we get the following arrangement of values above and below the median:

$$a \quad b \quad a \quad a \quad b \quad b \quad a \quad b \quad b \quad a \quad a \quad a \quad b \quad a \quad a \quad b \quad b \quad a \quad a \quad b \quad a \quad b \quad b \quad b$$

1. *Hypotheses*

$$H_0: \quad \text{arrangement is random}$$

$$H_A: \quad \text{arrangement is not random}$$

2. *Level of significance*　$\alpha = 0.01$

3. *Criterion*　Since $n_1 = 12$, $n_2 = 12$, and $\alpha = 0.01$, we get $u'_{0.005} = 6$ and $u_{0.005} = 20$ from Table VIII; thus, the null hypothesis must be rejected if $u \leq 6$ or $u \geq 20$; otherwise, accept it or reserve judgment.

4. *Calculations*   As can be seen from the arrangements of *a*'s and *b*'s above, there are $u = 14$ runs.

5. *Decision*   Since $u = 14$ falls between 6 and 20, the null hypothesis cannot be rejected; in other words, there is no real evidence to suggest that the data cannot be treated as if it constitutes a random sample.   ■

EXERCISES

Exercises 13.28, 13.32, and 13.37 are practice exercises; their complete solutions are given on pages 463 and 464.

✓ **13.28** The following sequence of *C*'s and *A*'s shows the order in which 25 cars with California or Arizona license plates crossed the Colorado River at Blyth, California, to enter Arizona.

$$C\ A\ A\ C\ A\ C\ C\ A\ C\ C\ C\ A\ A\ C\ A\ A\ A\ A\ C\ A\ C\ C\ A\ C\ C$$

Test for randomness at the level of significance $\alpha = 0.05$.

13.29 A prestidigitator draws 25 cards, with replacement from an ordinary deck of cards, and the sequence of red cards (R) and black cards (B) follows:

$$R\ B\ R\ B\ R\ B\ R\ B\ R\ B\ R\ B\ R\ B\ R\ B\ R\ B\ R\ B\ R\ B\ R\ B\ R$$

Test for randomness at the 0.05 level of significance.

13.30 The prestidigitator of Exercise 13.29 once again draws 25 cards, with replacement, but the sequence of red cards (R) and black cards (B) is now

$$R\ R\ R\ R\ R\ R\ R\ R\ R\ R\ R\ R\ R\ B\ B\ B\ B\ B\ B\ B\ B\ B\ B\ B\ B$$

Test for randomness at the 0.05 level of significance.

13.31 A driver buys gasoline either at a Shell station, S, or at a Chevron Texaco station, C, and the following arrangement shows where he purchased gasoline (in the given order) over a certain period of time:

$$C\ C\ C\ S\ C\ S\ C\ S\ S\ C\ C\ S\ C\ S\ C\ S\ C\ S\ S\ C\ S\ C$$

Test for randomness at the 0.05 level of significance.

✓ *13.32 Representing each 0, 2, 4, 6, and 8 by the letter E and each 1, 3, 5, 7, and 9 by the letter O, check at the 0.05 level of significance whether the arrangements of the 50 digits in the first column of the random-numbers table, Table XII, may be regarded as random.

*13.33 The following arrangement shows whether 50 persons interviewed consecutively in the given order are for, F, or against, A, an increase in the city sales tax:

$$A\ A\ A\ A\ A\ F\ A\ A\ F\ F\ A\ A\ A\ A\ A\ A\ A\ F\ A\ A\ A\ A\ F\ F\ F$$
$$\text{(cont.)}\ A\ A\ A\ F\ A\ A\ F\ A\ A\ A\ A\ F\ F\ A\ A\ A\ A\ A\ A\ A\ A\ F\ A\ A\ A$$

Test for randomness at the level of significance $\alpha = 0.05$.

*13.34 Use the large-sample u test to rework Exercise 13.28.

*13.35 To test whether a radio signal contains a message or constitutes random noise, an interval of time is subdivided into a number of very short intervals, and for each of these it is determined whether the signal strength exceeds, E, or does not exceed, N, a certain level of background noise. Test at the level of significance $\alpha = 0.05$

whether the following arrangement, thus obtained, may be regarded as random and hence that the signal contains no message and may be regarded as random noise:

$$E \quad N \quad N \quad N \quad E \quad N \quad E \quad N \quad E \quad N \quad N \quad N \quad E \quad E \quad N \quad N \quad N \quad E \quad E \quad N \quad E \quad N \quad E \quad N \quad N \quad N$$

(cont.) $E \quad E \quad N \quad N \quad N \quad N \quad E \quad E \quad N \quad E \quad N \quad N \quad E \quad N \quad N \quad N \quad E \quad E \quad E \quad N \quad N \quad N$

(cont.) $E \quad N \quad E \quad N \quad N \quad N \quad N \quad E \quad N$

*13.36 Mentally simulate 50 flips of a coin, and test at the 0.05 level of significance whether the resulting sequence of H's and T's (heads and tails) may be regarded as random.

✓ 13.37 The following are the numbers of students absent from a school on 24 consecutive days:

$$38 \quad 31 \quad 32 \quad 27 \quad 28 \quad 30 \quad 26 \quad 33 \quad 36 \quad 30 \quad 28 \quad 35$$
$$33 \quad 31 \quad 29 \quad 35 \quad 31 \quad 33 \quad 31 \quad 28 \quad 30 \quad 28 \quad 25 \quad 29$$

Test for randomness at the 0.05 level of significance.

13.38 The following are the numbers of business lunches that an insurance agent had in 30 consecutive months:

$$6 \quad 7 \quad 5 \quad 6 \quad 8 \quad 6 \quad 8 \quad 6 \quad 6 \quad 4 \quad 3 \quad 2 \quad 4 \quad 4 \quad 3$$
$$4 \quad 7 \quad 5 \quad 6 \quad 8 \quad 6 \quad 6 \quad 3 \quad 4 \quad 2 \quad 5 \quad 4 \quad 4 \quad 3 \quad 7$$

Discarding the three values that equal the median, 5, test for randomness at the level of significance $\alpha = 0.01$.

*13.39 The following are the examination grades of 40 students in the order in which they finished an examination:

$$75 \quad 95 \quad 77 \quad 93 \quad 89 \quad 83 \quad 69 \quad 77 \quad 92 \quad 88 \quad 62 \quad 64 \quad 91 \quad 72$$
$$76 \quad 83 \quad 50 \quad 65 \quad 84 \quad 67 \quad 63 \quad 54 \quad 58 \quad 76 \quad 70 \quad 62 \quad 65 \quad 41$$
$$63 \quad 55 \quad 32 \quad 58 \quad 61 \quad 68 \quad 54 \quad 28 \quad 35 \quad 49 \quad 82 \quad 60$$

Test for randomness at the level of significance $\alpha = 0.05$.

*13.40 The total number of retail stores opening for business and also quitting business within the calendar years 1971–2003 in a large city were

$$108 \quad 103 \quad 109 \quad 107 \quad 125 \quad 142 \quad 147 \quad 122 \quad 116 \quad 153 \quad 144$$
$$162 \quad 143 \quad 126 \quad 145 \quad 129 \quad 134 \quad 137 \quad 143 \quad 150 \quad 148 \quad 152$$
$$125 \quad 106 \quad 112 \quad 139 \quad 132 \quad 122 \quad 138 \quad 148 \quad 155 \quad 146 \quad 158$$

Making use of the fact that the median is 138, test at the 0.05 level of significance whether there is a significant trend. Work this exercise with and without the continuity correction.

13.41 The following are the weights in pounds of tomatoes harvested from 50 consecutive tomato plants:

$$5.1 \quad 5.0 \quad 5.2 \quad 5.8 \quad 6.6 \quad 4.9 \quad 5.2 \quad 6.1 \quad 7.1 \quad 5.8$$
$$6.4 \quad 6.1 \quad 5.2 \quad 7.0 \quad 5.4 \quad 5.6 \quad 6.9 \quad 5.0 \quad 5.9 \quad 5.3$$
$$5.9 \quad 6.1 \quad 7.2 \quad 5.5 \quad 6.2 \quad 6.4 \quad 6.0 \quad 7.4 \quad 6.7 \quad 6.0$$
$$5.3 \quad 6.5 \quad 6.7 \quad 5.5 \quad 6.9 \quad 6.5 \quad 6.8 \quad 5.8 \quad 6.4 \quad 7.4$$
$$7.2 \quad 7.6 \quad 7.8 \quad 6.6 \quad 7.8 \quad 5.9 \quad 6.7 \quad 7.0 \quad 6.8 \quad 7.2$$

Making use of the fact that the median is 6.3 pounds, test for randomness at the level of significance $\alpha = 0.05$ by using runs above and below the median.

13.42 A habitual coffee drinker consumed these numbers of cups on each of the 30 days of April:

$$
\begin{array}{cccccccccc}
5 & 3 & 4 & 3 & 5 & 4 & 2 & 2 & 4 & 4 \\
3 & 4 & 3 & 5 & 6 & 6 & 6 & 7 & 5 & 4 \\
3 & 4 & 3 & 3 & 3 & 2 & 2 & 4 & 4 & 5
\end{array}
$$

Test for randomness at the level of significance $\alpha = 0.05$ by using runs above and below the median.

13.10 RANK CORRELATION

Since the significance test for r of Section 12.6 is based on very stringent assumptions, we sometimes use a nonparametric alternative that can be applied under much more general conditions. This test of the null hypothesis of no correlation is based on the **rank-correlation coefficient**, often called **Spearman's rank-correlation coefficient** and denoted by r_S.

The calculation of the rank-correlation coefficient for a given set of n pairs of x's and y's requires several steps. We first rank the x's among themselves from low to high (or high to low). Then we rank the y's in the same way. Then we find the d's, the differences between the ranks, and substitute into the formula

RANK-CORRELATION
COEFFICIENT

$$
r_S = 1 - \frac{6\left(\sum d^2\right)}{n(n^2 - 1)}
$$

When there are ties in rank, we proceed as before and assign to each of the tied observations the mean of the ranks that they jointly occupy.

EXAMPLE 13.10 The following are the numbers of hours that 10 students studied for an examination and the grades that they received:

Number of hours studied x	Grade in examination y
9	56
5	44
11	79
13	72
10	70
5	54
18	94
15	85
2	33
8	65

Calculate r_S.

Solution Ranking the x's among themselves from low to high and also the y's, we get the ranks shown in the first two columns of the following table:

Rank of x	Rank of y	d	d^2
5	4	1.0	1.00
2.5	2	0.5	0.25
7	8	−1.0	1.00
8	7	1.0	1.00
6	6	0.0	0.00
2.5	3	−0.5	0.25
10	10	0.0	0.00
9	9	0.0	0.00
1	1	0.0	0.00
4	5	−1.0	1.00
		Total	4.50

Then, determining the d's and their squares and substituting $n = 10$ and $\sum d^2 = 4.50$ into the formula for r_S, we get

$$r_S = 1 - \frac{6(4.50)}{10(10^2 - 1)} \approx 0.97$$

■

As can be seen from this example, r_S is easy to compute manually, and this is why it is sometimes used instead of r when no calculator is available. When there are no ties, r_S actually equals the correlation coefficient r calculated for the two sets of ranks; when ties exist, there may be a small (but usually negligible) difference. Of course, by using ranks instead of the original data, we lose some information, but this is usually offset by the rank-correlation coefficient's computational ease.

When we use r_S to test the null hypothesis of no correlation between two variables x and y, we do not have to make any assumptions about the nature of the populations sampled. Under the null hypothesis of no correlation—indeed, the null hypothesis that the x's and y's are randomly matched—the sampling distribution of r_S has the mean 0 and the standard deviation

$$\sigma_{r_S} = \frac{1}{\sqrt{n-1}}$$

Since this sampling distribution can be approximated with a normal distribution even for relatively small values of n, we base the test of the null hypothesis of no correlation on the statistic

STATISTIC FOR TESTING THE SIGNIFICANCE OF r_S

$$z = \frac{r_S - 0}{1/\sqrt{n-1}} = r_S\sqrt{n-1}$$

which has approximately the standard normal distribution.

EXAMPLE 13.11 With reference to the preceding example, where we had $n = 10$ and $r_S = 0.97$, test the significance of this value of r_S at the 0.01 level of significance.

Solution   1. *Hypotheses*

$$H_0: \quad \rho = 0 \quad \text{(no correlation)}$$

$$H_A: \quad \rho \neq 0$$

2. *Level of significance* $\alpha = 0.01$
3. *Criterion*   Reject the null hypothesis if $z \leq -2.575$ or $z \geq 2.575$, where

$$z = r_S \sqrt{n - 1}$$

Otherwise, accept it or reserve judgment.

4. *Calculations*   For $n = 10$ and $r_S = 0.97$, we get

$$z = 0.97\sqrt{10 - 1} \approx 2.91$$

5. *Decision*   Since $z = 2.91$ exceeds 2.575, the null hypothesis must be rejected; in other words, we conclude that there is a relationship between study time and examination grades in the populations sampled.   ▪

Exercises 13.43 and 13.44 are practice exercises; their complete solutions are given on pages 464 and 465.

✓ **13.43**   Calculate r_S for the following data representing the statistics grades, x, and psychology grades, y, of a sample of 15 students:

x	y	x	y
75	70	70	83
69	64	93	91
73	70	77	79
96	89	85	73
73	84	93	84
52	70	87	85
57	66		
61	63		
71	68		

✓ **13.44**   Test at the level of significance $\alpha = 0.05$ whether the value obtained for r_S in Exercise 13.43 is significant.

13.45   Calculate r_S for the following sample data representing the number of minutes it took 12 mechanics to assemble a piece of machinery in the morning, x, and in the late afternoon, y:

x	y
10.8	15.1
16.6	16.8
11.1	10.9
10.3	14.2
12.0	13.8
15.1	21.5
13.7	13.2
18.5	21.1
17.3	16.4
14.2	19.3
14.8	17.4
15.3	19.0

13.46   Use the level of significance $\alpha = 0.05$ to test whether the value obtained for r_S in Exercise 13.45 is significant.

EXERCISES

13.47 Ten weeks' sales of a downtown department store, x, and its suburban branch, y, are

x	y
71	49
64	31
58	24
80	68
63	30
69	40
76	62
60	22
66	35
55	16

where the units are $10,000. Calculate r_S.

13.48 Assuming that the data of Exercise 13.47 may be looked upon as random samples from the two stores' sales, use the level of significance $\alpha = 0.01$ to test whether the value obtained for r_S is significant.

13.49 In Exercise 12.27 we gave the percentages of the vote predicted by a poll for seven candidates for the U.S. Senate in different states, x, and the corresponding percentages of the vote that they actually received, y, as

x	y
42	51
34	31
59	56
41	42
53	58
40	35
55	54

Calculate r_S and compare it with the value of r obtained for these data in Exercise 12.27.

13.50 If a sample of $n = 20$ pairs of data yielded $r_S = 0.41$, is this rank-correlation coefficient significant at the 0.05 level of significance?

13.51 The following table shows how a panel of nutrition experts and a panel of heads of household ranked 15 breakfast foods on their palatability:

Breakfast food	Nutrition experts	Heads of household
I	7	5
II	3	4
III	11	8
IV	9	14
V	1	2
VI	4	6
VII	10	12
VIII	8	7
IX	5	1
X	13	9
XI	12	15
XII	2	3
XIII	15	10
XIV	6	11
XV	14	13

Calculate r_S as a measure of the consistency of the two rankings.

13.52 The following are the ranks that three judges gave to the work of 10 artists:

$$
\begin{array}{llllllllllll}
\textit{Judge A:} & 5, & 8, & 4, & 2, & 3, & 1, & 10, & 7, & 9, & 6 \\
\textit{Judge B:} & 3, & 10, & 1, & 4, & 2, & 5, & 6, & 7, & 8, & 9 \\
\textit{Judge C:} & 8, & 5, & 6, & 4, & 10, & 2, & 3, & 1, & 7, & 9 \\
\end{array}
$$

Calculate r_S for each pair of rankings and decide

(a) which two judges are most alike in their opinions about these artists;

(b) which two judges differ the most in their opinions about these artists.

✓ SOLUTIONS TO PRACTICE EXERCISES

13.1 1. *Hypotheses*

$$H_0: \quad \tilde{\mu} = 5$$

$$H_A: \quad \tilde{\mu} \neq 5$$

2. *Level of significance* $\alpha = 0.05$

3′. *Test statistic*   The statistic is x, the number of plus signs.

4′. *Calculations*   We find that x, the number of plus signs, is equal to 8.

5′. *Decision*   Using Table I for $n = 12$ and $p = 0.50$, the probability of $x \geq 8$ is 0.194. This corresponds to half the p-value, so that $p = 0.388$. Since 0.388 exceeds 0.05, the null hypothesis cannot be rejected.

13.7 Since three of the pairs (12 and 12, 9 and 9, and 15 and 15) have to be discarded, the sample size for the sign test is only $n = 17$.

1. *Hypotheses*

$$H_0: \quad \mu_1 = \mu_2$$

$$H_A: \quad \mu_1 \neq \mu_2$$

2. *Level of significance* $\alpha = 0.01$

3. *Criterion*   Reject the null hypothesis if $z \leq -2.575$ or $z \geq 2.575$, where

$$z = \frac{x - np_0}{\sqrt{np_0(1 - p_0)}}$$

with $p_0 = 0.50$. Otherwise, accept it or reserve judgment.

4. *Calculations*   Replacing each positive difference with a plus sign and each negative difference with a minus sign, and discarding the three pairs of equal values, we get

$$- - - + - - - - - + - - - + - - -$$

where the number of plus signs is 3. Thus, we get

$$z = \frac{3 - 17(0.50)}{\sqrt{17(0.50)(0.50)}} \approx -2.67$$

5. *Decision*   Since -2.67 is less than -2.575, the null hypothesis must be rejected; in other words, we conclude that, on average, the two policemen do not issue

equally many speeding tickets per day. Had we used the continuity correction, we would have obtained

$$z = \frac{3.5 - 17(0.50)}{\sqrt{17(0.50)(0.50)}} \approx -2.43$$

Since this value falls between -2.575 and 2.575, we now find that the null hypothesis should not have been rejected.

13.14 1. *Hypotheses*

$$H_0: \quad \text{populations are identical}$$

$$H_A: \quad \mu_1 \neq \mu_2$$

2. *Level of significance* $\alpha = 0.01$
3. *Criterion*   Reject the null hypothesis if $U \leq 2$, which is the value of U'_α for $n_1 = 6, n_2 = 6$, and $\alpha = 0.01$; otherwise, accept it or reserve judgment.
4. *Calculations*   Arranging the data jointly according to size, we get 25, 28, 30, 32, 35, 37, 38, 39, 42, 45, 46, and 48. Assigning the data in this order the ranks 1, 2, 3, ..., and 12, we find that the values of the first sample (spring) occupy ranks 6, 7, 9, 10, 11, and 12, while those of the second sample (fall) occupy ranks 1, 2, 3, 4, 5, and 8. Thus,

$$W_1 = 6 + 7 + 9 + 10 + 11 + 12 = 55$$

$$W_2 = 1 + 2 + 3 + 4 + 5 + 8 = 23$$

so that

$$U_1 = 55 - \frac{6 \cdot 7}{2} = 34$$

$$U_2 = 23 - \frac{6 \cdot 7}{2} = 2$$

and $U = 2$.

5. *Decision*   Since $U = 2$ is less than or equal to 2, the null hypothesis must be rejected; in other words, we conclude that, on average, there are not equally many assaults per week in the spring and in the fall.

*13.19 1. *Hypotheses*

$$H_0: \quad \text{populations are identical}$$

$$H_A: \quad \mu_1 \neq \mu_2$$

2. *Level of significance* $\alpha = 0.05$
3. *Criterion*   Reject the null hypothesis if $z \leq -1.96$ or $z \geq 1.96$, where

$$z = \frac{U_1 - \mu_{U_1}}{\sigma_{U_1}}$$

otherwise, accept it or reserve judgment.

4. *Calculations*   Arranging the data jointly according to size, we get 23, 30, 41, 44, 47, 54, 55, 56, 58, 62, 64, 70, 71, 71, 72, 73, 78, 80, 81, 87, 91, 93, 94, and 96. Assigning the data in this order the ranks 1, 2, 3,..., and 24, we find that the values of the first sample (minority group 1) occupy ranks 4, 7, 10, 12, 15, 16, 17, 18, 20, 21, 23, and 24, so that

$$W_1 = 4 + 7 + 10 + 12 + 15 + 16 + 17 + 18 + 20 + 21 + 23 + 24$$

$$= 187$$

and

$$U_1 = 187 - \frac{12 \cdot 13}{2} = 109$$

Since

$$\mu_{U_1} = \frac{12 \cdot 12}{2} = 72 \quad \text{and} \quad \sigma_{U_1} = \sqrt{\frac{12 \cdot 12 \cdot 25}{12}} \approx 17.32$$

it follows that

$$z = \frac{109 - 72}{17.32} \approx 2.14$$

5. *Decision*   Since $z = 2.14$ exceeds 1.96, the null hypothesis must be rejected; in other words, students from the two minority groups cannot be expected to score equally well on the test.

*13.24 1. *Hypotheses*

H_0:   population are identical

H_A: population means are not all equal

2. *Level of significance* $\alpha = 0.05$

3. *Criterion*   Reject the null hypothesis if $H \geq 5.991$, which is the value of $\chi_{0.05}^2$ for $3 - 1 = 2$ degrees of freedom; otherwise, accept it or reserve judgment.

4. *Calculations*   Arranging the data jointly according to size, we get 14, 16, 17, 18, 19, 21, 21, 22, 23, 24, 26, 27, 28, 29, 31, 31, 31, and 32. Assigning the data in this order the ranks 1, 2, 3,..., and 18, we find that the values of the first sample occupy ranks 1, 9, 11, 13, 14, and 16, while those of the second sample occupy ranks 2, 5, 6.5, 12, 16, and 18, and those of the third sample occupy ranks 3, 4, 6.5, 8, 10, and 16. Thus,

$$R_1 = 1 + 9 + 11 + 13 + 14 + 16 = 64$$

$$R_2 = 2 + 5 + 6.5 + 12 + 16 + 18 = 59.5$$

$$R_3 = 3 + 4 + 6.5 + 8 + 10 + 16 = 47.5$$

and it follows that

$$H = \frac{12}{18 \cdot 19} \left(\frac{64^2}{6} + \frac{59.5^2}{6} + \frac{47.5^2}{6} \right) - 3 \cdot 19 \approx 0.85$$

5. *Decision*   Since $H = 0.85$ is less than 5.991, the null hypothesis cannot be rejected; in other words, the data tends to support the claim that there is no difference in the true average mileage yield of the three kinds of gasoline.

13.28 1. *Hypotheses*

H_0:   arrangement is random

H_A: arrangement is not random

2. *Level of significance* $\alpha = 0.05$

3. *Criterion*   Since $n_1 = 13$, $n_2 = 12$, and $\alpha = 0.05$, we get $u'_{0.025} = 8$ and $u_{0.025} = 19$ from Table VIII; thus, the null hypothesis must be rejected if $u \leq 8$ or $u \geq 19$.

4. *Calculations* $u = 15$

5. *Decision* Since $u = 15$ falls between 8 and 19, the null hypothesis cannot be rejected; in other words, there is no evidence to suspect that the arrangement is not random.

*13.32 1. *Hypotheses*

$$H_0: \quad \text{arrangement is random}$$

$$H_A: \quad \text{arrangement is not random}$$

2. *Level of significance* $\alpha = 0.05$

3. *Criterion* Reject the null hypothesis if $z \leq -1.96$ or $z \geq 1.96$, where

$$z = \frac{u - \mu_u}{\sigma_u}$$

Otherwise, accept it or reserve judgment.

4. *Calculations* Since there are $n_1 = 28$ even digits and $n_2 = 22$ odd digits, and $u = 23$, we get

$$\mu_u = \frac{2 \cdot 28 \cdot 22}{28 + 22} + 1 \approx 25.64$$

$$\sigma_u = \sqrt{\frac{2 \cdot 28 \cdot 22(2 \cdot 28 \cdot 22 - 28 - 22)}{(28 + 22)^2(28 + 22 - 1)}} \approx 3.45$$

and hence $z = \dfrac{23 - 25.64}{3.45} \approx -0.77$.

5. *Decision* Since $z = -0.77$ falls between -1.96 and 1.96, the null hypothesis cannot be rejected; there is no reason to suspect any lack of randomness.

13.37 1. *Hypotheses*

$$H_0: \quad \text{arrangement is random}$$

$$H_A: \quad \text{arrangement is not random}$$

2. *Level of significance* $\alpha = 0.05$

3. *Criterion* Since the median is $30.5, n_1 = 12, n_2 = 12,$ and $u'_{0.025} = 7$ and $u_{0.025} = 19$, according to Table VIII; thus, the null hypothesis must be rejected if $u \leq 7$ or $u \geq 19$. Otherwise, accept the null hypothesis or reserve judgment.

4. *Calculations* The arrangement of values above and below the median is

$$a\,a\,a\,b\,b\,b\,a\,a\,b\,b\,a\,a\,a\,b\,a\,a\,a\,a\,b\,b\,b\,b\,b$$

so that $u = 8$.

5. *Decision* Since $u = 8$ falls between 7 and 19, the null hypothesis cannot be rejected; in other words, there is no significant deviation from randomness.

13.43 Ranking the x's among themselves and also the y's, we get the ranks shown in the first two columns of the following table:

Rank of x	Rank of y	d	d^2
9	6	3	9
4	2	2	4
7.5	6	1.5	2.25
15	14	1	1
7.5	11.5	−4	16
1	6	−5	25
2	3	−1	1
3	1	2	4
6	4	2	4
5	10	−5	25
13.5	15	−1.5	2.25
10	9	1	1
11	8	3	9
13.5	11.5	2	4
12	13	−1	1
		Total	108.5

Then, determining the d's and their squares and substituting $n = 15$ and $\sum d^2 = 108.5$ into the formula for r_S, we get

$$r_S = 1 - \frac{6(108.5)}{15(15^2 - 1)} \approx 0.81$$

13.44 1. *Hypotheses*

$$H_0: \quad \rho = 0 \quad \text{(no correlation)}$$

$$H_A: \quad \rho \neq 0$$

2. *Level of significance* $\alpha = 0.05$
3. *Criterion*   Reject the null hypothesis if $z \leq -1.96$ or $z \geq 1.96$, where

$$z = r_S \sqrt{n - 1}$$

Otherwise, accept it or reserve judgment.
4. *Calculations*   For $n = 15$ and $r_S = 0.81$, we get

$$z = 0.81 \sqrt{15 - 1} \approx 3.03$$

5. *Decision*   Since $z = 3.03$ exceeds 1.96, the null hypothesis must be rejected; in other words, there is a relationship between students' grades in the two subjects.

12 & 13

Achievements

Having read and studied these chapters and having worked a good proportion of the exercises, you should be able to

1. Explain the method of least squares.
2. Fit a least-squares line by solving the two normal equations.
3. Fit a least-squares line by using the special formulas for a and b.
4. Use the equation of a least-squares line to predict a value of y.
5. Explain what is meant by *regression line*.
*6. State the assumptions underlying normal regression analysis.
*7. Calculate the standard error of estimate.
*8. Test hypotheses about the regression coefficient β.
*9. Construct confidence intervals for the regression coefficient β.
*10. Construct confidence limits for the mean of y when $x = x_0$.
11. Define the correlation coefficient in terms of the residual and total sums of squares.
12. Explain the difference between positive and negative correlation.
13. Use the computing formula to calculate r.
14. Use r to judge the strength of a linear relationship.
15. Use values of r to compare the strength of two relationships.
16. Test the significance of a value of r.
17. Explain what is meant by *nonparametric tests*.
18. Perform a one-sample sign test.
19. Apply the sign test to paired data.
20. Perform a large-sample sign test.
21. Use the U test as a nonparametric alternative to the two-sample t test (concerning the difference between two means).
*22. Perform a large-sample U test.
*23. Use the H test as a nonparametric alternative to a one-way analysis of variance.
24. Test for randomness on the basis of the total number of runs.
*25. Perform a large-sample test of randomness based on the total number of runs.
26. Test for randomness on the basis of runs above and below the median.
27. Calculate the rank-correlation coefficient.
28. Test the significance of a rank-correlation coefficient.

CHECKLIST OF KEY TERMS (with page references to their definitions)

R.156 If $r = 0.56$ for one set of paired data and $r = 0.97$ for another set of paired data, compare the strengths of the two relationships.

R.157 The following are the high school averages, x, and first-year-college grade-point averages, y, of 10 students:

x	y
3.0	2.6
2.7	2.4
3.8	3.9
2.6	2.1
3.2	2.6
3.4	3.3
2.8	2.2
3.1	3.2
3.5	2.8
3.3	2.5

Fit a least-squares line that will enable us to predict first-year-college grade-point averages in terms of high school averages, and use it to predict the first-year-college grade-point average of a student with a high school average of 3.5.

*__R.158__ With reference to Exercise R.157, construct a 95% confidence interval for the regression coefficient β.

R.159 If a set of $n = 42$ paired data yields $r = 0.33$, test the null hypothesis of no correlation

(a) at the 0.05 level of significance;

(b) at the 0.01 level of significance.

R.160 A sample of 30 suitcases carried by an airline on transoceanic flights weighed the following (in pounds):

$$32 \quad 46 \quad 48 \quad 27 \quad 35 \quad 52 \quad 66 \quad 41 \quad 49 \quad 36$$
$$50 \quad 44 \quad 48 \quad 36 \quad 40 \quad 35 \quad 63 \quad 42 \quad 52 \quad 40$$
$$38 \quad 36 \quad 43 \quad 41 \quad 49 \quad 52 \quad 44 \quad 60 \quad 31 \quad 35$$

Use the sign test at the 0.01 level of significance to test the null hypothesis $\tilde{\mu} = 37$ pounds against the alternative hypothesis $\tilde{\mu} > 37$ pounds.

R.161 The following are measurements of the strength of samples of two kinds of fishing lines (in pounds):

| Fishing line 1: | 12.0, | 11.5, | 11.8, | 11.3, | 12.2, | 11.7, | 11.9, |
| Fishing line 2: | 11.1, | 11.6, | 11.4, | 12.1, | 10.5, | 10.0, | 10.8 |

Use the U test at the 0.05 level of significance to test the null hypothesis that the two samples come from identical populations against the alternative hypothesis that the two populations have unequal means.

R.162 The following sequence shows whether a certain member was present, P, or absent, A, at 20 consecutive meetings of a fraternal organization:

$$P \ P \ P \ P \ P \ P \ P \ P \ A \ P \ P \ P \ P \ P \ P \ P \ P \ A \ A \ A \ A$$

Test for randomness at the 0.05 level of significance.

R.163 The following are the scores of 12 golfers on two consecutive Sundays:

68 and 71	73 and 76	70 and 73	74 and 71	69 and 72	72 and 74
67 and 70	72 and 68	71 and 72	73 and 74	68 and 69	70 and 72

Use the sign test at the 0.05 level of significance to test the null hypothesis $\mu_1 = \mu_2$ against the alternative hypothesis $\mu_2 > \mu_1$ (perhaps due to heavy winds on the second Sunday).

R.164 The following are the numbers of hours that 10 persons (interviewed as part of a sample survey) spent reading books or magazines, x, and watching television, y, during the preceding week:

x	y
4	18
8	12
9	9
3	15
2	27
10	12
5	12
1	19
5	25
7	18

Calculate the coefficient of correlation.

R.165 With reference to Exercise R.164, test the significance of r at the 0.05 level of significance.

R.166 The following are the closing prices of a commodity on 20 consecutive trading days (in dollars):

378	379	379	378	377	376	374	374	373	373
374	375	376	376	376	375	374	374	373	374

Test for randomness at the 0.01 level of significance.

R.167 State in each case whether you would expect a positive correlation, a negative correlation, or no correlation (a zero correlation):

(a) the price of gasoline and the occupancy rate of motels;

(b) exposure to the sun and the incidence of skin cancer;

(c) shoe size and years of education;

(d) mintage figures and the values of old coins;

(e) blood pressure and eye color;

(f) baseball players' batting averages and their salaries.

R.168 The following are the numbers of minutes it took two ambulance services to reach the scenes of accidents:

Ambulance service 1: 9.3, 5.5, 13.1, 10.0, 7.6, 9.2, 11.2, 6.4, 14.0, 10.3,
Ambulance service 2: 12.7, 6.6, 9.1, 4.5, 7.2, 6.4, 7.5

Use the U test at the 0.05 level of significance to test the null hypothesis that the two samples come from identical populations against the alternative hypothesis that the two populations have unequal means.

R.169 The following are the numbers of persons who attended a "singles only" dance on 12 consecutive Saturdays:

$$152 \quad 188 \quad 149 \quad 212 \quad 103 \quad 146 \quad 177 \quad 158 \quad 201 \quad 175 \quad 188 \quad 162$$

Use the sign test based on Table I and the 0.05 level of significance to test the null hypothesis $\tilde{\mu} = 149$ against the alternative hypothesis $\tilde{\mu} > 149$.

*R.170 The following sequence shows whether a television news program had at least 12% of the viewing audience, A, or less than 12%, L, on 36 consecutive weekday evenings:

$$L \quad L \quad L \quad L \quad A \quad A \quad L \quad L \quad L \quad A \quad L \quad L \quad L \quad A \quad A \quad A \quad A \quad L$$
(cont.) $\quad A \quad L \quad L \quad L \quad A \quad A \quad L \quad L \quad L \quad L \quad L \quad A \quad L \quad L \quad L \quad L \quad L \quad A$

Test for randomness at the 0.05 level of significance.

R.171 The following data was obtained in a study of the relationship between the resistance (ohms), x, and the failure time (minutes), y, of certain overloaded resistors:

x	y
33	38
41	45
48	40
46	44
35	33
30	32
50	48

Fit a least-squares line and use it to predict the failure time of such a resistor when the resistance is 40 ohms.

*R.172 With reference to Exercise R.171, construct a 95% confidence interval for the mean failure time of such a resistor when the resistance is 40 ohms.

*R.173 The following are the numbers of minutes that patients had to wait in the offices of four doctors:

Doctor 1:	20,	25,	38,	31,	23
Doctor 2:	19,	12,	21,	24,	9
Doctor 3:	8,	10,	27,	25,	14
Doctor 4:	17,	25,	28,	21,	15

Use the H test at the 0.05 level of significance to test the null hypothesis that the four samples come from identical populations against the alternative hypothesis that the means of the populations are not all equal.

R.174 Calculate r for the data of Exercise R.171 and test whether it is significant at the 0.01 level of significance.

R.175 The following are the numbers of burglaries committed in two suburbs of a large city on 24 days:

20 and 17	12 and 9	25 and 21	18 and 14	12 and 15	27 and 23
22 and 18	20 and 15	18 and 18	19 and 21	24 and 22	17 and 23
20 and 11	16 and 22	13 and 9	18 and 15	20 and 17	13 and 18
22 and 16	25 and 21	20 and 13	12 and 6	18 and 9	17 and 25

Use the large-sample sign test at the 0.05 level of significance to test the null hypothesis that on the average there are equally many burglaries in the two suburbs against the alternative hypothesis that there are not equally many.

R.176 The following are the batting averages, x, and home runs hit, y, by 15 baseball players during the first half of a baseball season:

x	y
0.252	12
0.305	6
0.299	4
0.303	15
0.285	2
0.191	2
0.283	16
0.272	6
0.310	8
0.266	10
0.215	0
0.211	3
0.272	14
0.244	6
0.320	7

Calculate the rank-correlation coefficient and check whether it is significant at the 0.05 level of significance.

R.177 Test at the 0.05 level of significance whether the following sequence of cars observing, O, and exceeding, E, the 65-mph speed limit may be regarded as random:

$$O\ O\ O\ E\ E\ O\ E\ O\ O\ O\ O\ E\ E\ E\ O\ E\ O\ O\ O\ O\ E\ O$$

*R.178 The following is data on the percentage kill of two kinds of insecticides used against mosquitos:

Insecticide A: 41.9, 46.9, 44.6, 43.9, 42.0, 44.0, 41.0, 43.1, 39.0, 45.2, 44.6, 42.0
Insecticide B: 45.7, 39.8, 42.8, 41.2, 45.0, 40.2, 40.2, 41.7, 37.4, 38.8, 41.7, 38.7

Use the large-sample U test at the 0.05 level of significance to test the null hypothesis that the two samples come from identical populations against the alternative hypothesis that on the average the first insecticide is more effective than the second.

R.179 Rework Exercise R.178 using Table VII and the alternative hypothesis that on the average the two insecticides are not equally effective.

R.180 The following are the processing times (minutes), x, and hardness readings, y, of certain machine parts:

x	y
20	282
34	275
19	171
10	142
24	145
31	340
25	282
13	105
29	233

Fit a least-squares line and use it to predict the hardness reading of a part that has been processed for 25 minutes.

*R.181 With reference to Exercise R.180, test the null hypothesis $\beta = 9.5$ against the alternative hypothesis $\beta < 9.5$ at the 0.01 level of significance.

R.182 With reference to Exercise R.180, what percentage of the variation among the hardness readings can be attributed to differences in processing times?

R.183 If $r_s = 0.34$ for a set of $n = 50$ paired data, is this rank-correlation coefficient significant at the 0.01 level of significance?

APPENDIX A: TI-83 TIPS

■ **KEYS**   Each button (key) on the TI-83 can perform up to three functions.

The primary key function is written in white on the key. To use the primary function, simply press the key.

2nd key   The secondary function is written in yellow above the key. To use a secondary function, you must first press the yellow 2nd key.

■ **EXAMPLE**   To turn the calculator off, press 2nd ON for OFF.

alpha key   The alpha function is written in green above the key. To use an alpha function, you must first press the green ALPHA key. Most alpha functions simply print a letter or variable on the screen.

■ **SCREENS**   When you first turn on the calculator you are in the home screen, where you
home screen   can perform computations and other commands. You can use the calculator keys to bring up other screens and menus that allow you to do more than is directly available on the keyboard. You can return to the home screen at any time by pressing 2nd MODE for QUIT.

■ **EXAMPLE**   To get to the graphing screen, press GRAPH.
graph screen   (The GRAPH key is in the upper right-hand corner of the keyboard.)

■ **EXAMPLE**   To return to the home screen, press 2nd MODE for QUIT.

■ **MENUS**   One type of screen you will use often is a menu. Menus allow you to do more than is directly available on the keyboard. You can move around these menus using the arrow keys located in the upper right-hand corner of the keyboard.

Some of the most common statistics menus are:

STAT EDIT	(press STAT)
STAT CALC	(press STAT ▶)
STAT TESTS	(press STAT ▶▶)
DISTR	(press 2nd VARS)
DISTR DRAW	(press 2nd VARS ▶)
STAT PLOT	(press 2nd Y =)
VARS Statistics	(press VARS 5)
MATH PRB	(press MATH ◀)

CLEARING LISTS   To clear a list press STAT ENTER. You should see

Move the cursor so that it is flashing on the label at the top of a list.

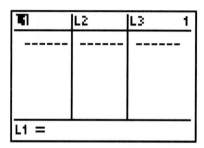

Press CLEAR ENTER and all data in the list will be deleted.

CLEARING PLOTS   Graphical representation of data is important in statistics. Before plotting a new set of data, you should clear old plots.

If you want to turn off statistics plots only, without affecting anything else, press 2nd Y = for STAT PLOT.
Press 4 for 4 : PlotsOff; then press ENTER.

You can also turn off plots by resetting the calculator defaults (see Resetting defaults, below).

RESETTING DEFAULTS   You can restore the initial or default settings on the TI-83 without losing your data or programs.

To reset defaults, press 2nd+ for MEM.
Press 5 for 5 : Reset ....
Press 2 for 2 : Defaults...; then press 2 for 2 : Reset .

CAUTION: Be careful not to reset all memory or you will lose all data and programs!

Resetting defaults will turn off all plots and reset the graphing screen to its standard scale (see Setting graph dimensions, page 476). Other defaults that you probably never changed will also be reset.

trouble
shooting   Any time you find that your calculator seems to be acting strange, try resetting defaults. You may have previously changed some of the defaults accidentally. Resetting defaults may just fix the problem.

■ **DATA ENTRY**   To enter data on the TI-83, you need to use the List Editor. You can access the List Editor by pressing STAT ENTER for EDIT.

■ **EXAMPLE**   To enter the following data set

91   59   75   66   60   69   61   72   90   89

1. Press STAT ENTER for STAT EDIT.
   You should see

   This screen is the List Editor. The lists should be empty. If not, clear the lists (see Clearing lists, page 474).

2. Press 91.
   You will see L1(1) = 91 at the bottom of the screen.

3. Press ENTER.
   91 is entered into L1.

4. Press 59 ENTER.
   59 is entered into L1.

5. Enter the rest of the data by typing in each number and pressing ENTER after each number. When you are done, you should see

L1	L2	L3	1
60			
69			
61			
72			
90			
89			

L1(11) =

   Notice that you may not see all of the data. When you reach the bottom of the screen, the data scrolls off the screen at the top. You can use the up arrow key to move back up the data list and see the numbers that scrolled off the screen. You can use the down arrow key to move back down through the list. You can use the left and right arrow keys to move between lists.

correcting mistakes   If you entered any incorrect numbers, use the arrow keys to highlight the incorrect list entry, retype the entry, and press ENTER. Note that the highlighted entry is always shown at the bottom of the screen.

negative sign   If your data includes a negative number, you need to use the gray negative sign (−) key to enter the number rather than the blue subtraction key.

■■ **EXAMPLE**   Change the first data entry in the above data set to -91.

1. Use the arrow keys to highlight the 91.
2. Press the (−) key, then press 91 ENTER.
   The first entry in the list should change to -91.

If you got an error message, you probably used the subtraction key. Return to the List Editor and try again.

■■ **SETTING GRAPH DIMENSIONS**   The standard graphing screen shows the rectangular coordinate system with x-values going from −10 to 10 and y-values going from −10 to 10.

window settings   To see the graphing screen settings, press WINDOW. You will see the current scale settings for the graphing screen.

Xmin = smallest x-value shown in the graphing screen.

Xmax = largest x-value shown in the graphing screen.

Xscl = distance between successive scale marks on the x-axis.

Ymin = smallest y-value shown in the graphing screen.

Ymax = largest y-value shown in the graphing screen.

Yscl = distance between successive scale marks on the y-axis.

Xres = resolution of graph with Xres = 1 being the most improved.

You can change the settings using the down arrow key to access each line and making a new entry on that line.

graphs of data   If you would like to make a graph of a dataset, select the appropriate graph, press ZOOM, and scroll down to 9: ZoomStat and press ENTER.
This will automatically set up the correct Window to display the graph.

graphs of distributions   The TI-83 allows you to draw common probability distributions. These graphs and their dimensions are based on information about the distribution rather than actual data. The most common of these distributions is the normal distribution. For a normal distribution, virtually the entire distribution falls within 5 standard deviations of the mean. You can use the following window settings for graphs of the standard normal distribution:

```
WINDOW
  Xmin=-5
  Xmax=5
  Xscl=1
  Ymin=-.1
  Ymax=.5
  Yscl=.1
  Xres=1
```

■■ **EXAMPLE**   To sketch the standard normal distribution:

1. Set the graph dimensions.
2. Press 2nd VARS ▶ for DISTR DRAW.
3. Press ENTER for 1:ShadeNorm(.
4. To get just the distribution with no shading, press 0, 0).
   (You are telling the calculator to shade from 0 to 0.)
   You should see 1:ShadeNorm(0, 0).
5. Press ENTER.
   You should see

■■ **INFINITY**   In statistics, you can indicate that there is no highest number in a function by using infinity. Since infinity is not a number, there is no key for infinity on the calculator. However, you can use instead the largest number the calculator can hold, which is 1×10^{99}, meaning 1 followed by 99 zeros.

To enter 1×10^{99}, press 1.
Press 2nd, then press the comma key for EE, and then press 99.
You should see 1E99.

1E99 is engineering notation for 1×10^{99}.

minus infinity   You can use -1×10^{99} for negative infinity. Press the negative sign (−) key, then press 1; press 2nd followed by the comma key for EE, and then press 99.

EXPRESSIONS WITH PARENTHESES   Many of the functions on the calculator automatically give an opening parenthesis. Be sure to include a closing parenthesis at the end of the expression.

■ **MOST**
COMMON
STATISTICS
MENUS

STAT EDIT
(press STAT)

```
EDIT CALC TESTS
1:Edit…
2:SortA(
3:SortD(
4:ClrList
5:SetUpEditor
```

STAT CALC
(press STAT)

```
EDIT CALC TESTS
1:1-Var Stats
2:2-Var Stats
3:Med-Med
4:LinReg(ax+b)
5:QuadReg
6:CubicReg
7↓QuartReg
```

```
EDIT CALC TESTS
7:QuartReg
8:LinReg(a+bx)
9:LnReg
0:ExpReg
A:PwrReg
B:Logistic
C:SinReg
```

STAT TESTS
(press STAT ▶▶)

```
EDIT CALC TESTS
1:Z-Test…
2:T-Test…
3:2-SampZTest…
4:2-SampTTest…
5:1-PropZTest…
6:2-PropZTest…
7↓ZInterval…
```

```
EDIT CALC TESTS
7:ZInterval…
8:TInterval…
9:2-SampZInt…
0:2-SampTInt…
A:1-PropZInt…
B:2-PropZInt…
C↓X²-Test…
```

```
EDIT CALC TESTS
C:X²-Test…
D:2-SampFTest…
E:LinRegTTest…
F:ANOVA(
```

DISTR
(press 2nd VARS)

```
DISTR DRAW
1:normalpdf(
2:normalcdf(
3:invNorm(
4:tpdf(
5:tcdf(
6:X²pdf(
7↓X²cdf(
```

```
DISTR DRAW
7:X²cdf(
8:Fpdf(
9:Fcdf(
0:binompdf(
A:binomcdf(
B:poissonpdf(
C↓poissoncdf(
```

```
DISTR DRAW
C:poissoncdf(
D:geometpdf(
E:geometcdf(
```

DISTR DRAW
(press 2nd VARS ▶)

```
DISTR DRAW
1:ShadeNorm(
2:Shade_t(
3:ShadeX²(
4:ShadeF(
```

STAT PLOT	VARS Statistics	MATH PRB
(press 2nd Y=)	(press VARS 5)	(press MATH ◄)

```
STAT PLOTS
1:Plot1…Off
   L1   L2    ▫
2:Plot2…Off
   L1   L2    ▫
3:Plot3…Off
   L1   L2    ▫
4↓PlotsOff
```

```
XY Σ EQ TEST PTS
1:n
2:x̄
3:Sx
4:σx
5:ȳ
6:Sy
7↓σy
```

```
MATH NUM CPX PRB
1:rand
2:nPr
3:nCr
4:!
5:randInt(
6:randNorm(
7:randBin(
```

```
STAT PLOTS
2↑Plot2…Off
   L1   L2    ▫
3:Plot3…Off
   L1   L2    ▫
4:PlotsOff
5:PlotsOn
```

```
XY Σ EQ TEST PTS
7:σy
8:minX
9:maxX
0:minY
A:maxY
```

BIBLIOGRAPHY

A. PROBABILITY AND STATISTICS FOR THE LAYMAN

Brook, R. J., Arnold, G. C., Hassard, T. H., and Pringle, R. M., eds., *The Fascinati*
Statistics. New York: Marcel Dekker, Inc., 1986.

Campbell, S. K., *Flaws and Fallacies in Statistical Thinking.* Englewood Cliffs, N.J.: Pre
Hall, 1974.

Federer, W. T., *Statistics and Society: Data Collection and Interpretation.* New York: M
Dekker, Inc., 1991.

Gonick, L., and Smith, W., *The Cartoon Guide to Statistics.* New York: Ha
Collins, 1993.

Hollander, M., and Proschan, F., *The Statistical Exorcist: Dispelling Statistics Anxiety.*
York: Marcel Dekker, Inc., 1984.

Hooke, R., *How to Tell the Liars from the Statisticians.* New York: Marcel De
Inc., 1983.

Huff, D., *How to Lie with Statistics.* New York: W. W. Norton & Company, Inc., 199.

Jaffe, A., and Spirer, H. F., *Misused Statistics: Straight Talk for Twisted Numbers*,
York: Marcel Dekker, Inc., 1987.

Mosteller, F., Pieters, R. S., Kruskal, W. H., Rising, G. R., Link, R. F., Carlson, R.
Zelinka, M., *Statistics by Example.* Reading, Mass.: Addison-Wesley Publishing
Inc., 1973.

Paulos, J. A., *Innumeracy: Mathematical Illiteracy and Its Consequences.* New York:
and Wang, 2001.

Wang, C., *Sense and Nonsense of Statistical Inference.* New York: Marcel De
Inc., 1993.

Weaver, W., *Lady Luck: The Theory of Probability.* New York: Dover Publicat
Inc., 1982.

B. SOME BOOKS ON THE THEORY OF PROBABILITY A STATISTICS

Barr, D. R., and Zehna, P. W., *Probability: Modeling Uncertainty.* Reading, M
Addison-Wesley Publishing Company, Inc., 1983.

Bennett, D. J., *Randomness.* Cambridge, Mass.: Harvard University Press, 1998.

Chernoff, H., and Moses, L. E., *Elementary Decision Theory.* New York: Dover Pul
tions, Inc., 1987 reprint.

Freund, J. E., *Introduction to Probability.* New York: Dover Publications, Inc., 1993 rep

Goldberg, S., *Probability: An Introduction.* New York: Dover Publications, Inc., 198

Hodges, J. L., and Lehmann, E. L., *Basic Concepts of Probability and Statistics*, 2nd
ed. San Francisco: Holden-Day, 1970.

Koopmans, L. H., *An Introduction to Contemporary Statistics.* North Scituate, M
Duxbury Press, 1981.

Kotz, S., and Stroup, D. E., *Educated Guessing: How to Cope in an Uncertain World.*
York: Marcel Dekker, Inc., 1983.

Mendenhall, W., Scheaffer, R. L., and Wackerly, D. D., *Mathematical Statistics*
Applications, 6th ed. North Scituate, Mass.: Duxbury Press, 2001.

Miller, I., and Miller, M., *John E. Freund's Mathematical Statistics*, 6th ed. Upper Sa
River, N.J.: Prentice Hall, 1999.

Packel, E. W., *The Mathematics of Games and Gambling.* Washington, D.C.: Mathem:
Association of America, 1981.

Rice, J. A., *Mathematical Statistics and Data Analysis*, 2nd ed. Belmont, Calif.: Wadsworth Publishing Co., 1995.

Ross, S., *A First Course in Probability*, 6th ed. Upper Saddle River, N.J.: Prentice Hall, 2002.

Scheaffer, R. L., Mendenhall, W., and Ott, L., *Elementary Survey Sampling*, 5th ed. Boston, Mass.: Duxbury Press, 1996.

C. SOME BOOKS DEALING WITH SPECIAL TOPICS

Box, G. E. P., Hunter, W. G., and Hunter, J. S., *Statistics for Experimenters*. New York: John Wiley & Sons, Inc., 1978.

Chatterjee, S., Hadi, A. S., and Price, B., *Regression Analysis by Example*, 3rd ed. New York: John Wiley & Sons, Inc., 2000.

Daniel, C., and Wood, F., *Fitting Equations to Data*, 2nd ed. New York: John Wiley & Sons, Inc., 1980.

Draper, N. R., and Smith, H. *Applied Regression Analysis*, 3rd ed. New York: John Wiley & Sons, Inc., 1998.

Everitt, B. S., *The Analysis of Contingency Tables*, 2nd ed. New York: Chapman and Hall, 1992.

Fleiss, J. L., *The Design and Analysis of Clinical Experiments*, New York: John Wiley & Sons, Inc., 1986.

Gibbons, J. D., and Chakraborti, S., *Nonparametric Statistical Inference*, 3rd ed. New York: Marcel Dekker, 1992.

Goodman, L. A., *The Analysis of Cross-Classified Data Having Ordered Categories*. Cambridge, Mass.: Harvard University Press, 1984.

Hartwig, F., and Dearing, B. E., *Exploratory Data Analysis*. Beverly Hills, Calif.: Sage Publications, Inc., 1979.

Hicks, C. R., and Turner, K. V., *Fundamental Concepts in the Design of Experiments*, 5th ed. New York: Oxford University Press, 1999.

Lehmann, E. L., *Nonparametrics: Statistical Methods Based on Ranks*, rev. ed. Upper Saddle River, N.J.: Prentice Hall, 1998.

Morgan, L., *Statistics Handbook for the TI-83*. Austin, Tex.: Texas Instruments, 1997.

Mosteller, F., and Rourke, R. E. K., *Sturdy Statistics*. Reading, Mass.: Addison Wesley Publishing Company, Inc., 1973.

Mosteller, F., *Fifty Challenging Problems in Probability with Solutions*. New York: Dover Publications, Inc., 1987.

Noether, G. E., *Introduction to Statistics: The Nonparametric Way*. New York: Springer-Verlag, 1991.

Weisberg, S., *Applied Linear Regression*, 2nd ed. New York: John Wiley & Sons, Inc., 1985.

Williams, W. H., *A Sampler on Sampling*. New York: John Wiley & Sons, Inc., 1978.

Wonnacott, T. H., and Wonnacott, R. J., *Regression: A Second Course in Statistics*. New York: John Wiley & Sons, Inc., 1981.

D. SOME GENERAL REFERENCE WORKS AND TABLES

Freund, J. E., and Williams, F. J., *Dictionary/Outline of Basic Statistics*. New York: Dover Publications, Inc., 1991 reprint.

Hauser, P. M., and Leonard, W. R., eds., *Government Statistics for Business Use*, 2nd ed. New York: John Wiley & Sons, Inc., 1956.

Kendall, M. G., and Buckland, W. R., *A Dictionary of Statistical Terms*, 4th ed. London: Longman Group Ltd., 1982.

National Bureau of Standards, *Tables of the Binomial Probability Distribution*. Washington, D. C.: Government Printing Office, 1950.

Owen, D. B., *Handbook of Statistical Tables*. Reading, Mass.: Addison-Wesley Publishing Company, Inc., 1962.

Rand Corporation, *A Million Random Digits with 100,000 Normal Deviates*. New York: The Free Press, 1955.

Romig, H. G., *50–100 Binomial Tables*. New York: John Wiley & Sons, Inc., 1953.

STATISTICAL TABLES

TABLE I   Binomial Probabilities[†]

n	x	0.05	0.1	0.2	0.3	0.4	*p* 0.5	0.6	0.7	0.8	0.9	0.95
2	0	0.902	0.810	0.640	0.490	0.360	0.250	0.160	0.090	0.040	0.010	0.002
	1	0.095	0.180	0.320	0.420	0.480	0.500	0.480	0.420	0.320	0.180	0.095
	2	0.002	0.010	0.040	0.090	0.160	0.250	0.360	0.490	0.640	0.810	0.902
3	0	0.857	0.729	0.512	0.343	0.216	0.125	0.064	0.027	0.008	0.001	
	1	0.135	0.243	0.384	0.441	0.432	0.375	0.288	0.189	0.096	0.027	0.007
	2	0.007	0.027	0.096	0.189	0.288	0.375	0.432	0.441	0.384	0.243	0.135
	3		0.001	0.008	0.027	0.064	0.125	0.216	0.343	0.512	0.729	0.857
4	0	0.815	0.656	0.410	0.240	0.130	0.062	0.026	0.008	0.002		
	1	0.171	0.292	0.410	0.412	0.346	0.250	0.154	0.076	0.026	0.004	
	2	0.014	0.049	0.154	0.265	0.346	0.375	0.346	0.265	0.154	0.049	0.014
	3		0.004	0.026	0.076	0.154	0.250	0.346	0.412	0.410	0.292	0.171
	4			0.002	0.008	0.026	0.062	0.130	0.240	0.410	0.656	0.815
5	0	0.774	0.590	0.328	0.168	0.078	0.031	0.010	0.002			
	1	0.204	0.328	0.410	0.360	0.259	0.156	0.077	0.028	0.006		
	2	0.021	0.073	0.205	0.309	0.346	0.312	0.230	0.132	0.051	0.008	0.001
	3	0.001	0.008	0.051	0.132	0.230	0.312	0.346	0.309	0.205	0.073	0.021
	4			0.006	0.028	0.077	0.156	0.259	0.360	0.410	0.328	0.204
	5				0.002	0.010	0.031	0.078	0.168	0.328	0.590	0.774
6	0	0.735	0.531	0.262	0.118	0.047	0.016	0.004	0.001			
	1	0.232	0.354	0.393	0.303	0.187	0.094	0.037	0.010	0.002		
	2	0.031	0.098	0.246	0.324	0.311	0.234	0.138	0.060	0.015	0.001	
	3	0.002	0.015	0.082	0.185	0.276	0.312	0.276	0.185	0.082	0.015	0.002
	4		0.001	0.015	0.060	0.138	0.234	0.311	0.324	0.246	0.098	0.031
	5			0.002	0.010	0.037	0.094	0.187	0.303	0.393	0.354	0.232
	6				0.001	0.004	0.016	0.047	0.118	0.262	0.531	0.735
7	0	0.698	0.478	0.210	0.082	0.028	0.008	0.002				
	1	0.257	0.372	0.367	0.247	0.131	0.055	0.017	0.004			
	2	0.041	0.124	0.275	0.318	0.261	0.164	0.077	0.025	0.004		
	3	0.004	0.023	0.115	0.227	0.290	0.273	0.194	0.097	0.029	0.003	
	4		0.003	0.029	0.097	0.194	0.273	0.290	0.227	0.115	0.023	0.004
	5			0.004	0.025	0.077	0.164	0.261	0.318	0.275	0.124	0.041
	6				0.004	0.017	0.055	0.131	0.247	0.367	0.372	0.257
	7					0.002	0.008	0.028	0.082	0.210	0.478	0.698
8	0	0.663	0.430	0.168	0.058	0.017	0.004	0.001				
	1	0.279	0.383	0.336	0.198	0.090	0.031	0.008	0.001			
	2	0.051	0.149	0.294	0.296	0.209	0.109	0.041	0.010	0.001		
	3	0.005	0.033	0.147	0.254	0.279	0.219	0.124	0.047	0.009		
	4		0.005	0.046	0.136	0.232	0.273	0.232	0.136	0.046	0.005	
	5			0.009	0.047	0.124	0.219	0.279	0.254	0.147	0.033	0.005
	6			0.001	0.010	0.041	0.109	0.209	0.296	0.294	0.149	0.051
	7				0.001	0.008	0.031	0.090	0.198	0.336	0.383	0.279
	8					0.001	0.004	0.017	0.058	0.168	0.430	0.663

[†]All values omitted in Table I are 0.0005 or less.

TABLE I	Binomial Probabilities (*Continued*)										
						p					
n	*x*	0.05	0.1	0.2	0.3	0.4	0.5	0.6	0.7	0.8	0.9	0.95
9	0	0.630	0.387	0.134	0.040	0.010	0.002					
	1	0.299	0.387	0.302	0.156	0.060	0.018	0.004				
	2	0.063	0.172	0.302	0.267	0.161	0.070	0.021	0.004			
	3	0.008	0.045	0.176	0.267	0.251	0.164	0.074	0.021	0.003		
	4	0.001	0.007	0.066	0.172	0.251	0.246	0.167	0.074	0.017	0.001	
	5		0.001	0.017	0.074	0.167	0.246	0.251	0.172	0.066	0.007	0.001
	6			0.003	0.021	0.074	0.164	0.251	0.267	0.176	0.045	0.008
	7				0.004	0.021	0.070	0.161	0.267	0.302	0.172	0.063
	8					0.004	0.018	0.060	0.156	0.302	0.387	0.299
	9						0.002	0.010	0.040	0.134	0.387	0.630
10	0	0.599	0.349	0.107	0.028	0.006	0.001					
	1	0.315	0.387	0.268	0.121	0.040	0.010	0.002				
	2	0.075	0.194	0.302	0.233	0.121	0.044	0.011	0.001			
	3	0.010	0.057	0.201	0.267	0.215	0.117	0.042	0.009	0.001		
	4	0.001	0.011	0.088	0.200	0.251	0.205	0.111	0.037	0.006		
	5		0.001	0.026	0.103	0.201	0.246	0.201	0.103	0.026	0.001	
	6			0.006	0.037	0.111	0.205	0.251	0.200	0.088	0.011	0.001
	7			0.001	0.009	0.042	0.117	0.215	0.267	0.201	0.057	0.010
	8				0.001	0.011	0.044	0.121	0.233	0.302	0.194	0.075
	9					0.002	0.010	0.040	0.121	0.268	0.387	0.315
	10						0.001	0.006	0.028	0.107	0.349	0.599
11	0	0.569	0.314	0.086	0.020	0.004						
	1	0.329	0.384	0.236	0.093	0.027	0.005	0.001				
	2	0.087	0.213	0.295	0.200	0.089	0.027	0.005	0.001			
	3	0.014	0.071	0.221	0.257	0.177	0.081	0.023	0.004			
	4	0.001	0.016	0.111	0.220	0.236	0.161	0.070	0.017	0.002		
	5		0.002	0.039	0.132	0.221	0.226	0.147	0.057	0.010		
	6			0.010	0.057	0.147	0.226	0.221	0.132	0.039	0.002	
	7			0.002	0.017	0.070	0.161	0.236	0.220	0.111	0.016	0.001
	8				0.004	0.023	0.081	0.177	0.257	0.221	0.071	0.014
	9				0.001	0.005	0.027	0.089	0.200	0.295	0.213	0.087
	10					0.001	0.005	0.027	0.093	0.236	0.384	0.329
	11							0.004	0.020	0.086	0.314	0.569
12	0	0.540	0.282	0.069	0.014	0.002						
	1	0.341	0.377	0.206	0.071	0.017	0.003					
	2	0.099	0.230	0.283	0.168	0.064	0.016	0.002				
	3	0.017	0.085	0.236	0.240	0.142	0.054	0.012	0.001			
	4	0.002	0.021	0.133	0.231	0.213	0.121	0.042	0.008	0.001		
	5		0.004	0.053	0.158	0.227	0.193	0.101	0.029	0.003		
	6			0.016	0.079	0.177	0.226	0.177	0.079	0.016		
	7			0.003	0.029	0.101	0.193	0.227	0.158	0.053	0.004	
	8			0.001	0.008	0.042	0.121	0.213	0.231	0.133	0.021	0.002
	9				0.001	0.012	0.054	0.142	0.240	0.236	0.085	0.017
	10					0.002	0.016	0.064	0.168	0.283	0.230	0.099
	11						0.003	0.017	0.071	0.206	0.377	0.341
	12							0.002	0.014	0.069	0.282	0.540

TABLE I Binomial Probabilities (*Continued*)

n	*x*	0.05	0.1	0.2	0.3	0.4	0.5	0.6	0.7	0.8	0.9	0.95
13	0	0.513	0.254	0.055	0.010	0.001						
	1	0.351	0.367	0.179	0.054	0.011	0.002					
	2	0.111	0.245	0.268	0.139	0.045	0.010	0.001				
	3	0.021	0.100	0.246	0.218	0.111	0.035	0.006	0.001			
	4	0.003	0.028	0.154	0.234	0.184	0.087	0.024	0.003			
	5		0.006	0.069	0.180	0.221	0.157	0.066	0.014	0.001		
	6		0.001	0.023	0.103	0.197	0.209	0.131	0.044	0.006		
	7			0.006	0.044	0.131	0.209	0.197	0.103	0.023	0.001	
	8			0.001	0.014	0.066	0.157	0.221	0.180	0.069	0.006	
	9				0.003	0.024	0.087	0.184	0.234	0.154	0.028	0.003
	10				0.001	0.006	0.035	0.111	0.218	0.246	0.100	0.021
	11					0.001	0.010	0.045	0.139	0.268	0.245	0.111
	12						0.002	0.011	0.054	0.179	0.367	0.351
	13							0.001	0.010	0.055	0.254	0.513
14	0	0.488	0.229	0.044	0.007	0.001						
	1	0.359	0.356	0.154	0.041	0.007	0.001					
	2	0.123	0.257	0.250	0.113	0.032	0.006	0.001				
	3	0.026	0.114	0.250	0.194	0.085	0.022	0.003				
	4	0.004	0.035	0.172	0.229	0.155	0.061	0.014	0.001			
	5		0.008	0.086	0.196	0.207	0.122	0.041	0.007			
	6		0.001	0.032	0.126	0.207	0.183	0.092	0.023	0.002		
	7			0.009	0.062	0.157	0.209	0.157	0.062	0.009		
	8			0.002	0.023	0.092	0.183	0.207	0.126	0.032	0.001	
	9				0.007	0.041	0.122	0.207	0.196	0.086	0.008	
	10				0.001	0.014	0.061	0.155	0.229	0.172	0.035	0.004
	11					0.003	0.022	0.085	0.194	0.250	0.114	0.026
	12					0.001	0.006	0.032	0.113	0.250	0.257	0.123
	13						0.001	0.007	0.041	0.154	0.356	0.359
	14							0.001	0.007	0.044	0.229	0.488
15	0	0.463	0.206	0.035	0.005							
	1	0.366	0.343	0.132	0.031	0.005						
	2	0.135	0.267	0.231	0.092	0.022	0.003					
	3	0.031	0.129	0.250	0.170	0.063	0.014	0.002				
	4	0.005	0.043	0.188	0.219	0.127	0.042	0.007	0.001			
	5	0.001	0.010	0.103	0.206	0.186	0.092	0.024	0.003			
	6		0.002	0.043	0.147	0.207	0.153	0.061	0.012	0.001		
	7			0.014	0.081	0.177	0.196	0.118	0.035	0.003		
	8			0.003	0.035	0.118	0.196	0.177	0.081	0.014		
	9			0.001	0.012	0.061	0.153	0.207	0.147	0.043	0.002	
	10				0.003	0.024	0.092	0.186	0.206	0.103	0.010	0.001
	11				0.001	0.007	0.042	0.127	0.219	0.188	0.043	0.005
	12					0.002	0.014	0.063	0.170	0.250	0.129	0.031
	13						0.003	0.022	0.092	0.231	0.267	0.135
	14							0.005	0.031	0.132	0.343	0.366
	15								0.005	0.035	0.206	0.463

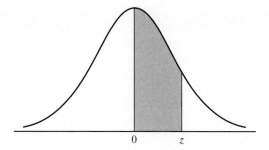

The entries in Table II are the probabilities that a random variable having the standard normal distribution will take on a value between 0 and z. They are given by the area of the gray region under the curve in the figure.

TABLE II	Normal-Curve Areas									
z	.00	.01	.02	.03	.04	.05	.06	.07	.08	.09
0.0	.0000	.0040	.0080	.0120	.0160	.0199	.0239	.0279	.0319	.0359
0.1	.0398	.0438	.0478	.0517	.0557	.0596	.0636	.0675	.0714	.0753
0.2	.0793	.0832	.0871	.0910	.0948	.0987	.1026	.1064	.1103	.1141
0.3	.1179	.1217	.1255	.1293	.1331	.1368	.1406	.1443	.1480	.1517
0.4	.1554	.1591	.1628	.1664	.1700	.1736	.1772	.1808	.1844	.1879
0.5	.1915	.1950	.1985	.2019	.2054	.2088	.2123	.2157	.2190	.2224
0.6	.2257	.2291	.2324	.2357	.2389	.2422	.2454	.2486	.2517	.2549
0.7	.2580	.2611	.2642	.2673	.2704	.2734	.2764	.2794	.2823	.2852
0.8	.2881	.2910	.2939	.2967	.2995	.3023	.3051	.3078	.3106	.3133
0.9	.3159	.3186	.3212	.3238	.3264	.3289	.3315	.3340	.3365	.3389
1.0	.3413	.3438	.3461	.3485	.3508	.3531	.3554	.3577	.3599	.3621
1.1	.3643	.3665	.3686	.3708	.3729	.3749	.3770	.3790	.3810	.3830
1.2	.3849	.3869	.3888	.3907	.3925	.3944	.3962	.3980	.3997	.4015
1.3	.4032	.4049	.4066	.4082	.4099	.4115	.4131	.4147	.4162	.4177
1.4	.4192	.4207	.4222	.4236	.4251	.4265	.4279	.4292	.4306	.4319
1.5	.4332	.4345	.4357	.4370	.4382	.4394	.4406	.4418	.4429	.4441
1.6	.4452	.4463	.4474	.4484	.4495	.4505	.4515	.4525	.4535	.4545
1.7	.4554	.4564	.4573	.4582	.4591	.4599	.4608	.4616	.4625	.4633
1.8	.4641	.4649	.4656	.4664	.4671	.4678	.4686	.4693	.4699	.4706
1.9	.4713	.4719	.4726	.4732	.4738	.4744	.4750	.4756	.4761	.4767
2.0	.4772	.4778	.4783	.4788	.4793	.4798	.4803	.4808	.4812	.4817
2.1	.4821	.4826	.4830	.4834	.4838	.4842	.4846	.4850	.4854	.4857
2.2	.4861	.4864	.4868	.4871	.4875	.4878	.4881	.4884	.4887	.4890
2.3	.4893	.4896	.4898	.4901	.4904	.4906	.4909	.4911	.4913	.4916
2.4	.4918	.4920	.4922	.4925	.4927	.4929	.4931	.4932	.4934	.4936
2.5	.4938	.4940	.4941	.4943	.4945	.4946	.4948	.4949	.4951	.4952
2.6	.4953	.4955	.4956	.4957	.4959	.4960	.4961	.4962	.4963	.4964
2.7	.4965	.4966	.4967	.4968	.4969	.4970	.4971	.4972	.4973	.4974
2.8	.4974	.4975	.4976	.4977	.4977	.4978	.4979	.4979	.4980	.4981
2.9	.4981	.4982	.4982	.4983	.4984	.4984	.4985	.4985	.4986	.4986
3.0	.4987	.4987	.4987	.4988	.4988	.4989	.4989	.4989	.4990	.4990

Also, for $z = 4.0, 5.0$, and 6.0, the areas are 0.49997, 0.4999997, and 0.499999999.

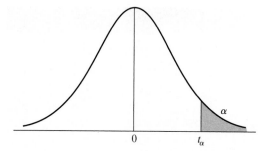

The entries in Table III are values for which the area to their right under the t distribution with given degrees of freedom (the gray area in the figure) is equal to α.

TABLE III   Values of $t^{\dagger}$

d.f.	$t_{.050}$	$t_{.025}$	$t_{.010}$	$t_{.005}$	d.f.
1	6.314	12.706	31.821	63.657	1
2	2.920	4.303	6.965	9.925	2
3	2.353	3.182	4.541	5.841	3
4	2.132	2.776	3.747	4.604	4
5	2.015	2.571	3.365	4.032	5
6	1.943	2.447	3.143	3.707	6
7	1.895	2.365	2.998	3.499	7
8	1.860	2.306	2.896	3.355	8
9	1.833	2.262	2.821	3.250	9
10	1.812	2.228	2.764	3.169	10
11	1.796	2.201	2.718	3.106	11
12	1.782	2.179	2.681	3.055	12
13	1.771	2.160	2.650	3.012	13
14	1.761	2.145	2.624	2.977	14
15	1.753	2.131	2.602	2.947	15
16	1.746	2.120	2.583	2.921	16
17	1.740	2.110	2.567	2.898	17
18	1.734	2.101	2.552	2.878	18
19	1.729	2.093	2.539	2.861	19
20	1.725	2.086	2.528	2.845	20
21	1.721	2.080	2.518	2.831	21
22	1.717	2.074	2.508	2.819	22
23	1.714	2.069	2.500	2.807	23
24	1.711	2.064	2.492	2.797	24
25	1.708	2.060	2.485	2.787	25
26	1.706	2.056	2.479	2.779	26
27	1.703	2.052	2.473	2.771	27
28	1.701	2.048	2.467	2.763	28
29	1.699	2.045	2.462	2.756	29
inf.	1.645	1.960	2.326	2.576	inf.

†Richard A. Johnson and Dean W. Wichern, *Applied Multivariate Statistical Analysis*, © 1982, p. 582. Adapted by permission of Prentice Hall, Upper Saddle River, N.J.

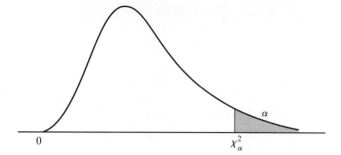

The entries in Table IV are values for which the area to their right under the chi-square distribution with given degrees of freedom (the gray area in the figure) is equal to α.

TABLE IV	Values of $\chi^{2\dagger}$		
d.f.	$\chi^2_{0.05}$	$\chi^2_{0.01}$	d.f.
1	3.841	6.635	1
2	5.991	9.210	2
3	7.815	11.345	3
4	9.488	13.277	4
5	11.070	15.086	5
6	12.592	16.812	6
7	14.067	18.475	7
8	15.507	20.090	8
9	16.919	21.666	9
10	18.307	23.209	10
11	19.675	24.725	11
12	21.026	26.217	12
13	22.362	27.688	13
14	23.685	29.141	14
15	24.996	30.578	15
16	26.296	32.000	16
17	27.587	33.409	17
18	28.869	34.805	18
19	30.144	36.191	19
20	31.410	37.566	20
21	32.671	38.932	21
22	33.924	40.289	22
23	35.172	41.638	23
24	36.415	42.980	24
25	37.652	44.314	25
26	38.885	45.642	26
27	40.113	46.963	27
28	41.337	48.278	28
29	42.557	49.588	29
30	43.773	50.892	30

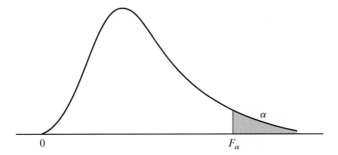

The entries in Table V are values for which the area to their right under the F distribution with given degrees of freedom (the gray area in the figure) is equal to α.

TABLE V Values of $F_{0.05}^{\dagger}$

Values of $F_{0.05}^{\dagger}$

Degrees of freedom for numerator

Degrees of freedom for denominator	1	2	3	4	5	6	7	8	9	10	12	15	20	24	30	40	60	120	∞
1	161	200	216	225	230	234	237	239	241	242	244	246	248	249	250	251	252	253	254
2	18.5	19.0	19.2	19.2	19.3	19.3	19.4	19.4	19.4	19.4	19.4	19.4	19.4	19.5	19.5	19.5	19.5	19.5	19.5
3	10.1	9.55	9.28	9.12	9.01	8.94	8.89	8.85	8.81	8.79	8.74	8.70	8.66	8.64	8.62	8.59	8.57	8.55	8.53
4	7.71	6.94	6.59	6.39	6.26	6.16	6.09	6.04	6.00	5.96	5.91	5.86	5.80	5.77	5.75	5.72	5.69	5.66	5.63
5	6.61	5.79	5.41	5.19	5.05	4.95	4.88	4.82	4.77	4.74	4.68	4.62	4.56	4.53	4.50	4.46	4.43	4.40	4.37
6	5.99	5.14	4.76	4.53	4.39	4.28	4.21	4.15	4.10	4.06	4.00	3.94	3.87	3.84	3.81	3.77	3.74	3.70	3.67
7	5.59	4.74	4.35	4.12	3.97	3.87	3.79	3.73	3.68	3.64	3.57	3.51	3.44	3.41	3.38	3.34	3.30	3.27	3.23
8	5.32	4.46	4.07	3.84	3.69	3.58	3.50	3.44	3.39	3.35	3.28	3.22	3.15	3.12	3.08	3.04	3.01	2.97	2.93
9	5.12	4.26	3.86	3.63	3.48	3.37	3.29	3.23	3.18	3.14	3.07	3.01	2.94	2.90	2.86	2.83	2.79	2.75	2.71
10	4.96	4.10	3.71	3.48	3.33	3.22	3.14	3.07	3.02	2.98	2.91	2.85	2.77	2.74	2.70	2.66	2.62	2.58	2.54
11	4.84	3.98	3.59	3.36	3.20	3.09	3.01	2.95	2.90	2.85	2.79	2.72	2.65	2.61	2.57	2.53	2.49	2.45	2.40
12	4.75	3.89	3.49	3.26	3.11	3.00	2.91	2.85	2.80	2.75	2.69	2.62	2.54	2.51	2.47	2.43	2.38	2.34	2.30
13	4.67	3.81	3.41	3.18	3.03	2.92	2.83	2.77	2.71	2.67	2.60	2.53	2.46	2.42	2.38	2.34	2.30	2.25	2.21
14	4.60	3.74	3.34	3.11	2.96	2.85	2.76	2.70	2.65	2.60	2.53	2.46	2.39	2.35	2.31	2.27	2.22	2.18	2.13
15	4.54	3.68	3.29	3.06	2.90	2.79	2.71	2.64	2.59	2.54	2.48	2.40	2.33	2.29	2.25	2.20	2.16	2.11	2.07
16	4.49	3.63	3.24	3.01	2.85	2.74	2.66	2.59	2.54	2.49	2.42	2.35	2.28	2.24	2.19	2.15	2.11	2.06	2.01
17	4.45	3.59	3.20	2.96	2.81	2.70	2.61	2.55	2.49	2.45	2.38	2.31	2.23	2.19	2.15	2.10	2.06	2.01	1.96
18	4.41	3.55	3.16	2.93	2.77	2.66	2.58	2.51	2.46	2.41	2.34	2.27	2.19	2.15	2.11	2.06	2.02	1.97	1.92
19	4.38	3.52	3.13	2.90	2.74	2.63	2.54	2.48	2.42	2.38	2.31	2.23	2.16	2.11	2.07	2.03	1.98	1.93	1.88
20	4.35	3.49	3.10	2.87	2.71	2.60	2.51	2.45	2.39	2.35	2.28	2.20	2.12	2.08	2.04	1.99	1.95	1.90	1.84
21	4.32	3.47	3.07	2.84	2.68	2.57	2.49	2.42	2.37	2.32	2.25	2.18	2.10	2.05	2.01	1.96	1.92	1.87	1.81
22	4.30	3.44	3.05	2.82	2.66	2.55	2.46	2.40	2.34	2.30	2.23	2.15	2.07	2.03	1.98	1.94	1.89	1.84	1.78
23	4.28	3.42	3.03	2.80	2.64	2.53	2.44	2.37	2.32	2.27	2.20	2.13	2.05	2.01	1.96	1.91	1.86	1.81	1.76
24	4.26	3.40	3.01	2.78	2.62	2.51	2.42	2.36	2.30	2.25	2.18	2.11	2.03	1.98	1.94	1.89	1.84	1.79	1.73
25	4.24	3.39	2.99	2.76	2.60	2.49	2.40	2.34	2.28	2.24	2.16	2.09	2.01	1.96	1.92	1.87	1.82	1.77	1.71
30	4.17	3.32	2.92	2.69	2.53	2.42	2.33	2.27	2.21	2.16	2.09	2.01	1.93	1.89	1.84	1.79	1.74	1.68	1.62
40	4.08	3.23	2.84	2.61	2.45	2.34	2.25	2.18	2.12	2.08	2.00	1.92	1.84	1.79	1.74	1.69	1.64	1.58	1.51
60	4.00	3.15	2.76	2.53	2.37	2.25	2.17	2.10	2.04	1.99	1.92	1.84	1.75	1.70	1.65	1.59	1.53	1.47	1.39
120	3.92	3.07	2.68	2.45	2.29	2.18	2.09	2.02	1.96	1.91	1.83	1.75	1.66	1.61	1.55	1.50	1.43	1.35	1.25
∞	3.84	3.00	2.60	2.37	2.21	2.10	2.01	1.94	1.88	1.83	1.75	1.67	1.57	1.52	1.46	1.39	1.32	1.22	1.00

†Reproduced from M. Merrington and C. M. Thompson, "Tables of percentage points of the inverted beta (F) distribution," *Biometrika*, Vol. 33 (1943), by permission of the *Biometrika* trustees.

TABLE V Values of $F_{0.01}^{+}$ (Continued)

Values of $F_{0.01}$

Degrees of freedom for numerator

Denom.	1	2	3	4	5	6	7	8	9	10	12	15	20	24	30	40	60	120	∞
1	4,052	5,000	5,403	5,625	5,764	5,859	5,928	5,982	6,023	6,056	6,106	6,157	6,209	6,235	6,261	6,287	6,313	6,339	6,366
2	98.5	99.0	99.2	99.2	99.3	99.3	99.4	99.4	99.4	99.4	99.4	99.4	99.4	99.5	99.5	99.5	99.5	99.5	99.5
3	34.1	30.8	29.5	28.7	28.2	27.9	27.7	27.5	27.3	27.2	27.1	26.9	26.7	26.6	26.5	26.4	26.3	26.2	26.1
4	21.2	18.0	16.7	16.0	15.5	15.2	15.0	14.8	14.7	14.5	14.4	14.2	14.0	13.9	13.8	13.7	13.7	13.6	13.5
5	16.3	13.3	12.1	11.4	11.0	10.7	10.5	10.3	10.2	10.1	9.89	9.72	9.55	9.47	9.38	9.29	9.20	9.11	9.02
6	13.7	10.9	9.78	9.15	8.75	8.47	8.26	8.10	7.98	7.87	7.72	7.56	7.40	7.31	7.23	7.14	7.06	6.97	6.88
7	12.2	9.55	8.45	7.85	7.46	7.19	6.99	6.84	6.72	6.62	6.47	6.31	6.16	6.07	5.99	5.91	5.82	5.74	5.65
8	11.3	8.65	7.59	7.01	6.63	6.37	6.18	6.03	5.91	5.81	5.67	5.52	5.36	5.28	5.20	5.12	5.03	4.95	4.86
9	10.6	8.02	6.99	6.42	6.06	5.80	5.61	5.47	5.35	5.26	5.11	4.96	4.81	4.73	4.65	4.57	4.48	4.40	4.31
10	10.0	7.56	6.55	5.99	5.64	5.39	5.20	5.06	4.94	4.85	4.71	4.56	4.41	4.33	4.25	4.17	4.08	4.00	3.91
11	9.65	7.21	6.22	5.67	5.32	5.07	4.89	4.74	4.63	4.54	4.40	4.25	4.10	4.02	3.94	3.86	3.78	3.69	3.60
12	9.33	6.93	5.95	5.41	5.06	4.82	4.64	4.50	4.39	4.30	4.16	4.01	3.86	3.78	3.70	3.62	3.54	3.45	3.36
13	9.07	6.70	5.74	5.21	4.86	4.62	4.44	4.30	4.19	4.10	3.96	3.82	3.66	3.59	3.51	3.43	3.34	3.25	3.17
14	8.86	6.51	5.56	5.04	4.70	4.46	4.28	4.14	4.03	3.94	3.80	3.66	3.51	3.43	3.35	3.27	3.18	3.09	3.00
15	8.68	6.36	5.42	4.89	4.56	4.32	4.14	4.00	3.89	3.80	3.67	3.52	3.37	3.29	3.21	3.13	3.05	2.96	2.87
16	8.53	6.23	5.29	4.77	4.44	4.20	4.03	3.89	3.78	3.69	3.55	3.41	3.26	3.18	3.10	3.02	2.93	2.84	2.75
17	8.40	6.11	5.19	4.67	4.34	4.10	3.93	3.79	3.68	3.59	3.46	3.31	3.16	3.08	3.00	2.92	2.83	2.75	2.65
18	8.29	6.01	5.09	4.58	4.25	4.01	3.84	3.71	3.60	3.51	3.37	3.23	3.08	3.00	2.92	2.84	2.75	2.66	2.57
19	8.19	5.93	5.01	4.50	4.17	3.94	3.77	3.63	3.52	3.43	3.30	3.15	3.00	2.92	2.84	2.76	2.67	2.58	2.49
20	8.10	5.85	4.94	4.43	4.10	3.87	3.70	3.56	3.46	3.37	3.23	3.09	2.94	2.86	2.78	2.69	2.61	2.52	2.42
21	8.02	5.78	4.87	4.37	4.04	3.81	3.64	3.51	3.40	3.31	3.17	3.03	2.88	2.80	2.72	2.64	2.55	2.46	2.36
22	7.95	5.72	4.82	4.31	3.99	3.76	3.59	3.45	3.35	3.26	3.12	2.98	2.83	2.75	2.67	2.58	2.50	2.40	2.31
23	7.88	5.66	4.76	4.26	3.94	3.71	3.54	3.41	3.30	3.21	3.07	2.93	2.78	2.70	2.62	2.54	2.45	2.35	2.26
24	7.82	5.61	4.72	4.22	3.90	3.67	3.50	3.36	3.26	3.17	3.03	2.89	2.74	2.66	2.58	2.49	2.40	2.31	2.21
25	7.77	5.57	4.68	4.18	3.86	3.63	3.46	3.32	3.22	3.13	2.99	2.85	2.70	2.62	2.53	2.45	2.36	2.27	2.17
30	7.56	5.39	4.51	4.02	3.70	3.47	3.30	3.17	3.07	2.98	2.84	2.70	2.55	2.47	2.39	2.30	2.21	2.11	2.01
40	7.31	5.18	4.31	3.83	3.51	3.29	3.12	2.99	2.89	2.80	2.66	2.52	2.37	2.29	2.20	2.11	2.02	1.92	1.80
60	7.08	4.98	4.13	3.65	3.34	3.12	2.95	2.82	2.72	2.63	2.50	2.35	2.20	2.12	2.03	1.94	1.84	1.73	1.60
120	6.85	4.79	3.95	3.48	3.17	2.96	2.79	2.66	2.56	2.47	2.34	2.19	2.03	1.95	1.86	1.76	1.66	1.53	1.38
∞	6.63	4.61	3.78	3.32	3.02	2.80	2.64	2.51	2.41	2.32	2.18	2.04	1.88	1.79	1.70	1.59	1.47	1.32	1.00

Degrees of freedom for denominator

TABLE VI   Critical Values of r

n	$r_{0.025}$	$r_{0.005}$	n	$r_{0.025}$	$r_{0.005}$
3	0.997		18	0.468	0.590
4	0.950	0.999	19	0.456	0.575
5	0.878	0.959	20	0.444	0.561
6	0.811	0.917	21	0.433	0.549
7	0.754	0.875	22	0.423	0.537
8	0.707	0.834	27	0.381	0.487
9	0.666	0.798	32	0.349	0.449
10	0.632	0.765	37	0.325	0.418
11	0.602	0.735	42	0.304	0.393
12	0.576	0.708	47	0.288	0.372
13	0.553	0.684	52	0.273	0.354
14	0.532	0.661	62	0.250	0.325
15	0.514	0.641	72	0.232	0.302
16	0.497	0.623	82	0.217	0.283
17	0.482	0.606	92	0.205	0.267

TABLE VII   Critical Values of $U^{\dagger}$

n_1 \ n_2	2	3	4	5	6	7	8	9	10	11	12	13	14	15
2							0	0	0	0	1	1	1	1
3			0	1	1	2	2	3	3	4	4	5	5	
4			0	1	2	3	4	4	5	6	7	8	9	10
5		0	1	2	3	5	6	7	8	9	11	12	13	14
6		1	2	3	5	6	8	10	11	13	14	16	17	19
7		1	3	5	6	8	10	12	14	16	18	20	22	24
8	0	2	4	6	8	10	13	15	17	19	22	24	26	29
9	0	2	4	7	10	12	15	17	20	23	26	28	31	34
10	0	3	5	8	11	14	17	20	23	26	29	30	36	39
11	0	3	6	9	13	16	19	23	26	30	33	37	40	44
12	1	4	7	11	14	18	22	26	29	33	37	41	45	49
13	1	4	8	12	16	20	24	28	30	37	41	45	50	54
14	1	5	9	13	17	22	26	31	36	40	45	50	55	59
15	1	5	10	14	19	24	29	34	39	44	49	54	59	64

Values of $U'_{0.05}$

n_1 \ n_2	3	4	5	6	7	8	9	10	11	12	13	14	15
3							0	0	0	1	1	1	2
4			0	0	1	1	2	2	3	3	4	5	
5		0	1	1	2	3	4	5	6	7	7	8	
6		0	1	2	3	4	5	6	7	9	10	11	12
7		0	1	3	4	6	7	9	10	12	13	15	16
8		1	2	4	6	7	9	11	13	15	17	18	20
9	0	1	3	5	7	9	11	13	16	18	20	22	24
10	0	2	4	6	9	11	13	16	18	21	24	26	29
11	0	2	5	7	10	13	16	18	21	24	27	30	33
12	1	3	6	9	12	15	18	21	24	27	31	34	37
13	1	3	7	10	13	17	20	24	27	31	34	38	42
14	1	4	7	11	15	18	22	26	30	34	38	42	46
15	2	5	8	12	16	20	24	29	33	37	42	46	51

Values of $U'_{0.01}$

$\dagger$ This table in based on Table 11.4 of Donald B. Owen, *Handbook of Statistical Tables*, © 1962, Addison-Wesley, Reading, Massachusetts. Reprinted with permission.

TABLE VIII   Critical Values of $u^\dagger$

n_1 \ n_2	Values of $u_{0.025}$											
	4	**5**	**6**	**7**	**8**	**9**	**10**	**11**	**12**	**13**	**14**	**15**
4		9	9									
5	9	10	10	11	11							
6	9	10	11	12	12	13	13	13	13			
7		11	12	13	13	14	14	14	14	15	15	15
8		11	12	13	14	14	15	15	16	16	16	16
9			13	14	14	15	16	16	16	17	17	18
10			13	14	15	16	16	17	17	18	18	18
11			13	14	15	16	17	17	18	19	19	19
12			13	14	16	16	17	18	19	19	20	20
13				15	16	17	18	19	19	20	20	21
14				15	16	17	18	19	20	20	21	22
15				15	16	18	18	19	20	21	22	22

n_1 \ n_2	Values of $u'_{0.025}$													
	2	**3**	**4**	**5**	**6**	**7**	**8**	**9**	**10**	**11**	**12**	**13**	**14**	**15**
2											2	2	2	2
3				2	2	2	2	2	2	2	2	2	2	3
4			2	2	2	3	3	3	3	3	3	3	3	3
5		2	2	3	3	3	3	3	4	4	4	4	4	4
6		2	2	3	3	3	3	4	4	4	4	5	5	5
7		2	2	3	3	3	4	4	5	5	5	5	5	6
8		2	3	3	3	4	4	5	5	5	6	6	6	6
9		2	3	3	4	4	5	5	5	6	6	6	7	7
10		2	3	3	4	5	5	5	6	6	7	7	7	7
11		2	3	4	4	5	5	6	6	7	7	7	8	8
12	2	2	3	4	4	5	6	6	7	7	7	8	8	8
13	2	2	3	4	5	5	6	6	7	7	8	8	9	9
14	2	2	3	4	5	5	6	7	7	8	8	9	9	9
15	2	3	3	4	5	6	6	7	7	8	8	9	9	10

†This table is adapted, by permission, from F. S. Swed and C. Eisenhart, "Tables for testing randomness of grouping in a sequence of alternatives," *Annals of Mathematical Statistics*, Vol. 14.

TABLE VIII	Critical Values of u (*Continued*)										

n_2		Values of $u_{0.005}$									
n_1	5	6	7	8	9	10	11	12	13	14	15
5		11									
6	11	12	13	13							
7		13	13	14	15	15	15				
8		13	14	15	15	16	16	17	17	17	
9			15	15	16	17	17	18	18	18	19
10			15	16	17	17	18	19	19	19	20
11			15	16	17	18	19	19	20	20	21
12				17	18	19	19	20	21	21	22
13				17	18	19	20	21	21	22	22
14				17	18	19	20	21	22	23	23
15					19	20	21	22	22	23	24

n_2			Values of $u'_{0.005}$									
n_1	3	4	5	6	7	8	9	10	11	12	13	14	15
3										2	2	2	2
4						2	2	2	2	2	2	2	3
5				2	2	2	2	3	3	3	3	3	3
6			2	2	2	3	3	3	3	3	3	4	4
7			2	2	3	3	3	3	4	4	4	4	4
8		2	2	3	3	3	3	4	4	4	5	5	5
9		2	2	3	3	3	4	4	5	5	5	5	6
10		2	3	3	3	4	4	5	5	5	5	6	6
11		2	3	3	4	4	5	5	5	6	6	6	7
12	2	2	3	3	4	4	5	5	6	6	6	7	7
13	2	2	3	3	4	5	5	5	6	6	7	7	7
14	2	2	3	4	4	5	5	6	6	7	7	7	8
15	2	3	3	4	4	5	6	6	7	7	7	8	8

TABLE IX	Factorials
n	$n!$
0	1
1	1
2	2
3	6
4	24
5	120
6	720
7	5,040
8	40,320
9	362,880
10	3,628,800
11	39,916,800
12	479,001,600
13	6,227,020,800
14	87,178,291,200
15	1,307,674,368,000

TABLE X   Binomial Coefficients

n	$\binom{n}{0}$	$\binom{n}{1}$	$\binom{n}{2}$	$\binom{n}{3}$	$\binom{n}{4}$	$\binom{n}{5}$	$\binom{n}{6}$	$\binom{n}{7}$	$\binom{n}{8}$	$\binom{n}{9}$	$\binom{n}{10}$
0	1										
1	1	1									
2	1	2	1								
3	1	3	3	1							
4	1	4	6	4	1						
5	1	5	10	10	5	1					
6	1	6	15	20	15	6	1				
7	1	7	21	35	35	21	7	1			
8	1	8	28	56	70	56	28	8	1		
9	1	9	36	84	126	126	84	36	9	1	
10	1	10	45	120	210	252	210	120	45	10	1
11	1	11	55	165	330	462	462	330	165	55	11
12	1	12	66	220	495	792	924	792	495	220	66
13	1	13	78	286	715	1287	1716	1716	1287	715	286
14	1	14	91	364	1001	2002	3003	3432	3003	2002	1001
15	1	15	105	455	1365	3003	5005	6435	6435	5005	3003
16	1	16	120	560	1820	4368	8008	11440	12870	11440	8008
17	1	17	136	680	2380	6188	12376	19448	24310	24310	19448
18	1	18	153	816	3060	8568	18564	31824	43758	48620	43758
19	1	19	171	969	3876	11628	27132	50388	75582	92378	92378
20	1	20	190	1140	4845	15504	38760	77520	125970	167960	184756

If necessary, use the identity $\binom{n}{r} = \binom{n}{n-r}$.

TABLE XI	Values of e^{-x}						
x	e^{-x}	x	e^{-x}	x	e^{-x}	x	e^{-x}
0.0	1.000	2.5	0.082	5.0	0.0067	7.5	0.00055
0.1	0.905	2.6	0.074	5.1	0.0061	7.6	0.00050
0.2	0.819	2.7	0.067	5.2	0.0055	7.7	0.00045
0.3	0.741	2.8	0.061	5.3	0.0050	7.8	0.00041
0.4	0.670	2.9	0.055	5.4	0.0045	7.9	0.00037
0.5	0.607	3.0	0.050	5.5	0.0041	8.0	0.00034
0.6	0.549	3.1	0.045	5.6	0.0037	8.1	0.00030
0.7	0.497	3.2	0.041	5.7	0.0033	8.2	0.00028
0.8	0.449	3.3	0.037	5.8	0.0030	8.3	0.00025
0.9	0.407	3.4	0.033	5.9	0.0027	8.4	0.00023
1.0	0.368	3.5	0.030	6.0	0.0025	8.5	0.00020
1.1	0.333	3.6	0.027	6.1	0.0022	8.6	0.00018
1.2	0.301	3.7	0.025	6.2	0.0020	8.7	0.00017
1.3	0.273	3.8	0.022	6.3	0.0018	8.8	0.00015
1.4	0.247	3.9	0.020	6.4	0.0017	8.9	0.00014
1.5	0.223	4.0	0.018	6.5	0.0015	9.0	0.00012
1.6	0.202	4.1	0.017	6.6	0.0014	9.1	0.00011
1.7	0.183	4.2	0.015	6.7	0.0012	9.2	0.00010
1.8	0.165	4.3	0.014	6.8	0.0011	9.3	0.00009
1.9	0.150	4.4	0.012	6.9	0.0010	9.4	0.00008
2.0	0.135	4.5	0.011	7.0	0.0009	9.5	0.00008
2.1	0.122	4.6	0.010	7.1	0.0008	9.6	0.00007
2.2	0.111	4.7	0.009	7.2	0.0007	9.7	0.00006
2.3	0.100	4.8	0.008	7.3	0.0007	9.8	0.00006
2.4	0.091	4.9	0.007	7.4	0.0006	9.9	0.00005

TABLE XII   Random Numbers[†]

04433	80674	24520	18222	10610	05794	37515
60298	47829	72648	37414	75755	04717	29899
67884	59651	67533	68123	17730	95862	08034
89512	32155	51906	61662	64130	16688	37275
32653	01895	12506	88535	36553	23757	34209
95913	15405	13772	76638	48423	25018	99041
55864	21694	13122	44115	01601	50541	00147
35334	49810	91601	40617	72876	33967	73830
57729	32196	76487	11622	96297	24160	09903
86648	13697	63677	70119	94739	25875	38829
30574	47609	07967	32422	76791	39725	53711
81307	43694	83580	79974	45929	85113	72268
02410	54905	79007	54939	21410	86980	91772
18969	75274	52233	62319	08598	09066	95288
87863	82384	66860	62297	80198	19347	73234
68397	71708	15438	62311	72844	60203	46412
28529	54447	58729	10854	99058	18260	38765
44285	06372	15867	70418	57012	72122	36634
86299	83430	33571	23309	57040	29285	67870
84842	68668	90894	61658	15001	94055	36308
56970	83609	52098	04184	54967	72938	56834
83125	71257	60490	44369	66130	72936	69848
55503	52423	02464	26141	68779	66388	75242
47019	76273	33203	29608	54553	25971	69573
84828	32592	79526	29554	84580	37859	28504
68921	08141	79227	05748	51276	57143	31926
36458	96045	30424	98420	72925	40729	22337
95752	59445	36847	87729	81679	59126	59437
26768	47323	58454	56958	20575	76746	49878
42613	37056	43636	58085	06766	60227	96414
95457	30566	65482	25596	02678	54592	63607
95276	17894	63564	95958	39750	64379	46059
66954	52324	64776	92345	95110	59448	77249
17457	18481	14113	62462	02798	54977	48349
03704	36872	83214	59337	01695	60666	97410
21538	86497	33210	60337	27976	70661	08250
57178	67619	98310	70348	11317	71623	55510
31048	97558	94953	55866	96283	46620	52087
69799	55380	16498	80733	96422	58078	99643
90595	61867	59231	17772	67831	33317	00520
33570	04981	98939	78784	09977	29398	93896
15340	93460	57477	13898	48431	72936	78160
64079	42483	36512	56186	99098	48850	72527
63491	05546	67118	62063	74958	20946	28147
92003	63868	41034	28260	79708	00770	88643
52360	46658	66511	04172	73085	11795	52594
74622	12142	68355	65635	21828	39539	18988
04157	50079	61343	64315	70836	82857	35335
86003	60070	66241	32836	27573	11479	94114
41268	80187	20351	09636	84668	42486	71303

[†]Based on parts of *Tables of 105,000 Random Decimal Digits*. Interstate Commerce Commission, Bureau of Transport Economics and Statistics, Washington, D.C.

TABLE XII   Random Numbers (*Continued*)

48611	62866	33963	14045	79451	04934	45576
78812	03509	78673	73181	29973	18664	04555
19472	63971	37271	31445	49019	49405	46925
51266	11569	08697	91120	64156	40365	74297
55806	96275	26130	47949	14877	69594	83041
77527	81360	18180	97421	55541	90275	18213
77680	58788	33016	61173	93049	04694	43534
15404	96554	88265	34537	38526	67924	40474
14045	22917	60718	66487	46346	30949	03173
68376	43918	77653	04127	69930	43283	35766
93385	13421	67957	20384	58731	53396	59723
09858	52104	32014	53115	03727	98624	84616
93307	34116	49516	42148	57740	31198	70336
04794	01534	92058	03157	91758	80611	45357
86265	49096	97021	92582	61422	75890	86442
65943	79232	45702	67055	39024	57383	44424
90038	94209	04055	27393	61517	23002	96560
97283	95943	78363	36498	40662	94188	18202
21913	72958	75637	99936	58715	07943	23748
41161	37341	81838	19389	80336	46346	91895
23777	98392	31417	98547	92058	02277	50315
59973	08144	61070	73094	27059	69181	55623
82690	74099	77885	23813	10054	11900	44653
83854	24715	48866	65745	31131	47636	45137
61980	34997	41825	11623	07320	15003	56774
99915	45821	97702	87125	44488	77613	56823
48293	86847	43186	42951	37804	85129	28993
33225	31280	41232	34750	91097	60752	69783
06846	32828	24425	30249	78801	26977	92074
32671	45587	79620	84831	38156	74211	82752
82096	21913	75544	55228	89796	05694	91552
51666	10433	10945	55306	78562	89630	41230
54044	67942	24145	42294	27427	84875	37022
66738	60184	75679	38120	17640	36242	99357
55064	17427	89180	74018	44865	53197	74810
69599	60264	84549	78007	88450	06488	72274
64756	87759	92354	78694	63638	80939	98644
80817	74533	68407	55862	32476	19326	95558
39847	96884	84657	33697	39578	90197	80532
90401	41700	95510	61166	33757	23279	85523
78227	90110	81378	96659	37008	04050	04228
87240	52716	87697	79433	16336	52862	69149
08486	10951	26832	39763	02485	71688	90936
39338	32169	03713	93510	61244	73774	01245
21188	01850	69689	49426	49128	14660	14143
13287	82531	04388	64693	11934	35051	68576
53609	04001	19648	14053	49623	10840	31915
87900	36194	31567	53506	34304	39910	79630
81641	00496	36058	75899	46620	70024	88753
19512	50277	71508	20116	79520	06269	74173

TABLE XII   Random Numbers (*Continued*)

24418	23508	91507	76455	54941	72711	39406
57404	73678	08272	62941	02349	71389	45605
77644	98489	86268	73652	98210	44546	27174
68366	65614	01443	07607	11826	91326	29664
64472	72294	95432	53555	96810	17100	35066
88205	37913	98633	81009	81060	33449	68055
98455	78685	71250	10329	56135	80647	51404
48977	36794	56054	59243	57361	65304	93258
93077	72941	92779	23581	24548	56415	61927
84533	26564	91583	83411	66504	02036	02922
11338	12903	14514	27585	45068	05520	56321
23853	68500	92274	87026	99717	01542	72990
94096	74920	25822	98026	05394	61840	83089
83160	82362	09350	98536	38155	42661	02363
97425	47335	69709	01386	74319	04318	99387
83951	11954	24317	20345	18134	90062	10761
93085	35203	05740	03206	92012	42710	34650
33762	83193	58045	89880	78101	44392	53767
49665	85397	85137	30496	23469	42846	94810
37541	82627	80051	72521	35342	56119	97190
22145	85304	35348	82854	55846	18076	12415
27153	08662	61078	52433	22184	33998	87436
00301	49425	66682	25442	83668	66236	79655
43815	43272	73778	63469	50083	70696	13558
14689	86482	74157	46012	97765	27552	49617
16680	55936	82453	19532	49988	13176	94219
86938	60429	01137	86168	78257	86249	46134
33944	29219	73161	46061	30946	22210	79302
16045	67736	18608	18198	19468	76358	69203
37044	52523	25627	63107	30806	80857	84383
61471	45322	35340	35132	42163	69332	98851
47422	21296	16785	66393	39249	51463	95963
24133	39719	14484	58613	88717	29289	77360
67253	67064	10748	16006	16767	57345	42285
62382	76941	01635	35829	77516	98468	51686
98011	16503	09201	03523	87192	66483	55649
37366	24386	20654	85117	74078	64120	04643
73587	83993	54176	05221	94119	20108	78101
33583	68291	50547	96085	62180	27453	18567
02878	33223	39199	49536	56199	05993	71201
91498	41673	17195	33175	04994	09879	70337
91127	19815	30219	55591	21725	43827	78862
12997	55013	18662	81724	24305	37661	18956
96098	13651	15393	69995	14762	69734	89150
97627	17837	10472	18983	28387	99781	52977
40064	47981	31484	76603	54088	91095	00010
16239	68743	71374	55863	22672	91609	51514
58354	24913	20435	30965	17453	65623	93058
52567	65085	60220	84641	18273	49604	47418
06236	29052	91392	07551	83532	68130	56970

TABLE XII   Random Numbers (*Continued*)

94620	27963	96478	21559	19246	88097	44926
60947	60775	73181	43264	56895	04232	59604
27499	53523	63110	57106	20865	91683	80688
01603	23156	89223	43429	95353	44662	59433
00815	01552	06392	31437	70385	45863	75971
83844	90942	74857	52419	68723	47830	63010
06626	10042	93629	37609	57215	08409	81906
56760	63348	24949	11859	29793	37457	59377
64416	29934	00755	09418	14230	62887	92683
63569	17906	38076	32135	19096	96970	75917
22693	35089	72994	04252	23791	60249	83010
43413	59744	01275	71326	91382	45114	20245
09224	78530	50566	49965	04851	18280	14039
67625	34683	03142	74733	63558	09665	22610
86874	12549	98699	54952	91579	26023	81076
54548	49505	62515	63903	13193	33905	66936
73236	66167	49728	03581	40699	10396	81827
15220	66319	13543	14071	59148	95154	72852
16151	08029	36954	03891	38313	34016	18671
43635	84249	88984	80993	55431	90793	62603
30193	42776	85611	57635	51362	79907	77364
37430	45246	11400	20986	43996	73122	88474
88312	93047	12088	86937	70794	01041	74867
98995	58159	04700	90443	13168	31553	67891
51734	20849	70198	67906	00880	82899	66065
88698	41755	56216	66852	17748	04963	54859
51865	09836	73966	65711	41699	11732	17173
40300	08852	27528	84648	79589	95295	72895
02760	28625	70476	76410	32988	10194	94917
78450	26245	91763	73117	33047	03577	62599
50252	56911	62693	73817	98693	18728	94741
07929	66728	47761	81472	44806	15592	71357
09030	39605	87507	85446	51257	89555	75520
56670	88445	85799	76200	21795	38894	58070
48140	13583	94911	13318	64741	64336	95103
36764	86132	12463	28385	94242	32063	45233
14351	71381	28133	68269	65145	28152	39087
81276	00835	63835	87174	42446	08882	27067
55524	86088	00069	59254	24654	77371	26409
78852	65889	32719	13758	23937	90740	16866
11861	69032	51915	23510	32050	52052	24004
67699	01009	07050	73324	06732	27510	33761
50064	39500	17450	18030	63124	48061	59412
93126	17700	94400	76075	08317	27324	72723
01657	92602	41043	05686	15650	29970	95877
13800	76690	75133	60456	28491	03845	11507
98135	42870	48578	29036	69876	86563	61729
08313	99293	00990	13595	77457	79969	11339
90974	83965	62732	85161	54330	22406	86253
33273	61993	88407	69399	17301	70975	99129

ANSWERS TO ODD-NUMBERED EXERCISES

Note: Answers that are given in the Solutions to Practice Exercises at the end of each chapter of the text are not repeated here.

CHAPTER 1

1.1 Practice exercise.

1.3 Single, 9; married, 7; widow or widower, 3; divorced, 5.

1.5 **(a)** The data is viewed as a population if we are interested in the tips received by waitresses employed in this restaurant during this week.

(b) The data is viewed as a sample if we are studying waitresses in different restaurants or in different time periods.

1.7 Practice exercise.

1.9 **(a)** If the filling machine was defective and failed to fill (for example) alternate cans completely, the inspector would see only the cans that were filled correctly, or only the cans that were not filled correctly. He would not see a random sample of correctly and incorrectly filled cans.

(b) Randomly selected telephone directory pages are often not representative of the population of the city. For example, if the page is filled with names beginning with Mac or Mc, their ethnic backgrounds would differ from those of Hispanic and other backgrounds. Or if the page is opened to the name Saint or St., the telephone calls would reach a disproportionate number of schools, hospitals, and churches with differing views on some subjects than the general population of the city. In any event the use of the telephone to conduct interviews risks missing a disproportionate number of low-income people who cannot afford telephone service, but nevertheless go to the polls and vote.

(c) Some products, including grains, settle in moving freight cars. The heaviest portion moves to the bottom of the car and the lightest remains on top. To sample the product we must also take small quantities from the sides, bottom, and throughout the car.

1.11 **(a)** Descriptive. Numbers can be read directly from the data.

(b) Descriptive. Numbers can be calculated from the data. $21 - 19 = 2$ thousand.

(c) Descriptive. Numbers can be calculated from the data. $34 - 26 = 8$ thousand.

(d) Generalization. There may be other reasons for the increase in unemployment.

1.13 Practice exercise.

1.15 **(a)** Descriptive. This conclusion can be read from line 1 of the table.

(b) Generalization. This conclusion cannot be determined from the data.

(c) Descriptive. This conclusion can be obtained by reading columns 2 and 3 of the table.

(d) Descriptive. This conclusion can be obtained by reading columns 2 and 3 of the table.

(e) Generalization. This conclusion cannot be determined from the data.

(f) Generalization. This conclusion cannot be determined from the data.

CHAPTER 2

2.1 Practice exercise.

2.3

     0   1   2   3

2.5

Coffee	● ● ● ● ● ● ● ● ● ● ● ● ● ● ●
Soda pop	● ● ● ● ● ● ● ● ● ●
Juice	● ● ● ● ● ● ●
Tea	● ● ● ● ● ●
None	● ● ● ●

2.7

16	84
17	613
18	9481
19	53981757
20	7302
21	64
22	7

2.9 **(a)** 10, 11, 11, 12, 15, 17, 18.

     (b) 120, 122, 123, 123, 125.

     (c) 0.60, 0.62, 0.63, 0.66.

     (d) 1.50, 1.51, 1.51, 1.52, 1.53, 1.54, 1.56.

2.11

5*	4 2
5·	6 9 7 6
6*	2 2 0 0 1 3 4 0
6·	6 7 8 7 5 9 5 8
7*	1 3 1
7·	8 6
8*	0

2.13

2·	9 8
3*	3 1 1 1 4 0
3·	5 7 6 8 9 8 7 6
4*	3 3 0 1 0 4 4
4·	6 6 9 7
5*	1 0
5·	6

2.15   Practice exercise.

2.17   One possibility is 0–249,999; 250,000–499,999; 500,000–749,999; 750,000–999,999; 1,000,000–1,249,999;   1,250,000–1,499,999;   1,500,000–1,749,999;   1,750,000–1,999,999.

2.19   **(a)** No;   **(b)** yes;   **(c)** no;   **(d)** yes.

2.21   Practice exercise.

2.23   **(a)**   0, 20, 40, 60, 80, 100, 120, and 140.
   (b)   19, 39, 59, 79, 99, 119, 139, and 159.
   (c)   9.5, 29.5, 49.5, 69.5, 89.5, 109.5, 129.5, and 149.5.
   (d)   20.

2.25   Practice exercise.

2.27   **(a)**   31.5, 40.5, 49.5, 58.5, 67.5, 76.5, 85.5, 94.5.
   (b)   32–40, 41–49, 50–58, 59–67, 68–76, 77–85, 86–94.

2.29   There is no provision for 11 and 31. Also, there is ambiguity because 23 can be put into the fourth or fifth class.

2.31   **(a)**

Justices of the U. S. Supreme Court	
Years of service	Percentage
0–4	9
5–9	25
10–14	15
15–19	25
20–24	11
25–29	7
30–34	5
35–39	2
Total	100 (Approx.)

   (b)

Justices of the U. S. Supreme Court	
Years of service	Cumulative percentage
Less than 5	9
Less than 10	34
Less than 15	49
Less than 20	74
Less than 25	85
Less than 30	92
Less than 35	98
Less than 40	100

(c)

Justices of the U. S. Supreme Court	
Years of service	Cumulative percentage
More than 0	100
More than 5	91
More than 10	66
More than 15	51
More than 20	26
More than 25	15
More than 30	8
More than 35	2
More than 40	0

2.33 (a)

Age	Cumulative frequency
Less than or equal to 19	0
Less than or equal to 24	129
Less than or equal to 29	350
Less than or equal to 34	660
Less than or equal to 39	823
Less than or equal to 44	928
Less than or equal to 49	990
Less than or equal to 54	1,000

(b)

Age	Cumulative frequency
20 or more	1,000
25 or more	871
30 or more	650
35 or more	340
40 or more	177
45 or more	72
50 or more	10
55 or more	0

2.35 In 7, 8, 8, and 17 World Series, the defeated team won 0, 1, 2, and 3 games, respectively.

2.37 The frequencies corresponding to excellent, very good, good, fair, poor, and very poor are 3, 9, 20, 6, 1, and 1, respectively.

2.39

Dollars	Cumulative frequency
Less than 1	0
Less than 21	18
Less than 41	80
Less than 61	143
Less than 81	186
Less than 101	200

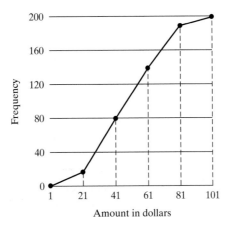

2.41 (a)

Thousands of dollars	Cumulative frequency
Less than 50	0
Less than 75	3
Less than 100	24
Less than 125	43
Less than 150	46
Less than 175	47
Less than 200	50

(b)

2.43   The central angles are $16.7°, 230.5°, 53.2°, 37.4°, 19.1°, 3.2°$.

2.45   The central angles for the degrees are Bachelor's, $192.4°$; Doctor's, $7.5°$; Master's, $69.8°$; Associate's, $90.4°$.

CHAPTER 3

3.1   Practice exercise.

3.3   **(a)**   The information is a parameter if the research department is interested only in the effect of the pill on the voluntary subjects but will not use the information for other purposes.

(b)   The information is a statistic (sample) if the research department assumes that the pill will have the same effect on the general population.

3.5   75.0 years.

3.7 $641.

3.9   The mean is 2, but as used here, the term *average criminal* is too vague.

3.11   10,752, which is less than the maximum load of 12,000 pounds.

3.13 $\bar{x} = 7.3$ hours. It is useful but it conceals the variation of the numbers around their average.

3.15 $\bar{x} =$ a comfortable $85°$, but the mean conceals the variation in the temperatures which include some very hot days.

3.17   Practice exercise.

3.19   20.1 thousands of pounds.

3.21   0.295.

3.23 $6,108.

3.25   Practice exercise.

3.27   **(a)**   20, so that the median is the 20th value;

(b)   75.5, so that the median is the mean of the 75th and 76th values.

3.29   Median is 5 meals.

3.31   The median is 59.5 power failures.

3.33   The median is 2.

3.35   The median is 68.5.

3.37   Practice exercise.

3.39   Experiments will produce various results.

3.41   Since the three midranges are 29.8, 30.0, and 30.3, the manufacturers of car C can use the midranges to substantiate the claim that their car performed best in the test.

3.43   The median is the mean of the ninth and tenth values. Since $\dfrac{9+1}{2} = 5$, Q_1 is the 5th value and Q_3 is the fifth value counting from the other end, namely, the 14th value counting from left to right.

3.45   Practice exercise.

3.47

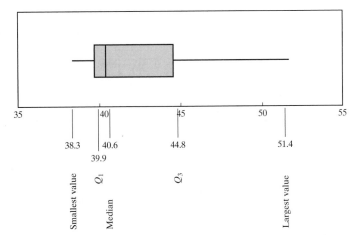

3.49 (a) The mean is $4\frac{1}{3}$ eggs.

(b) The median is 4 eggs.

(c) Q_1 is 3 eggs; Q_3 is 5 eggs.

(d) The midquartile is 4 eggs.

(e) The midrange is 5 eggs.

3.51 (a) The mode is 4%.

(b) The median is 4%.

(c) The mean is 3.9%.

3.53 The mode is 5.

3.55 Practice exercise.

3.57 The mode is 13 purchases with credit card.

3.59 Practice exercise.

3.61 For stock A, the range is $18.41 - 17.03 = \$1.38$.
For stock B, the range is $20.60 - 19.55 = \$1.05$.
Thus, stock B is less variable than stock A.

3.63 The range is 2.84 kilograms per cubic meter.

3.65 (a) $s = 4.06$;   **(b)** $s = 4.06$. s is the same by both methods.

3.67 $s = \$3.8$ thousand.

3.69 $s = 19.7$ vetoes.

3.71 17.62 billions of kWh.

3.73 $s = 2.06$, and the range is $7 - 3 = 4$. This supports the claim that for samples of size $n = 4$ the range should be roughly twice as large as the standard deviation.

3.77 Practice exercise.

3.79 For nearly all sets of data, the actual percentage of data lying between the limits is much greater using the empirical rule than with Chebyshev's theorem.

3.81 (a) At least 96%;

(b) at least 98.44%;

(c) at least 99%;

(d) at least 99.75%.

3.83 (a) About 68% of the flights will arrive within this interval;

(b) about 95% of the flights will arrive within this interval;

(c) about 99.7% of the flights will arrive within this interval.

3.85 (a) At least $\dfrac{3}{4}$ of the values must fall between 33 and 57;

(b) at least $\dfrac{15}{16}$ of the values must fall between 21 and 69.

3.87 (a) Since $z = 1.48$ and $z = 1.39$ for the two universities, the student is in a relatively better position with respect to the first university.

(b) Since $z = 3.10$ and $z = 3.18$ for the two universities the student is in a relatively better position with respect to the second university.

3.89 (a) Mean age is 22 years, s is 3.16 years.

(b) Mean weight is 136.4 pounds, s is 16.06 pounds.

(c) V for age is 14.4%, V for weight is 11.8%.
The age data is relatively more variable than the weight data.

3.91 Practice exercise.

3.93 $\bar{x} = 114.1°F$, $S_{xx} = 2{,}004.5$, and $s = 6.4°F$.

3.95 Practice exercise.

3.97 **(a)** $\bar{x} = 490.50$ sq. miles. **(b)** Median is 469.5 sq. miles.

3.99 Practice exercise.

3.101 (a) $\bar{x} = 12.9$;

(b) $S_{xx} = 1{,}594.375$, $s = 6.4$;

(c) $\tilde{x} = 12.4$;

(d) $Q_1 = 8.0$;

(e) $Q_3 = 17.3$;

(f) $P_{80} = 18.4$.

3.103 (a) Symmetrical.

(b) Positive skewness.

(c) Negative skewness.

3.105 $P_{20} = 64.12\%$.
$P_{80} = 86.17\%$.

3.107 Practice exercise.

3.109 $s = 0.48$, and $SK = 1.81$.

3.111 The distribution is U shaped because the number of H's is apt to stay ahead of the number of T's once it gets ahead, and vice versa.

3.113 (a) $x_1 + x_2 + x_3 + x_4 + x_5 + x_6$;

(b) $y_1 + y_2 + y_3 + y_4 + y_5$;

(c) $x_1 y_1 + x_2 y_2 + x_3 y_3$;

(d) $x_1 f_1 + x_2 f_2 + x_3 f_3 + x_4 f_4 + x_5 f_5 + x_6 f_6 + x_7 f_7 + x_8 f_8$;

(e) $x_3^2 + x_4^2 + x_5^2 + x_6^2 + x_7^2$;

(f) $(x_1 + y_1) + (x_2 + y_2) + (x_3 + y_3) + (x_4 + y_4)$.

3.115 Practice exercise.

3.117 (a) 27;

(b) 35;

(c) 137;

(d) 587;

3.119 **(a)** 8, −1, 0. 8 − 1 + 0 = 7

(b) 4, 5, 2, −4. 4 + 5 + 2 − 4 = 7

(c) 7.

3.121 No.

REVIEW EXERCISES FOR CHAPTERS 1, 2, AND 3

R.1 **(a)** $\bar{x} = 184.5$ home runs.

(b) $\tilde{x} = 185$.

R.3 Statements (a) and (d) are descriptive, while statements (b) and (c) are generalizations.

R.5 **(a)** The data would constitute a *population* if the CPA wanted to determine the average Federal Income Tax withheld from all the paychecks of residents of New York City in the year 2003 only.

(b) The data would constitute a *sample* if the CPA wants to predict the average Federal Income Tax which will be withheld in subsequent years from the residents of New York City or other major cities.

R.7 The mean is not a good average in this case. The calculated value of 14.2 is greatly enlarged by one large value from California and is not at all like the values of the other states.

R.9 Aside from the two families with incomes of $180,000 and $175,000, the mean of $72,900 is higher than the incomes of all other families. The mean is not a central value in this case.

R.11

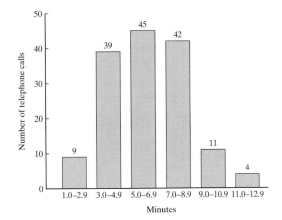

R.13 **(a)** $\bar{x} = 0.0638$ p.p.m.; $s \approx 0.0044$ p.p.m.

(b) The median is 0.0625 p.p.m., and the range is 0.012 p.p.m.

R.15   3 | 2 5 5 5 7
          4 | 1 8 4 7 3 9 1 1 1 1 4 9
          5 | 7 2 4 9 4 9 2 7
          6 | 2 8 4 8
          7 | 3

R.17 **(a)** Few persons will give honest answers about their personal habits.

(b) Because of the holiday season, December spending is not typical of spending patterns throughout the year.

R.19 By Chebyshev's theorem, the probability is at least $1 - \dfrac{1}{4^2} = \dfrac{15}{16}$.

R.21 **(a)** $\tilde{x} = 1{,}572.5$ feet.

(b) Since the median is the 19.5 value, the location of Q_1 is $\dfrac{19+1}{2} = 10$, and its value is 1,150. The location of Q_3 is 10 from the other end, and its value is 2,150. Since the values are arranged from the largest to the smallest, the designations Q_1 and Q_3 can be reversed.

R.23 **(a)** 50, 100, 150, 200, 250, 300, and 350;

(b) 99, 149, 199, 249, 299, 349, and 399;

(c) 49.5, 99.5, 149.5, 199.5, 249.5, 299.5, 349.5, and 399.5;

(d) 74.5, 124.5, 174.5, 224.5, 274.5, 324.5, and 374.5;

(e) 50.

R.25 $s = 7.6$ work stoppages.

R.27 **(a)** It is possible to find both the mean and the median,

(b) It is possible to find the median which falls into the 100–109 class, but not the mean.

(c) It is impossible to find either the median or the mean.

R.29 **(a)** This is the histogram:

R.35 (a)

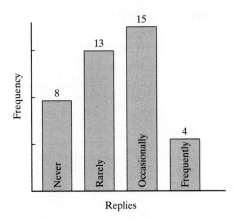

(b) The modal reply is "occasionally."

R.37

```
21 | 6
22 | 4 8 9 1 8 3 2
23 | 5 5 0 6 7 3 1 0 0
24 | 0 4 9 0 5
25 | 3 6
26 | 9
```

With rearranged leaves

```
21 | 6
22 | 1 2 3 4 8 8 9
23 | 0 0 0 1 3 5 5 6 7
24 | 0 0 4 5 9
25 | 3 6
26 | 9
```

R.39 $\bar{x} = \$17.5$ million. $Q_1 = \$12$ million. $Q_3 = \$26$ million.

CHAPTER 4

4.1 Practice exercise.

4.3 The tree diagram for this exercise is

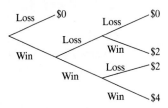

In two cases he will be exactly $1.00 ahead.

4.5 (a) 15,625.

(b) 1.

(c) 4,096.

4.7 Practice exercise.

4.9 60 ways.

(b) This is the cumulative "less than" distribution:

Number of vehicles	Cumulative frequency
Less than 20	4
Less than 40	15
Less than 60	43
Less than 80	67
Less than 100	80

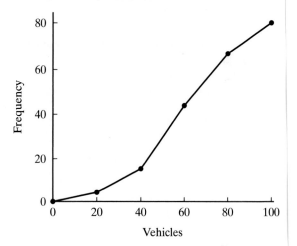

R.31 Here is a possible drawing for the box-and-whisker plot.

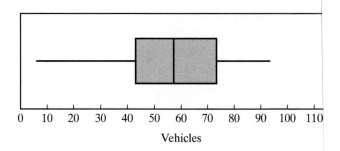

R.33 (a) $\bar{x} = \$189.4$ thousand.

$\tilde{x} = \$182.1$ thousand.

(b) $\bar{x} = \$308.2$ thousand and $\tilde{x}$ is unchanged.

(c) The printing error seriously affected the mean, but the me
unaffected.

4.11 500 different ways.

4.13 120 ways.

4.15 67,600,000 identifiable charge cards.

4.17 60 ways.

4.19 Here is one possible version of this diagram. The symbols are used as follows:

$$SK = \text{spackle the kitchen}$$

$$PK = \text{paint the kitchen}$$

$$SD = \text{spackle the dinning room}$$

$$PD = \text{paint the dining room}$$

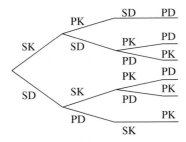

There are six different ways in which these tasks can be done.

4.21 Practice exercise.

4.23 **(a)** Since $\dfrac{2}{6} + \dfrac{1}{6} = \dfrac{1}{2}$, the answer is false.

  (b) Since $\dfrac{15!}{12!} = 15 \cdot 14 \cdot 13$, the answer is true.

  (c) Since $\dfrac{6 \cdot 5 \cdot 4 \cdot 3 \cdot 2 \cdot 1}{3 \cdot 2 \cdot 1 \cdot 1 \cdot 2 \cdot 1} = 60$, the answer is false.

4.25 Practice exercise.

4.27 **(a)** 720 permutations.

  (b) 6 permutations.

  (c)

1. Miziuko	Citigroup	Allianz
2. Miziuko	Allianz	Citigroup
3. Citigroup	Miziuko	Allianz
4. Citigroup	Allianz	Miziuko
5. Allianz	Miziuko	Citigroup
6. Allianz	Citigroup	Miziuko

4.29 For (a), (b), and (c) , the solutions are all $6 \cdot 5 \cdot 4 \cdot 3 \cdot 2 \cdot 1 = 720$ ways.

4.31 **(a)** The number of circular permutations is $(n-1)!$.

  (b) $5! = 120$;

  (c) $3! = 6$.

4.33 **(a)** 840;

  (b) 3,360;

  (c) 90,720;

  (d) 34,650.

4.35 Practice exercise.

4.37 **(a)** 364;
   (b) identical to (a);
   (c) 3,003;
   (d) identical to (c) .

4.39 **(a)** $\binom{25}{5}$;

   (b) $\binom{14}{4}$;

   (c) $\binom{10}{3}$;

   (d) $\binom{18}{6}$.

4.41 **(a)** 56;
   (b) 112;
   (c) 48;
   (d) 4.

4.43 Reading the answers directly from Table X, Binomial Coefficients, we get carpenters, 10 ways; plumbers, 6 ways; and electricians, 15 ways. The overall result is obtained by multiplying the 3 values, obtaining $10 \cdot 6 \cdot 15 = 900$ ways.

4.45 Reading directly from the Table of Binomial Coefficients, we get, for parts (a) and (b),
   (a) 6 ways;
   (b) 10 ways;
   (c) $6 \times 10 = 60$

4.47 The next three rows of Pascal's triangle are

		1		6		15		20		15		6		1	
	1		7		21		35		35		21		7		1
1		8		28		56		70		56		28		8		1

4.49 Practice exercise.

4.51 **(a)** $\dfrac{2}{52} = \dfrac{1}{26}$;

   (b) $\dfrac{26}{52} = \dfrac{1}{2}$;

   (c) $\dfrac{16}{52} = \dfrac{4}{13}$;

   (d) $\dfrac{13}{52} = \dfrac{1}{4}$;

   (e) $\dfrac{39}{52} = \dfrac{3}{4}$;

   (f) $\dfrac{48}{52} = \dfrac{12}{13}$.

4.53 The probabilities are $\dfrac{1}{8}, \dfrac{3}{8}, \dfrac{3}{8}$, and $\dfrac{1}{8}$.

4.55 (a) $\dfrac{1}{6}$;

(b) The even numbers are 2, 4, and 6, so the probability is $\dfrac{3}{6}$ or $\dfrac{1}{2}$.

4.57 2,730 ways.

4.59 (a) $\dfrac{6}{203}$;   (b) $\dfrac{95}{203}$.

4.61 Practice exercise.

4.63 0.55.

4.65 0.20.

4.69 Practice exercise.

4.71 $0.10 or 10¢.

4.73 Begin by noting that the probability of getting a red number is $\dfrac{18}{38} = \dfrac{9}{19}$, the probability of getting a black number is $\dfrac{18}{38} = \dfrac{9}{19}$, and the probability of getting a green number is $\dfrac{2}{38} = \dfrac{1}{19}$.

(a) −$0.0526;   (b) −$0.0526;   (c) −$0.0526.

4.75 $81,500.

4.77 1.92 times in any given year.

4.79 Practice exercise.

4.81 $p = 0.16$.

4.83 1.35 lawsuit filings.

4.85 Practice exercise.

4.87 The expected profit is $4,500,000 \cdot \dfrac{1}{2} - 2,700,000 \cdot \dfrac{1}{2} = \$900,000$ if they continue the operation and $-1,800,000 \cdot \dfrac{1}{2} + 450,000 \cdot \dfrac{1}{2} = -\$675,000$ if they do not continue the operation. Thus, continuing the operation will maximize the company's expected profit.

4.89 The expected profit is $451,000 \cdot \dfrac{1}{6} - 110,000 \cdot \dfrac{5}{6} = -\$16,500$ if the new factory is built, and $220,000 \cdot \dfrac{1}{6} + 22,000 \cdot \dfrac{5}{6} = \$55,000$ if the new factory is not built. Thus, not building the new factory will maximize the expected profit.

4.91 The best strategy calls for furnishing two stores, and the expected profit is $18,000.

CHAPTER 5

5.1 Practice exercise.

5.3 $D' = \{1, 2, 6\}$ which is the event we roll a 1, 2, or 6.
$D \cup E = \{2, 3, 4, 5, 6\}$ which is the event we do not roll a 1.
$D \cap E = \{4\}$ which is the event we roll a 4.

5.5 (a) $M \cup N = \{Q\clubsuit, K\clubsuit, Q\diamondsuit, K\diamondsuit, Q\heartsuit, K\heartsuit, 10\spadesuit, J\spadesuit, Q\spadesuit, K\spadesuit\}$ is the event that we draw a queen, a king, or the 10 or jack of spades.

(b) $M \cap N = \{Q\spadesuit, K\spadesuit\}$ is the event that we draw the queen or king of spades.

(c) $M' = (A\clubsuit, 2\clubsuit, \ldots, J\clubsuit, A\diamondsuit, 2\diamondsuit, \ldots, J\diamondsuit, A\heartsuit, 2\heartsuit, \ldots, J\heartsuit, A\spadesuit, 2\spadesuit, \ldots, J\spadesuit\}$ is the event that we do not draw a queen or a king.

(d) $N' = \{A\clubsuit, \ldots, K\clubsuit, A\diamondsuit, \ldots, K\diamondsuit, A\heartsuit, \ldots, K\heartsuit, A\spadesuit, \ldots, 9\spadesuit\}$ is the event that we do not draw the 10, jack, queen, or king of spades.

(e) $M' \cup N' = \{A\clubsuit, \ldots, K\clubsuit, A\diamondsuit, \ldots, K\diamondsuit, A\heartsuit, \ldots, K\heartsuit, A\spadesuit, \ldots, J\spadesuit\}$ is the event that we do not draw the queen or king of spades.

(f) $M' \cap N' = \{A\clubsuit, \ldots, J\clubsuit, A\diamondsuit, \ldots, J\diamondsuit, A\heartsuit, \ldots, J\heartsuit, A\spadesuit, \ldots, 9\spadesuit\}$ is the event that we do not draw a queen, a king, the 10 of spades or the jack of spades.

5.7   Since $A \cap B = \{3, 4\}$, A and B are not mutually exclusive.
Since $A \cap C = \emptyset$, A and C are mutually exclusive.
Since $B \cap C = \{5, 6, 7\}$, B and C are not mutually exclusive.

5.9   **(a)**   Mutually exclusive;
(b)   not mutually exclusive;
(c)   not mutually exclusive;
(d)   mutually exclusive;

5.11   **(a)**   Not mutually exclusive;
(b)   mutually exclusive;
(c)   not mutually exclusive;
(d)   not mutually exclusive;
(e)   not mutually exclusive;
(f)   mutually exclusive.

5.13   **(a)**   Regions 1 and 2 together represent the event that a person vacationing in Southern California visits Disneyland.
(b)   Regions 2 and 3 together represent the event that a person vacationing in Southern California visits Disneyland or Universal Studios, but not both.
(c)   Regions 2 and 4 together represent the event that a person vacationing in Southern California does not visit Universal Studios.

5.15   **(a)**   Regions 1 and 3 together represent the event that the murder suspect is guilty.
(b)   Regions 1 and 4 together represent the event that either the murder suspect is guilty and allowed out on bail, or is not guilty and not allowed out on bail.
(c)   Regions 3 and 4 together represent the event that the murder suspect is not allowed out on bail.

5.17   **(a)**   The laboratory facilities are adequate, the financial support is adequate, and it is staffed by competent personnel.
(b)   The financial support is adequate, it is staffed by competent personnel, but the laboratory facilities are not adequate.
(c)   It is staffed by competent personnel, but the laboratory facilities and financial support are not adequate.
(d)   The laboratory facilities are not adequate, the financial support is not adequate, and it is not staffed by competent personnel.
(e)   The laboratory facilities are adequate and it is staffed by competent personnel.
(f)   The financial support is adequate, but the laboratory facilities are not adequate.
(g)   The personnel are competent.
(h)   The personnel are not competent.

5.19   Practice exercise.

5.21   **(a)** 0.70;   **(b)** 0.75;   **(c)** 0.45.

5.23 **(a)** The probability cannot be greater than 1, and it is given as 1.09.

(b) The sum of the two probabilities cannot be greater than 1, and it is 1.10.

(c) The sum of the two probabilities is $0.30 + 0.40 = 0.70$, and not 0.90.

(d) The sum of the two probabilities should be 1, but it is given as $0.55 + 0.35 = 0.90$.

5.25 **(a)** $1 - 0.45 = 0.55$;

(b) $1 - 0.30 = 0.70$;

(c) $0.45 + 0.30 = 0.75$;

(d) 0;

(e) $P(Q' \cap R) = P(R) = 0.30$;

(f) $P(Q \cap R') = P(Q) = 0.45$.

5.27 **(a)** 0.35;   **(b)** 0.45;   **(c)** 0.65.

5.29 1. Since $0 \le s \le n$, division by n yields $0 \le \dfrac{s}{n} \le 1$.

2. If an event is certain to occur, $s = n$, so that its probability is $\dfrac{s}{s} = \dfrac{n}{n} = 1$; if an event is certain not to occur, then $s = 0$ and the probability is 0.

3. The respective probabilities are $\dfrac{s_1}{n}$, $\dfrac{s_2}{n}$, and $\dfrac{s_1 + s_2}{n}$, and the sum of the first two is equal to the third.

4. $\dfrac{s}{n} + \dfrac{n-s}{n} = \dfrac{n}{n} = 1$.

5.31 Practice exercise.

5.33 **(a)** The odds are $\dfrac{6}{10}$ to $\left(1 - \dfrac{6}{10}\right) = 6$ to 4 or 3 to 2.

(b) The odds are $\dfrac{8}{24}$ to $\left(1 - \dfrac{8}{24}\right) = 8$ to 16 or 1 to 2, which is 2 to 1 against the occurrence.

(c) The odds are $\dfrac{6}{16}$ to $\left(1 - \dfrac{6}{16}\right) = 6$ to 10 or 3 to 5, which is 5 to 3 against the occurrence.

(d) The odds are $\dfrac{13}{52}$ to $\left(1 - \dfrac{13}{52}\right)$ or 13 to 39 or 1 to 3, which is 3 to 1 against the occurrence.

5.35 The probabilities that he will run for the House of Representatives, the Senate, or either one, are $\dfrac{1}{1+5} = \dfrac{1}{6}$, $\dfrac{1}{1+4} = \dfrac{1}{5}$, and $\dfrac{1}{1+2} = \dfrac{1}{3}$, and since $\dfrac{1}{6} + \dfrac{1}{5} = \dfrac{11}{30}$, which is more than equal to $\dfrac{1}{3}$, they are not consistent.

5.37 The demonstration proceeds as follows:

$a(1 - p) = bp$

$a - ap = bp$

$a = ap + bp$

$a = p(a + b)$

$p = \dfrac{a}{a + b}$

5.39 **(a)** The even money bet is attractive.

(b) Attractive.

(c) Unattractive.

5.41   0.24.

5.43   **(a)** 0.60;   **(b)** 0.25;   **(c)** 0.34.

5.45   Practice exercise.

5.47   **(a)** 0.34;   **(b)** 0.44;   **(c)** 0.22.

5.49   Practice exercise.

5.51   0.11.

5.53   Filling the numbers associated with the various regions, we get

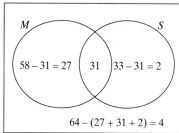

and it can be seen that the probability is $\dfrac{4}{64} = \dfrac{1}{16}$.

5.55   Practice exercise.

5.57   **(a)** 0.25;   **(b)** 0.125;   **(c)** 0.625;   **(d)** 0.375.

5.59 $P(A) \cdot P(B) = (0.70)(0.40) = 0.28 \neq 0.25$, so that A and B are not independent.

5.61   Practice exercise.

5.63   **(a)**   0.706;

   (b)   0.338;

   (c)   0.212;

   (d)   0.168;

   (e)   0.747;

   (f)   0.570.

5.65   **(a)** $\dfrac{5}{12}$;

   (b) $\dfrac{3}{23}$;

   (c) $\dfrac{3}{20}$;

   (d) $\dfrac{4}{11}$;

   (e) $\dfrac{19}{42}$;

   (f) $\dfrac{8}{23}$.

5.67   **(a)**   The probability that a very rare sword is in good condition.

   (b)   The probability that an antique sword which is not a reproduction is in good condition.

   (c)   The probability that an antique sword which is not very rare is a reproduction.

   (d)   The probability that a very rare antique sword is in good condition, and is a reproduction.

(e) The probability that an antique sword which is not in good condition is very rare.

(f) The probability that an antique sword which is very rare and a reproduction is in good condition.

5.69 **(a)** $\dfrac{1}{32}$; **(b)** $\dfrac{1}{2,197}$; **(c)** 0.013; **(d)** 0.16.

5.71 **(a)** In the first game, Abel's proportion of completions was $\dfrac{11}{20} = 0.55$, while Baker's was $\dfrac{3}{5} = 0.60$. Baker did better.

(b) In the second game, Abel's proportion of completions was $\dfrac{1}{5} = 0.20$, while Baker's was $\dfrac{6}{16} = 0.375$. Baker did better again.

(c) For the season-to-date, Abel threw 25 passes, of which 12 were completions. His proportion of completions was 0.48. Baker threw 21 passes, with 9 completions, so his proportion of completions was approximately 0.4286. Thus, even though Baker did better in each game, his overall performance was worse!

5.73 $P(L|A) = \dfrac{0.63}{0.75} = 0.84 = P(L)$.

5.75 **(a)** $\dfrac{1}{6} \cdot \dfrac{1}{6} = \dfrac{1}{36}$;

(b) $\dfrac{1}{6} \cdot \dfrac{1}{6} = \dfrac{1}{36}$.

The results of rolling a single die 2 times, or 2 dice 1 time are the same.

5.77 **(a)** $\dfrac{64}{225}$; **(b)** $\dfrac{8}{29}$.

5.79 0.45.

5.81 **(a)** 0.06; **(b)** 0.18.

5.83 **(a)** 0.196; **(b)** 0.084.

5.85 Practice exercise.

5.87 Let A, B, and C correspond to Arthur, Beatrice, and Carla in the following diagram.

and

(a) $\dfrac{0.0025}{0.0054} \approx 0.46$;

(b) $\dfrac{0.0009}{0.0054} \approx 0.17$;

(c) $\dfrac{0.0020}{0.0054} \approx 0.37$.

5.89 $P(B|A) = 0.810$.

5.91 $P(Y|T) = 0.136$.

REVIEW EXERCISES FOR CHAPTERS 4 AND 5

R.41   19,600 ways.

R.43   The 15 points of the sample space are shown in the following diagram:

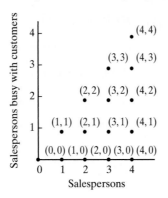

R.45   1.55 alarms.

R.47   **(a)** $\{A, D\}$;
(b) $\{C, E\}$;
(c) $\{B\}$.

R.49   Since a ticket pays 0.65, on average, and the ticket costs $1.00, it is not a wise purchase.

R.51   **(a)** $\dfrac{1}{15}$;   **(b)** $\dfrac{7}{15}$.

R.53   **(a)** 24;   **(b)** 9;   **(c)** 8;   **(d)** 6;   **(e)** 0;   **(f)** 1.

R.55   **(a)** 48;   **(b)** 12;   **(c)** 24.

R.57   **(a)** 0.60;   **(b)** 0.65;   **(c)** 0.35.

R.59   210 ways.

R.61 $\dfrac{2}{3}$ or 0.67.

R.63   Observe that $P(A|R') = \dfrac{160}{310}$, $P(A \cap R') = \dfrac{160}{380}$, $P(R') = \dfrac{310}{380}$, and that

$$\frac{\dfrac{160}{380}}{\dfrac{310}{380}} = \frac{160}{310}.$$

R.65   **(a)** $\dfrac{5}{7}$;   **(b)** 2 to 5;   **(c)** $\dfrac{2}{7}$;   **(d)** 1.

R.67   Region 1 represents the event that the school's football team is rated among the top twenty by both AP and UPI. Region 2 represents the event that the school's football team is rated among the top twenty by AP but not by UPI. Region 3 represents the event that the school's football team is rated among the top twenty by UPI but not by AP. Region 4 represents the event that the school's football team is rated among the top twenty by neither AP nor UPI.

R.69   **(a)** $\dfrac{1}{256}$ or 0.0039;   **(b)** $\dfrac{17,160}{6,497,400}$ or 0.0026.

R.71   **(a)** 15 ways;   **(b)** 225 ways;   **(c)** 495 ways.

R.73   **(a)**   Region 5.
(b)   Regions 1 and 2 together.

(c) Regions 3, 5, and 6 together.

(d) Regions 1, 3, 4, and 6 together.

R.75 **(a)** $\dfrac{10}{10+1} = \dfrac{10}{11}$ or approximately 0.91;

(b) $\dfrac{5}{5+3} = \dfrac{5}{8}$ or approximately 0.62;

(c) $\dfrac{1}{1+1} = \dfrac{1}{2}$ or 0.50.

R.77 **(a)** 0.35;   **(b)** 0.43;   **(c)** 0.42;   **(d)** 0.45.

R.79 $\dfrac{1}{16}$.

R.81 **(a)** $P(A|B) = \dfrac{0.13}{0.20} = 0.65 = P(A)$;

(b) $P(A|B') = \dfrac{0.65 - 0.13}{1 - 0.20} = \dfrac{0.52}{0.80} = 0.65 = P(A)$;

(c) $P(B|A) = \dfrac{0.13}{0.65} = 0.20 = P(B)$;

(d) $P(B|A') = \dfrac{0.20 - 0.13}{1 - 0.65} = \dfrac{0.07}{0.35} = 0.20 = P(B)$.

CHAPTER 6

6.1   Practice exercise.

6.3   **(a)** No;   **(b)** no;   **(c)** yes;   **(d)** no.

6.5   Practice exercise.

6.7   Practice exercise.

6.9   0.147.

6.11   0.478.

6.13   0.168.

6.15   **(a)** 0.942 or 0.941;   **(b)** 0.004;   **(c)** 0.497.

6.17   **(a)** 0.196;   **(b)** 0.304;   **(c)** 0.153;   **(d)** 0.151.

6.19   **(a)** 0.099;   **(b)** 0.545;   **(c)** 0.179.

6.21   **(a)** 0.9596;   **(b)** 0.3414.

6.23   **(a)** 0.2968 (no);   **(b)** 0.8866 (no);   **(c)** 0.9940 (yes);   **(d)** $n = 21$.

6.25   Practice exercise.

6.27   0.051.

6.29   0.412.

6.31   **(a)** 0.292;   **(b)** 0.525;   **(c)** 0.175;   **(d)** 0.008.

6.33   Practice exercise.

6.35   0.005.

6.37   0.206.

6.39   **(a)** No;   **(b)** no;   **(c)** yes.

6.41   0.171.

6.43 $f(2) = 0.0450$, $f(3) = 0.0900$, and the sum of the probabilities is $0.0450 + 0.0900 = 0.1350$.

6.45   **(a)** $n \le (0.05)(a+b)$, $n \ge 100$, and $n\dfrac{a}{a+b} < 10$.

(b) Use $n\dfrac{a}{a+b}$ for np in the Poisson calculation.

6.47   0.145.

6.49 $f(4) = 0.090$, $f(5) = 0.128$, and the sum of the probabilities is $0.090 + 0.128 = 0.218$.

6.51   0.139.

6.53   **(a)** 0.843645;   **(b)** 0.640552.

6.55   Practice exercise

6.57   0.117.

6.59   **(a)** 0.207;   **(b)** 0.207;   **(c)** the multinomial distribution with $k = 2$ is the binomial distribution.

6.61   3.27.

6.63   **(a)** 2.40;   **(b)** 1.24.

6.65   **(a)** 1.7078;   **(b)** 1.7078.

6.67   Practice exercise.

6.69   **(a)** 1.25;   **(b)** 1.25.

6.71   **(a)** 0.9413;   **(b)** 0.9487.

6.73   **(a)** 2.5 heads;   **(b)** 0.5 no shows;   **(c)** 0.6 seconds.

6.75   0.9375.

6.77   Practice exercise.

6.79   **(a)** At least 0.94;   **(b)** at least 0.84;   **(c)** at least 0.87.

CHAPTER 7

7.1   Practice exercise.

7.3   Practice exercise.

7.5   Practice exercise.

7.7   **(a)** Process B is preferable;   **(b)** no "right" answer.

7.9   **(a)** 0.3023;   **(b)** 0.7422;   **(c)** 0.0222;   **(d)** 0.0230.

7.11   Practice exercise.

7.13   **(a)** 2.00;   **(b)** 1.11;   **(c)** 0.55.   **(d)** 3.08

7.15   **(a)** 0.9977;   **(b)** 0.0918;   **(c)** 0.9059.

7.17   **(a)**   Since $0.5000 - 0.0250 = 0.4750$ corresponds to $z = 1.96$ in Table II, we get $z_{0.025} = 1.96$.

   (b)   Since $0.5000 - 0.0050 = 0.4950$, and 0.4949 and 0.4951 are the entries in Table II corresponding to $z = 2.57$ and $z = 2.58$, we get $z_{0.005} = \dfrac{2.57 + 2.58}{2} = 2.575$.

7.19   20.

7.21 $\mu = 258.47$ and $\sigma = 30.61$.

7.23   **(a)** 0.0239;   **(b)** 0.1554;   **(c)** 0.5384.

7.25   0.0582.

7.27   4.90 inches.

7.29   Practice exercise.

7.31   0.0455.

7.33   646 patients.

7.35   **(a)**   Conditions are satisfied. $np = 10.8$ and $n(1 - p) = 7.2$.

   (b)   Conditions are not satisfied. $np = 4.5$.

   (c)   Conditions are not satisfied. $np = 5$ and $n(1 - p) = 5$.

7.37 0.0001.

7.39 (a) 0.0601;   (b) 0.2389.

7.41 0.1492.

7.43 (a) 0.2358;   (b) 0.4908;   (c) 0.9556.

7.45 (a) −0.0056;   (b) −0.0034.

7.47 (a) 0.2514;   (b) 0.4701.

CHAPTER 8

8.1 Practice exercise.

8.3 (a) 4;   (b) 20;   (c) 220;   (d) 4,555,100.

8.5 (a) 0.2500;   (b) 0.0500;   (c) 0.0045;   (d) 0.0012.

8.7 *abc, bcd, abd, acd.*

8.9 (a) $M A U T, A U T O, U T O M, T O M A, O M A U$;

    (b) 1/5;

    (c) 4/5.

8.11 03, 11, 22, 13, 01, 08, 24, 21, 10, 17, 04, 09.

8.13 Students 2997, 1487, 3852, 0372, 2705, 1005, 3113, 0732, 3780, 3815, 2742, 1764, 3247, 3375, and 3700.

8.15 (a) $\dfrac{1}{6}$;

    (b) $\dfrac{1}{6}$;

    (c) 0.0254;

    (d) $\dfrac{3}{118} < \dfrac{1}{36}$ since $3 \cdot 36 = 108 < 118$.

8.17 Practice exercise.

8.19 (a) $\mu = 5$, and $\sigma = 2.236$.

    (b) The sixteen possible samples are 2 and 2, 2 and 4, 2 and 6, 2 and 8, 4 and 2, 4 and 4, 4 and 6, 4 and 8, 6 and 2, 6 and 4, 6 and 6, 6 and 8, 8 and 2, 8 and 4, 8 and 6, and 8 and 8. Their means are 2, 3, 4, 5, 3, 4, 5, 6, 4, 5, 6, 7, 5, 6, 7, and 8.

    (c)

Sample mean	Probability
2	$\dfrac{1}{16}$
3	$\dfrac{2}{16}$
4	$\dfrac{3}{16}$
5	$\dfrac{4}{16}$
6	$\dfrac{3}{16}$
7	$\dfrac{2}{16}$
8	$\dfrac{1}{16}$

    (d) The mean of the sampling distribution is five and the variance of the sampling distribution is 2.5. The standard deviation is $\sqrt{2.5} \approx 1.581$. Since the sampling is with replacement, the standard deviation formula is 1.581. The calculations agree.

8.21 **(a)** $\frac{1}{3}$; **(b)** $\frac{3}{1}$; **(c)** $\frac{1}{3}$; **(d)** $\frac{3}{1}$.

8.23 0.995.

8.25 **(a)** 0.889 or 0.866;

   (b) 0.871 or 0.866;

   (c) 0.975 or 0.975;

   (d) 0.912 or 0.910.

8.31 We can assert with the probability $1 - \frac{1}{3^2} = \frac{8}{9} \approx 0.89$ that the error will be less than $k \cdot \sigma_{\bar{x}} = (3)(0.75) = 2.25$.

8.33 **(a)** We get $k = 1.80$, so that the probability is at least $1 - \frac{1}{(1.80)^2} \approx 0.691$.

   (b) The probability is $2(0.4641) = 0.9282$.

8.35 The probability is at least 0.36.

8.37 **(a)** The probability is $2(0.4192) = 0.8384$.

   (b) The probability is $2(0.2580) = 0.5160$.

8.39 Practice exercise.

8.41 $n = 50^2 = 2{,}500$.

REVIEW EXERCISES FOR CHAPTERS 6, 7, AND 8

R.83 **(a)** 0.0467; **(b)** 0.7667; **(c)** 0.2333.

R.85 **(a)** 0.1736; **(b)** $0.3264 + 0.2324 = 0.5588$; **(c)** $0.5000 - 0.2324 = 0.2676$.

R.87 No.

R.89 **(a)** 0.003; **(b)** 0.151; **(c)** 0.153; **(d)** 0.349.

R.91 **(a)** 0.125; **(b)** 0.24.

R.93 22, 13, and 01.

R.95 **(a)** 0.030; **(b)** 0.185; **(c)** 0.189; **(d)** 0.077.

R.97 **(a)** 0.65; **(b)** 0.64.

R.99 0.8413.

R.101 **(a)** 0.420; **(b)** 0.160.

R.103 **(a)** They cannot be the values of a probability distribution.

   (b) They cannot be the values of a probability distribution.

   (c) They can be the values of a probability distribution.

R.105 $\mu = 365$, $\sigma = 9.927$, and the probability is 0.0721.

R.107 **(a)** The conditions are satisfied.

   (b) The conditions are not satisfied.

   (c) The conditions are not satisfied.

   (d) The conditions are satisfied

R.109 0.711.

R.111 **(a)** The condition is not satisfied.

   (b) The condition is satisfied.

   (c) The condition is not satisfied.

R.113 The error is less than 8.4.

R.115 **(a)** The conditions for the Poisson approximation are satisfied.

   (b) The conditions for the normal approximation are satisfied.

R.117 0.1692.

CHAPTER 9

9.1 Practice exercise.

9.3 0.008 inches.

9.5 0.1288 minutes.

9.7 0.9979 chair, or 1.0 chair.

9.9 **(a)** 0.1225 ounces;   **(b)** 0.1609 ounces.

9.11 1 minute.

9.13 96.04, which we move up to the next integer, 97 children.

9.15 37.14 which we move up to the next integer, 38 cars.

9.17 At least 42.68, but we should use $n = 43$.

9.19 **(a)** 22.72 tons $< \mu <$ 27.48 tons.

     (b) 21.97 tons $< \mu <$ 28.23 tons.

9.21 5,853.96 pounds $< \mu <$ 6,746.04 pounds.

9.23 83.10 $< \mu <$ 86.90 blood tests.

9.25 **(a)** 1.812;   **(b)** 3.106;   **(c)** 2.179;   **(d)** 2.650.

9.27 1.13 pounds.

9.29 5.68 $< \mu <$ 12.32.

9.31 18.1 $< \mu <$ 21.9 minutes.

9.33 **(a)** 1.44;   **(b)** 1.37;   **(c)** 1.35.

9.35 Computer or graphing calculator exercise.

9.37 Practice exercise.

9.39 4.14 $< \sigma <$ 6.82.

9.41 1.58 minutes $< \sigma <$ 2.21 minutes.

9.43 Practice exercise.

9.45 0.21 $< p <$ 0.39.

9.47 0.041.

9.49 1.93%.

9.51 0.019.

9.53 Practice exercise.

9.55 3,482.

9.57 246.

CHAPTER 10

10.1 Practice exercise.

10.3 **(a)** We would commit a Type I error if we erroneously reject the hypothesis that the average number of gallons of gasoline purchased is 9.3.

     (b) We would commit a Type II error if we erroneously accept the null hypothesis that the number of gallons of gasoline purchased is 9.3.

10.5 Test the null hypothesis that the use of telemarketing is not effective in increasing automobile sales

10.7 **(a)** 0.02;   **(b)** 0.5000.

10.9 **(a)** 0.78 and also 0.78;

     (b) 0.5000, and by symmetry it is also 0.5000;

     (c) 0.06, and by symmetry it is also 0.06.

10.11   Practice exercise.

10.13   **(a)**   Since $z = 2.04 \geq 1.96$, the null hypothesis must be rejected.
   (b)   Since $z = 2.04$ falls between -2.575 and 2.575, the null hypothesis cannot be rejected.

10.15   Since $z = -0.86$ lies between -1.96 and 1.96, the null hypothesis cannot be rejected.

10.17   Since $z = 3.50$ is greater than 2.33, the null hypothesis must be rejected.

10.19   **(a)** $z = 1.69$.
   (b) $z = 1.69$. The null hypothesis cannot be rejected.
   (c) $z = 1.69$. Reject the null hypothesis.
   (d) $z = 1.69$. The null hypothesis cannot be rejected.
   (e)   The p-value is $2(0.5 - 0.4545) = 0.0910$.

10.21   Practice exercise.

10.23   Since $z = -2.03$ is less than -1.833, the null hypothesis must be rejected.

10.25   Since $t = 1.05$ falls between -3.250 and 3.250, the null hypothesis cannot be rejected.

10.27   Since $t = 3.09$ is greater than 2.998, the null hypothesis must be rejected. Since $t_{.01} = 2.998$ and $t_{.005} = 3.499$, and $t = 3.09$ falls between these two values, we can write $0.005 < p < 0.01$.

10.29   **(a)** $t = -2.33$;   **(b)** $0.01 < p < 0.025$.

10.31   **(a)** 0.8784;   **(b)** 0.6082.

10.35   Practice exercise.

10.37   Since $z = 2.40$ falls between -2.575 and 2.575, the null hypothesis cannot be rejected. Since $p = 0.0164$, which is greater than $\alpha = 0.01$, we cannot reject the null hypothesis.

10.39   **(a)** 0.0434;
   (b) 0.0434;
   (c) 0.0478;
   (d) 0.0478;
   (e) 0.5824;
   (f) 0.0052;
   (g) 1.0000;
   (h) $2(0.50000 - 0.4999997) = 0.0000006$.

10.41   The probability that none will reject is 0.95^{20}. The probability that at least one will get a spurious result is 0.64.

10.43   Practice exercise.

10.45   Practice exercise.

10.47 $t = -1.99$. Since $t = -1.99$ falls between -2.878 and 2.878, the null hypothesis cannot be rejected.

10.49   Practice exercise.

10.51   **(a)** $34, 5.2$, and $F = 6.54$.
   (b)   The differences among the sample means are significant.

10.53

Source of Variation	Degrees of Freedom	Sum of Squares	Mean Square	F
Treatments	2	31.5833	15.7916	0.08
Error	21	4,182.3750	199.1607	
Total	23	4,213.9583		

Since $F = 0.08$ does not exceed 5.78, the null hypothesis cannot be rejected; the differences among the average weekly earnings may be attributed to chance.

10.55

Source of Variation	Degrees of Freedom	Sum of Squares	Mean Square	F
Treatments	3	412.6447	137.5482	6.84
Error	27	543.0972	20.1147	
Total	30	955.7419		

Since $F = 6.84$ exceeds 4.68, the null hypothesis must be rejected; we conclude that the secretary does not type with equal speed on the four typewriters.

CHAPTER 11

11.1 Practice exercise.

11.3 Since 0.033 is less than 0.05, the null hypothesis must be rejected.

11.5 Since neither probability is less than or equal to 0.05, the null hypothesis cannot be rejected.

11.7 Table I shows that the probability of 12 or fewer successes is 0.184. Since 0.184 is not less than or equal to 0.05, the null hypothesis cannot be rejected.

11.9 Since -2.12 is greater than -2.33, the null hypothesis cannot be rejected.

11.11 Since 1.85 does not exceed 2.33, the null hypothesis cannot be rejected.

11.13 Since -2.10 is less than -1.645, the null hypothesis must be rejected.

11.15 Since $z = 2.89$ exceeds 2.33, the null hypothesis must be rejected.

11.17 Since $z = 1.09$ falls between $-z_{0.025} = -1.96$ and $z_{0.025} = 1.96$, we find that the null hypothesis cannot be rejected.

11.19 Practice exercise.

11.21 Since $\chi^2 = 3.75$ is less than 5.991, the null hypothesis cannot be rejected.

11.23 Since $\chi^2 = 16.55$ exceeds 13.277, the null hypothesis must be rejected.

11.25 Since $\chi^2 = 2.19$ does not exceed 3.841, the null hypothesis cannot be rejected.

11.27, 11.29, 11.31 Practice exercises.

11.33, 11.35, 11.37 Computer exercises.

11.39, 11.41, 11.43, 11.45 Practice Exercises.

REVIEW EXERCISES FOR CHAPTERS 9, 10, & 11

R.119 Since $t = 1.84$ falls between -2.110 and 2.110, the null hypothesis cannot be rejected. The difference is not significant.

R.121 The interval may be written as $(0.82, 0.90)$.

R.123 (a) 1.812; (b) 2.201; (c) 2.681; (d) 3.012.

R.125 $2.50 < \mu < 2.70$.

R.127 $0.34 < \sigma < 0.48$.

R.129 $E = 0.541$ hours.

R.131 **(a)** 121.33 minutes $< \mu <$ 131.67 minutes;

  (b) 119.70 minutes $< \mu <$ 133.30 minutes.

R.133 $\$114.94 < \mu < \$137.96$.

R.135 1.11 minutes $< \sigma <$ 1.54 minutes.

R.137 $E = 0.137$.

R.139 $0.658 < p < 0.722$.

R.141 **(a)** She is testing the null hypothesis that the average value is not $\$195,000$.

  (b) She is testing the null hypothesis that the average value is $\$195,000$.

R.143 Since $z = -2.47 < -1.96$, we reject the null hypothesis.
Since $z = -2.47$ and since the test is two sided, we find that the p-value is 0.0136.
This is well below 5% and it indicates that we should reject the null hypothesis.

R.145 Since 2.08 is less than 2.33, the null hypothesis cannot be rejected. The p-value for 2.08 (which is not asked for) is 0.0188. Since this is greater than 0.01, the conclusion is the same as before.

R.147 Since $t = 1.89$ is greater than 1.796, the null hypothesis must be rejected.

R.149 Using Table I for $n = 15$ and $p = 0.80$ the probability of 9 or fewer successes is 0.061. The null hypothesis cannot be rejected.

R.151 Table I shows that the probability of 6 or fewer successes is 0.094, and the probability of 6 or more successes is 0.967. The null hypothesis cannot be rejected.

R.153 **(a)** The probability of 11 or fewer successes is 0.055, and the probability of 11 or more successes is 0.988. Since neither probability is less than or equal to 0.025, the null hypothesis cannot be rejected.

  (b) This is similar to (a), but with $x = 15$ the probability of 15 or fewer success is 1.000 and the probability of 15 or more successes is 0.206, and the null hypothesis cannot be rejected. Thus, even getting a 100% success rate in the experiment does not lead us to conclude that $p \neq 0.90$.

R.155 Since $z = 7.119$ exceeds 1.645, the null hypothesis must be rejected.

CHAPTER 12

12.1 Practice exercise.

12.3 **(a)** $n = 6, \sum x = 102, \sum y = 276, \sum xy = 5,160, \bar{x} = 17, \bar{y} = 46$;

  (b) $S_{xx} = 122$, and $S_{xy} = 468$;

  (c) $b = 3.836$, and $a = -19.212$;

  (d) $-19.212 + 3.836x$;

  (e) 57.508 cups of coffee.

12.5 59.07 inches

12.7 $\hat{y} = 4.26$, or $\$4,260,000$. The equation is $\hat{y} = 0.66 + 0.6x$.

12.9 $\hat{y} = 70.950 + 1.126x$.

12.11 Practice exercise.

12.13 $t = -2.14$. Since $t = -2.14$ is not less than -3.747, the null hypothesis cannot be rejected.

12.15 $2.108 < \beta < 5.564$.

12.17 (92.897, 98.283).

12.19 $t = 0.250$. The null hypothesis cannot be rejected.

12.21 $0.177 < \beta < 2.076$.

12.23 Practice exercise.

12.25 $r = 0.352$.

12.27 $r = 0.885$.

12.29 $r = 0.916$.

12.31 Practice exercise.

12.33 45%.

12.35 **(a)** We get $r = 1$ since the larger value of x corresponds to the larger value of y.

   (b) We get $r = -1$ since the larger value of x corresponds to the smaller value of y.

12.37 **(a)** $r = 0.62$, significant;   **(b)** $r = -0.47$, significant;   **(c)** $r = -0.58$, not significant;   **(d)** $r = 0.63$, not significant.

12.39 $r = 0.514$.

12.41 $r = 0.916$, so $(0.916)^2(100\%) \approx 83.9\%$.

CHAPTER 13

13.1 Practice exercise.

13.3 Since 0.113 exceeds 0.050, the null hypothesis cannot be rejected.

13.5 Since 0.212 exceeds 0.01, the null hypothesis cannot be rejected.

13.7 Practice exercise.

13.9 Since 0.073 and 0.981 both exceed 0.05, the null hypothesis cannot be rejected.

13.11 Since $z = 2.75$ exceeds 2.33, the null hypothesis must be rejected.

13.13 Since $z = -2.04$ is less than -1.645, the null hypothesis must be rejected.

13.15 Since $U = 5.5$ is less than 8, the null hypothesis must be rejected.

13.17 Since $U = 6$ is greater than 5, the null hypothesis cannot be rejected.

13.19 Practice exercise.

13.21 Since $z = 3.20$ exceeds 2.33, the null hypothesis must be rejected.

13.23 Since $z = 1.51$ is less than 1.645, the null hypothesis cannot be rejected.

13.25 Since $H = 0.39$ is less than 5.991, we cannot reject the null hypothesis.

13.27 Since $H = 0.245$ does not exceed 9.210, the null hypothesis cannot be rejected.

13.29 Since $u = 25$ is greater than 19, the null hypothesis of randomness must be rejected.

13.31 Since $u = 17$ equals 17, the null hypothesis must be rejected.

13.33 Since $z = -0.88$ falls between -1.96 and 1.96, the null hypothesis cannot be rejected.

13.35 Since $z = 0.38$ falls between -1.96 and 1.96, the null hypothesis cannot be rejected.

13.37 Practice exercise.

13.39 Since $z = -2.88$ is less than -1.96, the null hypothesis must be rejected.

13.41 Since $z = -0.57$ falls between -1.96 and 1.96, the null hypothesis cannot be rejected.

13.43 Practice exercise.

13.45 $r_s = 0.65$.

13.47 $r_s = 0.99$.

13.49 $r_s = 0.893$.

13.51 $r_s = 0.75$.

REVIEW EXERCISES FOR CHAPTERS 12 AND 13

R.157 $\hat{y} = -1.07 + 1.22x$; 3.2.

R.159 **(a)** $r = 0.33$ is significant;   **(b)** $r = 0.33$ is not significant.

R.161 $U = 9$. Since $U = 9$ is greater than 8, the null hypothesis cannot be rejected.

R.163 $x = 2$; since 0.019 is less than 0.05, the null hypothesis must be rejected.

R.165 $r = -0.65$. Since $r = -0.65$ is less than -0.632, the null hypothesis must be rejected.

R.167 **(a)** Negative correlation;   **(b)** positive correlation;   **(c)** zero correlation;   **(d)** negative correlation;   **(e)** zero correlation;   **(f)** positive correlation.

R.169 From Table I, the probability of 9 or more successes is 0.032. Since 0.032 is less than 0.05, the null hypothesis must be rejected.

R.171 $y = 14.247 + 0.637\ (40) = 39.727$ minutes.

R.173 Since $H = 5.03$ does not exceed 7.815, the null hypothesis cannot be rejected.

R.175 Since $z = 2.29$ exceeds 1.96, the null hypothesis must be rejected.

R.177 Since $u = 11$ falls between 6 and 16, the null hypothesis cannot be rejected.

R.179 Since $U = 35$ is less than 37, the null hypothesis must be rejected.

R.181 Since $t = -0.805$ is greater than -2.998, the null hypothesis cannot be rejected.

R.183 Since $z = 2.38$ falls between -2.575 and 2.575, the null hypotheses cannot be rejected.

INDEX

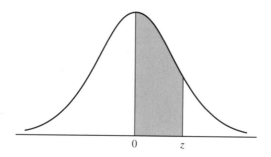

TABLE II   Normal-Curve Areas

z	.00	.01	.02	.03	.04	.05	.06	.07	.08	.09
0.0	.0000	.0040	.0080	.0120	.0160	.0199	.0239	.0279	.0319	.0359
0.1	.0398	.0438	.0478	.0517	.0557	.0596	.0636	.0675	.0714	.0753
0.2	.0793	.0832	.0871	.0910	.0948	.0987	.1026	.1064	.1103	.1141
0.3	.1179	.1217	.1255	.1293	.1331	.1368	.1406	.1443	.1480	.1517
0.4	.1554	.1591	.1628	.1664	.1700	.1736	.1772	.1808	.1844	.1879
0.5	.1915	.1950	.1985	.2019	.2054	.2088	.2123	.2157	.2190	.2224
0.6	.2257	.2291	.2324	.2357	.2389	.2422	.2454	.2486	.2517	.2549
0.7	.2580	.2611	.2642	.2673	.2704	.2734	.2764	.2794	.2823	.2852
0.8	.2881	.2910	.2939	.2967	.2995	.3023	.3051	.3078	.3106	.3133
0.9	.3159	.3186	.3212	.3238	.3264	.3289	.3315	.3340	.3365	.3389
1.0	.3413	.3438	.3461	.3485	.3508	.3531	.3554	.3577	.3599	.3621
1.1	.3643	.3665	.3686	.3708	.3729	.3749	.3770	.3790	.3810	.3830
1.2	.3849	.3869	.3888	.3907	.3925	.3944	.3962	.3980	.3997	.4015
1.3	.4032	.4049	.4066	.4082	.4099	.4115	.4131	.4147	.4162	.4177
1.4	.4192	.4207	.4222	.4236	.4251	.4265	.4279	.4292	.4306	.4319
1.5	.4332	.4345	.4357	.4370	.4382	.4394	.4406	.4418	.4429	.4441
1.6	.4452	.4463	.4474	.4484	.4495	.4505	.4515	.4525	.4535	.4545
1.7	.4554	.4564	.4573	.4582	.4591	.4599	.4608	.4616	.4625	.4633
1.8	.4641	.4649	.4656	.4664	.4671	.4678	.4686	.4693	.4699	.4706
1.9	.4713	.4719	.4726	.4732	.4738	.4744	.4750	.4756	.4761	.4767
2.0	.4772	.4778	.4783	.4788	.4793	.4798	.4803	.4808	.4812	.4817
2.1	.4821	.4826	.4830	.4834	.4838	.4842	.4846	.4850	.4854	.4857
2.2	.4861	.4864	.4868	.4871	.4875	.4878	.4881	.4884	.4887	.4890
2.3	.4893	.4896	.4898	.4901	.4904	.4906	.4909	.4911	.4913	.4916
2.4	.4918	.4920	.4922	.4925	.4927	.4929	.4931	.4932	.4934	.4936
2.5	.4938	.4940	.4941	.4943	.4945	.4946	.4948	.4949	.4951	.4952
2.6	.4953	.4955	.4956	.4957	.4959	.4960	.4961	.4962	.4963	.4964
2.7	.4965	.4966	.4967	.4968	.4969	.4970	.4971	.4972	.4973	.4974
2.8	.4974	.4975	.4976	.4977	.4977	.4978	.4979	.4979	.4980	.4981
2.9	.4981	.4982	.4982	.4983	.4984	.4984	.4985	.4985	.4986	.4986
3.0	.4987	.4987	.4987	.4988	.4988	.4989	.4989	.4989	.4990	.4990

Also, for z = 4.0, 5.0, and 6.0, the areas are 0.49997, 0.4999997, and 0.499999999.